Physics: The First Science

Physics: The First Science

Peter Lindenfeld and Suzanne White Brahmia
Rutgers, The State University of New Jersey

RUTGERS UNIVERSITY PRESS
New Brunswick, New Jersey, and London

Library of Congress Cataloging-in-Publication Data

Lindenfeld, Peter.
Physics : the first science / Peter Lindenfeld and Suzanne White Brahmia.
p. cm.
Includes bibliographical references and index.
ISBN 978-0-8135-4937-8 (pbk. : alk. paper)
1. Physics—Textbooks.   I. Brahmia, Suzanne White, 1963–   II. Title.
QC23.2.L55 2011
530—dc22
2010021009

A British Cataloging-in-Publication record for this book is available from the British Library.

Copyright © 2011 by Peter Lindenfeld and Suzanne White Brahmia

All rights reserved

No part of this book may be reproduced or utilized in any form or by any means, electronic or mechanical, or by any information storage and retrieval system, without written permission from the publisher. Please contact Rutgers University Press, 100 Joyce Kilmer Avenue, Piscataway, NJ 08854–8099. The only exception to this prohibition is "fair use" as defined by U.S. copyright law.

Visit our website: http://rutgerspress.rutgers.edu

Manufactured in the United States of America

Contents

Preface ix

Additional Notes, Primarily for the Instructor xi

Acknowledgments xv

CHAPTER 1
Atoms to Stars: Scales of Size, Energy, and Force 1

1.1 The microbe and the elephant: the hierarchy of size 2
1.2 Energy and stability 5
1.3 The four forces 7
1.4 Atoms and the periodic table of elements 8
1.5 Seeing atoms: the scanning tunneling microscope 11
1.6 Summary 12
1.7 Review activities and problems 12

CHAPTER 2
Some Tools of the Trade: Numbers, Quantities, and Units 16

2.1 The language of physics: symbols and formulas 17
2.2 Once more the four forces, this time quantitatively 26
2.3 Summary 28
2.4 Review activities and problems 28

CHAPTER 3
There Is No Rest: Describing Motion 33

3.1 Getting started: simplification and approximation: models 34
3.2 Keep your eye on the ball: where is it and where is it going? 35
3.3 The mathematics of change 46
3.4 Summary 49
3.5 Review activities and problems 49

CHAPTER 4
Forces and Motion: Newton's Framework 57

4.1 Newton's laws of motion 58
4.2 Adding forces: vectors 62
4.3 Momentum and its conservation: action, reaction, and Newton's third law 71
4.4 One more motion that is everywhere: rotation 77
4.5 Summary 81
4.6 Review activities and problems 83

CHAPTER 5
Laws and Their Limits: The Organization of Scientific Knowledge 90

5.1 How do we know? Reality and interpretation 91
5.2 The Newtonian model and its limitations 97
5.3 Mechanics and beyond 99
5.4 Summary 100
5.5 Review activities and problems 101

CHAPTER 6
Energy and Its Conservation 103

6.1 Work: not always what you think 104
6.2 Energy of motion: kinetic energy 105
6.3 Energy of position: potential energy 106
6.4 Friction and the loss of mechanical energy 113
6.5 Internal energy and the law of conservation of energy 116
6.6 Work and energy revisited 118
6.7 Power: not what the power company sells 125
6.8 Summary 125
6.9 Review activities and problems 126

CHAPTER 7
Materials and Models   135

7.1 Ideal systems and models: the ideal gas   136
7.2 Other systems: adding pieces from reality   146
7.3 Summary   153
7.4 Review activities and problems   154

CHAPTER 8
Electricity: It Is Everywhere   158

8.1 The electric force   158
8.2 The electric field   160
8.3 Field lines and flux   165
8.4 Summary   171
8.5 Review activities and problems   172

CHAPTER 9
More on Electricity: From Force to Energy, from Field to Potential   176

9.1 Electric potential energy and electric potential   177
9.2 Energy transformations and electric circuits   187
9.3 Summary   192
9.4 Review activities and problems   192

CHAPTER 10
Magnetism: Electricity's Traveling Companion   198

10.1 Again—force, field, and motion   199
10.2 The electron: an old friend turns out to be the elemental magnet   210
10.3 Generating electricity: motional emf and Faraday's law   212
10.4 Summary   217
10.5 Review activities and problems   218

CHAPTER 11
Waves: Mechanical and Electromagnetic   223

11.1 What is a wave?   224
11.2 What can waves do? Describing waves and their properties   226
11.3 Sound and musical scales   229
11.4 Maxwell's great contribution: electromagnetic waves   233
11.5 Observing interference of light   237
11.6 Reflection and refraction   241
11.7 Where Einstein started: electromagnetism and relativity   247
11.8 Summary   252
11.9 Review activities and problems   254

CHAPTER 12
Quantum Physics   260

12.1 The "old" quantum physics   261
12.2 The new synthesis   270
12.3 Order in the universe: the elements   281
12.4 Summary   286
12.5 Review activities and problems   288

CHAPTER 13
The Nucleus: Heart of the Atom   293

13.1 Henri Becquerel, Marie Curie, and the beginning of nuclear physics   294
13.2 What is the universe made of? The stable nuclei and their binding energy   295
13.3 Radioactivity   298
13.4 Biological effects   303
13.5 The nuclear force   305
13.6 Observing radioactive radiations   306
13.7 Nuclear reactions   308
13.8 Particles   314
13.9 Summary   315
13.10 Review activities and problems   317

CHAPTER 14
Energy in Civilization   322

14.1 The flow of energy   324
14.2 Electric energy: what is it and what does it do for us?   325
14.3 DC and AC: transformer and generator   325
14.4 Energy storage   326
14.5 Entropy and the second law of thermodynamics: the limits of energy transformation   328
14.6 Our addiction to fossil fuels   331
14.7 Other sources of energy   333

14.8 Summary   337

14.9 Review activities and problems   338

CHAPTER 15
Atomic Physics Pays Off: Solar Cells, Transistors, and the Silicon Age   339

15.1 The real solid: metals, insulators, and semiconductors   340

15.2 Tiny changes and vast consequences: impurities in semiconductors   342

15.3 The transistor and the information revolution   344

15.4 Summary   347

15.5 Review activities and problems   348

CHAPTER 16
There Is No End   349

Bibliography   351

Some Constants, Astronomical Quantities, and Masses   353

Index   355

To Lore, who saw this book come to life,
and was the first to listen to some of its passages.

Preface

The world is a wonderful place. Our aim is that working with this book will help you to be more aware of it than you were before and to see it more fully, both the aspects that are visible and those that are hidden.

Let's look at the different parts of this book. First there is the text. We hope that you find it clear, interesting, and sometimes surprising. We want this to be a book that you *read*. In between the narrative passages are the examples. They turn the words into action. They are essential parts of the book, and it is very important that you read and study them carefully. Each example has questions, usually with numbers and quantities. Working through them is not difficult since the answers are there too.

You might think that leaves nothing for you to do. That's where the "Guided Review" comes in. There is one at the end of each chapter. For each example in the text there is a Guided Review question that is closely related to it. The idea is that as you think about the question at the end of the chapter you will go back to the example that goes with it and study it more closely, and also read and reread the part of the text that goes with it.

The other questions at the end of the chapters are for review, and to let you try out the ideas and the different ways of finding answers. Some of them also take the story further. They are grouped in sections called "Problems and reasoning skill building," "Multiple choice questions," and "Synthesis problems and projects." For simplicity all quantities in the examples and other questions are assumed to be known to three figures unless otherwise noted.

A particularly interesting feature consists of the simulations that have been developed at the University of Colorado. They give you an even closer look at the phenomena that are discussed in the text. In most of them you are asked to make specific measurements, but in addition they invite free explorations that take the concepts further.

We should say a little about the math. That's the language of physics, and it is important to know how to use it. We don't expect you to come knowing more than you learned in high school. But there may be ways of using math that are new to you, and symbols that are different from those that you are used to. We explain them as we go along. We don't use calculus in the problems, but we talk about the ideas of calculus where they are helpful.

The important part of learning physics is not to memorize facts and formulas. It is, rather, to approach and analyze unfamiliar situations. It uses ways of thinking that are accessible to everyone. The study of physics can be frustrating at times, but also exhilarating and rewarding. We hope that you will find it to be an adventure that enriches your view of the world. There may be hard work ahead, but we want you to enjoy the trip that we take together!

Additional Notes, Primarily for the Instructor

We want to change the way physics is taught and the way it is perceived.

We think there is too much rote learning, too many formulas, with too little thought, meaning, and overview. There are too many problems that have no existence outside the elementary physics class, too many that are more like games that the instructors like to play than searching investigations. We want students to see the subject as a vital part of their world, and as a foundation for all of the sciences.

Physics: The First Science is an algebra-based text that is both research based and different from the many textbooks that are currently available. Our book is less than half as long as the tomes that have become the norm, and correspondingly less expensive. It is designed to be read by the students, we hope with interest, and even with pleasure. We would like the students to think of it as contributing to their learning and outlook in an important way, and to remember it long after the course has ended.

The book is supported by the website firstscience.rutgers.edu. It allows us to post comments, additional material, and corrections, as well as answers and solutions to some of the questions and problems.

We hope that our book will lead to new patterns of teaching. The best courses are transforming experiences. We want to help to create such a course, where the student's view of the world is enriched by a new awareness and understanding of its phenomena.

Some cherished topics are not there. In particular, we have tried to leave out problems that might be called "puzzles." We believe that the teaching of physics should be clear and direct, and we have made great efforts to keep it so.

The topics follow the usual sequence reasonably closely, but we have approached them with a fresh mind, asking ourselves why each is included and how it should be presented. Most topics are there because we think of them as fundamental physics; sometimes because of their importance for technology, their societal relevance, or their historical significance; occasionally we add something that we find beautiful and illuminating, or fun and surprising.

MCAT

Many of the students will want to take the Medical College Admission Test (MCAT). According to the website of the Association of American Medical Colleges, which administers the test, it not only asks for mastery of basic concepts in subjects including physics, but also "assesses capacity for problem solving and critical thinking", and "evaluates your ability to understand, evaluate and apply information and arguments presented in prose style." We have kept the requirements of the test in mind, and believe that our text provides excellent preparation for it.

Examples and end-of-chapter material

The worked examples are an essential part of the reading. They are part of the story line and can't be skipped. They are linked to the *Guided Review* at the end of the chapter, which consists of problems that are alternate versions or minor extensions of the examples. They are fairly simple to do if you read and work through the examples. In that way the Guided Review really forces the student to take the examples and the associated sections of the book seriously. It guides the students' reading and helps them focus on the important concepts. The Guided Review was a very popular feature with the students of the class in which the text was tested.

The end-of-chapter material includes questions beyond the Guided Review in sections called "Problems and reasoning skill building,"

"Multiple choice questions," and "Synthesis problems and projects." These sections provide plenty of practice. Some of the questions also require thought and creativity well beyond the typical array found in most textbooks and test banks. Note that all quantities in the examples and problems are assumed to be known to three significant figures unless otherwise noted.

Many simulations are incorporated in the text, as examples, as parts of the development, and in the end-of-chapter questions. They provide an excellent way to bring the concepts to life. They were produced by the PhET (Physics Education Technology) project at the University of Colorado, initiated by Carl Wieman, a physicist who decided to devote himself to education after winning the Nobel Prize for his work on atomic physics. Accessing the simulations is quite straightforward, but the students may at first need some guidance. Log onto http://phet.colorado.edu (no www). Then either "play with sims" or "on line" gets you to the list. Click on "physics," then on one of the subdivisions, choose a simulation, and finally "run now." We provide explicit questions, but hope that the students (and the instructor!) will go further and try different parts and possibilities. The PhET group occasionally makes changes to the online simulations so that the instructions may have to be modified, but that should not create major obstacles.

Mathematics

There is no requirement for previous mathematical knowledge beyond high-school algebra. We introduce other mathematics when it is appropriate. Mathematics is part of the language that we use, and we foster the development of mathematical reasoning and physics numeracy. We avoid formulaic use and boxed equations. Most often the mathematical development becomes part of the narrative.

Calculus is not used in the problems, but its ideas are introduced in terms of slopes and areas, and then used where they are relevant, as in the development of kinematics. Chapter 2 reinforces mathematical skills that are both vital in our book and often poorly understood, even by students who are skilled in "doing problems." We encourage the students to use mathematics as a descriptive tool rather than as an algorithmic puzzle solver. We avoid problems that are primarily mathematical exercises.

The road map

Atoms, nuclei, and a modern microscopic view are introduced at the start and referred to frequently. This deviates from the practice of leaving twentieth-century physics ideas until the end of the course. While we think mechanics is important, we are eager to get to other subjects for which there is often too little time. We include an introduction to quantum properties that is earlier and more comprehensive than is usual. We sometimes discuss phenomena on a microscale and sometimes on a macroscale, depending on which is more appropriate.

The first chapter is an overview that introduces atoms and nuclei, scales of size and energy, and the four fundamental forces. It is followed by a review of some of the essential mathematics. The next two chapters are on kinematics and Newton's laws, and make frequent use of multiple representations, such as motion diagrams, graphs, and energy bar charts.

Chapter 5 discusses the nature of models and theories in some detail. It provides an opportunity to think about scientific evidence and the models and theories based on it. Chapter 6 is on energy, not only mechanical but also electric energy and internal energy. Chapter 7 on materials and models is organized somewhat nontraditionally. It includes some kinetic theory, fluids, and the first law of thermodynamics, with remarks on quantum theory and other modern developments.

Chapters 8, 9, and 10 are on electricity and magnetism. Some traditional topics and many traditional problems are left out, but there is some material that is not usually included. The same is true for Chapter 11 on waves, which, in the spirit of the book, deals with sound, light, and electomagnetic waves so as to show their common features and their differences. Chapter 12 is intellectually perhaps the most important chapter. It begins with the "old" quantum theory, including the Bohr model, primarily to introduuce some basic quantities and concepts, while making clear from the outset where it is inadequate. The chapter proceeds to the modern synthesis of particles and waves, both for photons and particles, and discusses this fundamental feature of modern physics without resort to mystery or paradox in a way that we think is more straightforward and understandable than most others that we have seen. It is followed by the chapter on nuclear physics, which is quite extensive.

Chapters 14 and 15 represent major departures from the usual subject matter. The chapter *Energy in Civilization*, more than any other, demonstrates the importance of physics in societal issues. It also includes an introduction to entropy and the second law of thermodynamics. We think that it makes these topics more accessible than is usually the case. Finally, we felt that we could not leave out the concepts and devices underlying our electronic age, and end with a chapter that describes some of what is behind the gadgetry that every child now grows up with.

Highways and byways

Even with our drastic cuts, there remains more material in the book than is manageable in the usual one-year course. There are sections, examples, and problems that an instructor may wish to leave out, and others that can be assigned to be read without spending much class time. We are very aware of the need to deal with unfamiliar concepts in some depth and with sufficient time. Each instructor will need to construct his or her own road map. For each part you can choose the level of intensity that you wish to use. The first is to read so as to become familiar with the concepts. The second is to study and be able to make straight-forward applications. Finally, there is the intense involvement that leads to deeper understanding and allows the use of the material in unfamiliar situations. Here are some suggestions for variations.

We like our first chapter, but wouldn't expect to spend a lot of time with it. Similarly, Chapter 2 can be dealt with rather quickly with moderately experienced students. Chapter 5 is short, and much of it can be a reading assignment.

The last section of Chapter 7 consists of topics that are not generally included, and may be assigned as extra reading. It's an advantage to get to the end of Chapter 7 with time to spare, because Chapter 8 on electricity can then be part of the first semester. This allows a fast start to the second semester and additional time for the more sophisticated sections to come. Chapters 8, 9, and 10 are basic, but not everyone will want to deal with each part in the same depth. This applies, for instance, to Gauss's law (in Chapter 8), to the brief part where calculus is used in Chapter 9, and to Section 10.2 on the microscopic aspects of magnetism. The sections on sound (11.3) and light (11.5, 11.6) offer options, and so does the section on relativity in Chapter 11.

Even though Chapter 12 is important, there can be variations in depth and emphasis. We regard most of the first part and the qualitative introduction to the Schrödinger equation as essential, but we realize that the following sections have parts that are more difficult and go far beyond the usual subject matter of a one-year introductory course.

Chapter 13 on nuclear physics is more extensive than usual, but uses only simple algebra. Candidates for bypasses are the semiempirical binding energy relation, and Section 13.6 on the observation of radioactive radiations.

The review of present-day energy use in Chapter 14 may be the first part of the book to be overtaken by what happens. Since it comes near the end it is in danger of being left out. It would be better to assign it as extra reading. It should then be possible to spend some class time on the section on the second law of thermodynamics. Finally, the instructor will have to decide how much time is left for Chapter 15.

In summary, we believe that we have a thoroughly modern course that is open, inviting, and accessible. Some sections are easier and more basic than usual, and some are more sophisticated. Each instructor will have to decide what to include and what to emphasize. Earlier versions of the text were tested during three successive summers at Rutgers University in a course for premed and science majors. The instructor was not one of the authors, but it contributed to the success of the book that he was sympathetic to its aims and ideas, and that the procedures and exams reflected the spirit of the book.

Acknowledgments

We would like to thank our colleagues at Rutgers University for their support and encouragement, especially Premala Chandra, Eugenia Etkina, Charles Glashausser, Noémie Koller, Paul Leath, and Alan Van Heuvelen. Alan Van Heuvelen pioneered the use of multiple representations, including energy bar charts. Some of the examples and problems had their origin in the collaboration of S.W.B with him and Eugenia Etkina.

We would like to acknowledge the role of George Horton (1926–2009) in making our collaboration possible. His was a lifetime effort to help to create a culture in which caring about student learning has a very high value.

P. L. would like to acknowledge discussions with Art Hobson, whose writings influenced the section on the synthesis of wave and particle phenomena. The section on the *sustainability transition* was inspired by the work of his friend and mentor Robert W. Kates.

Helpful comments on early versions of parts of the manuscript were made by E. L. Jossem, Arthur Kosowsky, Mario Iona, and Christian Römelsberger.

Special thanks go to Lonni Sue Johnson for her drawings, including the one on the cover.

We have incorporated activities based on the PhET interactive Simulations developed by the *Physics Education Technology* group at the University of Colorado, licensed under a Creative Commons Attribution United States License. We thank them for their openness and cooperation.

The book has profited greatly from the contributions of Michael J. Gentile, who was the first to use it as a text in a course. On the basis of this experience he made many thoughtful and helpful comments and suggestions that have led to numerous improvements in the present version.

P. L. and S.W.B.
August 2010

CHAPTER 1

Atoms to Stars: Scales of Size, Energy, and Force

The microbe and the elephant: the hierarchy of size

Energy and stability

The four forces

Atoms and the periodic table of elements

Seeing atoms: the scanning tunneling microscope

As we look around us we see a world that is marvelously ordered and organized. Outward from the earth we see the moon, the planets, and the stars. Light comes to us from them, and other kinds of radiation, signals that tell us how they look, how big they are, and what they are made of.

We can also go in the other direction, down to smaller and smaller sizes, until we come to pieces that are too small to see directly. They too can be studied by the radiation that they give off, or through microscopes, or, more indirectly, by bouncing other particles off them. As we continue, we get to the atoms that were once thought to be the smallest, the ultimate building blocks. Today we know that atoms have their own structure, each with an even tinier nucleus in the center, with electrons racing around it.

How is the world organized, what can we know and understand about its order? What is it made of, how are its pieces held together, on the earth, out to the stars, and down to the tiniest pieces that we know? How do we find out and how do we learn more? These are some of the questions that we will explore.

The sun, the planets, the atoms, and the nuclei are very different, most obviously in their size. That allows us to study them quite separately, almost as if each existed alone. But no part of the universe is alone. Each is acted on by *forces*, as its neighbors push and pull. In spite of the enormous variety that we observe, it seems that there are just four kinds of fundamental forces. They are the gravitational force, the electromagnetic force, and two kinds of nuclear forces. Each reigns supreme in its own realm. Together they cooperate to create the world that we know, from nuclei to stars, with our own, the human scale, right in the middle.

1.1 The microbe and the elephant: the hierarchy of size

Cut an apple in two. See the flesh, the peel, the seeds. Each is separate, each has its own existence and its own function. The peel provides a barrier against the outside; the flesh protects the seed, which, in turn, waits to play its part when its time comes.

The parts may be separate, but they are not independent. Each depends on the others, each develops from the same seed.

There is an order in space, and an order in the sequence of time. This is what gives order and sequence to our perception and our understanding. We first see the different components separately, and only later ask how they change each other and interact with each other and their environment.

We can cut the apple further, into smaller pieces. How long can we continue before what we have is no longer recognizable as a piece of apple?

Democritus, more than 2000 years ago, asked that question, and thought that there must be a limit, a point beyond which we cannot continue, when we have arrived at a piece that cannot be cut further, the *atom*, the not-to-be-cut.

Today we know that he was both right and wrong. The atoms are the building blocks of which all materials are made. But we can go further. Each atom has its own structure, with its nucleus deep inside and its electrons around it.

The first one who thought in detail about what an atom might be like with a nucleus surrounded by electrons was Niels Bohr in 1913. He imagined a picture, a "model" of the atom, with precisely known forces and exactly predicted motions of the electrons. It was so successful that it still colors a lot of talk about atoms, even though some of its most important features are incorrect.

Bohr's picture of an atom had electrons circling the nucleus, pretty much as the planets travel around the sun. But an atom is not a planetary system, with electrons moving in fixed and predictable paths. One of the great insights stemming from the development of *quantum mechanics* in 1925 is that the laws governing atoms are different from those that are followed by the sun and the planets.

Electrons can be observed surrounding the nucleus, but we can no longer talk about them as moving in circles, or ellipses, or in any other definite path. With the methods of quantum mechanics we can, however, calculate how likely the electrons are to be observed at any particular distance from the nucleus.

The figure shows the kind of picture that we have. Here a high density of dots represents a high probability that an electron can be found there. There is a simple relation between Bohr's model and the quantum mechanical picture. Bohr said that the electrons had to circle the nucleus at certain definite distances. The new picture says that electrons can be found at other distances, but that the highest probability is that they are observed close to where Bohr said they had to be.

As with the apple, we can study the nucleus alone, and separately the dance of the electrons. The two are distinct, but each is linked to the other, and is rarely without it.

There is further structure within the nucleus, where there are two kinds of *nucleons*, the *protons* and the *neutrons*. In turn, each nucleon

consists of three *quarks*. As far as we know, however, the quarks are locked inside and cannot exist independently.

The figure shows schematic representations of an apple, a cell, a molecule, an atom, a nucleus, and a nucleon.

There is also further structure as we go outside the apple, to the tree, the earth, and the solar system in which the earth is just one of the planets, to the galaxy in which the sun is one of myriads of stars, and to the clusters of galaxies, as we come to the limit of the known.

EXAMPLE 1

(a) How much bigger is our planetary system than the sun at its center: what is the ratio of the diameter of the orbit of Pluto to that of the sun?

(b) What is the ratio of the diameter of a hydrogen atom to that of its nucleus?

Ans.:

(a) Pluto travels about the sun in an elliptical orbit. Its farthest distance from the sun is 7.38×10^{12} m and the closest is 4.45×10^{12} m. The average radius of the orbit is about 5.9×10^{12} m. (10^{12} is the number "1." with the decimal point shifted 12 steps to the right, i.e., followed by 12 zeros. If you're not sure about using this notation, look at the section called "Numbers, huge and tiny: powers of 10" in Chapter 2. In the same chapter there is also a section called "Quantities and units" and one called "Precision: significant figures." Please read them!)

The sun's diameter is 1.39×10^9 m. The ratio of the two diameters is

$$\frac{(2)(5.9 \times 10^{12})}{1.39 \times 10^9} = 8.5 \times 10^3 \text{ or about } 8500.$$

(b) The radius of the smallest orbit of the electron in a hydrogen atom, according to the Bohr model, is 0.53×10^{-10} m. (10^{-10} is the number 1.0 with the decimal point shifted 10 steps to the left, while filling the empty spaces with zeros.) Although the Bohr model is an obsolete representation of the atom, the Bohr orbit's size gives a good approximate number for the atom's size. The nucleus of an ordinary hydrogen atom is a proton. What we call the size of the proton depends on what property is measured, but it is about 10^{-15} m. The ratio of the radii, and therefore also the ratio of the diameters, is 5.3×10^4 or about 53,000.

Here is a table of some distances: from Pluto's orbit to the nucleon size. (The radii are averages.)

Radius of Pluto's orbit	5.9×10^{12} m
Radius of earth's orbit	1.5×10^{11} m
Radius of moon's orbit around the earth	3.84×10^8 m
Radius of sun	6.9×10^8 m
Radius of earth	6.38×10^6 m
Radius of moon	1.74×10^6 m
Human size	2 m
Diameter of human hair	2×10^{-4} m
Red blood cell	10^{-5} m
Escherichia coli bacterium	2×10^{-6} m
Rhinovirus	2×10^{-8} m
Radius of uranium atom	1.4×10^{-10} m
Radius of hydrogen atom	5.3×10^{-11} m
Radius of proton	0.9×10^{-15} m

We know what is between the sun and its planets. Not much. (There is about one atom per cubic centimeter in interstellar space, and perhaps 10 times as many within the planetary system.) The space between the nucleus and the electrons in an atom is very much smaller, but it is even emptier. When we hold something in our hands, a stone, a book, whatever is there, whatever we see or feel, the overwhelming amount is empty space.

Our planetary system as a whole is so much larger than the sun that we can think of the two quite separately. What happens to the earth or the other planets has almost no influence on the sun. It is like microbes on an elephant, or for that matter, in ourselves. They live in their own world, unseen, unthought of, by the world of their host.

4 / Atoms to Stars: Scales of Size, Energy, and Force

The same is true about the size difference between an atom and its nucleus. That's why we can think of the two as separate and distinct—not independent of each other, but each following its course on a vastly different scale. Similarly, each planet travels along its path, unaffected by anything that may happen to one or a few of its atoms, or even to one of us, walking on it.

We are somewhere between, so that on the one hand we think of atoms as invisibly small, and on the other hand of the sun, the planets, and the stars as vast and vastly far away. Each of these systems is to a great extent outside our direct experience, yet each is visible, at least indirectly. Each is known, at least in part, and is being explored in more and more detail.

When all the atoms in a material are of the same kind, it cannot be decomposed into other substances. It is an *element*. But atoms are not usually alone. One way for them to combine is to form molecules. The simplest are the combinations of two atoms of the same element, as in oxygen and nitrogen. These two elements make up most of the air that we breathe. In the air the oxygen and nitrogen molecules fly around separately at great speed. A water molecule consists of three atoms, one of oxygen and two of hydrogen. Other molecules can have many more atoms, especially the organic molecules of living matter, which often consist of hundreds or thousands of atoms.

It is the nucleus that decides which element we have. It does that by the number of protons in it. A nucleus with just one proton is a hydrogen nucleus, with two it is helium, with eight it is oxygen, with 29 it is copper, with 50 it is tin, and so on. We see a wonderful order, with families of elements that have similar properties, with each kind of atom, each element, in its place in the *periodic table of elements*.

Molecules can be separate, as in air and other gases. They can also combine to form liquids, as in water, or solids. But many solids do not consist of molecules. Their atoms can be in an irregular, or *amorphous* arrangement. More often the atoms form a *lattice*, in which each has its special place. This happens not only in visibly crystalline materials, such as diamond, but also in many others, where the crystal structure is too small to be easily apparent, for example, in most metals.

In metals the atoms in the lattice do not have all their electrons attached to them. One or more of the electrons from each atom leave their "home" atom, and roam freely throughout the piece of metal. These *free* electrons give the metal its characteristic properties, such as its appearance, its strength, and its ability to transport energy and charge.

The ordered hierarchy and separate identities of nuclei, atoms, planets, and stars provide a plan for our quest toward knowledge. When we study the structure or the motion of a planet we don't need to think of whether there are people, or animals, or even mountains and valleys on it. We can study a nucleus without thinking of what is happening to the atom of which it is a part. Each can be looked at separately. Just imagine how different it would be if the universe were more like an ocean, in which the constituents meld into one another, and are distinguishable only with much greater difficulty.

Where do we fit in? It is no accident that the human scale is huge compared to atomic sizes and tiny compared to the size of the earth. On the one side the human structure is so complicated that each of us must be made of a very large number of atoms and molecules. To have the varied and vastly subtle structures of skin and flesh, tissue and blood, the building blocks must be small enough to be capable of being put together in many different ways.

At the other end, as organisms get larger, there is another limit. Large animals struggle with the effect of their greater weight, which makes it more difficult for them to move and even to stand. We are fortunate not to have the massive legs of elephants.

We can see how the elephant's weight and legs are affected by changing the animal's size. Look at what happens when all lengths are increased by the same factor. If the factor is two, the volume and therefore the weight are multiplied by eight. (If the factor is f, they are multiplied by f^3.) But a leg's strength depends on its cross-sectional area, which goes up by only a factor of four (or f^2).

If all lengths increase by the same factor, the leg becomes less able to support the animal's weight. To support the increased weight the leg's cross section must increase by more than f^2. In other words, the leg must become thicker compared to other parts of the elephant's body.

Another factor that is affected by scaling is the energy requirement. Warm-blooded animals lose energy primarily through their surface. The energy that is lost is supplied by the food they eat. As the surface increases by f^2, the volume increases by f^3. The surface-to-volume ratio decreases by f.

EXAMPLE 2

Consider a giant, similar to you, but with linear dimensions that are each 10 times as large.

What are the factors by which each of the following is larger than for you?

length of leg	10
surface area	100
weight	1000
volume	1000
cross section of leg bones	100
weight supported by leg bones	1000
pressure on leg bones $= \dfrac{\text{weight}}{\text{cross section of bone}}$	10
food requirement	100
energy requirement per unit weight	1/10

What might be the giant's attitude toward a steak whose linear dimensions are also each 10 times those of a one-pound steak that you might eat?

"This 1000-pound steak is about 10 times as large as what I need for the rate at which I lose energy through my skin. Also, I don't move around much, since my bones are only a tenth as strong as yours compared to my weight. They are close to breaking when I stand up. My heart is 1000 times as heavy as yours, but since I don't use as much energy my heart rate is much less than yours. That may make me seem sluggish, and has given rise to the image of giants as not so smart. Actually I'm a lot smarter than you. My brain has 1000 times the volume of yours, and therefore has much more room for *neurons*, the threads that provide the pathways for the electric signals that are responsible for sensory perception and thought processes."

These features give only a partial indication of changes with scale. The actual changes in animals with different sizes and weights are more complex. For example, "Kleiber's law" is an empirical relation that says that the resting metabolic rate (the rate of energy expenditure) of similar animals is proportional to $M^{\frac{3}{4}}$, where M is their mass.

1.2 Energy and stability

So far we have concentrated on size. There is another measure that distinguishes the various realms. How hard is it to break apart a solid, a molecule, an atom, or a nucleus? The quantity that tells us how hard it is to carry out each of these transformations is the *energy*.

We can tear a piece of paper, we can break a twig of wood. With a saw we can divide a plate of iron. In each case the material remains the same.

We haven't changed from paper, wood, or iron. We haven't changed the molecules, the atoms, or the nuclei of which they are composed.

Can we change all of these? Yes, but it gets harder and harder, and takes more and more energy, as we go down the scale to smaller and smaller pieces. Changing the molecules is the easiest. That happens when we burn the paper or the wood, or when the iron rusts. In each of these cases oxygen combines with some of the substance that is there to begin with. In each case the atoms remain, but they combine in different ways. That's what happens in a chemical reaction.

Now look at an atom. It consists of a nucleus surrounded by electrons. The electrons are attracted to the nucleus. Usually the number of protons in the nucleus is the same as the number of electrons. The atom is then *neutral*. It is possible to pull off an electron or to add an extra one. The atom is then an *ion*. The number of protons is no longer the same as the number of electrons. That happens when electrons jump between our hair and a comb or between our shoes and a carpet. We can sometimes tell because the electrons will tend to jump back, and may then produce a spark that we can see or feel.

Transforming an atom into an ion (*ionizing it*) has changed the atom, but not the nucleus. That is much harder. If we can do that we can change one element into another. It takes energy to ionize an atom. It takes much more energy to change the composition of a nucleus.

The alchemists wanted to do that. They wanted to change the relatively common element mercury into gold. They failed, but in the process they helped to develop the field of chemistry. Today we know that they weren't so far off in their thinking. The mercury nucleus contains 80 protons, and the gold nucleus 79. We also know how to transform one into the other by a *nuclear reaction*. It requires expensive machinery (like a *cyclotron*) to produce a very small amount of gold, and therefore it isn't the way to get rich that the alchemists had hoped for.

A molecule can be changed by a chemical reaction. An atom can lose or gain electrons. A nucleus can be transformed by a nuclear reaction. But it takes 100,000 to a million times as much energy to change a nucleus as it does to change an atom. That's why we can, most of the time, talk about an atom as if it contained a stable, never-changing nucleus, and hence about the elements as the unchanging constituents of matter.

The binding energy of an object is the amount of energy that is required to take it apart. We can talk about the "total" binding energy, which is the energy required to separate it into all of its pieces. We can also consider the energy required to remove just one piece.

The binding energy of an electron in an atom is the amount of energy that has to be given to the atom to remove the electron. This is also called the *ionization energy*.

The *joule* (J) and the *electron volt* (eV) are units of energy: $1 \text{ eV} = 1.6 \times 10^{-19}$ J. (One joule is approximately the amount of energy that it takes to raise an object whose weight is one pound to a height of 9 inches. More exactly it is one newton-meter, i.e., the amount of energy used by a force of one newton to move an object through a distance of one meter. We will look at energy and its units in more detail later.) Electron binding energies are usually measured in eV, nuclear binding energies in millions of electron volts, MeV. (1 MeV = 10^6 eV.)

For the electron in a hydrogen atom in its usual condition or "state" (called the *ground state*) the binding energy (or the ionization energy) is 13.6 eV.

In atoms with more than two electrons the electrons are arranged in shells and subshells. The outer electrons are easiest to remove. In a neutral lead atom, for instance, with its 82 electrons, the easiest electron to remove has a binding energy of about 18 eV. The electrons nearest to the nucleus, which are the hardest to remove, have binding energies of about 88,000 eV.

The total binding energy of a nucleus is the amount of energy required to separate it into all of its nucleons. The binding energy per nucleon is the total binding energy divided by the number of nucleons.

If the total binding energy of a nucleus is higher, it takes more energy to decompose the nucleus into its constituents, and the nucleus is more stable.

For a group of nucleons the most stable arrangement is that which gives the largest binding energy per nucleon.

For uranium the binding energy per nucleon is about 7.5 MeV. The highest number is for iron and nickel, with about 8.8 MeV/nucleon.

The simplest nucleus with more than one nucleon is the deuteron, which consists of one proton and one neutron. It has the lowest binding energy, namely 2.2 MeV. The alpha particle consists of two protons and two neutrons and has a binding energy of 28 MeV.

EXAMPLE 3

A given number of nucleons can be arranged in various ways, as a group of deuterons, alpha particles, etc. Arrange the following in order of stability from the least stable to the most stable:

deuterons, alpha particles, iron nuclei, uranium nuclei, separated nucleons.

Ans.:
The quantity that needs to be considered is the binding energy per nucleon.

For separated nucleons the binding energy is zero. Therefore this is the least stable arrangement. In ascending order of binding energy per nucleon the others are deuterons, alpha particles, uranium nuclei, and iron nuclei. The values of the binding energy per nucleon are 0, 1.1, 7, 7.5, and 8.8.

1.3 The four forces

In our examination of the hierarchy from the very large to the very small, we started with the differences in size, and went on to the differences in the amounts of energy that are required for change or disruption. There is still another way to separate the various realms, and that is by the different kinds of forces that play the most important role in them.

We are most familiar with the gravitational force. It acts between us and the earth, with a force that we call our weight. It also acts between the earth and the sun and between the moon and the earth. It was Newton's great insight to realize that it acts between all objects and particles, attracting each to all others. We are normally aware of it only when one of the objects is of astronomical size, such as the earth, but already in 1798, a little more than 100 years after Newton's publication of the law of gravitation, Cavendish demonstrated that it also acts between objects light enough to be held in our hands.

On the astronomical scale it is the dominant force, and determines the paths of the moons and planets, as well as the large-scale structure of the universe, and, to a large extent, the fate of the stars.

It comes as a surprise to learn that there is a far stronger force that determines almost everything that we are aware of, including our very existence, and that of everything around us. That is the electric force. It is intimately connected to the magnetic force, and their combination is called the electromagnetic force.

Our direct awareness of the electric force is rare. We feel it when we get an electric shock, and we know that it is responsible for lightning. But its sweep goes enormously farther.

Our civilization depends, in some places almost entirely, on electricity for light, for heat and cold, for motors, for the many forms of communication from telephone to radio and television, and for information technology in all its variety.

It is, however, in the microscopic realm that the electric force reigns supreme. It holds the electrons to the nuclei to form atoms, and atoms to each other to form molecules, liquids, and solids. Each time we push or pull, we exert electric forces. The pathways along our nerves, as we feel, or see, or hear, or smell, are electrical.

All chemical changes are the result of changes in the way electrons move around nuclei. Even further, this is true as well for all biological processes, at least those that we know sufficiently to come to any conclusions at all about their microscopic nature.

The electrical nature of our civilization is apparent all around us. Each time we switch on the light, or the toaster, or the vacuum cleaner, pick up the telephone, turn on the radio, television set, or computer, we affect the motion of electrons. Each electron is so light, so small, that it cannot possibly be detected by our senses. Yet they cooperate, move along wires, and bring about the large-scale, macroscopic effects that we experience.

The electric nature of matter is far less apparent. All nuclei contain protons. All atoms contain electrons. The protons and electrons attract each other. But protons repel other protons and electrons repel other electrons.

We describe that by saying that both kinds of particles are *electrically charged*, and that there are two different kinds of charge: *positive* and *negative*. Each proton has a positive charge and each electron has a negative charge. Positive charges experience forces away from each other. They repel. The same is true for negative charges. But positive charges and negative charges attract. The force of attraction between the protons and the electrons, and the force of repulsion between the protons, and between the electrons, is called the *electric force*. This is the force responsible for the existence of atoms. The force between atoms is weaker, but it is also a manifestation of the electric force.

If there are charges in every atom and nucleus, why are we not more aware of them? The answer lies in the very strength of the force. The protons and electrons attract each other so strongly that we rarely find them separately. We seldom find a piece of material that does not contain equal numbers of protons and electrons, and when we do, the difference is minute.

Although the numbers of protons and electrons are almost always equal, we can shift the particles around with respect to each other. We can cause the electrons to move, on average, a little further away from the protons, or a little closer to them. This is what happens each time there is a chemical reaction.

The energy that changes when atoms and molecules combine or separate is commonly called *chemical energy*. Each such change, whether it is as subtle as in biological processes or as violent as in burning and explosion, is, on the microscopic, atomic scale, a change in electric energy.

With the vast variety of phenomena in the universe, it is amazing that just four fundamental forces are responsible for them. In addition to the gravitational force and the electromagnetic force there are two more, and their domain is the nucleus. One is the strong nuclear force, usually called the nuclear force, and the other is the weak nuclear force, most often simply called the weak force.

As we think of the nucleus, made up of protons and neutrons, it is clear that a special force must hold these particles together. The gravitational force is too weak, the electric force does not act at all on the neutrons, and repels the protons from each other. There must be a force that acts on both kinds of particles, i.e., on all nucleons, and that can overcome the disruptive effect of the electric force between the protons. This is the nuclear force. It is the strongest force that we know, but its range is so short that it acts only between neighboring nucleons. It holds the nucleons together to form the nucleus.

That leaves the weak force. Although it acts on nucleons and electrons, it is usually far overshadowed by the other forces. Because of it, however, some nuclear phenomena can occur that would be impossible without it, and they turn out to have profound influences on the universe. The weak force determines the timing of the life cycle of the stars, hence the existence of planets, and ultimately the conditions that make it possible for life to exist.

The difference in scale between the solar system, the atom, and the nucleus is reflected in the differences between the forces that dominate each realm. Each planet is held in its orbit by the gravitational force between it and the sun. The electrons and the nucleus are held to each other in the atom by the vastly stronger electrical force. The still stronger nuclear force acts between the protons and neutrons in the nucleus. The feeble weak force comes into its own in some special situations, which include some that help to control the pulse of the universe.

This brings us to one more surprise, one more insight unimagined before the twentieth century. The path from nucleus to atom, through us and the planets to the stars, is joined at the ends. It is the reactions between nuclei in the sun and the other stars that cause them to release their energy, the energy that radiates to the planets, that has brought about life, and that sustains our existence.

1.4 Atoms and the periodic table of elements

Although the concept of indivisible atoms as the building blocks of matter has been around since the time of the ancient Greeks, until the nineteenth century it was not based on experiments, and had not developed into a theory capable of analysis of known facts or the prediction of new ones.

The closely related concept of *chemical elements* developed gradually during the seventeenth and eighteenth centuries. By the end of that time it was generally accepted that there were elements, like oxygen, hydrogen, sulfur, and a number of metals, that could not be decomposed into other substances.

The two concepts were combined by John Dalton (1766–1844), starting in 1808, into a coherent scheme with many of the properties that are fundamental to modern chemistry. These include that elements are composed of atoms that are alike, that chemical compounds consist of what we now call molecules, each composed of two or more atoms, and that in a chemical reaction atoms are rearranged, but not created or destroyed.

At the same time some of the features of Dalton's scheme show how far his understanding was from today's. He thought, for instance, that atoms were at rest, held in place by repulsive forces between them. He also thought that each was surrounded by a shell of "caloric," the hypothetical substance that was supposed to represent heat.

Dalton's atoms and molecules explained why compounds consist of definite fractions of elements, as measured by the weights of their constituents. For example, hydrogen and oxygen combine to produce water. In each water molecule there are always exactly two hydrogen atoms and one oxygen atom.

Dalton's scheme led to the concept of *atomic weights*. These were only relative weights, since there was at that time no means of knowing of how many atoms or molecules a piece of material was composed. Knowledge of the size and weight of individual atoms and molecules did not come until more than half a century later, with the development of the kinetic theory of gases.

In the meantime the idea of atomic weights had a striking influence on the progress of knowledge of elements and their atomic structure. Hydrogen was recognized as the lightest element, and in 1815 William Prout, an English physician, suggested that all other elements were multiples of hydrogen. That turned out not to be the case, and Prout's hypothesis was abandoned. From today's perspective it seems remarkably prophetic.

The same idea was carried to a much more fruitful level by Dmitri Ivanovich Mendeleev in 1869. He ordered the known elements in the sequence of their atomic weights. It was known by that time that there were families of elements like the *halogens*, fluorine, chlorine, bromine, and iodine, and the *alkalis*, lithium, potassium, sodium, rubidium, and cesium, each with a number of common properties. These were partly chemical properties, such as the way in which they formed compounds. *Salts*, for example, are composed of one alkali element and one halogen element, as in table salt, *sodium chloride*. They were also physical characteristics, as for example the sequence of melting points within the families.

When the elements are put in the order of their atomic weights, it becomes apparent that there is a striking periodicity. After the first group of about eight elements, the next ones, in order, belong to the same families.

H = 1							
Li = 7	Be = 9.4	B = 11	C = 12	N = 14	O = 16	F = 19	
Na = 23	Mg = 24	Al = 27.3	Si = 28	P = 31	S = 32	Cl = 35.5	
K = 39	Ca = 40	– = 44	Ti = 48	V = 51	Cr = 52	Mn = 55	Fe = 56, Co = 59 Ni = 59, Cu = 63
(Cu = 63)	Zn = 65	– = 68	– = 72	As = 75	Se = 78	Br = 80	
Rb = 85	Sr = 87	?Yt = 88	Zr = 90	Nb = 94	Mo = 96	– = 100	Ru = 104, Rh = 104 Pd = 106, Ag = 108
(Ag = 108)	Cd = 112	In = 113	Sn = 118	Sb = 122	Te = 125	J = 127	
Cs = 133	Ba = 137	?Di = 138	?Ce = 140	–	–	–	–
–	–	?Er = 178	?La = 180	Ta = 182	W = 184	–	Os = 195, Ir = 197 Pt = 198, Au = 199
(Au = 199)	Hg = 200	Tl = 204	Pb = 207	Bi = 208	–	–	
–	–	–	Th = 231	–	U = 240	–	–

Others had noticed that, and looked for further repetitions. Mendeleev's *periodic table of elements* (shown in the figure) goes further in two ways. He realized that the periodicity is more complicated than a repetition after every eight elements. In addition, he was so convinced of the correctness of his approach that when an element seemed to be missing, he confidently predicted that a new element would be discovered to fill the empty spot. He soon had some major successes. Gallium, germanium, and scandium were discovered and took their places in the periodic table. Others among the roughly 60 that were then known had their places confirmed or adjusted. The periodic table of elements remains a cornerstone of chemistry, and a bridge between chemistry and physics.

Mendeleev's table was entirely empirical. In other words, it rested only on observation, without any knowledge of the underlying atomic structure that it reflects. It was not until the twentieth century that the reasons for its existence, in terms of atomic and nuclear structure, were understood.

The fact that atoms have an internal structure, with a nucleus, was not known until the experiments of Ernest Rutherford and his coworkers in 1911. The realization that nuclei consist of protons and neutrons had to wait for the discovery of the neutron in 1932.

The atomic weight is not even the right quantity to order the elements. Their properties showed that iodine and tellurium, as well as argon and potassium, are not in the right order if their atomic weights are used. It turned out that it is the number of protons, Z, in the nucleus that determines the particular element, and its place in the periodic table. That is why it is called the *atomic number*.

The number of neutrons can vary. Each different neutron number represents a different *isotope*. Some elements have only a single stable neutron number (a single stable isotope) and others have several. When we write ^{16}O, it says that there are 16 nucleons in the nucleus. Eight of them are protons because oxygen is the eighth element in the periodic table. The rest (also eight) are neutrons.

The weight of the nucleus depends on the number of all the nucleons (protons and neutrons) in the nucleus. It is almost equal to the atomic weight, since the electrons are less massive by a factor of almost 2000. Occasionally the atomic number and the atomic weight don't increase in the same order. For example, potassium has one more proton in its nucleus than argon, so that its atomic number is higher, but its most abundant naturally occurring isotope has two less neutrons, and so its atomic weight is less than that of argon. (A modern table of the elements is shown in the figure.)

Here is one more fascinating question: why does the periodic table end where it does? In Mendeleev's day, and until 1940, it was thought

1 H																	2 He
3 Li	4 Be											5 B	6 C	7 N	8 O	9 F	10 Ne
11 Na	12 Mg											13 Al	14 Si	15 P	16 S	17 Cl	18 Ar
19 K	20 Ca	21 Sc	22 Ti	23 V	24 Cr	25 Mn	26 Fe	27 Co	28 Ni	29 Cu	30 Zn	31 Ga	32 Ge	33 As	34 Se	35 Br	36 Kr
37 Rb	38 Sr	39 Y	40 Zr	41 Nb	42 Mo	43 Tc	44 Ru	45 Rh	46 Pd	47 Ag	48 Cd	49 In	50 Sn	51 Sb	52 Te	53 I	54 Xe
55 Cs	56 Ba	57 La	72 Hf	73 Ta	74 W	75 Re	76 Os	77 Ir	78 Pt	79 Au	80 Hg	81 Tl	82 Pb	83 Bi	84 Po	85 At	86 Rn
87 Fr	88 Ra	89 Ac	104 Rf	105 Db	106 Sg	107 Bh	108 Hs	109 Mt	110	111	112		114		116		118

58 Ce	59 Pr	60 Nd	61 Pm	62 Sm	63 Eu	64 Gd	65 Tb	66 Dy	67 Ho	68 Er	69 Tm	70 Yb	71 Lu
90 Th	91 Pa	92 U	93 Np	94 Pu	95 Am	96 Cm	97 Bk	98 Cf	99 Es	100 Fm	101 Md	102 No	103 Lr

that uranium, with its atomic number of 92, was the last element. In that year neptunium and plutonium were made by bombardment of uranium, and identified as elements 93 and 94. Later other "transuranic" elements, with even higher atomic numbers, were made. That requires cyclotrons and other devices that allow protons and other nuclear particles to be accelerated and used as projectiles. As the atomic number increases beyond plutonium, the nuclei become more and more unstable, and live for shorter and shorter times before disintegrating. In fact, no nuclei beyond atomic number 83 (bismuth) are stable. They are *radioactive*, that is, they change into other nuclei by spontaneous nuclear reactions that change the number of protons in their nuclei. Each radioactive transformation proceeds at its own characteristic rate. Some disintegrate very slowly, especially some isotopes of uranium, whose "lifetime" is measured in billions of years.

The two kinds of forces that act between nucleons in the nucleus are the strong nuclear force and the electric force. (The gravitational force and the weak force are too weak to have a detectable effect.) The nuclear force attracts neighboring nucleons to each other. The electric force acts in the opposite direction, repelling protons from each other. While the nuclear force is only effective between neighboring nuclei, the electric force acts over much larger distances, so that all protons in a nucleus experience it. As we look at nuclei with more and more nucleons, the effect of both forces grows. But because the nuclear force acts only between neighbors, and the electric force acts between all protons, the disruptive action of the electric force grows faster, and eventually overcomes the nuclear force that holds the nucleus together. At some atomic number no combination of nucleons can hold together even for a short time, and the periodic table of elements has reached its upper end.

1.5 Seeing atoms: the scanning tunneling microscope

After Dalton's work in the early nineteenth century the existence of atoms was gradually accepted, but remained controversial until the twentieth century. The evidence was indirect, and actually *seeing* atoms remained impossible.

Eventually atomic sizes could be estimated to be from one to several tenths of nanometers. (A nanometer, 1 nm, equal to 10^{-9} m, is a billionth of a meter.)

Simple magnifying lenses were known to the ancient Egyptians, and the compound microscope, consisting of two lenses, has been known since the end of the sixteenth century. There is, however, an inherent limit to the magnification that can be achieved by optical means. The wavelength of light is between 400 and 800 nm, so that light is too coarse a probe to be able to detect atoms. Shining light on a surface to see atoms is like trying to see a needle on a football field from a helicopter.

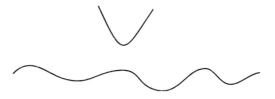

H. Rohrer and G. Binnig. Courtesy IBM Research-Zurich.

In 1981 Heinrich Rohrer and Gerd Binnig built a device based on an entirely different principle, capable of resolving distances of the order of atomic sizes. It depends on the fact that when a metal tip is very close to a metallic surface, electrons can flow from one to the other, as in a very small spark, even when the tip and the surface don't touch. This so-called "tunneling current" decreases very quickly when the distance is increased.

The realization of this instrument depended crucially on very steady and precise positioning of the tip, and on being able to move it in a controlled way along the surface that is being examined. Rohrer and Binnig used the *piezoelectric effect*, that is, the property of some materials, like quartz, to expand or contract when an electric voltage is applied across them.

Since the tunneling current varies so quickly with distance, they used an electric circuit to move the tip up or down so as to keep the current constant as the tip moves along the surface. This vertical motion then follows the details of the electron cloud on the surface, moving up and down at each atom.

Here is an example of a scanning tunneling microscope result. It shows a ring of iron atoms on a copper surface.

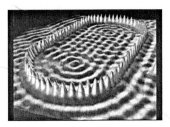

(Image originally created by IBM Corporation)

1.6 Summary

We looked at three ways to sort the different parts of the universe: their size, their binding energy, and the forces that predominate.

The sizes that we considered range from the *femtometers* (1 fm = 10^{-15} m) of the nucleons and the nucleus to the billions of kilometers across the solar system.

The binding energy of an object is the amount of energy that is required to take it apart. For atoms it is measured in *electron volts* (eV), for nuclei it is measured in millions of electron volts (MeV). 1 eV = 1.6×10^{-19} J. 1 MeV = 1.6×10^{-13} J.

The total binding energy of a nucleus is the amount of energy required to separate it into all its nucleons. We can also talk about the energy it takes to remove a single nucleon or other particle. It is called the binding energy of that particle.

The force that holds the atom together is the *electric force*. The force that holds the nucleus together is the *strong nuclear force*, sometimes just called the *strong force*, or the *nuclear force*. The *gravitational force* is much weaker. It is responsible for keeping the planets in their orbits around the sun. It is also the force that formed the planets and the stars.

The *weak nuclear force* is usually just called the *weak force*. Unlike the others, it doesn't hold anything together, but it regulates the timing of some important processes, including the life of the stars.

All materials are made up of the *elements*, each with its kind of atoms, starting with hydrogen. Each atom has a nucleus that consists of protons and neutrons. The number of protons in the nucleus is called the *atomic number*. The hydrogen atom has one proton in its nucleus and its atomic number is 1. At the other end of the list of naturally occurring elements is the uranium nucleus. It has 92 protons, so that its atomic number is 92. When the elements are put in order of their atomic number, they form the *periodic table of elements*. It is arranged in columns, each with a family of elements that shares a number of chemical and physical properties.

An element can have different *isotopes*, each with a different number of neutrons in the nucleus. Most properties, however, including the way elements combine (the *chemical* properties), depend only on the atomic number.

The numbers in the symbols ^{1}H, ^{16}O, ^{238}U, etc. represent the number of nucleons (protons and neutrons) in the nucleus. In a neutral atom the number of protons and electrons is the same. An atom with different numbers of protons and electrons is an *ion*.

Individual atoms can be separately detected ("seen") by using a *scanning tunneling microscope*.

1.7 Review activities and problems

Guided review

1. To make a large-scale representation of a hydrogen atom in its ground state, you start with a golf ball ($D = 4.27$ cm) to represent the nucleus. How far away (in km and miles— 1 mile = 1.61 km) will you have to put the representation of the electron?

2. Consider a giant similar to the one of Example 3, i.e., with body dimensions most of which are 10 times yours. The legs and leg bones, however, have a greater diameter, so that the giant's body weight, divided by the legs' cross-sectional area, is the same as for you.

(a) How much larger than yours is the cross section of the giant's leg?

(b) How much larger is the leg diameter?

(c) What are the answers to parts (a) and (b) in terms of the scale factor f (which is equal to 10 in this example)?

3. How much energy would it take to decompose a nucleus of ^{238}U, with its 238 nucleons into its separate nucleons? (This is the total binding energy of this nucleus.)

Problems and reasoning skill building

1. You want to draw a picture of the sun and the earth, to scale, on a sheet of paper. To fit it on the paper you start by representing the diameter of the earth's orbit around the sun by a distance of 25 cm.

(a) How large do you have to draw the diameters of the sun, the earth, and the orbit of our moon?

(b) How large a sheet of paper would you need to draw the earth's orbit if you had to represent the earth by a circle whose diameter is 1 mm?

2. The *order of magnitude* of a quantity is its size to the nearest factor of 10. For example, the order of magnitude of the size of an atom is 10^{-10} m.

What are the orders of magnitude of the following:

(a) sizes (in m) of the solar system, sun, earth, moon;

(b) binding energies (in eV) of the hydrogen atom and the deuteron.

3. What are the orders of magnitude of the sizes (in m) of the following:

(a) Highest mountain on earth

(b) Largest animal

(c) Smallest animal

(d) Hydrogen atom

(e) Nucleon

4. (a) How many people, standing on each other's shoulders, would it take to reach the sun from earth?

(b) Name an object so that the ratio of the size of a person to the size of the object is the same as the ratio of the radius of the earth's orbit to the size of a person.

5. For each of the following forces and combinations of forces name one or more objects that are affected by them.

(a) Nuclear but not electric

(b) Electric but not nuclear

(c) Electric and nuclear

(d) Electric and weak

(e) Weak but not electric

6. Which of the four fundamental forces predominates in each of the following cases?

(a) Motion of the planets

(b) Changing colors of leaves

(c) Radioactivity

(d) Digestion

(e) Friction

7. The "atomic" bomb dropped on Hiroshima contained about 60 kg of its explosive material (uranium "enriched" so as to contain more of the isotope ^{235}U than natural uranium). It released an amount of energy equivalent to that of about 15,000 tons of the "ordinary" chemical explosive TNT. Explain the magnitude of the ratio of these two weights in terms of the forces and energies involved in the explosion of the two materials.

8. Which two forces are in competition for the stability of nuclei? Which one holds the nucleus together and which one tends to disrupt it? Why is the disruptive force more important in large nuclei?

9. The atomic number of uranium is 92. How many neutrons are there in a nucleus of ^{235}U?

10. Hydrogen and helium are the first two elements in the periodic table of the elements. What would have to be changed in a nucleus of ^{3}He to convert it to a nucleus of ^{3}H?

11. What is the number of protons, neutrons, and electrons in a neutral atom of ^{23}Ne?

12. Gold and mercury are neighbors in the periodic table of elements. What has to be changed to change an atom of mercury into an atom of

gold? Explain why this is so difficult in terms of the forces and the binding energies.

13. The simplest stable nucleus consisting of more than one nucleon is the deuteron. It consists of a proton and a neutron. It is the nucleus of an atom of deuterium or heavy hydrogen, which is 0.015% of ordinary hydrogen. It takes an energy of 2.24 MeV to separate the deuteron into its two constituents.

(a) What is the ratio of this energy to the energy required to separate the proton and the electron in a hydrogen atom?

(b) What does this problem illustrate about the strength of the forces that are involved and about the relative stability of atoms and nuclei?

14. The binding energy of an electron in a helium atom is 24.6 eV. The binding energy of a neutron in an alpha particle (the helium nucleus) is 20.6 MeV. By what factor is it harder to remove the neutron than the electron? What general feature of nuclei and atoms does this example illustrate?

15. The nuclear force has such a short range that it acts only between neighboring nucleons. The electric force decreases with distance, but quite gradually (as $\frac{1}{r^2}$). Consider a nucleus with 100 protons and 120 neutrons. If one more proton is added, how many nucleons will experience an additional electric repulsion? How many nucleons will experience an added nuclear attraction? As nuclei get bigger, how does the importance of the electric force change, compared to that of the nuclear force?

Multiple choice questions

1. To take apart a nucleus is harder than to take apart an atom by a factor of about
    (a) 100
    (b) 10,000
    (c) 1,000,000
    (d) 100,000,000

2. The modern periodic table of elements is ordered by
    (a) Atomic weight
    (b) Atomic number
    (c) Number of nucleons in the nucleus
    (d) Number of neutrons in the nucleus

Synthesis problems and projects

1. About what fraction of the volume of the solar system is occupied by the sun and the planets?

2. About what fraction of the volume of an apple is empty space?
   Think of this question in steps.
   (a) What fraction of the hydrogen atom is empty space? (The electrons may be considered to be point particles that take up no space at all.)
   (b) As a rough approximation, what do you expect this ratio to be in other atoms?
   (c) What do you expect for the apple?

3. What is it that Mendeleev discovered that led to the first periodic table of the elements?

4. In the periodic table of elements the pairs argon and potassium, and iodine and tellurium are not "in the right order" if the elements are arranged in order of their atomic weights. What does it mean to say that they are not in the right order, and how can you tell?

5. Search the Internet under "Periodic Table of Elements." On several of the tables you can click on the symbol for an element to find its properties.
   What is the abundance of *deuterium*? This is the percentage of hydrogen where the nucleus of a hydrogen atom is not a proton, but a *deuteron*, consisting of a proton and a neutron.

6. What are some of the features of Mendeleev's original table that are incorrect? What features are missing?

7. Here is a statement from the text: "The weak force determines the timing of the life cycle of the stars, hence the existence of planets, and ultimately the conditions that make it possible for life to exist." Explain the connection between life on earth and the life cycle of stars.
   An excellent discussion of this question is in the article "Energy in the Universe" by Freeman J. Dyson in the September 1971 issue of *Scientific American*. ("The proton-proton reaction proceeds about 10^{18} times more slowly than a strong nuclear reaction at the same density and temperature," implying that if the life cycle of the sun were governed by a strong nuclear reaction it would have burned up long ago.) This issue was reprinted as "Energy and Power"

(W. H. Freeman, 1971). It includes a number of outstanding articles that remain relevant today.

8. Search the Internet under "Scanning Tunneling Microscope." Find the Nobel-Prize lecture of its inventors. Read at least the first half dozen paragraphs.

Was their original aim to build a microscope?

How much bigger is the area that they hoped to get down to than the area of a hydrogen atom? (Other atoms have more electrons and you might expect them to be much larger. But the nucleus then also has more charge and holds the electrons more tightly. As a result the sizes of atoms do not vary very much.)

You may need to know what an Angstrom (Å) is. It is 10^{-10} m. (Also: "spectroscopy" is the study of the possible energies of a system.)

Find and look at the "STM image gallery" (from IBM). It includes the STM image in this chapter.

9. What would be the form of Kleiber's law if the simple scaling described in the giant's soliloquy in Example 2 were followed?

CHAPTER 2

Some Tools of the Trade: Numbers, Quantities, and Units

The language of physics: symbols and formulas
> *Positive and negative numbers*
> *Zero*
> *Numbers, huge and tiny: powers of 10*
> *Precision: Significant figures*
> *Quantities and units*
> *Ratios and proportional reasoning*
> *Tables, graphs, equations, and functions*
> *Right-angled triangles*

Once more the four forces, this time quantitatively
> *The gravitational force*
> *The electric force*
> *The other forces*

So far we have used words, almost exclusively. We have talked about size, force, and energy, but mostly without using numbers, although a few times they almost forced themselves on us. But most of the time we have to know how big a distance is, or a force, or any other quantity, and that requires numbers and units. We also want to describe relations between different quantities, and do that with symbols, such as E for energy, and M for mass.

All of that is *mathematics*. Mathematics is the language in which the ideas, facts, and relationships of physics are best expressed. Sometimes it's just shorthand. It is much easier to write $v = 32$ m/s than "the velocity is thirty-two meters per second," or $\bar{v} = \frac{\Delta x}{\Delta t}$, rather than "the average velocity of an object is equal to the displacement along the x-axis divided by the time it took to make that displacement."

Sometimes the relationships are more complex, like $F = Ma$: "the force, or, if there is more than one, the sum of all the forces acting on an object, is equal to the mass of the object multiplied by its acceleration."

You can see that just as a way of writing things down in shorthand notation with symbols like v and x, math is very helpful. But it does more than that. It lets us write down relationships between different quantities, and change them so that they lead to other relationships. To do the same with words would be cumbersome even for the simplest ones, and close to impossible for others.

Mathematicians can make up their own rules. All that is required is that they are not in conflict with one another, i.e., that they are *internally consistent*. Physics, and more generally science, does something quite different. It describes the world as it exists. We can make up rules and laws, but more than consistency is required of them. The more severe test is whether they help us in the primary task, the description of the dynamic, interacting, ever-changing, endlessly new world that we observe.

2.1 The language of physics: symbols and formulas

Look at the equation $y = 3x^2$. It could be seen as a "formula," or "recipe" that expresses the quantity y in terms of the quantity x: when $x = 1, y = 3$, when $x = 2, y = 12$, and so on. Given x we can follow the rules and find y. A physicist is more likely to see it as a *relationship*, for which the mathematical equation provides the description. The equation shows how y depends on x. In other words, it shows y as a *function* of x. It shows that y is proportional to x^2, and brings to mind a picture of the graph of y against x that shows how the two quantities are related.

Some symbols stand for quantities, like distance, L, and their units, like meters, m. Others describe a procedure or operation, like *plus* (+) or *times* (×), or a relation like *equals* (=) or "is proportional to" ($\propto$).

A page with unfamiliar symbols does not give us understanding or comprehension. It is like a page of musical notes that comes alive to the musician as he or she looks at it, but remains hidden to those who do not know musical notation. Moreover, to know what each written note means is far from knowing what an orchestral passage sounds like. It is the same with a page of physics filled with mathematical notation. Our intention is to help you see, follow, and "hear" what is being described.

When we speak or write we use the "parts of speech," the nouns, verbs, adjectives, and so on. Their proper use leads to understanding and communication, while their improper use can lead to confusion. Similarly, to "speak physics" we need mathematical components, such as numbers (positive, negative, and zero), graphs, proportionalities, and equations. In the following sections we review some of the mathematical procedures and ideas that we will use to communicate.

Positive and negative numbers

In physics we need to quantify. We ask *How much? How far? How large?* The answers are expressed in numbers. In many cases we want to distinguish between opposites such as *right and left, past and future, speeding up and slowing down*. We do that by using positive and negative numbers.

Here is an example. I walk six steps to the right and then four steps to the left. How far am I from where I started?

We let the starting point be represented by zero, "to the right" by positive numbers, and "to the left" by negative numbers. We can then describe the motion mathematically as (+6 steps) + (−4 steps) = (+2 steps). We use *plus* and *minus* signs, but we use them in two entirely different ways. One (within the parentheses) is to indicate which way is to the right and which to the left. The other (between the parentheses) is to indicate the *operation* of addition.

To distinguish between the two uses we can reserve "plus" and "minus" for the operations of addition and subtraction, and use "positive" and "negative" to indicate positive and negative numbers. Our relation then reads "positive six steps plus negative four steps equals positive two steps." "Positive" and "negative" tell us what kind of numbers we have, and "plus" and "minus" tell us what we do with them.

The choice of which direction to call positive is up to us. If we decide to let "to the left" be positive, we would write (−6 steps) + (+4 steps) = (−2 steps). Both choices lead to the same result, namely that we end up two steps to the right.

Sometimes we don't care whether a number is positive or negative; we just want to know how big it is. That's its *magnitude*. The magnitude of both +15 (positive 15) and −15 (negative 15) is 15. When we look at the speedometer in a car we see how fast the car is going. That tells us its *speed*: the speed is 30 miles per hour. To incorporate the direction, we need the *velocity*: the velocity is 30 miles per hour, north. The speed is the magnitude of the velocity.

EXAMPLE 1

Zahra and Mona pull in opposite directions on a cart (*A*) that is initially at rest. Laila and Yasmine also pull in opposite directions on a second identical cart (*B*). The figure shows the forces that they apply in force units.

(a) Represent these actions by mathematical statements, using numbers.

(b) In which direction does each cart start to move?

(c) Which cart is being pulled with the greatest net force?

Ans.:
(a) Let the positive direction be to the left.
 Cart *A*: $(+3 \text{ units}) + (-2 \text{ units}) = (+1 \text{ unit})$.
 Cart *B*: $(+3 \text{ units}) + (-5 \text{ units}) = (-2 \text{ units})$.

(b) Cart *A* starts to move in the positive direction, to the left, as a result of the net pull (or force) to the left of one unit.
 Cart *B* starts to move in the negative direction, to the right, as a result of the net pull (or force) in that direction of 2 units.

(c) The size or amount of the net force on cart *A* is 1 unit, the size of the net force on cart *B* is 2 units. Cart *B* is being pulled with twice the net force of cart *A*. The negative sign tells us the direction and the number tells us how hard the pull is.

Zero

Zero can have several meanings. It can mean the absence of a quantity, as in "the car's speed is zero." Here the car has no speed at all. It is not moving.

A second meaning is for zero to denote a starting point. From this point we can move in one direction or the other, or the temperature can change up or down. More generally, zero can be a position, or a temperature or the value of some other quantity, from which we count positions, temperatures, etc. This is what we do when we use the *Celsius* scale of temperature, where *zero degrees* refers to the temperature at which water freezes.

A third use of *zero* is to denote a balance or neutral condition between opposites. Two equal forces in opposite directions produce a *zero net force*. A neutral atom has just as many protons (with their positive charge) as electrons (with their negative charge). Its *net electric charge* is zero.

EXAMPLE 2

You go to bed and the temperature is 18°C. Overnight the temperature drops by 18°C.

Write a mathematical expression for what happens, first in the C-scale and then in the F-scale. What is the temperature in the morning and in the evening in degrees Celsius and Fahrenheit?

Ans.:
$t = (18°C) + (-18°C) = 0°C$. The final temperature is 0°C.

The size of a degree C is $\frac{9}{5}$ times the size of a degree F. The magnitude of the drop in temperature is therefore $(18)(\frac{9}{5}) = 32.4$ Fahrenheit degrees.

0°C or 32°F is the freezing temperature. This is also the morning temperature. The evening temperature of 18°C is equal to $32°F + 32.4°F = 64.4°F$.

The final (morning) temperature is 0°C. It is also the evening temperature plus the change in temperature, $(64.4°F) + (-32.4°F) = (32°F)$.

EXAMPLE 3

Two teams are engaged in a tug of war. Each team pulls on the rope with 25 units of force. The rope does not move. Write a mathematical description of the forces with which the rope is being pulled.

Ans.:
Let the positive direction be to the right. $(25 \text{ units}) + (-25 \text{ units}) = (0 \text{ units})$.

The rope is being pulled from both sides as each team exerts a force, but the net force on it is zero.

EXAMPLE 4

The atomic number of sodium is 11. In its most common ionized state each atom loses one electron. Represent the charges of the neutral atom and the ion using a mathematical description of the numbers of electronic charges.

Ans.:
The proton and the electron have charges of the same magnitude, e, positive for the proton and negative for the electron. The neutral sodium atom contains 11 protons and 11 electrons. The net amount of charge is $(+11e)+(-11e) = 0$.

When the atom is ionized and loses one electron, the negative charge is $(-10e)$. The amount of charge of the ion is then $(+11e)+(-10e) = (+1e)$.

(The number of protons in a sodium nucleus is 11, regardless of whether the atom is neutral or whether it has gained or lost electrons.)

Numbers, huge and tiny: powers of 10

When we use atomic or astronomical distances we need to use very small and very large numbers. It is helpful to use a notation that avoids long strings of zeros.

We can write the number 100 as 10^2 and the number 1000 as 10^3. The "2" and the "3" are called "exponents." You can think of them as the number of steps that the decimal point is shifted to the right of the "1." This way of looking at the exponent also works in the other direction: 0.01 is written as 10^{-2}, where the negative sign means that we move the decimal point to the left from its position after the "1."

To multiply two numbers written in this power-of-ten notation we need to add the exponents: $100 \times 1000 = 100,000$, can be written as $10^2 \times 10^3 = 10^5$.

Since 0.01 is equal to $\frac{1}{100}$, we see that 10^{-2} is equal to $\frac{1}{10^2}$.

We can extend the notation to other numbers by writing 6.2×10^3 instead of 6200. Again, the exponent gives the number of steps that the decimal point is moved, either to the right if the exponent is positive, or to the left if it is negative.

EXAMPLE 5

The earth is about 150,000,000,000 m from the sun. Pluto is about 6,000,000,000,000 m from the sun.

(a) What is the ratio of the distances?
(b) What is the radial distance from Pluto's orbit to the orbit of the earth?

Ans.:
Sun to earth: 1.5×10^{11} m.
Sun to Pluto: 6.0×10^{12} m.
Ratio: $\frac{6.0 \times 10^{12}}{1.5 \times 10^{12}} = \frac{6.0}{1.5} \times 10^{(12-11)} = 4.0 \times 10^1 = 40$, so that Pluto is 40 times as far from the sun as is the earth.

Difference: $6.0 \times 10^{12} - 1.5 \times 10^{11}$. This is easiest if both exponents are the same: $6.0 \times 10^{12} = 60 \times 10^{11}$ and $60 \times 10^{11} - 1.5 \times 10^{11} = 58.5 \times 10^{11}$, which can be rounded off to 58 or 59×10^{11} m.

Precision: significant figures

You may describe the distance from the edge of a desk to the nearest wall as 5 feet and 4 inches, which is about 5.3 feet. But if someone tells you that 5 feet 4 inches is really 5.33333 feet, with the threes going on forever, you have to ask yourself how many of these threes to write.

It is true that 4 inches equal $\frac{1}{3}$ ft, or 0.333... ft, with an unlimited number of threes, but we are looking at the results of a measurement, not at an abstract number. To write down that third "3" implies that we know that it is not a "2," a "4," or some other number. If we really know, because the distance was measured sufficiently precisely, then this number should be there, because it is *significant*. Otherwise it should be left out.

If we write the distance as 5.3 ft, we imply that it is bigger than 5.2 feet and smaller than 5.4 feet. That's not a bad measurement. It says that we know the distance to about one part in 53. That's about two parts in 100, or 2%. If we write 5.33 ft, the implication is that it is not 5.32 or 5.34, so that we know the distance to about one part in 533, or 0.2%. You have to ask yourself whether you really know the distance that well.

A calculator can give numbers with many figures, and you have to decide how many of them are significant. If the number is the result of a measurement, the number of significant digits depends on the precision of the measurement. It is most often three figures or less. We usually "round" the last digit that we keep, either down, by leaving out the next one, if it is less than 5, or up, by increasing it by 1 if it is 5 or more. 5.333 becomes 5.33 and 5.336 becomes 5.34.

Sometimes we only know the exponent, that is, we may know only that a number is 10^{23}, and not 10^{22} or 10^{24}. We then say that we know the *order of magnitude*.

There are rules about what happens to the number of significant figures when numbers are combined. For example, if you multiply or divide two numbers, each with three significant figures, the result should also have three significant figures. In general, the least precise number in a combination will determine how precise the result is. On the other hand, if you are not quite sure, it is better to keep an extra significant figure until the end of a calculation, and only then to round it. Rules are useful, but common sense may also be necessary.

For simplicity we will assume that the quantities in our examples and problems are known to three significant figures unless we say something different. In other words, when we say 2 m, it will imply 2.00 m, and when we give a force as 5.3 N, that will imply 5.30 N. The calculations and answers should therefore also be carried out to three significant figures.

EXAMPLE 6

You have a rope that is 52.3 feet long, which you want to cut into three equal parts. How long should each piece be?

Ans.:
You put the numbers in your calculator and get $\frac{52.3}{3} = 17.43333333$ ft. You now look at the appropriate precision, and see that the length of the rope is given to three significant figures. You therefore round the answer to 17.4 ft. You check that this is reasonable by considering the precision of the measurement. 0.1 ft is about an inch, and this seems about right for the precision of the measurement of the length of a 50 ft rope.

Quantities and units

We have already talked about *energy*. You know that energy is what you get from the gasoline that you put in the car and from the electric company when you plug in a light. It is what you get from the food you eat and what you use when you lift this book. Another term that we have used is *force*. It describes our interaction with a wagon when we pull it and with a wall when we push on it. The baseball bat or the tennis racket interacts with a ball by means of forces, and so do you and the earth when you jump, and even when you stand still.

Terms like *force* and *energy* are used in everyday language, but they are given much more precise meanings in science. We need more than a general idea of what we are talking about. We want to be quantitative, and for that we need definitions that provide recipes for measuring the various quantities. Only then can we say how big a force is and how much energy is required for a particular task.

We will talk about the connections between distance, force, and energy that lead to their precise definitions in the next chapters. In the meantime we will continue to use them, since we already have a rough idea of what these quantities are. Let's add one more, the *mass*. It is the property that tells you how hard it is to get an object to change its velocity. From it you can also see how hard the earth pulls on it. That's the *weight* of the object.

To measure something we need units. Inches, meters, and miles are among the units that are used to measure distance, and only when we know how big each is can we make sense of what is meant by 3 inches, or 5 meters, or 10 miles. If we know how many kilometers are equivalent to 1 mile, we can change from describing a distance as 15 miles to saying that it is some number of kilometers.

For energy we need other units, such as the *kilowatt-hour*, the *calorie*, the *joule*, and the *electron-volt*. These are among the units that let us describe and compare the energy needed to lift this book from the floor to the table, to keep a light on for three hours, or to remove an electron from a hydrogen atom. They also let us describe how much energy a person uses in a day, or how much all of us in this country, or on the whole earth, use in a year.

We start by comparing different units for the same kind of quantity, for example distance. It is helpful to have a fundamental unit, and then to express the others in terms of it. Most of the time we will use the units of the "SI System" (from the French *Système Internationale*) as our fundamental units. The SI unit for distance is the *meter*, m, but distance can also be measured in many other units (miles, inches, centimeters, etc.). Other SI units include the *joule*, J, for energy, the *newton*, N, for force, the *kilogram*, kg, for mass, and the *second*, s, for time.

Although we will wait with the exact definitions of energy, force, and mass, we can anticipate some of what we will get to. The *kilogram* is the fundamental (SI) unit of mass. The *newton* is the SI unit for force. The weight of an object whose mass is 1 kg (the force that the earth exerts on it near the earth's surface) is about 2.2 pounds, or about 9.8 N.

Suppose you want to express a distance of 6.71 m in feet, using the fact that 1 m = 3.28 ft. The easiest way is to multiply the 6.71 m by a fraction that is equal to one, with the meters in the denominator and the same distance, in feet, in the numerator. 6.71 m is the same as $\frac{6.71 \text{ m}}{1}$, so that we can write $(\frac{6.71 \text{ m}}{1})(\frac{3.28 \text{ ft}}{1 \text{ m}})$. Since units that appear both on the top and on the bottom cancel, the meters cancel, and we are left with (6.71)(3.28) ft, or 22.0 ft.

EXAMPLE 7

How long (in days and years) would it take you to count 100 million one-dollar bills if you counted them at the rate of two per second?

Ans:
$(2\frac{\$}{s})(t) = 10^8$, or $t = 0.5 \times 10^8$ s.

$(0.5 \times 10^8 \text{ s}) \frac{1 \text{ h}}{3600 \text{ s}} \frac{1 \text{ d}}{24 \text{ h}} = \frac{0.5 \times 10^8}{(3600)(24)} \text{ d} = 579 \text{ d}$

$= \frac{579}{365} \text{ y} = 1.6 \text{ y}$

At this rate it will take 1.6 years!

Ratios and proportional reasoning

The ratio of the circumference of a circle to its diameter is π. This is so regardless of the size of the circle. (π is a number with an infinite number of decimal places. To three figures it is 3.14.) We can write $\frac{c}{d} = \pi$, and $c = d\pi$, where c is the circumference and d is the diameter. We can also write "c is proportional to d" ($c \propto d$), which means that c is equal to a constant times d. If d is multiplied by 2 (or some other number) then c is also. This is an example of "proportional reasoning."

Here is another example. The energy of motion (or "kinetic energy," K) of a car is $\frac{1}{2}mv^2$, where m is its mass and v is its speed. What happens to K when v doubles?

When the speed has the value v_1, K has the value K_1, equal to $\frac{1}{2}mv_1^2$. When the speed is v_2, the energy is $K_2 = \frac{1}{2}mv_2^2$. $\frac{K_2}{K_1} = \frac{\frac{1}{2}mv_2^2}{\frac{1}{2}mv_1^2}$, which is equal to $\frac{v_2^2}{v_1^2}$ or $(\frac{v_2}{v_1})^2$. When the speed is doubled, $\frac{v_2}{v_1} = 2$, and $(\frac{v_2}{v_1})^2 = 4$, so that the energy is then four times as large.

More simply, we can write $K \propto v^2$ (K is proportional to v^2) since the mass does not change. This shows that when v is multiplied by some number (here 2), K is multiplied by the same number squared. Proportional reasoning allows us to say this without knowing the values of m, v, or K. The equation $K = \frac{1}{2}mv^2$ describes the relation between K and v. The proportionality, $K \propto v^2$, is part of that relation. It allows us to see how K changes when v changes.

Suppose that $P = \rho g h$, where ρ and g are constant. We don't need to know anything about ρ and g to see how P and h depend on each other: we see that $P \propto h$. If h is tripled, so is P. Proportional reasoning allows us to predict the effect on P of a change in h, while all other quantities remain unchanged. In this relation P is the pressure in a liquid whose density is ρ (greek *rho*), at a depth h, and g is the weight in *newtons* of a body whose mass is one *kilogram*, but regardless of what the letters represent, as long as ρ and g are constant, $P \propto h$.

EXAMPLE 8

An approximate relation between the force of air resistance, F_a, on a moving car and its speed v is $F_a = kv^2$. k is a constant quantity that depends on the size and shape of the car. The car speeds up from 40 miles per hour to 60 miles per hour. By what factor will the force of air resistance increase?

Ans.:
$F_a \propto v^2$. v increases from the initial speed $v_i = 40$ mph to the final speed $v_f = 60$ mph, while the force of air resistance changes from its initial value F_{ai} to its final value F_{af}. $\frac{F_{af}}{F_{ai}} = \frac{kv_f^2}{kv_i^2} = (\frac{v_f}{v_i})^2 = (\frac{60}{40})^2 = (1.5)^2 = 2.25$. The final value F_{af} is 2.25 times the initial value F_{ai}, i.e., the force of air resistance increases by a factor of 2.25.

Tables, graphs, equations, and functions

The results of an experiment or an observation can be shown in a table of data. The table might show values of some quantity x, and for each value of x the corresponding value of some other quantity y. We can also represent the data on a graph of y against x. The table and the graph show how y changes when x changes. They show y as a *function* of x, or in shorthand notation they show $y(x)$, pronounced "y of x."

We try to find patterns in the data. For example, the graph might be a straight line, a parabola, or some other shape, which we can then describe by an equation.

EXAMPLE 9

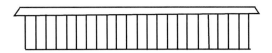

You skate down the sidewalk past a long building with regularly spaced windows. You count the number of windows that you pass. You can then make a table. Let x be the time in seconds and y the number of windows passed.

x time in s	y windows passed
0	0
1	4
2	8
3	12
4	16
5	20

(a) Make a graph of the data.

(b) Represent the data by an equation.

(c) Why is it physically meaningful to replace the data with a function?

Ans.:

(a)

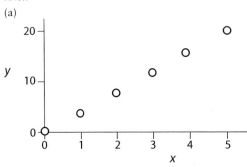

(b and c) You can draw a line that passes through the points. What do the parts of the line between the points represent? You didn't count anything there, but the line tells you what the measurements would be if the same trend is followed.

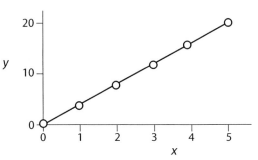

You can express the line on the graph by an equation: the equation that describes the line is $y = 4x$.

Each of the three descriptions (the table, the graph, and the equation) describes the same data. Each gives its particular perspective. Each is a different *representation*. Each illuminates what is happening in a different way.

EXAMPLE 10

x (days)	y ($ spent)
0	0
1	5
2	11
3	20
4	23
5	31
6	35

You start the month with $100 budgeted for eating out and entertainment. The table shows your expenses as they accumulate day by day for the first week.

(a) Make a graph of the data.

(b) Write the mathematical function that describes the line on the graph.

Ans.:
(a)

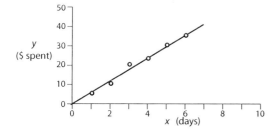

The points on the graph lie quite closely, but not exactly, along a straight line. By drawing a straight line through them you assume that the relationship is linear. The line represents your estimate of the "best fit" to the data. It *averages* over the deviations from the straight-line relationship.

(b) The graph shows the amount of money spent (y) plotted against the number of days that have passed (x). It shows y as a *function* of x, or $y(x)$. The equation of the line that we have drawn is $y = 6x$. The slope, the number of dollars spent per day, is 6 $/day. It is positive, showing that y increases as x increases.

Alternatively we can use the same data to plot the amount of money that remains against the number of days:

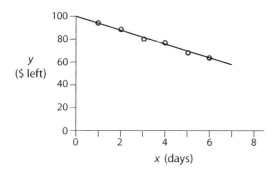

The line starts with $100 at time 0 and decreases by $6 each day, on average. The equation of the line is $y = 100 - 6x$. This time the slope is negative, equal to -6 $/day. The negative sign shows that y is decreasing as x increases.

Both graphs give a description of the same data, but they do so differently. Each is a different representation.

One value of the different representations is that they allow us to find patterns: the skater passes four windows in each second; the money is spent at the constant rate of six dollars per day.

A second value is that they allow us to see what happens at points in between the recorded data. You can look at the graph and see how many windows you have passed after 3.5 seconds. This is called *interpolation*. You can also see how many windows you would pass if you kept going for 10 seconds. This is called *extrapolation*. It gives the correct answer only if the pattern continues unchanged. In this example it will do so only if the windows continue to be equally spaced for a sufficient distance, and if you continue skating in exactly the same way.

For the example in which you spend money at the rate of six dollars a day, you can extrapolate to see when you will run out of money if you continue to spend it at the same rate.

What is the mathematical description of this question? The equation is $y = 100 - 6x$, where y is the money left and x is the number of days. You can then ask what x is when y is zero.

The equation then becomes $0 = 100 - 6x$, and it leads to the solution $6x = 100$, or $x = \frac{100}{6} = 16.7$ days. This shows that with the given rate the amount of money left will go to zero in 16.7 days. You can conclude that unless you change the rate at which you spend money you will not be able to go out to eat or go to a movie for almost half of the month.

Here is another example. A mouse comes out of its hole and moves along the floor in a straight line. That's the event. That's what happens. The words, spoken or on paper, are one representation. The drawing, and a photograph, are other representations. So are graphs of the mouse's position or speed as a function of time. Each representation gives some partial information about what happens.

Here is a table of measurements of the mouse's position. x is the distance from the starting point at the mouse hole, in centimeters. t is the time in seconds, in half-second time intervals, measured from the moment the mouse emerges.

x (cm)	t (s)
0	0
0.5	26
1	52
1.5	61
2	64
2.5	65
3	63
3.5	59
4	53
4.5	51
5	51

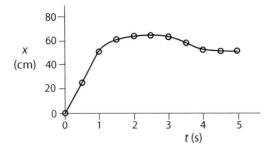

The graph shows the position of the mouse as a function of the time. It shows the *data*, i.e., the results of measurements. To make this graph, a number of decisions had to be made. What quantities do we plot on each axis? What are the units on each axis? What is the scale, i.e., how far apart are the numbers on each axis? Where is each quantity measured from, i.e., where is each of the quantities equal to zero? Is that also the place where the graph has its origin, i.e., where the axes (plural of *axis*) cross? Each of these decisions affects the way the graph looks. Together they describe the *coordinate system*. The facts (the data) are not changed. But the way the facts are represented by a graph is different for each coordinate system.

The graph starts out as a straight line. During this part of its travel the mouse covers equal distances in equal times. In other words its speed is constant: 26 cm in the first half second and 26 cm in the second half second, or 52 cm/s. It then slows down, going only 13 cm (from $x = 52$ cm to $x = 65$ cm in the next 1.5 s.

What happens then? We see that x, the distance from the starting point, decreases during the next two seconds. The mouse is moving back toward the starting point. During the last half second x does not change. The mouse is not moving.

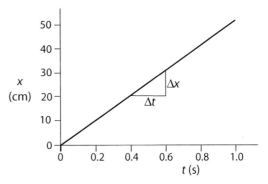

Now zoom in on the part of the graph on the left, near the origin. There the points lie along a straight line. The small triangle represents a change along the x-axis that we call Δx (capital Greek *delta*) and a change along the t-axis that we call Δt. Here Δx is 10.4 cm, and Δt is 0.2 s.

The ratio $\frac{\Delta x}{\Delta t}$ is the *slope* of the line. Here it is $\frac{10.4}{0.2}$ cm/s, or 52 cm/s. If the graph describes physical quantities (not just numbers), each axis must include *units*. Here they are cm and s. The slope has units of cm/s. For this line the slope is 52 cm/s. We see that the slope of the graph of $x(t)$ is the velocity.

On the graph of $y(x)$, y is a function of x. Just numbers are given this time, no units. The graph

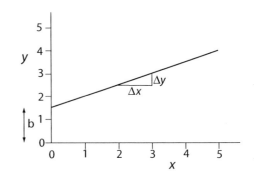

is a straight line. The place where the line crosses the y-axis is called the y-intercept. The symbol used to represent the y-intercept is b. Here it is 1.5. Δy is a change parallel to the y-axis (here 0.5), and Δx is a change parallel to the x-axis (here 1.0). The slope of the line is $\frac{\Delta y}{\Delta x}$. The symbol used for the slope is m. Here it is 0.5. We can choose a different triangle of Δx and Δy along the line, but their ratio, the slope, remains the same. In fact, we could define a straight line as one whose slope is constant. The equation of this line is $y = 0.5x + 1.5$. In general, the equation of a straight line is $y = mx + b$.

The main part of an equation is the *equal* sign. It tells you that what is on the left of it, the left-hand side of the equation (lhs), has the same value and sign as what is on the right-hand side (rhs). $3 + 8 = 4 + 7$, for instance. Most often there are not just numbers, but also symbols, as in $3x = 12$. Now x has to be equal to 4, or the lhs is not equal to the rhs. We can see what x must be by dividing each side by 3. This leaves x by itself on the left, and $\frac{12}{3}$ or 4 on the right.

We have changed both sides in the same way by dividing each side by the same number, 3. We can also multiply each side by the same number or quantity, add the same thing to each, or subtract the same from each side. In each case the lhs will still be equal to the rhs. The equal sign will still *hold*.

Equations can be much less complicated than expressing what they say in words. In fact, symbols and equations give us a marvelous shorthand way to express complex relationships and to describe a great deal of information.

Each equation describes a relationship: the lhs is equal to the rhs. We can change it in a lot of ways as long as each change preserves that basic relationship. Often we want to make these changes, manipulate the equation, so that one quantity (the *unknown*) is on the lhs, with other quantities, presumably known, on the rhs. If we can do this, we say that we have *solved* the equation.

EXAMPLE 11

The sales at a lemonade stand can be described by the function $y = -(x-2)^2 + 9$, where x is the number of hours of operation after opening and y is the rate of sales in cups per hour.

(a) Make a table of the values of x and y at hourly intervals.

(b) Plot the points from your table and sketch $y(x)$.

(c) After how many hours does the number of cups per hour go to zero?

Ans.:

(a)

x (h)	$y \frac{\text{cups}}{\text{h}}$
0	5
1	8
2	9
3	8
4	5
5	0

(b)

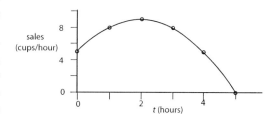

(c) The rate is seen to go to zero after 5 hours.
From the equation: when $y = 0$ it reduces to $0 = -(x-2)^2 + 9$, or $(x-2)^2 = 9$, and, if we take the square root of each side, $x - 2 = \pm 3$, so that $x = 5$ or -1.

Only positive values of t are appropriate here. But you can see on the graph that if the line is continued to the left, it crosses the x-axis at $x = -1$. This shows that it is not enough to follow the mathematics blindly. The solution has to be looked at to see whether it represents the physical situation correctly.

Right-angled triangles

Positive and negative numbers are sufficient to describe positions, distances, and directions as long as we stay along one line, i.e., in *one dimension*. On a plane (in *two dimensions*) or in three-dimensional space we need further tools.

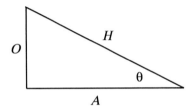

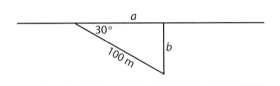
Ans.:

For example, we often make use of the properties of right-angled triangles.

The figure shows a triangle with one right angle (90°) and another angle, θ. One of its sides is marked "O" (for *opposite*) because it is opposite the angle θ and one side is marked "A" (for *adjacent*) because it is adjacent to this angle. The third side is the *hypotenuse*, "H," whose length is given by *the theorem of Pythagoras* (or the Pythagorean theorem) $A^2 + O^2 = H^2$.

The triangle shows the initial part of your path of 100 m, and the second part, to the shore, which is marked b, as well as the distance from there back to the starting point along the shore, which is marked a.

(a) $\sin \theta = \frac{\text{Opposite}}{\text{Hypotenuse}} = \frac{b}{100}$. $b = 100 \sin \theta = 50$ m.

(b) The distance back is a.
$\cos \theta = \frac{\text{Adjacent}}{\text{Hypotenuse}} = \frac{b}{100}$. $b = 100 \cos \theta = 87$ m.

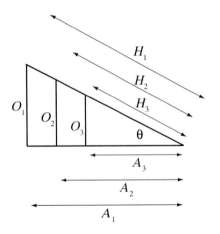

The ratio $\frac{O}{H}$ is the same as long as θ does not change. ($\frac{O_1}{H_1} = \frac{O_2}{H_2}$, etc.) This ratio is called the *sine* of the angle θ: $\sin \theta = \frac{O}{H}$. The ratio $\frac{A}{H}$ is the *cosine* of θ (cos θ). The ratio $\frac{O}{A}$ is the tangent: $\tan \theta = \frac{O}{A}$.

If you divide each term in the relation $A^2 + O^2 = H^2$ by H^2, you get $\frac{A^2}{H^2} + \frac{O^2}{H^2} = 1$, or $\sin^2 \theta + \cos^2 \theta = 1$, for any angle θ.

EXAMPLE 12

You swim from shore for 100 m at an angle of 30° with respect to the coastline.

(a) How far are you from the shore when you stop?

(b) If you swim directly back to the shore, how far will you then be from where you started?

2.2 Once more the four forces, this time quantitatively

The gravitational force

The relation that describes the gravitational force was first published by Isaac Newton in 1687 in his book *Philosophiae Naturalis Principia Mathematica*, "The Mathematical Principles of Natural Philosophy," usually referred to as *Principia*. It is known as Newton's law of universal gravitation:

$$F_g = G \frac{M_1 M_2}{r^2} \qquad (2.1)$$

It describes the gravitational force of attraction between any two bodies with masses M_1 and M_2, separated by a distance r. G is a proportionality constant. In the SI system, where M is in kg, r in m, and F_g in N, G is equal to 6.67×10^{-11} Nm²/kg².

The law is for small "point" objects, but we will see later that it holds also for large spherically symmetric bodies if for r we use the distance between their centers.

EXAMPLE 13

Calculate the gravitational force between the earth and a 1 kg object at its surface.

Ans.:

The first mass, M_1, becomes the mass of the earth, which we can write as M_e, equal to 5.97×10^{24} kg.

M_2 is the 1 kg mass. For r we have to use the distance from the center of the earth to the mass at the earth's surface. This is the radius of the earth, R_e, equal to 6.38×10^6 m. Then $F_g = G\frac{(M_e)(1)}{R_e^2}$, or $\frac{(6.67 \times 10^{-11})(5.97 \times 10^{24})}{(6.38 \times 10^6)^2}$, which comes out to be 9.78 N. Instead of calculating the answer directly it is helpful to collect the exponents separately. Here this is $\frac{(6.67)(5.97)}{(6.38)^2} \times 10^{-11+24-12}$, which is 0.978×10^1 or 9.78 N as before. This makes it easy to check the exponents and to see whether the answer is in accord with a rough estimate. Because the earth is not a homogeneous sphere, nor exactly spherically symmetric, the magnitude of g varies somewhat over the surface of the earth. We usually use the value 9.8 N/kg.

The result shows that the gravitational force that the earth exerts on an object whose mass is 1 kg is 9.8 N. Another way of saying that is that an object whose mass is 1 kg has a weight of 9.8 N at the surface of the earth.

EXAMPLE 14

At what distance from the surface of the earth does an object weigh half as much as it does on earth?

Ans.:
The gravitational force on an object whose mass is M is $F = G\frac{MM_e}{r^2}$. We see that this force (the weight) is proportional to $\frac{1}{r^2}$. At the surface of the earth $r = R_e$. We will call that distance R_1, and the force there F_1.

At some distance that we will call R_2, the force is F_2, so that $\frac{F_2}{F_1} = \frac{1}{2}$. The proportionality shows that $\frac{F_1}{F_2} = \frac{R_2^2}{R_1^2}$, which we can solve for R_2 to give $\sqrt{2}R_1$, or $1.41\ R_e$. This is the distance from the center of the earth. If we want the distance from the earth's surface, we have to subtract R_e, to get $0.41R_e$ as the answer.

The electric force

Charles Augustin Coulomb did the experiments that led to the law describing the force between charges in the years 1785 to 1787. It is called Coulomb's law, and the force is often referred to as the Coulomb force,

$$F_e = k\frac{Q_1 Q_2}{r^2} \quad (2.2)$$

Here F_e is the electric force between two point charges, Q_1 and Q_2, a distance r apart.

The SI unit for electric charge is the coulomb. k is a proportionality constant, which in the SI system is 9.00×10^9 Nm²/C². When both charges are positive or both negative, they repel; when one is positive and one is negative, they attract.

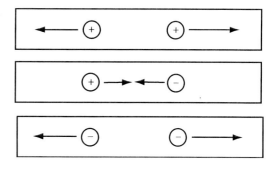

EXAMPLE 15

Here is an example that illustrates how enormously strong the electric force is. There are about 6.0×10^{23} atoms in 1 g of hydrogen. Imagine that all of the electrons in 1 g of hydrogen are put on the earth's north pole, and all the protons on the south pole 12.8×10^6 m away. Calculate the force between them.

Ans.:
The charge on a single electron or proton has a magnitude of 1.6×10^{-19} C. In the 1 g of hydrogen the electric charge of the protons is therefore (6×10^{23}) (1.6×10^{-19}) or 9.6×10^4 C. The negative charge on the electrons from the same gram of hydrogen has the same magnitude. The force between them is therefore $F_e = \frac{(9.0 \times 10^9)(9.6 \times 10^4)^2}{(12.8 \times 10^6)^2} = 5.1 \times 10^5$ N, which is equal to the weight of a mass of 5.2×10^4 kg or 52 metric tons.

(Since the weight of a small apple is about 1 N, the force is equal to the weight of about half a million apples.)

The other forces

The gravitational and the electric forces are long-range forces, and each decreases as $\frac{1}{r^2}$, i.e., $F \propto \frac{1}{r^2}$. If the separation of two masses or charges is increased by a factor of two, the force between them decreases by a factor of four. The gravitational force gets smaller with distance, but it still acts over very large distances, such as those between the sun and the earth and the other planets. The electric force has the same distance dependence.

The nuclear force can not be described by a similarly simple formula. But we know that it acts only between neighboring nucleons in a nucleus, or in collisions when they get just as close to each other.

For the weak force there is also no simple formula. Its contribution is to allow some nuclear processes to take place that could not otherwise happen. It is not responsible for any structures, as is the gravitational force for the planetary system, the electric force for atoms, and the nuclear force for nuclei.

2.3 Summary

Mathematics is the language of physics. Here are some reminders and reviews of some parts that we are going to use.

A *symbol* is shorthand for a quantity (like x) or an operation (like +). Symbols allow us to express relationships between quantities by *equations*.

Numbers can be positive, negative, or zero. Very large and very small numbers are best expressed with powers of ten. 5×10^6 is the same as 5 followed by six zeros, or 5 million. The *exponent* (in this case 6) gives the number of spaces by which the decimal point is shifted—to the right for positive exponents, and to the left for negative ones.

The number of *significant figures* indicates how precisely a number is known. When we give no other information about it we will assume that the numbers in the examples and problems are known to three significant figures.

Physical quantities are expressed in terms of *units*. We will most often use the units of the SI system, which include meters, (m), kilograms (kg), and seconds (s).

Relations between quantities can be described by using *tables, graphs,* or *equations*. All of these are *representations* of what happens, as are words and pictures.

A straight line on a graph of y against x is described by the equation $y = mx + b$. y is usually plotted along the vertical axis and x along the horizontal axis. m is the slope, equal to $\frac{\Delta y}{\Delta x}$, and b is the y-intercept, i.e., the value of y when x is zero.

In a triangle with a right angle and an angle θ, with O the side opposite θ, A the side adjacent to θ, and H the hypotenuse (whose length is $\sqrt{O^2 + A^2}$), $\sin\theta = \frac{O}{H}$, $\cos\theta = \frac{A}{H}$, and $\tan\theta = \frac{O}{A}$.

Newton's law of gravitation, $F_g = G\frac{M_1 M_2}{r^2}$, describes the gravitational force between two particles with masses M_1 and M_2, a distance r apart. Coulomb's law, $F_e = k\frac{Q_1 Q_2}{r^2}$, describes the electric force between two particles with charges Q_1 and Q_2 a distance r apart. The two expressions are similar in that both forces are inversely proportional to the square of the distance between the two interacting particles. Each includes the product of the characteristic quantities (mass or charge) of the particles. Each expression describes a very different physical property. Each has a very different strength, reflected in the respective proportionality constant (G or k).

2.4 Review activities and problems

Guided review

1. You walk to the right 50 yards, then to the left 17 yards.

(a) Represent the motions mathematically, letting "to the right" be positive. What does the result and its sign tell you?

(b) Now answer the question again, but this time letting "to the left" be positive. What does the result and its sign tell you?

2. The temperature drops from 15°F to 0°F.

(a) Write a mathematical description of what happens.

(b) Convert the temperatures to the Celsius scale and repeat.

3. A sign whose weight is 100 lbs is held up by a chain.

(a) With what force does the chain need to be pulled up to keep the sign in place?

(b) Write a mathematical description.

(c) What does the sign of each number represent?

4. Two properties of several elements are shown in the table. One is the atomic number and the

other is the most common ionization state. The number in the superscript shows how many electrons have been added to the neutral atom or removed from it. A "+" sign indicates that the resulting ion is positive and a "−" sign that it is negative.

(a) How much positive and how much negative charge is in each neutral atom? (Give the answer in multiples of e, the magnitude of the charge on an electron.)

(b) Similarly represent the charges in each of the ionized states.

Element	Al	Zn	N	C
Atomic number	13	30	7	6
Common ionization state	Al^{3+}	Zn^{2+}	N^{3-}	C^{4-}

5. The mass of a proton or a neutron is 0.000 000 000 000 000 000 000 000 001 67 kg. The mass of an electron is 0.000 000 000 000 000 000 000 000 000 000 911 kg. Put these numbers in power-of-ten notation.

(a) If you put a proton on one pan of a balance, how many electrons would you have to put on the other to balance it?

(b) What is the total mass of the particles that make up a neutral helium atom?

6. The circumference of a circle is 2.735 m. How big is the diameter? (How many significant figures do you know?)

7. Convert the following into SI units. (1 in = 2.54 cm, 1 mile = 5280 ft, 1 food calorie = 1000 cal = 4187 J, 9.8 N = 2.20 lb)

3.4 cm
5 in
140 lbs
30 $\frac{miles}{hour}$
14.7 $\frac{lb}{in^2}$
2000 $\frac{food\ cal}{day}$

8. The strength of attraction, F_m, between two magnets can be described as $F_m \propto \frac{1}{d^2}$.

You separate the two magnets from $d = 1$ cm to $d = 5$ cm. By what factor does F_m change? Does it increase or decrease?

9. Here is a table of the amount of water (in gallons) consumed by a soccer team during a practice session on a hot afternoon as a function of time in minutes.

Time	30	60	90	120	150	180
Water consumed	1.25	2.5	3.75	5	6.25	7.5

(a) Make a graph of the data.
(b) Write a mathematical function that describes the data.
(c) What is the rate at which the team drinks?
(d) They start with 10 gallons. How long can they practice before they run out of water?

10. Here is a plot of the number of leaves on a tree as a function of time, in days after the leaves have started to fall.

(a) Write an equation that describes the data.
(b) After how many days will all of the leaves have dropped if the equation continues to be valid?

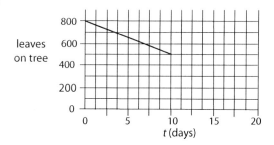

11. On a very hot day the sales at a lemonade stand can be represented by the equation $y = (x-2)^2 + 9$, where y is the rate of sales in cups per hour and x is the number of hours of operation after opening from $t = 0$ to $t = 5$, at which time all of the lemonade has been sold.

(a) Make a table of values of x and y at hourly intervals.
(b) Plot the points and sketch $y(x)$.
(c) What is the slope at $t = 2$ and 4?

12. A flagpole is mounted on the side of a house. It is 1 m long and makes an angle of 50° with the vertical.

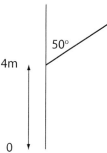

(a) If the sun shines straight down, how long is the flagpole's shadow?

(b) How far from the ground is the end of the flagpole?

13. What is the gravitational attraction between two 60 kg people standing 2 m apart?

14. What is the weight of a 1 kg mass on top of Mount Everest? ($H = 8850$ m)?

(a) Use proportional reasoning and the fact that the weight is 9.8 N at sea level.

(b) Check your answer by using Newton's law of gravitation.

15. Two spheres are electrically charged, one with $Q_1 = -15$ nC and the other with $Q_2 = +15$ nC. (1 nC $= 10^{-9}$ C.) This is the charge of about 100 billion electrons and protons. Find the electric attraction between the two spheres when they are 2 m apart.

Problems and reasoning skill building

1. You are out skiing, and in the course of the afternoon the temperature drops by 8°C from a beginning temperature of 4°C. Describe the temperature change using integers. What does the result and its sign tell you?

2. Two cylindrical buildings are side by side. One has a diameter of 100 m and a circumference of 314 m. The other has a diameter of 135 m. Use ratios only (not π!) to find the circumference of the second building.

3. The data in the table show how many pushups a young boy completes over a period of five days.

Day	Pushups
1	4
2	7
3	10
4	13
5	16

(a) Make a graph of the data.
(b) Write a function that describes the data.

4. The figure shows a pool table. Your ball is at the position marked X. $a = 12''$, $b = 36''$.

(a) How far is your ball from the left top corner pocket?

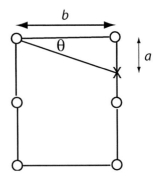

(b) What is the angle θ?

5. At the time of his famous studies on electricity Benjamin Franklin understood that there were two kinds of electric charge. He called one kind *positive* and the other *negative*. According to his (incorrect) model electric current in a wire is the movement of positive charge. The current, I, is defined as the amount of positive charge passing a cross section of the wire per second. The figure shows 15 positive charges in a wire, which move to the right, past the place marked by a dotted line in one second.

Today we know that it is actually the negative charges that move. What is the motion of charge for the same amount of current to the right, when negative charges move?

6. Your brother raises gerbils. At the end of each month you count the baby gerbils in the cage. The table shows how many there are for the first five months.

Month	Babies
1	0
2	3
3	8
4	15
5	24

(a) Make a graph of the data.
(b) Find a function that describes the data.

(c) If this trend continues, how many babies do you expect to find at the end of the sixth month? The seventh month?

(d) What factors does your mathematical model not consider?

7. Some people believe that the position of the planets at one's birth has a profound influence on the course of a person's life.

Calculate the gravitational interaction with a newborn of the planet Jupiter and of the attending physician. Which is larger?

8. The angle that the line to the sun makes with the horizontal can be used to estimate the time to sunset. (In the afternoon the angle can be estimated from the fact that $\sin 30° = 0.5$.) What is the time to sunset when the sun is $15°$ above the horizon?

9. A measure of the apparent size of the moon is the angle that it subtends, i.e., the small angle in the triangle of which the moon's diameter is one side and the two other sides are equal to the distance to the moon. The angle can be estimated by holding a pencil or piece of paper at arm's length so that part of it forms a triangle similar to the first, with the same small angle.

Draw a diagram with both triangles. Estimate the angle, and use the distance to the moon to estimate the diameter of the moon. Compare it to the value in a table of astronomical quantities.

10. Convert the following energy units to joules. Give the answer to three significant figures and use power-of-ten notation with one digit before the decimal point. Put the answers in the order of increasing energy.

cal
kwh
horsepower hours
eV
MeV
ft-lbs

11. The weight of a rock on earth is 50 N.
(a) What is its mass?
(b) Use Newton's law of gravitation to calculate the value of g on the moon.
(c) What are the mass and weight of the rock on the moon?

12. What can you conclude from the fact that the apparent sizes of the moon and the sun are approximately the same?

13. Use Newton's law of gravitation to determine the SI units of G. Use Coulomb's law to determine the SI units of k.

14. The tides are caused by both the moon and the sun. Which of the two exerts the greater force on the earth, and by what factor?

15. What is the ratio of the gravitational force between the electron and the deuteron in a deuterium atom to that in ordinary hydrogen (whose nucleus is a proton)? Answer the same question for the electric force and for the total force.

16. Use proportional reasoning to find by what factor the energy of motion of a runner, $\frac{1}{2}mv^2$ (where v is her speed), increases when she improves her racing time for a 5 km course from 22 min to 19 min.

17. A medieval alchemist had high hopes of turning lead into gold. The densities of the two metals are $\rho_{gold} = 19.3 \times 10^3$ kg/m^3 and $\rho_{lead} = 11.3 \times 10^3$ kg/m^3.

If he had succeeded, what would be the ratio of the space occupied by 1 kg gold to that occupied by 1 kg of lead? What would be the difference in the amounts of space?

18. An accused thief, named Stiles, implores Sherlock Holmes to help him, saying that he is wrongfully accused of stealing one million dollars in gold. The evidence against him is that he was seen placing the gold in a suitcase and then escaping, running down the street with the missing gold. Holmes asks Watson "How much does an ounce of gold go for these days?" and is told that it is 200 British pounds, with a value of 300 U.S. dollars. Why does Holmes agree to take the case?

Synthesis problems and projects

1. At "half moon" we see the moon illuminated by the sun from the side.

(a) Draw a diagram that shows the positions of the earth, the moon, and the sun at that time. What is the angle between the line EM from the earth to the moon and the line SM from the sun to the moon?

(b) Aristarchus of Samos (310 B.C.–230 B.C.) measured the angle θ between the line EM and the line ES from the earth to the sun. From this measurement he calculated the ratio of the length of

ES to the length of *EM*. How is this ratio related to the angle θ?

(c) From your knowledge of the modern values of these lengths (see the table at the end of the book) what are the values of the ratio and of the angle?

(d) Look up the value of the angle measured by Aristarchus and the ratio that he calculated.

(e) Why does a small error in the estimate of the angle lead to a large error in the ratio of the two distances?

(f) Aristarchus also noted that the angles subtended by the moon and the sun at the earth (i.e. their apparent sizes) are approximately the same. What does this observation lead to for the ratio of the radii of the two bodies? How does his value compare to the modern ratio?

2. Aristarchus used observations during an eclipse of the moon to determine the ratio of the diameters of the earth and the moon.

(a) Draw a diagram of the positions of the sun, the earth, and the moon during an eclipse of the moon. (The moon is in the earth's shadow so that it is not illuminated by the sun.)

(b) Aristarchus measured the time interval from the time when the moon first comes to the earth's shadow to the time when it is completely in the earth's shadow. He also measured the time of "totality," i.e., the time during which the moon is completely in the earth's shadow. This second time interval turned out to be approximately twice the first. Indicate the approximate size relation of the moon and the earth on your diagram.

The geometrical calculation of the ratio of the diameters from the observation is somewhat complicated. Several versions may be found on webpages devoted to Aristarchus. There you will also find discussions describing that he seems to have been the first to believe that the earth revolves around the sun rather than the other way around. Copernicus, more than 1700 years later, is usually given the main credit, but he refers to the work of Aristarchus.

3. Eratosthenes of Cyrene (273 B.C.–192 B.C.) determined the circumference of the earth as follows. At Aswan, on the Tropic of Cancer, at the summer solstice, at noon, there is no shadow in a vertical well, i.e., the sun is directly overhead. At that time, at a distance of 4900 stadia further north, in Alexandria, (with one stadium probably about 160 m) the sun was at an angle of 7° away from the vertical.

(a) Draw a diagram to show the earth and the sun at that time in the two places.

(b) What does his measurement lead to for the radius of the earth? What is the modern value?

4. The half lives of ^{238}U and ^{235}U are 4.5×10^9 y and 0.71×10^9 y, respectively. Consider the hypothesis that the two isotopes of uranium were present in equal amounts at a time near the beginning of the universe. Today ^{235}U is 0.72% of natural uranium. The rest is ^{238}U. Determine the time implied by this hypothesis as follows.

Make a graph of the amount of ^{235}U as a function of time on which the horizontal axis shows time in units of 10^9 y. On the vertical axis let equal factors be represented equally, i.e., one unit from 2 to 4, the next from 4 to 8, then from 8 to 16, and so on. (This is called a semilog plot.) On this graph the amount of uranium as a function of time is a straight line.

5. An electrically charged ball with a mass of 20 g floats 10 cm above a second one that carries the same charge, q, so that the net force on the floating ball is zero. What is the value of q?

CHAPTER 3

There Is No Rest: Describing Motion

Getting started: simplification and approximation: models

Keep your eye on the ball: where is it and where is it going?
How far? Distance and displacement
How fast? Speed, velocity, and acceleration
Constant velocity
Changing velocity: constant acceleration

The mathematics of change
Slopes and derivatives
Areas and integrals

There was a time when physics meant *mechanics*, and perhaps it should still. The word comes from the Greek for machine. We understand a piece of machinery when we know each of its parts, all the gears and levers, and their functions and interplay. That's really the program of all of natural science, to know the pieces of which the world is made and how they connect and work together.

The world is more complicated than any machine, and knowing it and understanding it is a work in progress. We learn more each day, we get to know some parts very well, but there is no end to the quest. We continue to be surprised by what we learn, sometimes by the complexity and intricacy, sometimes by the simplicity of the way in which the components act and interact.

The term *mechanics* has come to be used in a more restricted sense, as the science of motion and of the way that forces bring about motion. To start with, we will limit ourselves further by leaving out, at one end of the scale, the atoms and their constituents, and at the other end the stars and other massive astronomical objects, because they behave, in part, in ways that are different and outside our day-to-day experience. That leaves us with *classical* mechanics, the study of the realm between, the ordinary world of chairs and tables, baseballs and bullets.

We will look at the basic concepts that we need to describe motion, namely displacement, speed and velocity, and acceleration.

One of the features that has made physics so successful is that it asks the simplest questions, and so is able to look at them in great depth. That is also one of the dangers. Physicists love to go into more and more detail, until what they say is far from simple. That can easily happen when we go into the mathematical description of various kinds of motion, along straight lines, in circles, parabolas, ellipses, motion in one dimension or in many, rotational, spiraling, oscillating, and so on.

We will do only a little of that. The main reason we study mechanics is to see how it introduces basic ideas, such as force and energy, that are central to every part of science. These ideas will stay with us as we go on later to other realms, beyond the world that is immediately accessible to us through our senses. There we come to atoms and molecules, each one of them ceaselessly moving. It was one of the great insights in intellectual history when it was realized that it is their motion that underlies, on the larger scale, all of our experience with the phenomena of heat and temperature. It is also here, in the world of atoms and molecules, that we will find the forces between them, and within them, that shape the animate and inanimate world as we know it in its infinite variety.

3.1 Getting started: simplification and approximation: models

Look at yourself running. Your elbows move. Your feet move differently. In fact, every part of you moves, and moves differently from every other part. Usually you just see yourself as running in a straight line at a certain speed, and don't think about these much more complicated motions.

When you think only of the fact that you are running, concentrating on the whole of you, ignoring the fact that your knees are moving quite differently from your chin, the picture in your mind becomes much simpler. You can then describe your motion quite straightforwardly by saying "I am moving along this track with a speed of 6 miles per hour."

Now think of all that you have decided to ignore: not only the externally obvious parts of your body, all moving differently, but also the internal, vastly complex motions, as you breathe, swallow, digest, as your heart beats and your blood flows. On an even deeper level, each part of you consists of molecules, and each molecule is moving. Inside the molecules, the atoms are also moving, and inside the atoms the electrons and nuclei.

With each step that you take, you are pushing on the ground, changing its motion, and at least in some minute way you are changing the motion of the earth of which it forms a part, as it spins and travels around the sun, the earth and the sun each a collection of vastly complex constituents, all of them ceaselessly moving.

It's a good thing that you can simplify the story, ignoring most of it, and concentrating just on your motion around the track. That's what allows us to describe what is going on without being overwhelmed, without being paralyzed by the difficulty of dealing with every detail at once. This simplification is an essential part of science, and makes it possible for knowledge to progress, piece by piece.

You can study the motion of an object simpler than a human being, perhaps a baseball as it flies through the air. But here too you will have a tough time unless you begin by ignoring a large part of what happens. You may want to restrict yourself to still air, but even then, as the air molecules push against the ball, they will affect its motion. Internally, the ball is made of different materials, and they, like everything else, are made of moving molecules, atoms, and nuclei.

Can you imagine something even simpler, with no internal structure? Yes, at the smallest level there are "elementary particles," so called because we think that they have no internal structure at all. On our own much larger scale we can talk about balls and marbles as if they were particles. To do so we have to ignore a lot, and keep in mind that we are no longer talking about the ball or the marble as it exists. We are substituting something that we have invented, something that we will call a *model* of the ball. (This is a very different use of the word "model" than those that you are likely to be familiar with.)

We can talk about the model, in this case a particle, and how it moves and behaves. It is much simpler than the ball. It does not have all the properties of the ball. A particle can't be squeezed, and it can't be heated, because both of these processes represent internal changes. By using the model of the ball as a particle we have given up any possibility of describing what happens inside the ball.

We have to be careful when we simplify and approximate, i.e., when we use a model, because some part of the real story is then lost. But if we are clever, or lucky, the essential part will still be there.

3.2 Keep your eye on the ball: where is it and where is it going?

Here is what has been called *the* problem of mechanics: consider a particle at some instant of time. You know where it is, and how fast and in what direction it is moving. You also know the forces that are acting on it, at the beginning, and at all other times. Calculate where it will be and how fast and in what direction it will be moving at some given future moment.

Perhaps the most interesting point is that this problem *has* a solution. Always, as long as the relations of *classical mechanics* hold. What are these relations? Are they *right*? Or better, to what extent, and under what circumstances are they right? Should we think of them as *laws* that have to be obeyed? As we examine these questions and their implications we see already that it is just as important to look at the limits of validity of these or any other relations and "laws" as it is to look at the relations themselves.

We'll start with the description of the motion. This part of mechanics is called *kinematics*. It has to do with position and time, velocity, and acceleration. We will assume that we know what is meant by position and time. It took the genius of Einstein to show that even these concepts cannot be taken for granted. He looked more deeply at mechanics, and came to revolutionary conclusions, which we will discuss later. For now, we'll start at the beginning.

How far? Distance and displacement

When Jane walks to school, she changes her position. She first goes two blocks north along her street, then turns the corner and goes four blocks east. The distance she goes is therefore six blocks. If each block is 500 feet long, that distance is 3000 feet. To go there and back she has to traverse a distance of 6000 feet.

We will define another measure of how far she is from her starting point, called the *displacement*. Its magnitude (or size) is that of the straight-line distance from the starting position to the final position. In this case it is $\sqrt{2^2 + 4^2}$ blocks, or 4.47 blocks, which is equal to 2.22×10^4 ft. Besides this magnitude, the displacement has another attribute, namely its *direction*. In the present case it can be specified as 63.4° east of north.

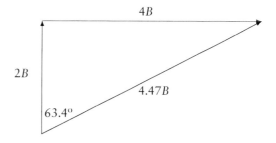

A quantity that has both magnitude and direction is called a *vector* quantity. A quantity that has only magnitude (as, for example, temperature, or volume) is called a *scalar* quantity.

When Jane goes to school and then back, she has gone a distance of 6000 feet. But since the displacement is the distance between the starting position and the final position, it is, in this case, zero.

EXAMPLE 1

Bob walks to the library. He walks one block north and then one block east. Find the magnitude and direction of his displacement.

36 / There Is No Rest: Describing Motion

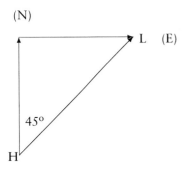

Bob starts from home (H) and goes to the library (L). His displacement is the vector from H to L. Its magnitude is $\sqrt{1^2 + 1^2}$, equal to $\sqrt{2}$, or 1.41 blocks. Its direction is 45° east of north.

How fast? Speed, velocity, and acceleration

Speed is what the speedometer shows. If it reads "20 miles per hour" the car will travel a distance of 20 miles in an hour, as long as the speed doesn't change. Usually, though, the speed does change. But if the car traverses 20 miles in one hour, even if it has stopped at a red light, and then speeded up to 35 miles per hour for part of the trip, the *average speed* is 20 miles per hour.

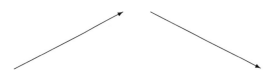

These are two different vectors. They have the same magnitude, but different directions.

If the car goes around a corner, and the speedometer still shows 20 miles per hour, its speed doesn't change. But to a passenger it feels quite different from what it would if the car continued in a straight line. To distinguish between these two situations, we define a different quantity, the *velocity*. Its magnitude is equal to the speed, but the velocity is constant only if the direction does not change. If, for example, a car goes first in one direction and then back in the opposite direction, both at 20 miles per hour, the speed is the same on the return trip, but the velocity is not. The speed has only magnitude. It is a scalar quantity. The velocity has magnitude and direction. It is a vector quantity.

As the car moves, its displacement from the starting point changes. On a graph the displacement, x, is measured from an *origin*, where $x = 0$. If x is plotted against t, we call this a graph of $x(t)$ (read "x as a function of t" or "x of t").

The quantity that describes how the displacement changes is the velocity. It is the rate of change with time of the displacement. When the velocity changes, there is an *acceleration*. The acceleration is the rate of change with time of the velocity. We will explore the relations between displacement, velocity, and acceleration in the rest of this chapter.

EXAMPLE 2

Bob walks one block north in five minutes.

(a) At what rate and in what direction does his displacement change?

(b) Draw a graph of $x(t)$ for Bob's motion.

(c) Find the slope of the graph and compare it to the answer to (a).

Ans.:

(a) The rate at which the displacement changes is the velocity. It is $v = \frac{1 \text{ block}}{5 \text{ min}} = 0.2 \frac{\text{blocks}}{\text{min}}$ to the north.

(b)

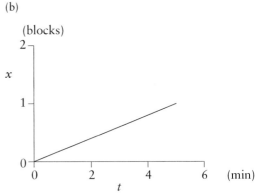

(c) The graph shows that the slope is $\frac{1 \text{ block}}{5 \text{ min}} = 0.2 \frac{\text{blocks}}{\text{min}}$. The slope of $x(t)$ is the velocity.

Constant velocity

Let's talk about a car that moves in a straight line with a velocity of 25 miles/hour south for 10 seconds. This is the *verbal representation* of the situation that we want to describe.

First of all, we prefer the SI units of m/s: 1 mile is about 1609 m, so that $\frac{25\,\text{miles}}{\text{h}} = \left(\frac{25\,\text{miles}}{\text{h}}\right)\left(\frac{1609\,\text{m}}{1\,\text{mile}}\right)\left(\frac{1\,\text{h}}{3600\,\text{s}}\right) = 11.2\,\frac{\text{m}}{\text{s}}$.

In addition to the verbal statement we can represent this motion in several other ways. We can draw a picture. This will be a snapshot at a particular moment. By itself it won't tell us anything about the motion, but we can write on it whatever we know about the motion. In this case we know that the velocity is $v = 11.2$ m/s south, during the time interval from $t = 0$ to $t = 10$ s.

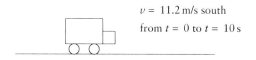

We can go further by drawing several snapshots at equal time intervals, as on a movie film. In this case, with its constant velocity, the car will move 11.2 m in each second, i.e., the same distance in each of the equal time intervals. We will call the sequence of such snapshots a *motion diagram*. To keep things simple we will represent the car by a dot. This is in keeping with the model of the car as a particle. Usually our motion diagrams will be qualitative, i.e., they will not have numbers or units, and just show whether the dots are, as in this case, equally spaced, or whether they get closer together or farther apart.

. . . .

EXAMPLE 3

B . . . .

A . . . .

(a) Which car moves faster, A or B?

(b) Are they ever side by side? If so, where?

Ans.:

(a) The points in A are farther apart. That means that the car moves through a greater distance in the same time. This car goes faster.

(b) They start out at the same place and are side by side at the beginning. After that, B goes more slowly, and the cars are never side by side again.

We can represent more information on a graph, or on a series of graphs. Look first on a graph of v against t. The velocity is constant. It has the same value for each value of t. Another way of saying the same thing is that the velocity as a function of time, $v(t)$, is constant. As t varies, v remains the same. The graph is therefore a horizontal line.

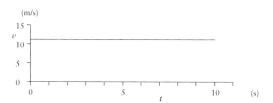

Note that the graph is marked with the quantities being plotted, v on the vertical axis and t on the horizontal axis. On each axis we also mark the units, m/s on the vertical axis and s on the horizontal axis.

What about the displacement, x? It starts at $x = 0$ when the time is $t = 0$. It then increases by 11.2 m every second. The graph is a straight line, but it is not horizontal. Again we mark the quantities and units, x and m on the vertical axis and t and s on the horizontal.

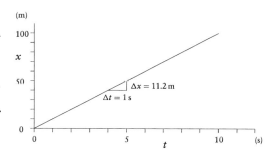

To see how fast x varies with t, we look at a change in t, for which we use the symbol Δt (read "*delta t*"). Let's make this time interval 1 s. During this time there is a change in x, which we call Δx, of 11.2 m. The slope of the graph is $\frac{\Delta x}{\Delta t}$, equal to 11.2 m/1 s, or 11.2 m/s. This is the velocity. The velocity has the units of x divided by the units of t, or m/s.

We can write a mathematical representation that describes this line. (We could also call it a "mathematical statement," or an "equation.") We can do that by starting with the standard expression for a straight line, $y = mx + b$. In this expression y is plotted along the vertical axis, x along the horizontal axis, m is the slope, and b is the y-intercept. (That's the value of y when x is equal to zero.) We can compare that with our present graph, where we plot the displacement along the vertical axis and the time along the horizontal axis. The slope is the velocity and the intercept is zero. The expression becomes $x = vt + 0$, or $x = vt$.

EXAMPLE 4

The graph describes Jean walking west.

(a) Where is she at $t = 2$ s?

(b) What does the intercept on the vertical axis represent?

(c) What is the slope of the line? What physical quantity does it represent?

(d) Write a mathematical representation of the line, first with symbols and then with numbers.

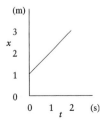

Ans.:

(a) The distance, x, is measured from a chosen origin, with "west" the positive direction. At $t = 2$ s she is 3 m west of the origin.

(b) The intercept is her position at $t = 0$, which is x_0. At that time she is 1 m west of the origin.

(c) The slope is her velocity. It is 1 m/s to the west.

(d) Comparing to the standard expression for a straight line,
$$y = mx + b$$
we have to change y to x, $(y \to x)$, $x \to t$, $m \to v$, $b \to x_0$, so that the mathematical expression becomes

$$x = vt + x_0$$

With the numbers, her displacement in meters, in terms of the time in seconds, is $x(\text{m}) = t(\text{m/s} \times \text{s}) + 1(\text{m})$, or $x = t + 1$, where the units of each term are meters (m).

EXAMPLE 5

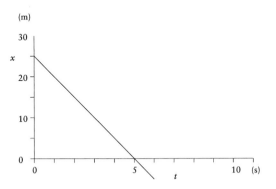

The graph shows the motion of a car as $x(t)$ for 6 s. The x direction is toward the right. The negative x or $(-x)$ direction is to the left.

(a) Where is the car at $t = 0$?

(b) Where is the car at $t = 5$ s?

(c) Where is the car at $t = 6$ s?

(d) In what direction does the car move?

(e) How far does it move between $t = 5$ s and $t = 6$ s?

(f) What is the car's velocity?

(g) What happens at $t = 5$ s?

Ans.:

(a) At $t = 0$ the car is 25 m from the origin in the $+x$ direction. The car is 25 m to the right of the origin.

(b) At $t = 5$ s the car is at the origin, $x = 0$.

(c) At $t = 6$ s the car is 5 m from the origin, in the negative $(-x)$ direction, or 5 m to the left of the origin.

(d) The car moves in the $-x$ direction, toward the left.

(e) It moves 5 m in the negative x-direction, to the left.

(f) It is moving with a constant speed of 5 m/s in the negative x direction, i.e., to the left. Its velocity is -5 m/s.

(g) At $t = 5$ s the car passes the point $x = 0$.

These graphs are quantitative. They show the magnitudes of v and x for each value of t. Sometimes we may want to draw a qualitative graph or *sketch* that just shows the shape of each function. Here, for example, is a sketch of x against t. The axes are still marked with x and t to show what quantities are being plotted, but without the units and numbers.

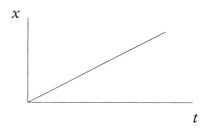

We can also draw a graph that shows how the acceleration, a, changes as the time, t, changes. Since the velocity does not change, the acceleration is zero. A graph of a against t is therefore a horizontal line at the value $a = 0$, right at the horizontal axis.

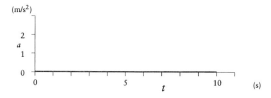

The units of t are seconds, but what are the units of the acceleration? If the velocity changes from 5 m/s to 11 m/s in 2 s, the change in the velocity (Δv) is 6 m/s. In each second the velocity changes by 3 m/s. The acceleration is 3 m/s per second, or $3\frac{m}{s}/s$, which we write as 3 m/s².

In symbols, the acceleration is the change in the velocity, Δv, divided by the time interval Δt during which the velocity changes by Δv: $a = \frac{\Delta v}{\Delta t}$. It therefore has the units of v divided by the units of t, or m/s divided by s, which is equal to m/s².

Finally, we can represent the motion by one or more mathematical relations. In this case they are $a = 0$, $v = 11.2$ m/s, and $x = vt = (11.2t)$ m.

To summarize: motion with constant velocity can be described in several ways:

(1) The verbal statement that the object moves through equal distances in equal time intervals.
(2) The motion diagram, which shows the equal distance intervals.
(3) The graphs: $a(t)$, horizontal at $a = 0$; $v(t)$, horizontal at the value (11.2 m/s) of the velocity; and $x(t)$, the straight line through the origin, with slope v.
(4) The mathematical expressions: $a = 0$, $v = +11.2$ m/s, and $x = vt$, or $x = (11.2)(t)$ m.

In this case, where the velocity is constant, the slope of the graph of x against t is constant, and the graph is a straight line. But the definition of the velocity as the slope on a graph of x against t still holds when the velocity is not constant.

As you look at the three graphs of a, v, and x against t, you see that each shows the slope of the next one. a, here equal to zero, is the slope of the horizontal line of $v(t)$. The value at which this horizontal line is drawn (11.2 m/s), in turn, is the slope of $x(t)$.

You can also see that the *area* under the graph of $v(t)$ gives the displacement, x. The area under the curve for the first second is (11.2 m/s)(1 s), or 11.2 m. This is the value of $x(t)$ when t is 1 s. Note that lines on this graph have units, m/s if they are vertical and s if they are horizontal. A rectangular area therefore has the units of m/s multiplied by s, (m/s)(s), or meters.

For the first five seconds the area is (11.2 m/s)(5 s), equal to 56 m.

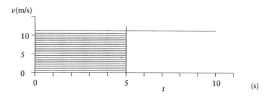

The relationships between the slopes and areas will always be the same, regardless of the actual values of a, v, and x, because they are the direct consequences of the definitions of these three quantities.

40 / There Is No Rest: Describing Motion

EXAMPLE 6

Go to the PhET website (http://phet.colorado.edu) and open the simulation "The Moving Man."

Play with the *Introduction*. Than go to *Charts*.

The position can be changed with the slider or by putting the desired position into the box and then pushing the button at the bottom. The scales can be changed with the buttons marked "+" and "−."

(a) Clear, put $x = -10$ m, $v = 2$ m/s (leave $a = 0$) and push to go. Stop the motion after the man hits the wall. The motion can be repeated by using *Playback*.

What does the graph of $v(t)$ show?

What happens to it when the man gets to the wall?

What is the equation for $v(t)$ before he hits the wall?

(b) Answer the same questions for $x(t)$.

What is the slope of the line?"

(c) What does the graph of $a(t)$ show?

(d) Experiment with other values. Go to "Special Features" and choose the "Expression Evaluator." It shows the graph for a mathematical expression that you put in the box. You have to use "*" for multiplication and "/" for division. Start with $x = 5 + 2*t$ and experiment with other expressions. Put in your expression for v and compare it to your result in (a).

Ans.:

(a) The graph is a straight horizontal line at $v = 2$ m/s until he gets to the wall, when it goes to zero. The equation is $v = 2$ m/s.

(b) The graph is a straight line. It starts at $x = -10$ m and goes to $x = 10$ m. Its slope is 2 m/s. This is equal to the constant velocity. The equation is $x = -10 + 2t$. To check, we see that at $t = 0$ $x = -10$ and at $t = 10$ $x = 10$.

(c) $a(t)$ stays at zero until it shows a negative spike where the man stops abruptly when he hits the wall.

EXAMPLE 7

The graph shows the velocity of a car for 10 s. It first moves at a steady velocity for 8 s.

(a) What is that velocity?

(b) What happens at $t = 8$ s?

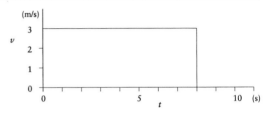

(c) How far is the car at $t = 0$ from its position at $t = 8$ s?

(d) How far does the car move between $t = 0$ and $t = 10$ s?

Ans.:

(a) The velocity is 3 m/s.

(b) At $t = 8$ s the velocity abruptly goes to zero. The car comes to an immediate stop. It apparently crashes into a wall.

(c) $x = vt = (3 \text{ m/s})(8 \text{ s}) = 24$ m.

(d) Between $t = 8$ s and $t = 10$ s the car does not move at all. During the 10 s it moves 24 m.

Changing velocity. Constant acceleration

We can use the same representations for other kinds of motion. Look next at what happens when the velocity changes, but the acceleration is constant.

An example is the motion of a freely-falling body near the surface of the earth. By "freely falling" we mean that we consider only the influence of the earth, and assume that the influence of air resistance and other extraneous retarding effects is so small that they can be neglected. That's a good approximation if we drop a metal object like a coin, but not for a piece of paper or a feather. It works very well on the moon, where there is almost no atmosphere.

It may only be legend that Galileo, around 1590, dropped objects with different weights from the leaning tower of Pisa, and showed that they all take the same time to fall to the ground. In any case, he showed in his later writings that free fall was motion with constant acceleration.

The magnitude of the acceleration of a freely-falling body is so important that we use a special symbol for it, g. Near the surface of the earth its magnitude is approximately 9.8 m/s^2, or 32 ft/s^2, and it is in the downward direction. (The speed increases by 32 ft/s in each second.) If you go up a mountain, g decreases. On the moon a freely-falling object falls with an acceleration of only 1.6 m/s².

EXAMPLE 8

In the process of organizing a colony on the moon, a group of astronauts has constructed a 100 m-high tower. One of them drops a hammer from the top of the tower to the ground.

(a) Use your knowledge of the value of g on the moon (1.6 m/s²) and the definition of acceleration to make a table of the velocity after 1, 2, 3... up to 12 s.

(b) Use the values from your table to make a graph of $v(t)$.

(c) From your knowledge that the displacement is the area under the curve of $v(t)$, construct a table of $x(t)$ for the same 12 s.

(d) From your table, estimate how long it takes for the hammer to fall to the ground.

(e) Estimate its velocity just before it hits the ground.

Ans.:
We can decide to let either the up or the down direction be positive. The opposite direction is then negative.

Let the downward direction be positive. That avoids having to use a lot of negative numbers while the hammer falls.

(a) The acceleration is the amount by which the velocity changes in each second: here the velocity changes by 1.6 m/s in each second. The velocity starts from zero. It is 1.6 m/s after 1 s, 2×1.6 m/s or 3.2 m/s after 2 s, 4.8 m/s after 3 s, and so on.

t	0	1	2	3	4	5	6
v	0	1.6	3.2	4.8	6.4	8.0	9.6

t	7	8	9	10	11	12	s
v	11.2	12.8	14.4	16.0	17.6	19.2	m/s

(b)
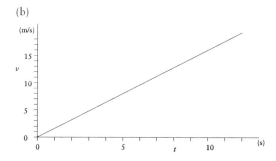

(c) After 1 s the area under the curve of $v(t)$ is that of a triangle with a horizontal side of 1 s and a vertical side of 1.6 m/s, i.e., ($\frac{1}{2}$)(1 s)(1.6 m/s), or 0.8 m. After 2 s the area is ($\frac{1}{2}$)(2 s)(3.2 m/s), or 3.2 m. After 3 s it is ($\frac{1}{2}$)(3 s)(4.8 m/s), or 7.2 m, and so on.

t	0	1	2	3	4	5	6
x	0	0.8	3.2	7.2	12.8	20.0	28.8

t	7	8	9	10	11	12	s
x	39.2	51.2	64.8	80.0	96.8	115.2	m

(d) From the table we see that after 11 s the hammer has fallen 96.8 m, so that it takes just a little longer to fall the 100 m.

(e) From the graph, after 11 s, $v = 17.6$ m/s. Since it takes a little longer, we can estimate the velocity to be about 18 m/s.

Let's look at the same representations as in the previous section. Here is the verbal statement: a car moves in the x direction with a constant acceleration of 2 m/s², starting from rest at $t = 0$, and continuing for 10 s.

The picture (or "pictorial representation") is the same as before, but with the information about the motion for the new situation.

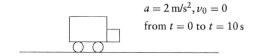

The motion diagram is a series of dots that represent the car's position at different times, separated by equal time intervals. This time the velocity increases, so that the distance between the dots gets larger.

. . .   .

The graph of $a(t)$ is the simplest: a remains the same as t increases. The graph is a horizontal line at the value given for a of 2 m/s².

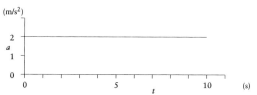

The acceleration is the rate at which the velocity changes with time. When the time increases by an amount Δt, the velocity increases by an amount Δv. The ratio $\frac{\Delta v}{\Delta t}$ is the acceleration. It is the slope of the graph of v against t. Since the car starts from rest, $v = 0$ at $t = 0$.

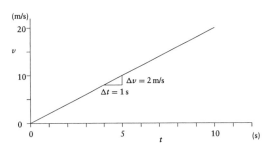

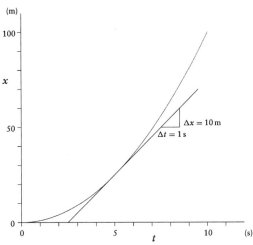

The figure shows $v(t)$, the relationship between v and t. The graph is a straight line, starting with $v = 0$ at $t = 0$, and with a slope of 2 m/s². After 10 s the velocity is $(2 \text{ m/s}^2)(10 \text{ s}) = 2$ m/s.

The next step is to go to $x(t)$, the graph of displacement against time. We can use our previous observation that x is given by the area under the curve of $v(t)$. It is therefore the area of the triangle, $(\frac{1}{2})(v)(t)$, which is equal to $(\frac{1}{2})(at)(t)$ or $\frac{1}{2}at^2$.

t	1	2	3	4	5	6	7	8	9	10
$v = at$	2	4	6	8	10	12	14	16	18	20
$x = \frac{1}{2}vt$	1	4	9	16	25	36	49	64	81	100

In this case the graph of x against t is not a straight line. (It is a parabola.) As usual, the displacement is measured from a starting point, or, on a graph, from the graph's origin. The *slope*, equal to the velocity, increases as the time increases.

Until now we have talked only about the slope of a straight line. What do we mean by the slope of a curve? To find the slope of a curved line at a point we draw a tangent to the curve, the straight line that just touches the curve at that point. The slope of this straight line is the slope of the curve.

Look, for example, at the point on the curve $x(t)$ where t is equal to 5 s, and draw a tangent there. The value of x at that point is 25 m. Draw the tangent line at this point. The slope of the tangent is 10 m/s.

We now also have the mathematical representations: $a = 2\frac{m}{s^2}$, $v = at = (2)(t)$ m/s, and $x = \frac{1}{2}at^2$.

We can go one step further by noting that for the same constant acceleration the motion might not start from rest. It could have some *initial* velocity v_0 at the time $t = 0$. Let's repeat our example, with $v_0 = 5$ m/s.

The picture is still the same.

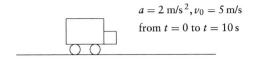

$a = 2$ m/s², $v_0 = 5$ m/s
from $t = 0$ to $t = 10$ s

The motion diagram shows the dots starting further apart, and the distance between them continuing to expand as the velocity increases.

. . . .

The graph of $a(t)$ remains the same, since the acceleration is the same as before.

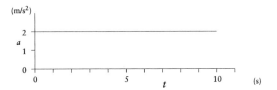

On the graph of $v(t)$ we start at the value $v_0 = 5$ m/s at $t = 0$. Since the acceleration is the same as before, the graph is still a straight line with a slope of $a = 2$ m/s^2. Its mathematical representation is $v = v_0 + at$.

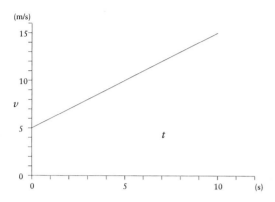

We can again compare this expression with the standard expression for a straight line, $y = mx + b$. This time we plot v along the vertical axis and t along the horizontal. The slope is a and the intercept on the vertical axis is v_0. The mathematical expression becomes $v = v_0 + at$.

To get the displacement $x(t)$ we look at the area under the curve. This time it is the sum of the rectangle whose area is $v_0 t$ and the triangle that we had before, whose area is $\frac{1}{2}at^2$, giving $x = v_0 t + \frac{1}{2}at^2$. (The shape is again that of a parabola, but the slope at the origin is no longer zero.)

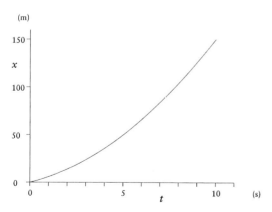

We have gone a long way toward solving the problem that we set ourselves at the beginning of the chapter for this special case. It was to take a particle at some starting point and starting time, with a known initial velocity at that place and time, to figure out where it would be at some later time, and how fast it would then be going.

Let our particle start out at $x = 0, t = 0$, with velocity v_0. The two mathematical statements,

$$x = v_0 t + \frac{1}{2}at^2 \qquad (3.1)$$

and

$$v = v_0 + at \qquad (3.2)$$

then do just what we were looking for. They tell us what x and v will be at any time, t. All we have to know is that the acceleration is constant, and what its value is.

That isn't quite what we promised. We talked about forces, not acceleration. But there is a close relationship between force and acceleration, which we will explore in the next chapter.

In the meantime these two relations describe what happens when a particle moves along a straight line with constant acceleration. It is often convenient to use a relation that does not contain the time explicitly. We can get this third relation by eliminating the time from the first two.

To do this, use the second relation to show what t is equal to, namely $t = \frac{v - v_0}{a}$, and substitute the right-hand side of this expression for t in the first relation: $x = (v_0)(\frac{v - v_0}{a}) + (\frac{1}{2})(a)(\frac{v - v_0}{a})^2$. If you now multiply every term by $2a$, you get $2ax = 2v_0(v - v_0) + (v - v_0)^2 = 2v_0 v - 2v_0^2 + v^2 - 2vv_0 + v_0^2 = v^2 - v_0^2$, which gives our third relation for constant acceleration:

$$v^2 = v_0^2 + 2ax \qquad (3.3)$$

Equations (3.1), (3.2), and (3.3) are three relations that give a complete kinematic description of motion in a straight line with constant acceleration.

EXAMPLE 9

Go to the PhET website (http://phet.colorado.edu) and open the simulation *The Moving Man*. (See the earlier example on this simulation.) Go to *Charts*.

(a) Put in $x = 0$, $v = 0$, and $a = 0.4$ m/s^2.
What is the shape of $v(t)$? What is its equation (up to where he hits the wall)?

What is the shape of $x(t)$? What is its equation? Check it with the Expression Evaluator.

(b) Put in $x = -10$ m, $v = 5$ m/s, $a = -1$ m/s^2.
Watch the velocity vector.
Watch how $x(t)$ follows the man's position.
What does the graph of $v(t)$ show? What is its equation? Check that the equation gives the values of v at $t = 0, 5$ s, and 10 s. What is the velocity at the point where the man turns around?
What does $a(t)$ show? What is its equation?

(c) What is the equation of $x(t)$? What is the value of x where the man turns around?

(d) Check your expressions with the Expression Evaluator.

Ans.:

(a) $v(t)$ is a straight line. Its slope is 0.4 m/s, which is equal to the acceleration. $v = 0.4t$ ($v = at$). $x(t)$ is a parabola: $x = 0.2t^2$ ($x = \frac{1}{2}at^2$).

(b) While the man moves to the right (in the positive direction), v is positive. It is zero where he turns around, and then negative while he moves to the left. $x(t)$ starts at $x = -10$ m/s and goes past zero and then back, decreasing toward -10 where the man hits a wall.

$v(t)$ is a straight line with the negative slope -1 m/s^2. Its equation is $v = 5 - t$ ($v = v_0 + at$). The graph and the equation show that at $t = 0$, $v = 5$ m/s, at $t = 5$ s, $v = 0$, and at $t = 10$ s, $v = -5$ m/s. At the point where the man turns around, $v = 0$.

a is constant at -1 m/s^2.

(c) The equation for x is more difficult to determine, because it does not follow the relation in the text, which is for a particle that starts out at $x = 0$. Here it starts at $x = -10$ m and the -10 m has to be added to the expression $x = v_0 t + \frac{1}{2}at^2$ to give $x = -10 + 5t - \frac{1}{2}t^2$.
At $t = 0$, $x = -10$ m, at $t = 10$ s, $x = -10 + 50 - 50 = -10$ m, and at $t = 5$ s when the man turns around, it is $-10 = 25 - 12.5$ or $x = 2.5$ m, as seen on the graph.

We have dealt with just two special cases of kinematics: first motion along a straight line with constant velocity and then motion along a straight line with constant acceleration. We could now examine motion with different kinds of accelerations, as in oscillatory motion, motion in a plane (two-dimensional motion), such as projectile motion and circular motion, and motion in space (three-dimensional motion), as in a spiral. The definitions of acceleration, velocity, and displacement remain the same, but the relation between these quantities is different for each kind of motion. Each fulfills the promise of a totally determined motion once the position and velocity at a given moment (the *initial conditions*) and the acceleration, $a(t)$, are given. In each case there are different characteristic features.

But no matter how complicated the calculations for a particular case are, the content of the calculations of kinematics is still the same: displacement, velocity, acceleration, time, and the relationships between them as they follow from the definitions. There are only the definitions: no other laws or principles are required. Once the acceleration $a(t)$ is known, there is no further recourse to nature or to observation. The definitions provide us with all we need to describe the motion.

EXAMPLE 10

A bicyclist, riding with a velocity of 4 m/s, slows down at a steady rate of 1.5 m/s^2 until she stops.

(a) Draw a graph of $v(t)$ and use it to determine how far she moves until she stops and how long it takes her to stop.

(b) Write a mathematical statement describing her displacement $x(t)$ while she is slowing down.

(c) Write a mathematical statement for $v(t)$ while she is slowing down.

(d) Use your mathematical statements to find the time and distance until she stops and compare them to your answers in part (a).

Ans.:

(a) Let's take the direction of the starting velocity to be positive. If v increases, a is then positive. Here it decreases, and is therefore negative: $a = -1.5$ m/s^2.

After 1 s the velocity has decreased by 1.5 m/s, from 4 m/s to $4 - 1.5$ or 2.5 m/s. After 2 s it has decreased by another 1.5 m/s, to 1 m/s.

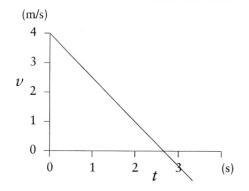

The line crosses the horizontal axis where the velocity has decreased to zero. That is after about 2.7 s.

The distance she has covered by that time is equal to the area under the curve, i.e., the area of the triangle whose sides are 4 m/s and 2.7 m. This area is $(\frac{1}{2})(4\text{ m/s})(2.7\text{ s}) = 5.4\text{ m}$.

(b) $x = v_0 t + \frac{1}{2}at^2$, where $v_0 = 4$ m/s and $a = -1.5$ m/s², so that $x = 4t - 0.75t^2$.

(c) $v = v_0 + at$, so that $v = 4 - 1.5t$.

(d) From the statement in part (c), $v = 0$ when $0 = 4 - 1.5t$, or $1.5t = 4$, or $t = 4/1.5 = 2.67$. When $t = 2.67$, the answer to part (b) shows that $x = (4)(2.67) - (.75)(2.67^2) = 10.67 - 5.33 = 5.33$ m. (This answer is more precise than that for [a], which was just read off the graph.)

EXAMPLE 11

A ball is thrown upward with an initial velocity whose magnitude is 20 m/s. It moves upward and then returns to the place where it started. We will assume that air resistance may be neglected, so that the magnitude of the downward acceleration is $g = 9.8$ m/s².

(a) Develop the mathematical statements for the displacement $y(t)$ and the velocity $v(t)$.

(b) Draw graphs of $a(t), v(t)$, and $y(t)$.

(c) What is the velocity when the ball returns to the starting point?

(d) What is the maximum height to which the ball rises?

(e) What is the length of time during which the ball is in the air?

Ans.:
We have to decide which direction we will take as positive. (It is up to us to decide, but we have to stick to the decision for the duration of the problem.) Here we will take the upward direction to be positive.

(a) We can use the first two relations for constant acceleration, $x = v_0 t + \frac{1}{2}at^2$ and $v = v_0 + at$, and use y instead of x, with $g = -9.8$ m/s² (negative, because we have chosen up to be positive, and the acceleration is down) and $v_0 = 20$ m/s, to get $y = 20t - 4.9t^2$ and $v = 20 - 9.8t$.

(c) There are different ways to get the answer. The simplest is to use the third of the equations that describe constant acceleration, $v^2 = v_0^2 + 2ax$. The displacement for the entire motion is now zero, so that we are left with $v^2 = v_0^2$, which has the solutions $v = \pm v_0$. Since the direction at the starting point is up, and therefore positive, it is negative on the way down, and we need to choose the negative value, $v = -v_0, = -20$ m/s.

We can also begin by answering the next two parts of the question (d and e): at the top $v = 0$. We can use $v = v_0 + at$ to find the time to reach the top: $0 = 20 - 9.8t$, which yields $t = \frac{20}{9.8} = 2.04$ s. We can then find y from $y = 20t - 4.9t^2 = 40.8 - 20.4 = 20.4$ m.

The motion downward can be analyzed in the same way: now the initial velocity (starting from the top) is zero, and the displacement to the starting point is -20.4 m (down, and therefore negative). We can use $v^2 = v_0^2 + 2ax$, now $v^2 = 0 + (2)(-9.8)(-20.4) = 400$, so that $v = \pm 20$ m/s. We know that the direction is negative, so that we have to choose the negative sign, as before, $v = -20$ m/s.

We can find t from $v = v_0 + at$, or $t = \frac{v - v_0}{a} = \frac{-20-0}{-9.8} = 2.04$ s, as before.

(d) The graphs of part (b) are well worth contemplating.

$a(t)$ is constant and negative with magnitude $g = 9.8$ m/s². This is also the slope of the graph of $v(t)$.

The graph of $v(t)$ is positive for half of the flight of the ball, until it reaches its highest point. During this part of the flight the velocity points up. It starts from 20 m/s and decreases, reaching zero when the ball gets to its highest point. During the second half of the flight the velocity is negative, as the ball falls back down to the starting point, where it has the same speed as initially, but in the opposite direction, so that the velocity has the same magnitude, but opposite sign.

$y(t)$ starts and ends at $y = 0$. Half way between these two points the ball is at its highest point and y is at its maximum value. The slope of $y(t)$ starts out positive and decreases during the first half until it is zero when the ball is at its maximum height. It is then negative, increasing in magnitude as the ball returns to its initial position. The slope of $y(t)$ is equal to $v(t)$.

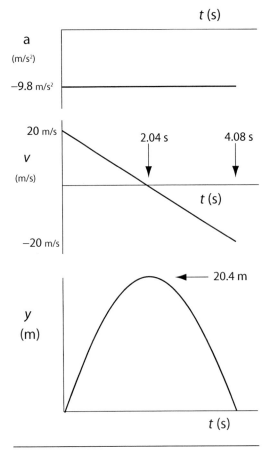

3.3 The mathematics of change

There is a branch of mathematics that deals with two of the concepts that we have just talked about: the slope of a curve and the area under a curve. It describes the relationships between changing quantities, i.e., how changes in one quantity affect another. It was developed to a large extent in the seventeenth century by Isaac Newton, Gottfried Wilhelm Leibniz, and others, and the primary motivation was to describe motion. We won't study this subject in detail, but we will look at some of its concepts.

We already know the relation between velocity and displacement and between acceleration and velocity. There are many other relationships between varying quantities, involving rates of change, and changes in the rate of change. They might deal with other mechanical quantities, or perhaps with the rate of transformation of electrical energy to internal energy. Other examples in which rates of change are important might be the way in which labor costs affect profits, or how the number of predators and the food supply affect animal populations.

The subject (the *calculus*) opens up a fascinating and wonderful world. We will look at some of its features and symbols, primarily to enhance and clarify topics that you already know.

Slopes and derivatives

Earlier we used the fact that the slope of the line of x plotted against t is $\frac{\Delta x}{\Delta t}$, the change of the displacement (Δx), divided by the time interval (Δt) during which the displacement changes by Δx. That was for a straight–line relationship between x and t. For a relationship represented by a curved line it was necessary to draw the tangent to the curve, and to use the changes in x and t, namely Δx and Δt, for the tangent line. We now introduce a shorthand notation for the slope of the tangent, namely $\frac{dx}{dt}$.

The quantity $\frac{dx}{dt}$ is the slope of the tangent line. It is also called "the derivative of x with respect to t." Instead of saying "the velocity is the slope of the graph of displacement against time," we can say that "the velocity is the derivative of the displacement with respect to time." The words may be unfamiliar, but what we mean to say is exactly the same. We can now use $v = \frac{dx}{dt}$ as the definition of the velocity.

We haven't done anything. We have only introduced a new word for "slope," namely "derivative," and a new symbol, $\frac{dx}{dt}$. Similarly, we can define the acceleration to be $a = \frac{dv}{dt}$.

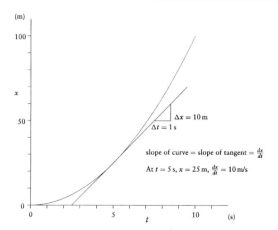

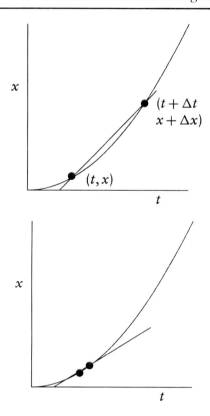

Of course this isn't the whole story. The methods of the calculus also tell us how to calculate derivatives. Here is an example of the kind of result that we can get: if the relationship between x and t is given by $x = t^2$, then the slope $\frac{dx}{dt}$ is equal to $2t$.

We can see how this comes about. Look at two points on the graph of $x(t)$, namely (t, x), and a point not too far away, $(t + \Delta t, x + \Delta x)$. Since both points are on the curve $x = t^2$, we can write $x = t^2$ and $(x + \Delta x) = (t + \Delta t)^2 = t^2 + 2t\Delta t + (\Delta t)^2$. Subtracting the first of these from the second, we get $\Delta x = 2t\Delta t + (\Delta t)^2$ or $\frac{\Delta x}{\Delta t} = 2t + \Delta t$. Now let the second point move toward the first. As Δt becomes smaller and smaller, the line through the two points becomes the tangent at (t, x), and $\frac{\Delta x}{\Delta t}$ becomes $\frac{dx}{dt}$, equal to $2t$.

More generally, for $x = A t^n$, where A and n are fixed numbers, it can be shown that the derivative $\frac{dx}{dt}$, the slope of the curve of x against t, is equal to nAt^{n-1}. [For $x = t^2$, we have $v = \frac{dx}{dt} = (2)(t^{2-1}) = 2t$.]

Let's see what this leads to for our old case of constant acceleration. Start with $x = v_0 t + \frac{1}{2}at^2$, and find the expressions for v and a.

We'll have to take one term at a time. For the term $v_0 t$, v_0 is a number, taking the place of A. The exponent n is one ($t^1 = t$). The derivative therefore becomes $(1)(v_0)(t^0)$. Since $t^0 = 1$, this is just v_0. Now look at the second term, $\frac{1}{2}at^2$. This time the exponent n is two, and the constant term (corresponding to A) is $\frac{1}{2}a$. The derivative is $(2)(\frac{1}{2}a)(t^1)$, or at. Taking the two terms together we see that $\frac{dx}{dt} = v_0 + at$. In other words, $v = v_0 + at$, just as we had it earlier.

We can apply the same procedure to $v = v_0 + at$, to find the derivative $\frac{dv}{dt}$, i.e., the slope of the graph of v against t, which we know to be the acceleration. The first term is v_0, which is constant, so that there is no change, and therefore no derivative. From the second term (where $n = 1$) we get $(1)(a)(t^0)$, or a, the constant value of the acceleration.

The particular relations between x, v, and a that we have used apply only when a is constant. But the definitions $v = \frac{dx}{dt}$ and $a = \frac{dv}{dt}$ are always valid.

Areas and integrals

Now let's look at the area under the curve of v against t. In the simplest case it can be a rectangle, but it can also have a more complicated shape. Consider two examples in which we look at the area between the values $t = 0$ and $t = 4$ s. Here is the calculus terminology for that: this same area is called "the integral of v with respect to t between the limits $t = 0$ and $t = 4$ s." The symbol for it is $\int_0^{4s} v \, dt$. This turns out to be the opposite of taking the derivative, and is sometimes called the antiderivative.

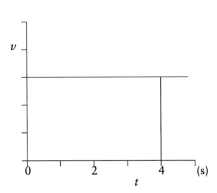
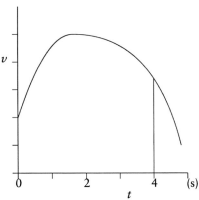

Again, we haven't done anything, except to use different symbols. If you are not familiar with them they look strange, perhaps forbidding. But then you realize that they are just symbols, just new ways to write down something that you already know.

For the particular case where $v = At^n$, $\int v\, dt = A\frac{t^{n+1}}{n+1}$. You can see that going backward, taking the derivative, gives you $(n+1)(A)\frac{t^{n+1-1}}{n+1}$, or At^n.

You may not have had a course in calculus, but here you see its basic ideas and will be able to use them. We won't do any calculations using calculus in this book, but we will use the ideas and the symbols occasionally.

Go to the PhET website (http://phet.colorado.edu) and open the simulation *Calculus Grapher*.

It allows you to explore derivatives and integrals. Start by checking "derivative" and choosing the triangle. Use the mouse to drag the f(x) line up from the x-axis. The vertical scale can be changed with the buttons on the left and the horizontal scale with the slider under the top figure on the right. Each time you want a different horizontal scale or to go to a new function, press zero and start again.

Try the various other functions. In each case see that the derivative graph ($\frac{df}{dx}$) shows the slope of the graph of f(x). Pay particular attention to the sinusoidal function (third down on the left) to which we will come back later: get f(x) to start upward at the origin so that it represents the sine function and continues for about one complete oscillation (up, down, and back). The derivative then represents the cosine function, starting at its maximum. What is the slope when f(x) is zero and going up, at its maximum, at its minimum, and at zero going down?

EXAMPLE 12

The astronaut revisited.

(a) Write mathematical statements for the astronaut's hammer in Example 8 for $v(t)$ and $x(t)$.

(b) Use your statement to find the answers to parts (d) and (e) of Example 8, and compare them to your previous answers.

Ans.:

(a) This time assume that upward directions are positive. $v = at$, where $a = -1.6$ m/s^2, so that $v = -1.6t$.
$$x = \tfrac{1}{2}at^2,\ \text{or}\ x = -0.8t^2.$$

(b) Use the second relation first: $x = -100$ m where we continue to use upward as positive so that the downward direction is negative. $-100 = -0.8t^2$, or $t^2 = \frac{100}{.8} = 125$, so that $t = 11.2$ s. (Note that, as usual, we have used SI units for all the data, so that the answer automatically comes out in the appropriate SI unit also.)

$v = at = (-1.5)(11.2) = -16.8$ m/s. The velocity just before the hammer hits the ground has the magnitude 16.8 m/s, and is in the negative or downward direction.

(We could have chosen the downward direction to be positive. All downward-directed vectors would then be positive. The calculation would look different, but the result would be the same.)

EXAMPLE 13

The graph shows the acceleration of an object.

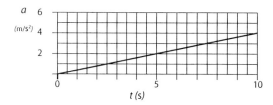

(a) Write down a mathematical statement that describes how the acceleration changes with time.

(b) Find $v(t)$.

(c) Previous examples have been for constant velocity or constant acceleration. How would you describe the motion of this graph? Explain.

Ans.:

(a) $a = 0.4t$.

(b) The velocity is the area under the curve: After 1 s it is $(\frac{1}{2})(1)(0.4)$. After 2 s it is $(\frac{1}{2})(2)(0.8)$. After t s it is $(\frac{1}{2})(t)(0.4t)$ or $0.2t^2$.
Note that if $v = 0.2t^2$, $a = \frac{dv}{dt} = 0.4t$.

(c) Neither v nor a is constant here. The acceleration increases linearly with time, i.e., it is proportional to t.

3.4 Summary

Everything around us is in motion. Even an object that appears to be at rest consists of atoms in motion, with their electrons and their nuclei also in motion. To describe an object in motion we use a number of concepts, primarily *displacement*, *velocity*, and *acceleration*. The displacement shows how far the object is from the start or from a reference point. The velocity describes how fast it moves. The acceleration measures how fast the velocity changes.

Most objects or systems are too complex to be analyzed completely. We therefore invent a *model*, a simplified version of the real system. It retains some of the essential features and allows an approximate analysis by neglecting some of the complicating aspects. For example, in this and the following chapters we often treat an object as a particle.

The displacement is the distance to a point, from a starting point or from a reference point. The velocity is the rate at which the displacement changes. If the displacement is in meters, we can measure the velocity in meters per second, m/s. If the displacement changes by Δx meters in a time Δt s, the velocity is $v = \frac{\Delta x}{\Delta t}$. On a graph of x against t this is the slope.

The displacement, the velocity, and the acceleration are *vector* quantities. They have both magnitude and direction. A vector can be represented by a line whose length indicates the magnitude, with the direction indicated by an arrow. The magnitude of the velocity (represented by the length of the velocity vector) is the *speed*. If the motion is along a straight line we need only positive and negative values for $x, v,$ and a to express the magnitude and the direction.

If a graph of x against t is curved, the *slope* at a point is the slope of the line that is tangent to the curve at that point. It is also called the *derivative* of x with respect to t, and is written $\frac{dx}{dt}$. The definitions can be written as $v = \frac{dx}{dt}$ and $a = \frac{dv}{dt}$. Since a is the derivative with respect to t of the derivative of x with respect to t, it is also called the *second derivative* of x with respect to t, and is written $\frac{d^2x}{dt^2}$. For $x = At^n$, $\frac{dx}{dt} = nAt^{n-1}$.

We go from x to v and to a by taking derivatives. We can go in the opposite direction with areas under the curves or *integrals*.

3.5 Review activities and problems

Guided review

1. Cindy bicycles to the grocery store. She first goes four blocks west and then three blocks south. Find the magnitude and direction of her displacement.

2. Cindy bicycles along the path described in Question 1. It takes her 5 minutes for the first part (going west) and 4 minutes for the second part (going south).

(a) What is her average velocity in each of these two parts? Assume that each block is 150 m and express each of the two velocities in m/s.

(b) Draw a graph of x(t), using meters and seconds, for the first part of her path.

3. Draw motion diagrams for a car that is
    (a) accelerating with a positive acceleration to the right,
    (b) slowing down as it moves to the right.

4. The graph shows Cindy again as she bicycles along a straight path.

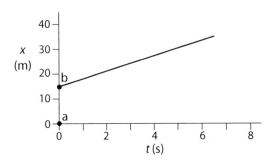

(a) How far does she move in 5 s?
(b) What do points a and b on the graph represent about Cindy's path?
(c) What is the slope of the line?
(d) What is her velocity?
(e) Write the equation for the line, first using only symbols, and then using only numbers except for x and t.

5. The graph shows Harry's motion. x is positive to the right.

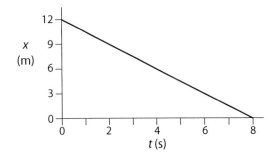

(a) Where does Harry's motion begin?
(b) At what time does he reach the point represented by x = 0?
(c) In what direction does he move?

(d) What are the magnitude and direction of his velocity? Does the direction of the velocity change? How can you tell? How can you tell whether his velocity is constant?

6. Go to the PhET website and open the simulation *The Moving Man*. Put in $x = 0$ and $v = -2$ m/s.

What does the graph of $v(t)$ show?
What happens to it when the man gets to the wall?
What is the equation for $v(t)$ before he gets to the wall?
What does the graph of $x(t)$ show?
What happens to it when the man gets to the wall?
What is the equation for $x(t)$ before he gets to the wall?
What does the graph of $a(t)$ show?
Put your expression for $x(t)$ into the Expression Evaluator and compare what you see to the "experimental" graph.

7. The graph shows the motion of your car for 16 s. The positive direction is to the right.

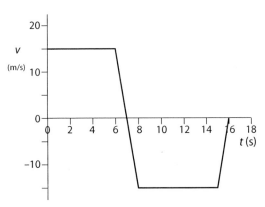

What is the type of motion of the car between
(a) 0 and 6 s
(b) 8 and 15 s
(c) 6 and 8 s. What happens at $t = 7$ s?
(d) 15 and 16 s

8. Tehmina's car accelerates from rest with $a = 5$ m/s^2.
(a) Make a table of v and t at 1 s intervals for 10 s.
(b) Make a graph of $v(t)$ for the 10 s.
(c) Make a table of x and t for the 10 s. (Let $x = 0$ when $t = 0$.)

(d) How long does it take for the car to reach a velocity of 40 km/h? How far has it then traveled?

9. Go to the PhET website and open the simulation *The Moving Man*.
  Put in $x = 10$, $v = -5$, and $a = 1$.
  (a) Watch the velocity vector and how the graph of $x(t)$ follows the man's position.
  What does the graph of $v(t)$ show? What is its equation? Check that the equation gives the values on the graph. What is the velocity when the man turns around? Check your equation with the Expression Evaluator.
  What does $a(t)$ show? What is its equation?
  (c) What is the equation of $x(t)$? What is the position where the man turns around? Check your equation with the Expression Evaluator.

10. Tony, riding a motorcycle, with a velocity of 21 m/s slows down with an acceleration whose magnitude is 3 m/s².
  (a) Draw a graph of $v(t)$ to the time when the motorcycle stops.
  (b) Use your graph to determine how far he moves before the motorcycle stops and how much time that takes.
  (c) Write mathematical statements for $x(t)$ and $v(t)$. Use your mathematical statements to find the answers to part (b), and compare them to the values that you obtained graphically.

11. A stone is thrown vertically upward from outside a window, with $v_0 = 18$ m/s. It moves up, and then moves down to the ground 420 m below the point where it started. Let "up" be positive.
  (a) Write down the mathematical statements for $a(t), v(t)$, and $y(t)$.
  (b) Draw graphs of these functions.
  (c) What is the maximum height above the window to which the stone rises?
  (d) How long is the rock in the air before it reaches the ground?
  (e) What is its velocity just before it hits the ground?

12. (a) Write down the mathematical statements for $v(t)$ and $x(t)$ of the car in Question 8 of this Guided Review.
  (b) Use your statement to find how long it takes for the car to reach 40 kilometers per hour and how far it has then traveled. Compare your answers to those of Question 8.

13.

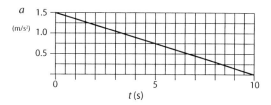

(a) Describe a procedure for finding $v(t)$ graphically for the $a(t)$ shown in the figure.
(b) Find $v(t)$. ($v_0 = 0$)
(c) Check to see what $v(t)$ is by using the relation for the integral of $a(t)$ and compare the result to your answer to part (b).

Problems and reasoning skill building

1. The figure shows three qualitative position-time graphs.

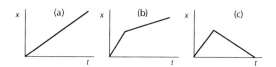

For each one
(a) draw a qualitative velocity-time graph,
(b) describe in words what happens, i.e., state what changes occur in the velocity and in the direction of motion.
(c) In the second and third graphs there is a kink in $x(t)$. What happens to the velocity at the kink? What is the acceleration at the kink? What is your comment on how realistic this is?

2. Draw qualitative graphs of $x(t)$ and $v(t)$ for an object that moves with constant speed away from the origin, stops briefly, and then continues in the same direction as before at twice the original speed. It stops again briefly, and then moves toward the origin at the original speed until it returns to the starting position.

3. A car moves toward the origin at 20 m/s for 5 s, stops, and then (nearly instantly) moves away from the origin at 20 m/s for the next 5 s.
  (a) Draw a graph of $v(t)$.
  (b) What is the change in the speed at $t = 5$ s?
  (c) What is the change in the velocity at $t = 5$ s?

4. The following are qualitative graphs of $v(t)$. For each one sketch the corresponding graph of $a(t)$.

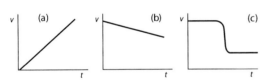

5. Sketch qualitative graphs of $x(t)$, $v(t)$, and $a(t)$ for a car that moves
   (a) away from the origin, in the positive direction, with constant speed,
   (b) toward the origin with constant speed,
   (c) away from the origin, starting from rest, with a constant positive acceleration,
   (d) past the origin with a positive velocity, with a constant negative acceleration until it stops.

6. Describe the motion corresponding to the graph. Where does the object start and end? What are the magnitude and direction of the initial velocity, and how does the velocity change? What is the sign of the acceleration?

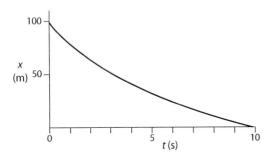

7. Two football players run toward each other with constant velocities, Bill to the right with twice the speed of Tom, who runs to the left, starting 25 m from Bill.
   (a) Draw a motion diagram for the two players.
   (b) Draw graphs of $v(t)$ for both on the same coordinate system.
   (c) Draw graphs of $x(t)$ for both on the same coordinate system.

8. The graph shows $x(t)$ for a marble. The positive direction is to the right.

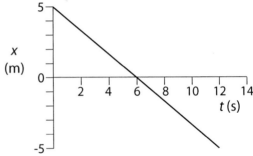

(a) What is the magnitude and direction of the velocity?
(b) What is the acceleration?
(c) How far does the marble travel?
(d) What is its average speed?

9. For the $x(t)$ of the diagram, what are the approximate values of $v(t)$ at points $a, b, c, d,$ and e? Sketch a diagram of $v(t)$.

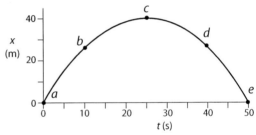

10. For the $x(t)$ of the diagram, what are the approximate values of $v(t)$ at 1, 2, 3,... to 12 s? Sketch the corresponding diagram of $v(t)$.

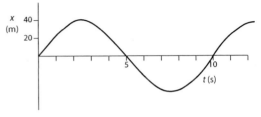

11. Draw a graph of $x(t)$ for a bicyclist who rides at 5 m/s to the right for 15 s, stops for 10 s, and then turns around to go back to the starting point, which takes her 25 s more.

12. Describe (in words) the motion represented by the following graph of $x(t)$:

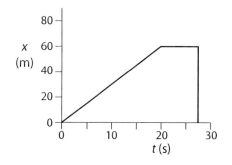

13. Describe (in words) the motion represented by the following graph of $x(t)$:

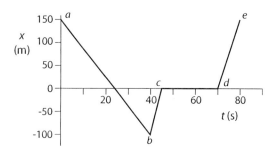

14. For the diagram of the previous question, what is the average speed between
    (a) a and b
    (b) a and c
    (c) a and d
    (d) a and e

15. A molecule travels to a wall with a constant speed of 100 m/s. It bounces back, and then travels in the opposite direction with a speed of 90 m/s.
    Describe what happens at the wall to its speed, its velocity, and its acceleration.

16. You run half-way around a circular track whose radius is 60 m at a steady 5 m/s. What are the following for this part of your path:
    (a) displacement
    (b) average speed
    (c) change in the velocity

17. The figure shows a graph of $v(t)$. How can you find the average velocity between points a and b from this graph?

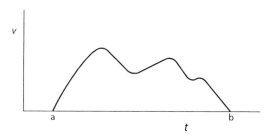

18. You stand at the edge of a cliff and throw a rock straight up with an initial velocity of 15 m/s. It reaches a maximum height, y_{max}, above the starting point, and then falls past you to a point 60 m below where it started.
    (a) What is the maximum height of the rock from the starting point?
    (b) How much time has elapsed to the time when the rock passes its starting point on the way down?
    (c) What is its speed just before it hits the ground 60 m below you?
    (d) What is the total time that the rock is in the air?

19. What are the displacement and the velocity of the rock of the previous question when the elapsed time is
    (a) 1 s
    (b) 2 s
    (c) 5 s?

20. (a) My house is 16 miles from my office. On Monday it took me 40 min to get from my house to the office. What was the magnitude of my average velocity in miles per hour?
    (b) On Tuesday I traveled the same route at 40 miles per hour for 20 minutes, and at 30 miles per hour for 15 minutes. I was stopped for 5 minutes at stoplights. What was the magnitude of my average velocity in miles per hour?
    (c) What is missing in the description of part (b) so that it is unrealistic?

21. The figure shows a graph of $v(t)$. What is the average velocity for the trip?

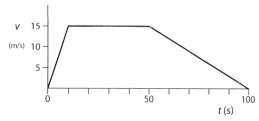

22. A car moves in a straight line at 40 m/s for 20 s and then at 30 m/s for 30 s. What is the average velocity?

23. What is the sum of the following displacements? First find the magnitude and direction

graphically. Then calculate the magnitude from the vector diagram.

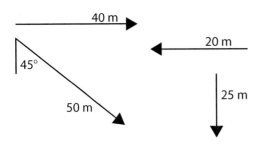

24. The following problems are about a ball moving straight up or down. We will assume that air resistance may be neglected, so that the acceleration is downward and has a magnitude of 9.8 m/s².

(a) The ball starts from rest and falls for 5 s. State your decision as to whether you wish up or down to be positive. Write down the expressions for $v(t)$ and $x(t)$ using numbers for every quantity except t. Draw graphs of $y(t)$ and $v(t)$. What are v and y when $t = 5$ s, from your graph and analytically?

(b) Answer the questions of part (a) for a ball that is thrown downward with an initial speed of $5 \frac{m}{s}$.

(c) Answer the questions of part (a) for a ball with an initial velocity of 5 m/s upward.

(d) Write down expressions for $v(x)$ for parts (a), (b), and (c).

25. A squirrel running at 3 m/s sees that it is pursued by a cat and accelerates to 8 m/s in 2 s. It travels 14 m in that time.

(a) What is its average acceleration?

(b) Is the acceleration constant? How can you tell?

26. As you drive along a highway at 50 miles per hour you suddenly see a deer coming into your path. Your reaction time is 0.25 s and your acceleration has a magnitude of 14 m/s². How far do you travel before your car stops?

27. (a) What is the average speed, in miles per hour, of a world class sprinter who runs 100 m in 10 s?

(b) What is the average speed (in miles per hour) of a runner who runs a mile in 4 minutes?

(c) What percentage is this of the answer to part (a)?

28. You are driving at 65 miles per hour, following a car 20 feet ahead of you, which is moving at the same speed. The car ahead of you crashes into a car that has drifted into its lane and comes to rest almost instantaneously. Your reaction time is 0.3 s.

(a) Describe your subsequent motion.

(b) If you step hard on your brakes, so that your acceleration has a magnitude of 15 m/s², how far behind the other car (when it crashes) do you need to be to avoid a collision?

(c) You have had two beers, and your reaction time has doubled. What is the answer to part (b) now?

Multiple choice questions

1. A ball is thrown upward from $y = 0$ and then falls back to its starting point. Neglect air resistance and let up be positive. Which of the following changes sign during the flight?

(a) y
(b) v
(c) a
(d) none of the above

2. For the ball of the previous question, which of the following is zero at the highest point?

(a) v
(b) a
(c) v and a
(d) $y, v,$ and a

3. For the ball of Question 1, which of the following changes sign at the highest point?

(a) x
(b) v
(c) x and v
(d) $x, v,$ and a

4. Which among the following pairs of graphs are corresponding graphs of $x(t)$ and $v(t)$?

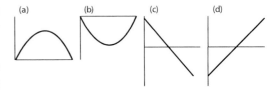

(a) a and d and b and c
(b) a and c and b and d
(c) a and d only
(d) b and d only

3.5 Review activities and problems

5. A toy car moves to the left with constant speed, then turns around and moves to the right with constant speed until it reaches its starting point. Which of the diagrams of $x(t)$ represents this motion? (Positive is to the right.)

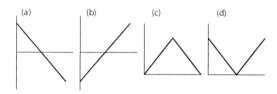

6. A toy car starts from rest, speeds up gradually to a constant speed, and slows down until it comes to rest. It speeds up gradually in the opposite direction, slows down, and comes to rest at its starting point. Which of the following graphs of $x(t)$ represents this motion?

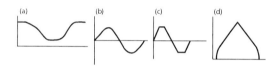

7. Which of the following graphs represents the velocity of a bouncing ball?

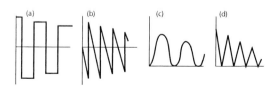

Synthesis problems and projects

1. You and your friends Tom and Jerry are standing at the edge of a cliff with two identical rocks.

(a) Tom says "If I throw one of these rocks up and the other down, both with the same initial speed, they will hit the ground below the cliff at the same time, because in the relation $y = v_0 t + \frac{1}{2}at^2$ the three quantities y, v_0, and a are the same for both cases." Do you agree? Explain.

(b) Jerry says: "Both rocks will have the same velocity just before hitting the ground." Do you agree? Explain.

2. A state trooper in a police car is parked on a highway. A car moving at 80 miles per hour (or 35.8 m/s) passes the police car. The trooper spends 5 s to call in a report and then accelerates to 100 miles per hour (or 44.7 m/s) to pursue the speeder.

(a) Draw graphs of $x(t)$ for both cars on the same coordinate system. Make the (not very realistic) assumption that the trooper changes his speed instantaneously. Let the origin of your graph ($x = 0, t = 0$) represent the place and time at which the speeding car passes the police car.

Use your graph to find the time when the trooper catches up with the speeder and the distance from the origin where this happens.

(b) Write down the mathematical statements for $x_1(t)$ for the speeding car. Develop a mathematical statement for $x_2(t)$ of the police car. (Check your expression by seeing where the police car is at $t = 5$ s.)

(c) From the two expressions answer the questions of part (a) analytically.

3. As you drive home with Leila she records the motion of your car by writing down the speedometer reading every second for 15 seconds, with the results shown in the following table.

t	v
1	17
2	17
3	17
4	17
5	16
6	15
7	14
8	13
9	12
10	11
11	10
12	9
13	8
14	7
15	6

(a) Plot the data for $v(t)$.
(b) Write two functions for $v(t)$. The first will be $v_1(t)$ for the time interval from $t = 0$ to $t = 4$ s.

The second will be $v_2(t)$ for $4\,\text{s} < t < 15\,\text{s}$. Let $(x = 0, t = 0)$ represent the position and time when she starts taking data. Use numbers for everything but t and the units.

(c) Write the corresponding functions for $a_1(t), a_2(t), x_1(t)$, and $x_2(t)$.

(d) How far does the car move in the first four seconds?

(e) If the car continued with the same acceleration as between $t = 4\,\text{s}$ and $t = 15\,\text{s}$, at what values of t and x would it come to rest?

4. A tortoise and a snake start a race together, the tortoise moving at 0.30 m/s and the snake at 0.80 m/s. The snake stops after a minute to talk to a fellow snake for 3 min before starting to go again at the same speed as before. The tortoise wins the race by 2 m.

(a) Draw graphs of $x(t)$ for both on the same coordinates.

(b) Where are they after 1, 2, 3, 4, and 5 minutes?

(c) What are the length and the time of the race between the two?

(d) What is the equation for $x(t)$ of the tortoise? What is the equation of the line that represents the graph of $x(t)$ for the snake after her rest? (Check your expression by seeing whether it gives the answers for part [b] for 4 and 5 minutes.)

(e) Use the two equations to determine the answers to part (c) analytically.

CHAPTER 4

Forces and Motion: Newton's Framework

Newton's laws of motion
> *When forces add up to zero: the first law*
> *What force really is: the second law*
> *Units*
> *Inertial mass, gravitational mass, and the principle of equivalence*

Adding forces: vectors
> *One dimension*
> *Two or more dimensions*
> *Force diagrams*
> *Vector components*
> *More on friction*
> *Object or system?*

Momentum and its conservation. Action, reaction, and Newton's third law

One more motion that is everywhere: rotation
> *Uniform circular motion*
> *Angular momentum and torque*
> *The angular momentum of particles*

What does it take to get something to move? You have to push a book to make it start to slide along the table. You have to throw a ball to make it leave your hand to fly through the air. The push on the book and that of your hand on the ball as you throw it are the forces that determine the motion. The book's motion depends not only on how hard you push, but also on the table and how smooth it is. The ball's motion also depends on forces other than that of your hand. Once the ball leaves your hand, the hand no longer exerts a force on it. The other forces continue to act: the earth pulls it down with the force of gravity. And on its way the air pushes against it and affects the path that it follows.

It's easy to think of more complicated examples. When you are on a bicycle, the downward push of your feet is linked to forces that make the bicycle move forward. And just think of all the forces that act in a moving car.

It took a long time for the relation between force and motion to be clarified. It was Isaac Newton, in the seventeenth century, who developed the framework that we still use today.

4.1 Newton's laws of motion

When the forces add up to zero: the first law

One of Newton's breakthrough contributions was to see that it takes no force at all just to keep an object moving in a straight line with constant speed. A nonzero *net* force is there only when the motion changes in speed or direction, in other words, when there is an acceleration.

Let's look at what happens when we slide a book along a table. At first it just sits there. We push it and it speeds up. We let go and it slides along by itself for a short distance. It slows down and comes to rest. On a smoother table it goes farther. On ice the same push makes it go quite far. In each case there is some friction, but the less friction there is, the farther the book moves. We can now imagine, as Newton did, that if there were no friction at all, the book would continue to move without losing speed. Today we can get quite close to that situation by letting an object move on a cushion of air, on an *air track* or *air table*. (You may also be familiar with a game called *air hockey*, in which a puck moves on a cushion of air, almost without friction.)

To make an object slide on a smoother and smoother surface is something we can do. It's an experiment. To make it move without any friction is something we cannot do. It's an *ideal* situation that we can only approach. Newton imagined what would happen in this ideal case, and concluded that if there were no friction, and no other horizontal force, the book would continue to move in a straight line with constant speed.

Are there any forces on the book when it just sits still on the table? Is the earth still pulling down on it? If the table were not there, the earth would pull the book down and it would fall to the floor. The table keeps it from falling, and while the earth pulls down, the table pushes the book up. The two forces, the force down by the earth and the force up by the table, are of equal size but in opposite directions. Their effects cancel each other out and the *net force* is equal to zero. Since there is no net force there is no change in the motion of the book.

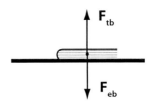

Each force is an *interaction*. It takes two! Whether it's the force of the earth on the book or the force of the hand on the ball, there are always two objects involved. The earth interacts with the book. The hand interacts with the ball.

When we write a symbol for *force*, we want it to tell us which two objects are interacting. We can write $F_{\text{earth on book}}$. To make that less clumsy we shorten it to F_{eb}. The second subscript stands for the object that we want to talk about, and the first for the other object that is exerting a force on it and so interacts with it.

EXAMPLE 1

A rope holds a tire as a swing on the playground. What are the forces on the tire?

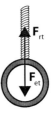

Ans.:

The tire is pulled down by the earth with a force $F_{\text{earth on tire}}$ or F_{et}.

The tire is pulled up by the rope with a force $F_{\text{rope on tire}}$ or F_{rt}.

In the ideal case, when we imagine the book to slide without friction or other horizontal forces, the two vertical forces are still there and add up to zero. After your hand is no longer in contact with the book and it no longer exerts a force, there are no horizontal forces, since we assume that there is no friction. Since the two vertical forces add up to zero, and since there are no other forces, the sum of all the forces acting on the book, the *net force*, or the *unbalanced* force, is zero.

This is the situation described by Newton's first law of motion. To have no force on an object is an ideal situation impossible to achieve. But we can talk about what happens when the *net force* (the sum of the forces on the object) is zero: the object remains at rest, or if it is moving, it continues to move with constant velocity, i.e., in a straight line with constant speed. In either case there is no acceleration. *If the sum of all the forces on an object is zero, its acceleration is zero.* This is Newton's first law of motion.

What force really is: the second law

We have talked about forces from the beginning of this book. We already know a good deal about different kinds of forces. We know that there are four fundamental forces, namely gravitational, electric, and two kinds of nuclear. We know something about the electric forces between atoms, which lead to the forces exerted by springs and ropes, to friction and air resistance, and to the forces exerted by our hands as we push and pull.

But we haven't really said exactly what we mean by *force*. To say that it is a push or a pull was enough to get us started. Now we will use our preliminary and intuitive knowledge to develop a precise and quantitative definition. In the process we will also define what is meant by mass, and get to Newton's second law of motion.

What happens when you step on a bathroom scale to weigh yourself? At least in an old-fashioned one there is a spring in it, which is compressed when you step on the scale.

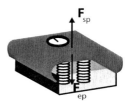

A pointer goes around a dial to tell you how much the spring is compressed. Two forces act on you as you stand on the scale: one is the force of the earth, pulling down on you ($F_{\text{earth on person}}$ or F_{ep}). This is the force that we call your *weight*. The other is the upward force of the scale with its spring ($F_{\text{scale on person}}$, or F_{sp}).

While you stand on the scale you have no acceleration. (Your velocity is constant and equal to zero.) That tells you that the net force on you is zero. The two forces on you must add up to zero.

The spring scale gives us a way to measure forces. We can also do that with a spring that is stretched. One end is attached to a fixed point, such as a hook on the ceiling or on a stand. From the other end we hang a pan on which we can place various objects to stretch the spring. A pointer is attached so that we can measure how far the spring has stretched.

Start with a set of identical metal blocks. Put one of them in the pan and mark a "1" on the scale next to the position of the pointer. With a second block the pointer moves further, and we mark a "2" where it stops, and so on. We then know how much any force stretches the spring.

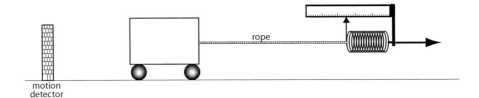

We say that the scale is now *calibrated* in units each of which is equal to the weight of one block.

This means that we now know what the pointer positions mean. When we take the blocks off and put on another object, such as a stone, the pointer moves to a new position. If it points to "4," we know that the weight of the stone is the same as the weight of four blocks. All we need to assume is that for a given weight on the scale, the pointer always returns to the same position. (This will be so as long as the spring is not stretched too far.)

Now let's do an experiment in which the object that is acted on by forces does not remain at rest. We can use a cart pulled with a rope, as on the diagram. If we attach our calibrated spring to the rope and pull on the spring, it will stretch and pull on the rope. The pointer position tells us the magnitude of the force with which the spring pulls on the rope and the magnitude of the force with which the rope pulls on the cart.

A sonic motion detector can measure the position of the cart at equal time intervals that are about 50 milliseconds apart. We can then use successive points to find the velocity, which can be plotted against time. Here is such a plot for a constant pulling force.

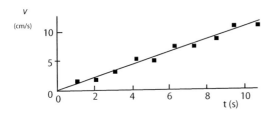

The graph is close to a straight line. Its slope is the acceleration, here equal to 1.11 in the units of the graph. If we repeat the measurements for different forces we find straight lines with different slopes, showing that the accelerations are different. We find that the relationship between the force and the acceleration is also represented by a straight line, showing us that the pulling force is proportional to the acceleration of the cart.

We can repeat the experiment with different numbers of blocks in the cart, but keeping the pulling force constant. As the number of blocks increases, the acceleration decreases.

To see what happens when we double the amount of material that is being pulled, we first determine the number of blocks that have the same weight as the cart. We can do this by using our spring scale. We find that doubling the amount of material being pulled by the same force leads to half the acceleration, tripling it to a third, and so on.

What property of the blocks determines how large the acceleration is? We call it their *mass*. More mass means less acceleration. We see that the acceleration is proportional to one over the mass (the reciprocal of the mass) that moves.

Now we're ready for a precise definition of force. We will take our preliminary and intuitive knowledge and the experimental results as guides. Only now we take the earlier statements to be exact: We saw that as the force was increased, the acceleration also increased. The graph showed that these two quantities are proportional. We also saw that as the mass was increased, the acceleration decreased. This time the graph showed that the acceleration is proportional to $\frac{1}{M}$, i.e., it is *inversely* proportional to the mass. We can combine these statements to say that the acceleration is proportional to $\frac{F}{M}$, i.e., that the net force is proportional to Ma.

We still need to choose the units for measuring force and mass. For mass we use a standard mass, that of a particular metal cylinder kept in a laboratory in Paris, as the mass of one unit in the SI system. We call this mass one kilogram (1 kg).

In the SI system the quantity Ma is then measured in $\frac{\text{kg m}}{\text{s}^2}$. Since the units on both sides of an equation have to be the same, we let that also be the SI unit for force. We give it its own name, the *newton*, N. We can now define the net force to be equal to Ma. *The sum of all the forces acting* on an object or system is equal to its mass times its acceleration. This is Newton's second law of motion.

Guided by the experiments we have refined our previously rough idea of the meaning of the term *force*, and defined both mass and force. A net force brings about an acceleration. The larger the net force, the larger the acceleration. The two are *proportional*: ($a \propto F_{net}$). A net force of 100 N on an object produces twice the acceleration of a net force of 50 N.

The amount of the acceleration also depends on the mass of the object on which the force acts. More mass means less acceleration ($a \propto \frac{1}{M}$). The same force of 100 N produces an acceleration on a 5 kg object, which is twice as large as that which it produces on a 10 kg object.

EXAMPLE 2

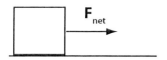

A block of ice has a mass of 5 kg. The net force on it is 100 N to the right. What is its acceleration?

Ans.:
The relation between the three quantities is $F = Ma$, so that $a = \frac{F}{M} = \frac{100\,\text{N}}{5\,\text{kg}} = 20\,\text{N/kg} = 20\,\text{m/s}^2$.

Since both the force and the mass are in SI units (newtons, N, for force, and kilograms, kg, for mass), the acceleration comes out automatically in SI units, m/s².

Units

It's really important to keep track of units. It helps to use a *system* of consistent units. There are different metric systems and various English systems. In this book we will stay, for the most part, with the SI system, which uses kg, m, s, and N. Most countries have adopted these units and multiples of them.

We have used the *kilogram* as a unit of mass. Even in the United States it (and the *gram*, equal to 10^{-3} kg) is used on food labels. But it is unlikely that you have seen the *newton* mentioned outside of physics class. In common, nonphysics language it is pretty much unknown.

This seems surprising, since we talk about forces frequently. The most common force is the *weight*, the force with which an object is attracted to the earth. We can measure our weight by stepping on a bathroom scale, but you won't find any that are graduated in newtons, even in countries that use the SI system exclusively. Instead the weight will be marked and referred to in kilograms. How is that possible, when kilograms measure mass?

It's a sort of shorthand. We know that if an object has a mass of 1 kg, its weight, the force with which the earth attracts it, is 9.8 N. (This value is approximately the same at all points on the surface of the earth.) We can say that the weight is Mg, where $g = 9.8$ N/kg. This relation for the weight as equal to Mg has the same form as $F = Ma$. g is the acceleration of an object when the only force on it is its weight. The units of g are the usual units of acceleration, m/s², which are the same as N/kg.

If people say (incorrectly) "the weight of this book is 1 kg," what they mean is "the weight is that of a book whose mass is 1 kg." If you step on a metric scale and it reads 70 kg, it means that your weight is that of any object whose mass is 70 kg.

You might say that it tells you that your mass is 70 kg, but that is not necessarily so. If you take your bathroom scale (the kind that has a spring inside) to the top of a mountain, it will read less because the force with which the earth attracts you is then smaller. The weight is still Mg, but the value of g is now smaller because you are farther away from the center of the earth.

But your mass will not change! It is the quantity that tells you what your acceleration is when a force is applied to you, and that doesn't change, whether you are on earth, on the moon, or anywhere else. In other words, a spring scale graduated in kg will no longer read correctly if g is no longer 9.8 m/s², or 9.8 N/kg.

In the commonly used English system the units are defined differently. Here the unit of force is the *pound*, equal to 4.445 N. In this system it is the unit of mass that is almost never used.

The weight (on earth) of an object whose mass is 1 kg is 9.8 N, or 2.205 lb. We can say, more briefly, that a kg weighs 9.8 N or 2.205 lb. It is important to remember that the kg is a unit of mass, while the N and the lb are units of force.

EXAMPLE 3

Bob's mass is 70 kg. He is on the surface of the earth, where $g = 9.8$ m/s^2.

(a) What is his weight in newtons?

(b) What does a metric scale register when he steps on it?

(c) What is his weight in pounds?

Ans.:

(a) His weight is $Mg = (70\,\text{kg})(9.8\,\text{N/kg}) = 686\,\text{N}$.

(b) The scale is graduated in kg, and registers 70 kg. His weight is that of an object whose mass is 70 kg. This weight is Mg or 686 N.

(c) Bob's weight in pounds is $(70)(2.205) = 154$ lbs.

Inertial mass, gravitational mass, and the principle of equivalence

Newton's law of gravitation says that there is a force of attraction between any two objects, i.e., there is a force on each, resulting from the gravitational attraction to the other. A curious feature now becomes apparent. We defined mass from $F = Ma$ as the quantity that tells us what the acceleration is when a given force is applied. Quite separately, Newton's law of gravitation says that mass is the quantity that tells us what the gravitational force on an object is. Two quite distinct physical phenomena are involved. Are we justified in using the same quantity (mass) and the same symbol (M) in both cases? It may be better to give them different names, *inertial mass* for the one based on $F = Ma$ and *gravitational mass* for the one based on gravitation. But no observation or experiment has ever been able to detect a difference.

In Newtonian, classical physics that's as far as we can go. The fact that the same quantity governs two seemingly unrelated phenomena remains a fortuitous quirk of nature.

Einstein, in 1915, saw more. Consider a closed box, or elevator, he said. Let go of an object in your hand, and it falls to the floor. That's not very mysterious. Presumably it is falling down because of the gravitational attraction of the earth. But you can't look out. Is there another explanation? Perhaps you are far from the earth and the elevator is accelerating. You feel the floor pushing up on you, and when you let go of the object in your hand, it accelerates with respect to the floor. There is no way, from inside the elevator, for you to tell the difference between the two explanations. The first case depends on gravitation and involves the gravitational mass. The second depends on acceleration and involves the inertial mass. But you can't tell the difference. They are the same.

This is the principle of equivalence, and it is the cornerstone and starting point of the *general theory of relativity*, the modern theory of gravitation.

To have unified the two aspects of mass would already be a great achievement. Beyond that, however, the general theory of relativity leads to results different from those described by Newton's law of gravitation. Einstein predicted three observational differences when he first described the theory in 1915. They are three astronomical effects. The first is a difference in the orbit of *Mercury* (the planet closest to the sun) from the Newtonian calculation. The second is the bending of light when it gets close to the sun. The third is a change in the wavelength of light emitted from a source where the gravitational force is very large.

For each case the observations showed that Einstein's predictions were correct. In other words, the general theory of relativity was seen to describe the observed gravitational effects better than Newton's law of gravitation. For decades they remained the only observable results of the theory that were different, and came to be regarded as curiosities. With advances in observational astronomy in the last part of the twentieth century, however, a whole new era began for the general theory. After lying dormant for a long time, it is now a vital component of modern physical science.

One widespread application is in Global Positioning Systems (GPS). The great precision of today's global positioning devices relies on calculations using the general theory of relativity.

4.2 Adding forces: vectors

One dimension

Let's talk again about the book at rest on the table. Suppose its weight is 10 N. That's the downward gravitational force attracting it to the earth. There is also the force with which the table pushes up. There is no acceleration, and therefore there is no net force. That means that the two forces (one up and one down) cancel. They are each 10 N, and they add up to zero. We see that we are dealing with quantities that behave very differently from numbers.

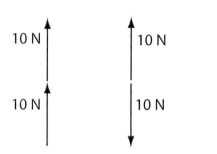

The direction matters. If two 10 N forces are in the same direction, their sum is 20 N. But if they are in opposite directions, they add up to zero. A quantity that has not only a size or magnitude, but also a direction, is a *vector* quantity. Examples are displacement, velocity, acceleration, and force. Quantities that have no direction, such as temperature, time, money, and energy, are scalar quantities. They add just like numbers.

When all the forces are vertical, we can say that we will let all upward forces be positive and all downward forces negative. (It could be the other way around. It's a choice we make. Either way, the physical result is the same.) The weight (the downward force that the earth exerts) is then -10 N, the upward force of the table is 10 N, and their sum is zero. The net force is zero even though there are two forces.

EXAMPLE 4

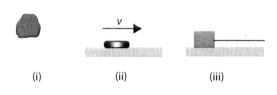

The figure shows (i) a freely falling rock, (ii) a hockey puck on ice, moving with constant velocity, and (iii) a box being pulled on a frictionless table.

For each of the objects in the figure,
(a) represent the forces on the object by vectors.
(b) Write the mathematical and the verbal statement of Newton's second law with these forces.

Ans.:

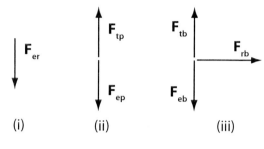

(i) $F_{er} = Ma$. The force of the earth on the rock (its weight) is equal to the rock's mass times its acceleration.

(ii) $F_{ep} - F_{tp} = 0$. The magnitude of the force of the earth on the puck (the weight of the puck) minus the magnitude of the force of the table on the puck is zero. There is no horizontal acceleration, and therefore no net horizontal force.

(iii) $F_{eb} - F_{tb} = 0$. $F_{rb} = Ma$.
Vertical forces: the magnitude of the weight of the box minus the magnitude of the force of the table on the box is zero.
Horizontal forces: There is only the force of the rope on the box. It is the net force and it is equal to the box's mass times its acceleration.

Two or more dimensions

Using only positive and negative numbers works as long as the forces are along the same line, up and down, or right and left. But what do we do if the forces are at some other angle to each other?

We start by representing each vector quantity by a line, whose length represents the magnitude, and whose orientation, with its arrow, shows the direction.

It's easiest to see the general method if we use displacement vectors. You just make a map. Suppose we look at the displacements from A to B, then from B to C, from C to D, and finally from D to E. The first vector goes from point A to point B, with its arrow pointing toward B. The other

64 / Forces and Motion: Newton's Framework

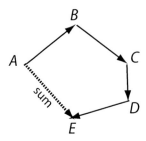

vectors follow, "tail" to "head," until we get to point *E*. The sum of the four displacements is the single displacement from the beginning point (*A*) to the endpoint (*E*).

That's the way it works for all kinds of vectors. The most straightforward way to add them is to draw them so that they touch, head to tail. Their sum is the single vector from the beginning (tail) of the first to the end (head) of the last. (If you make a mistake and let two heads touch, you'll get the wrong answer!)

Go to the PhET website (http://phet.colorado.edu) and open the simulation *Vector Addition*.

Check "Show Sum" and "None" under Component Display. The arrows representing vectors can be dragged from the basket. Their length can be changed and they can be tilted by grabbing their heads. Explore different angles and different numbers of vectors. Look at the sum when you (correctly) put two vectors head to tail as well as when you put them tail to tail and head to head. Tilt a vector and observe what happens to the vector that represents the sum.

Come back to this simulation later after the discussion of vector components.

A vector is usually described by a boldface symbol (**A**) or a symbol with an arrow over it ($\vec{A}$). The plain symbol then refers to the magnitude (or length) only. If the vector nature of a quantity is obvious we may leave out the vector notation. But we have to be careful. For example, $A + B$ represents the sum of the magnitudes or lengths of the two vectors **A** and **B**. But **A** + **B** is a vector whose magnitude is less than $A + B$ unless the two vectors point in the same direction. For example, if **A** is 10 N to the right, and **B** is 10 N to the left, the vector sum **A** + **B** is equal to zero, but (*A* + *B*) is the sum of the magnitudes of the two vectors, and is equal to 20 N.

If the three vectors $\mathbf{F}_1, \mathbf{F}_2, \mathbf{F}_3$ represent all of the forces that act on an object, then their vector sum, $\Sigma\mathbf{F} = \mathbf{F}_1 + \mathbf{F}_2 + \mathbf{F}_3$, is the net force, and this is the quantity that is equal to $M\mathbf{a}$. (The symbol Σ is a capital Greek *sigma*. It is often used to represent the word *sum*.) $\Sigma\mathbf{F}$ is the vector sum of all the forces acting on an object.

EXAMPLE 5

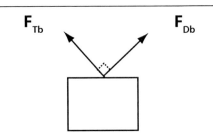

Tom and Dick pull on strings attached to a box with forces of 30 N each. (Call them F_{Tb} and F_{Db}.) They pull at right angles to each other, as shown in the diagram, which shows the view from above. What is the sum of the two forces?

Ans.:
Draw vectors representing the two forces head to tail:

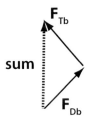

Their sum is the vector from the tail of the first to the head of the second. The two vectors and their sum form a right-angle triangle. The sum of the squares of the sides next to the right angle is equal to the square of the third side (the hypotenuse), i.e., the sum is $\sqrt{30^2 + 30^2}$ or 42.4 N. The sum of the two forces is therefore a force whose magnitude is 42.2 N, acting at an angle of 45° to the direction in which Tom pulls, as shown in the diagram. Note that the answer is the same if you start with Dick's force.

EXAMPLE 6

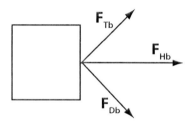

Tom, Dick, and Harry pull on strings attached to a box. Harry pulls to the right with $F_{Hb} = 40\,\text{N}$. Tom and Dick pull with forces of 30 N each at angles of 45° to the direction in which Harry pulls, as shown in the diagram. What is the sum of the three forces?

Ans.:
Start by adding Tom and Dick's forces, which we will call F_{Tb} and F_{Db}. Since they are at 45° to Harry's force, they are at 90° to each other. To add them we redraw the vectors head to tail.

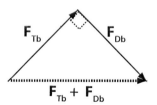

Their sum is the third side of the right-angled triangle formed by the two forces. It is to the right, with a magnitude of $\sqrt{30^2 + 30^2} = 42.4\,\text{N}$.

Harry's force is in the same direction, to the right. We are left with two forces, 42.4 N and 40 N, both to the right. Since they are in the same direction, we can just add their magnitudes, to get 82.4 N to the right.

Since two vectors of equal magnitude and opposite directions add up to zero, we see that to change the direction of a vector is to make it the negative of the original vector. If we want to subtract **B** from **A**, we can rewrite **A** − **B** as **A** + (−**B**), and add the vector **A** to the vector −**B**, which is in the opposite direction to the vector **B**.

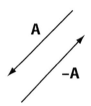

EXAMPLE 7

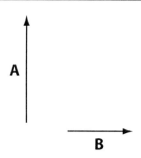

The vector **A** is 50 N north.
The vector **B** is 30 N to the east.
Draw the appropriate vector diagrams and

(a) find **A** + **B**

(b) find **A** − **B**.

Ans.:
(a) The magnitude of the sum is $\sqrt{50^2 + 30^2} = \sqrt{2500 + 900} = \sqrt{3400} = 58.3\,\text{N}$ in the direction shown in the diagram. (We can use the Pythagorean theorem, $A^2 + B^2 = C^2$, because the vectors are at right angles.)

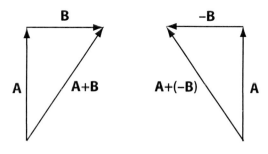

(b) The vector −**B** is 30 N west. Add **A** and −**B**.
The magnitude is the same as for part (a). The direction is now different, as shown in the diagram.

We have used the symbol F for force in Newton's second law, as if there were only one force acting on our object. In general, we should use the *net* force, i.e., the vector sum $\Sigma\mathbf{F}$ of all the forces acting on the object. Newton's second law can then be written as $\Sigma\mathbf{F} = M\mathbf{a}$: *When one or more forces are applied to an object whose mass is M, and the vector sum of all the forces acting on the object is* $\Sigma\mathbf{F}$, *the acceleration of the object is given by the relation* $\mathbf{a} = \frac{\Sigma\mathbf{F}}{M}$.

Force diagrams

When we want to apply Newton's second law to a particular situation, we first have to decide which object we want to apply it to. Then we have to look for all the forces acting on the object. That means that we are not looking for the acceleration, or for the velocity, or for forces acting on other objects! We make a diagram on which we represent the object by a dot, and draw vectors representing all of the forces on the object. The important thing is to make sure that we put on the diagram only the forces acting on that particular object, and that we put on all of them. The result is a *force diagram*.

If you are not sure whether a certain force should be included, ask yourself: what is the object that I am considering? What are the other objects that are interacting with it? Are they touching? The other object that exerts a force can be some distance away if the force is gravitational or electric, but it has to be touching for a force like the tension of a rope or the push of a hand or of a spring.

We can now add up the vectors representing the forces so as to find their sum, $\Sigma\mathbf{F}$, and use it in the expression for Newton's second law. If there are several forces in the x direction, and several forces in the y direction perpendicular to it, it is often helpful and convenient to apply Newton's second law separately in the two directions: $\Sigma F_x = Ma_x$ and $\Sigma F_y = Ma_y$.

EXAMPLE 8

A book whose mass is 1.2 kg is on a table. It is being pulled forward by a horizontal force of 7 N. There is also a force of friction of 2 N.

(a) Draw the force diagram, labeling all the forces.

(b) Find the book's acceleration.

Ans.:

(a) On the force diagram the dot represents the book. Four forces act on it: Its weight, $F_{eb} = Mg$ (the force of the earth on the book), F_{tb}, the upward force exerted by the table on the book, F_{pb}, (person on book) the horizontal pull on the book to the right of 7 N, and F_f, the force of friction.

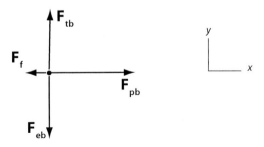

(b) Let the up direction be the y direction and the direction in which the 7-N force acts, the x direction.

The two forces along the y direction are F_{eb} and F_{tb}. There is no acceleration in the y direction. (The book does not jump up from the table.) Therefore the sum of the two forces along the y direction must be zero.

There are two forces along the x direction, the force of 7 N to the right and the force of friction (2 N) to the left, so that their sum is 5 N to the right. This is also the sum $\Sigma\mathbf{F}$ of all four forces, and it is therefore equal to $M\mathbf{a}$. $\mathbf{a} = \frac{\Sigma\mathbf{F}}{M} = \frac{5\,N}{1.2\,kg} = 4.17$ m/s². to the right. (Since the force and the mass are both given in SI units, the acceleration is automatically also in the same system of units.)

We have added four forces. Two are along the y-axis and add up to zero. There is a net force along the x-axis. It is also the sum of all the forces acting on the book, and is therefore equal to $M\mathbf{a}$.

Vector components

Together the x-axis and the y-axis are a *coordinate system*. It allows us to describe the positions of points, lines, or objects. It is an aid to describing a situation or solving a problem, and we can choose it any way we want. We choose one direction of each axis to be positive and the other to be negative, Many times we choose the axes to be horizontal and vertical, but that is not always so.

If a force is at an angle to the axes, we can replace it with two forces that are parallel to the two axes and that add up to it. The one parallel to the x-axis is called the x component, F_x, and the one parallel to the y-axis is called the y component, F_y. For an angle θ between the direction of F and the x-axis, $F_x = F \cos \theta$ and $F_y = F \sin \theta$.

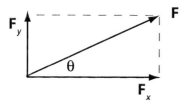

Adding vectors head to tail, as we have done so far, is convenient graphically. It also allows the calculation of numerical values of the magnitude and direction of the sum in simple cases, as, for example, when the vector diagram is a triangle with a right angle. In more complicated cases it may be better to use components.

To do this it is first necessary to choose a coordinate system, i.e., directions for the x- and y-axes. We can then find the x and y components of the vectors that are to be added.

All the x components can now be added. Since they are all along the same line, no angles need to be considered. It is important, however, to make sure which of the components are in the same direction as the x-axis, and are positive, and which are in the opposite direction and are negative. The x components are negative when the angle of the vector with the x-axis is between 90° and 270°, i.e., when the vector is in the second or the third quadrant. We can write the sum of the x components as ΣF_x. This quantity is the x component of the sum of the vectors.

Similarly the y components can be added to give ΣF_y, which is equal to the y component of the vector sum. A y component is negative when the angle of the vector with the x-axis is between 180° and 360°, i.e., when the vector is in the third or the fourth quadrant.

Knowing the x and y components of the vector sum gives us all of the information about the sum. We can calculate its magnitude, $\sqrt{(\Sigma F_x)^2 + (\Sigma F_y)^2}$. The angle that the sum makes with the x-axis is the angle whose tangent is $\frac{\Sigma F_y}{\Sigma F_x}$.

More on friction

Let's look again at the forces on the book on the table. When the book is at rest, without any horizontal forces, there are only two: one is the weight, Mg, for which we have also used the symbol F_{eb}, to indicate that it is the force of the earth on the book. It is always there, unless the object is far from the earth. Near the surface of the earth the value of g is about 9.8 m/s², or 9.8 N/kg. At some distance from the earth the weight is still Mg, but the value of g is then different. The direction of this force is toward the center of the earth.

The second force is the force of the table on the book, F_{tb}. It is upward and perpendicular to the surface of the table. This force is often called the *normal* force, F_n. (This does not mean that other forces are *abnormal*. The word *normal* here means the same thing as *perpendicular* or *at right angles*.) For a book sliding along an inclined plane, or for a car along a road or track at some angle, there is always a *normal* force, a force of the surface on the book or the car, at right angles to the surface along which it moves.

Now push the book horizontally with a force F_{pb} (person on book). At first the book does not move. There is a force of friction (F_f) in the opposite direction to F_{pb}, and it has the same magnitude as F_{pb}.

When we increase our push, F_{pb}, the force of friction initially grows with it, and the book remains at rest. There is, however, a maximum value beyond which F_f cannot grow, so that there is then a net force $F_{pb} - F_f^{max}$, and the book accelerates.

Once the book starts to move, the force of friction remains steady, but at its sliding value F_f^s, which is smaller than F_f^{max}.

The ratio $\frac{F_f^{max}}{F_n}$ is called the *coefficient of static friction*. The ratio $\frac{F_f^s}{F_n}$ is called the *coefficient of sliding friction*. Both of these coefficients depend on the nature of the two surfaces that are in contact. The symbol μ (Greek *mu*) is generally used for the coefficient of friction.

EXAMPLE 9

Go to the PhET website (http://phet.colorado.edu) and open the simulation *Forces and Motion*. Play with the *Introduction* and *Friction*. Look at the force diagram (here called "Free Body Diagram.") Then go to *Force Graphs*. Choose the small crate, friction off ("Ice"), check applied force, and click on all graphs (a, v, and x). Type in an applied force of 200 N, put the crate at -7 m, and record.

(a) What is the acceleration shown on the graph? Compare it to the value you expect from Newton's second law.

(b) What is the shape of the velocity graph? What is its equation? Calculate the velocity just before it hits the wall. Compare this value to the value on the velocity graph.

(c) What is the shape of the position graph? What is its equation? Calculate the position at the end. Compare it to the value on the position graph. (Don't forget where you started.)

(d) Clear. Check friction "wood." Check only the applied force graph. Press the "go" button.
   Increase the applied force until the file cabinet moves. Find the necessary force to the nearest 10 N.
   What is the coefficient of static friction?

(e) Once the object moves the force of friction decreases and the coefficient of friction decreases to the coefficient of sliding friction.
   What happens to the force of friction when the object moves?
   Find the coefficient of sliding friction.

Ans.:
(a) $a = 0.5 \frac{m}{s^2}$.

(b) $v = at$, $v = 0.5t$.

(c) Since he starts at $x = 7$ m, $x = -7 + v_0 t + \frac{1}{2}at^2$, i.e., $x = -7 + 0.25t^2$.

(d) The coefficient of static friction is $\frac{490}{100 \times 9.8} = 0.5$.

(e) The coefficient of sliding friction is $\frac{294}{100 \times 9.8} = 0.3$.

EXAMPLE 10

Find the x and y components of these vectors, using sines and cosines.

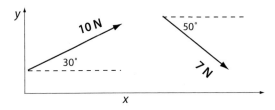

Ans.:
For the vector on the left, $F_x = 10 \cos 30° = (10)(0.866) = 8.66$ N.
$F_y = 10 \sin 30° = (10)(0.500) = 5.00$ N.
For the vector on the right, $F_x = 7 \cos 50° = (7)(.643) = 4.50$ N.
$F_y = 7 \sin 50° = (7)(.766) = 5.36$ N.
This is the *magnitude* of F_y. It points downward, in the $-y$ direction, and is therefore negative, equal to -5.36 N. (Similarly, an x component in the negative x direction would be negative.)

EXAMPLE 11

Go to the PhET website (http://phet.colorado.edu) and open the simulation *Ramp: Forces and Motion*. Play with the *Introduction*, Then go to *Force Graphs*,

(a) Check "wood." Choose the crate and use the position slider to put it on the ramp. Press the "go" button.
   Change the angle of the ramp (use the angle slider or drag the top of the ramp) until you find the largest angle at which the crate just rests without moving down.
   At this angle, what are the forces on the crate?
   What is their sum?
   One force is the weight. How are the other forces related to the weight?
   What is the magnitude of each of the forces?

(b) Set the applied force to 300 N. Do not change the angle. Answer all the other questions in part (a).

(c) Set the applied force to 1000 N. Answer all the questions in part (b).

Ans.:
(a) The angle is 26.6°. The vector sum of the forces is zero. The weight is $Mg = 980$ N. The normal force is $N = Mg \cos \theta = 876$ N. The force of friction is $\mu N = 438$ N.

(b) With an applied force of 300 N the crate still does not move. The sum of the forces is zero. The perpendicular components are N and $Mg \cos \theta$. The magnitude of the applied force is equal to the force of friction plus the component of the weight $Mg \sin \theta$. Note that the force of friction is not equal to μN. μN is the maximum force of friction.

(c) With the applied force of 1000 N the force of friction is not large enough to keep the crate from moving. The sum of the forces is now not zero. The perpendicular components are as before and add up to zero. In the direction parallel to the ramp, using "up the ramp" as the positive direction, the net force F is equal to the applied force, F_a, minus the force of friction, f, minus the component of the weight down the ramp: $F = F_a - f - Mg \sin \theta$. The crate accelerates up the ramp with $a = \frac{F}{M}$.

EXAMPLE 12

Repeat Example 8 for the same book, but this time the book is being pulled forward by a string at an angle of 15° with a force of 7 N.

(a) Draw the force diagram.

(b) What is the coefficient of friction in Example 8?

(c) Assume that the coefficient of friction remains the same, and find the force of friction.

(d) Find the book's acceleration.

Ans.:

(a) We choose the same coordinate system, the x-axis horizontal and the y-axis vertical. The force pulling the book ($F_{pb} = 7$ N) is now at an angle, and we can decompose it into its two components. Its x component is $7 \cos 15°$, or $(7)(.97) = 6.76$ N. The y component is $7 \sin 15° = (7)(.26) = 1.82$ N.

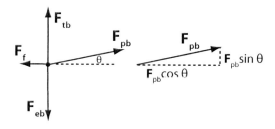

(b) The normal force, F_n, which is the force of the table on the book, F_{tb}, is different from its value in Example 8. It is $F_{eb} - F_{pb} \sin \theta$, i.e. $(1.2)(9.8) - 1.82$, which is $11.76 - 1.82$ or 9.94 N.

(c) In Example 8 the coefficient of friction is $\frac{F_f}{F_n} = \frac{2}{11.76} = 0.17$. In this example $\frac{F_f}{F_n}$ is again 0.17, but $F_n = 9.94$ N. Hence $F_f = (9.94)(0.17) = 1.69$ N.

The vertical forces add up to zero, as before, because there is no vertical acceleration. $\Sigma F_x = F_{pb} \cos \theta - F_f$, or $6.76 - 1.69$, which is 5.07 N.

(d) The acceleration is $\frac{5.07}{1.2} = 4.23$ m/s².

EXAMPLE 13

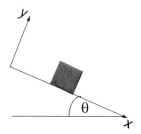

A box is at rest on a surface that slopes at an angle of 30° to the horizontal.

(a) List the forces on the box. Draw a force diagram.

(b) What forces are perpendicular to the surface? What forces are parallel to the surface?

(c) The weight is neither parallel nor perpendicular to the surface. Use a coordinate system with the x-axis parallel to the surface and down it, and the y-axis perpendicular to the surface and up. Find the weight's x component and its y component.

(d) Write down the relation between the y components of all the forces.

(e) Write down the relation between the x components of all the forces.

(f) The mass of the box is 2.5 kg. Find the magnitude of each of the forces.

Ans.:

(a,b) The weight, $F_{eb} = Mg$, is straight down, and therefore neither parallel nor perpendicular to the surface. The normal force, F_n, is perpendicular to the surface, and the force of friction, F_f, is parallel to the surface. (You can see that the two angles marked θ on the left are the same by imagining the two vectors F_{eb} and F_n rotating together through 90°.)

(c) The y component of the weight is $Mg \cos 30°$. The x component is $Mg \sin 30°$.

(d) $Mg \cos 30° - F_n = 0$.

70 / Forces and Motion: Newton's Framework

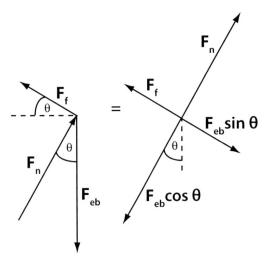

(e) $Mg \sin 30° - F_f = 0$

(f) $Mg = (2.5)(9.8) = 24.5$ N.
$F_n = Mg \cos 30° = 21.2$ N.
$F_f = Mg \sin 30° = 12.25$ N.

EXAMPLE 14

(a) The box of the previous example slides down the same surface without friction. What is its acceleration?

(b) The box slides down the surface with an acceleration of 2 m/s². What are the magnitude and direction of the force of friction?

Ans.:

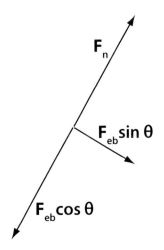

(a) This time F_f is zero. The other two forces are unchanged. There is again no acceleration along the y direction. The force along the x direction is the x component of the weight, 12.25 N. There is no other force with a component in this direction, so that $a = \frac{12.25}{2.5} = 4.9 \frac{m}{s^2}$.

(b) The force of friction is in the direction opposite to the direction of motion, in the $-x$-direction. The forces perpendicular to the surface are unchanged. The forces parallel to the plane, in the x-direction, are $Mg \sin 30°$ in the positive direction and F_f in the negative direction. $Mg \sin 30° - F_f = Ma$. $F_f = Mg \sin 30° - Ma = 12.25 - 5 = 7.25$ N.

EXAMPLE 15

You jump from a table straight down to the floor. Neglect air resistance. Assume that your mass is 65 kg.

(a) Draw a force diagram for the time that you are in the air.

(b) What objects interact with you? What is the net force on you?

(c) What is your acceleration?

Ans.:
(a)

(b) Since we are neglecting air resistance, the only force on you is your weight, F_{ep}. It is equal to Mg or $(65 \text{ kg})(9.8 \text{ m/s}^2) = 637$ N.

(c) Since the only force on you is your weight, your acceleration is $g = 9.8$ m/s².

EXAMPLE 16

You jump from a table, as in the previous example, but this time you push off so that you jump away from the table and not straight down. Repeat parts a, b, and c of the previous example.

Ans.:
This time your path is not along a straight vertical line. However, as long as we continue to neglect air resistance, the only force that acts on you is still your weight. The answers to all three parts of this question

are therefore the same as before. Your weight is the same, and your acceleration is g.

Object or system?

We often apply Newton's second law to a combination or *system* of more than one object. In fact, unless we are talking about a single particle without internal structure, every object consists of more than one particle. In the previous section we said that we first have to decide to which object we want to apply the law. Now we see that it is really the *system* that we have to choose. Only then can we decide what the forces are that we need to consider, namely all the forces that act on the system that we have chosen. To make clear what the system that we choose is, we can draw a dotted line around it on our sketch.

EXAMPLE 17

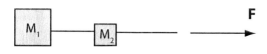

Two blocks are connected by a string. Their masses are 5 kg and 3 kg. The less massive one is being pulled by a second string, with a force, F, so that its acceleration is 2 m/s^2. (Neglect friction and all other forces. The vertical forces add up to zero and need not be considered.)

(a) Draw force diagrams for the system consisting of both blocks, and for each block separately (for the horizontal forces only).

(b) What is the force (the "tension") of the second string?

(c) What is the tension in the string connecting the two blocks?

Ans.:
(a)

Let F_{s1} be the force of the string on M_1 and F_{s2} the force of the same string on M_2. These two forces have the same magnitude but are in opposite directions. (This is an approximation. We are neglecting the mass of the string.) The force diagrams are for the system consisting of both blocks, the system with only M_1, and the system with only M_2.

(b) Apply Newton's second law to the system consisting of both blocks. There is only one force on this system, the force F. The mass of the system is the sum of the two masses, $5 + 3 = 8$ kg. Since the acceleration is 2 m/s^2, $F = Ma = (8)(2) = 16 \text{ N}$.

(c) The only force acting on the 5 kg mass is the tension in the string that connects the two masses. It is the force exerted by the string on the 5 kg mass, and we have called it F_{s1}. $F_{s1} = Ma = (5)(2) = 10 \text{ N}$.

The string connecting the two blocks pulls to the right on the 5 kg block with the force F_{s1} and to the left on the 3 kg block with a force F_{s2} of the same magnitude.

We know from part (b) that the force to the right on the 3 kg block is 16 N. We can use the forces on this block to find F_{s2} (and F_{s1}) in a second way. The net force is $16 - F_{s2}$ to the right. It is equal to Ma.

$16 - F_{s2} = Ma = (3)(a) = (3)(2) = 6$. $F_{s2} = 16 - 3a = 16 - 6 = 10 \text{ N}$, as before.

4.3 Momentum and its conservation. Action, reaction, and Newton's third law

When two objects collide, there are two quantities that we need to focus on. One is how fast they are going, the other is their mass. We define a new quantity, the *momentum*, equal to the product of the mass and the velocity. Like velocity, it is a vector quantity. A small car, moving slowly, has a much smaller momentum than a truck barreling along at great speed. A table-tennis ball coming at you fast is not likely to hurt you. A car with the same speed is vastly more dangerous.

Acceleration is the quantity that tells us how fast the velocity is changing. It is the rate of change with time of the velocity. Force (mass times acceleration) is therefore the quantity that tells us how fast the momentum (mass times velocity) is changing. It is the rate of change with time of the momentum. In symbols we can write $\mathbf{a} = \frac{d}{dt}(\mathbf{v})$, $\mathbf{F} = M\mathbf{a} = \frac{d}{dt}(M\mathbf{v})$. We can rephrase Newton's second law to say that the force on an object is equal to the rate at which the object's momentum changes.

If there is no net force on an object then its momentum does not change, it is *conserved*. At first sight this hardly seems like a new or important statement. Isn't that just Newton's first law, $a = 0$ when $F = 0$? Yes, that's true if we talk about a single particle, but look at what happens if we talk about two particles. Together we can think of them as a composite object, or a *system*. If there is no force on this system from outside it, its momentum is constant.

EXAMPLE 18

A ball whose mass is 0.2 kg is at rest on the ground. It is hit head-on by a second ball whose mass is 0.1 kg, and which is initially moving with a velocity of 3 m/s in the x direction. (Neglect all horizontal forces other than those that the two balls exert on each other.)

(a) Make a sketch and draw a dotted line around the system that you will consider. Show all momentum vectors.

(b) Is there a net force on the system?

(c) After the collision the two balls stick to each other and move off together. What is their velocity right after the collision?

(d) In a different collision, starting as before, the balls do not stick to each other, and the ball that was initially at rest is observed afterward to move with a velocity of 1.2 m/s in the x direction. What is the velocity of the other ball after the collision?

(e) In still another collision, again with the same start, the ball that was originally at rest is observed after the collision to be moving with a velocity of 0.8 m/s in a direction at 45° to the x direction. Draw a diagram that shows all momentum vectors. From your diagram find (graphically and analytically) the momentum of the other ball after the collision. What is its velocity?

Ans.:

(a)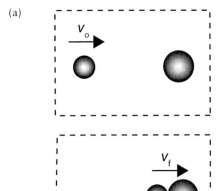

(b) No, the forces that the balls exert on each other are internal to the system. (There are vertical forces, but they add up to zero.)

(c) Before the collision the total momentum of the system is that of the moving ball, (0.1 kg)(3 m/s) = 0.3 kg m/s in the x direction.

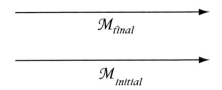

The momentum of the system consisting of the two particles is conserved, i.e., the momentum before the collision is equal to the momentum after the collision, still 0.3 kg m/s. (The diagram shows the initial momentum vector and the final momentum vector. They are equal in magnitude and direction.)

The mass of the system is the sum of the masses of the two particles, 0.2 kg + 0.1 kg = 0.3 kg. The velocity of the system of the two balls sticking together after the collision is therefore $\frac{0.3 \text{ kg m/s}}{0.3 \text{ kg}} = \frac{1 \text{ m}}{\text{s}}$ in the x direction.

(d) The initial total momentum is the same as in part (a), 0.3 kg m/s in the x direction. The total momentum is conserved, and is again the same after the collision.

The momentum of the 0.2 kg ball is (0.2 kg)(1.2 m/s) = 0.24 kg m/s in the x direction.

It is shown as M_2 on the diagram. The momentum of the 0.1 kg ball is therefore 0.3 kg m/s − 0.24 kg m/s = 0.06 kg m/s. It is shown as M_1 on the left-side diagram. Its velocity is $\frac{0.06 \text{ kg m/s}}{0.1 \text{ kg}} = \frac{0.6 \text{ m}}{\text{s}}$, in the x direction.

direction, so as to keep their combined momentum constant. The rate at which the momentum of one changes must be just as large as the rate at which the momentum of the other changes, and in the opposite direction. But the rate at which the momentum of a particle changes is the force on it. We see that the force on one particle is equal in magnitude to the force on the other, and is in the opposite direction.

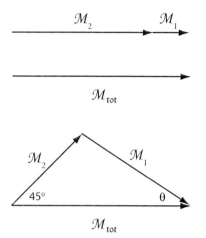

(e) The initial momentum is again the same, and since momentum is conserved, so is the momentum after the collision. But the directions are now not along the x direction. The diagram shows the momentum M_2 after the collision of the 0.2 kg ball at 45° to the x-direction. Together with the momentum M_1 of the 0.1 kg ball it must add up to the total momentum, which is again 0.3 kg m/s, both before and after the collision.

Graphically, from the momentum diagram, the momentum of the 0.1 kg ball (M_1) is 0.22 kg m/s, at an angle of 31° to the x direction.

Analytically: M_2 has two components, $M_2 \cos 45 = (0.16)(.707) = 0.113$ kg m/s and $M_2 \sin 45$, which has the same magnitude.

M_1 has two components: $(M)_x − (M)_{tot} − M_{2x} = 0.3 − 0.113 = 0.187$ and M_y whose magnitude is 0.113. The magnitude of M_1 is $\sqrt{.187^2 + .113^2} = 0.219$.

The tangent of the angle θ is $\frac{.113}{.187}$, so that $\theta = 31.1°$.

To find the velocity we divide by the mass to get 2.2 m/s.

There is no force from the outside. The only forces are the ones inside the system that the two particles exert on each other. These are the force that particle one exerts on particle two ($\mathbf{F}_{12}$) and the force that particle two exerts on particle one ($\mathbf{F}_{21}$). These forces must be equal in magnitude and opposite in direction. $\mathbf{F}_{12} = −\mathbf{F}_{21}$. This is Newton's third law.

But we knew this already! At least intuitively. Force is an interaction. The two objects that interact either attract or repel each other with forces that have the same magnitudes.

Newton's third law remains true regardless of any other forces. It says that all forces occur in pairs. If I push you, you experience a force. As I do that I experience a force on me from you, equal to my push and in the opposite direction. The same is true if I push against a wall. The wall pushes back on me with a force equal in magnitude and opposite in direction.

If the momentum of one of the two particles changes (still with no forces from outside the system), the other must change also, in the opposite

You may ask: "If all forces come in pairs that add up to zero, how can anything ever get going?" The two forces of a Newton's third law pair act on different objects! The net force on any one object is the sum of the forces that act on that one object, and this is what gives rise to its acceleration.

If there are two masses, M_1 and M_2, and you want to know how M_1 moves, you have to draw a force diagram for M_1. One of the forces on it will be F_{21}, the force exerted on M_1 by M_2. There may be other forces on it. The vector sum of all the forces on M_1 determines what happens to M_1. Similarly, if you want to know what happens to M_2, and draw a force diagram for it, one of the forces (among others) will be F_{12}.

You may also want to draw a force diagram for the system composed of both M_1 and M_2.

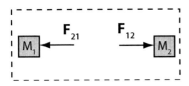

This time the pair of forces F_{12} and F_{21} are forces inside the system. They don't influence the motion of the system as a whole, only what happens inside it. They are *internal* forces that add up to zero.

Sometimes Newton's third law is described by saying that "for every action there is an equal and opposite reaction," but this is not a particularly helpful way to think about it. The words *action* and *reaction* are not defined, and they don't make it clear that the two forces act on different objects. Moreover a reaction is usually thought to come after the action, and this is not true here. The two forces that Newton's law refers to act simultaneously.

Next, think of a system or combined object consisting of many particles. Even the tiniest visible object consists of a huge number of microscopic pieces, its atoms and molecules. They exert forces on each other, each one on its neighbor, and to a lesser extent on those further away. Inside the atoms the nuclei and electrons also exert forces on each other. Each fragment of material is a seething mass of particles exerting internal forces on each other.

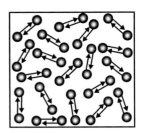

We can now see that there is a simple way to deal with these forces. After all, when we hold a ball in our hands we are not aware of all of the atoms and molecules that constitute it, nor of the forces between them. Because they come in pairs, each of which adds up to zero, we can just ignore them!

It is really important to be able to talk about a tennis ball or any other object without each time having to consider all the molecules (and their constituents) within it. We have always done that, and now we see that it is Newton's third law that allows us to do it.

Once we agree that the sum of the internal forces is zero, the net force acting on the system can be found without considering them, regardless of how many collisions the pieces make with each other, or how they separately change their velocity and their momentum. If there is no *external* force on the system then the total momentum of the system does not change. This is the *law of conservation of momentum*.

Look at just two particles or objects colliding with each other. Each will change its velocity, and its momentum. But the sum of the momenta (not the velocities!), i.e., the total momentum of the two combined, will be the same before the collision and after the collision.

This is the basis of our knowledge of all kinds of collisions, whether they are between cars or between the molecules of a gas.

EXAMPLE 19

The horse and the wagon

The way we use Newton's second and third laws is so important that we will go over it once more in this example.

A horse pulls on a wagon.

(a) Describe the force of interaction between the horse and the wagon.

(b) Describe the forces on the system containing both the horse and the wagon.

(c) Describe the forces on the system containing only the wagon.

(d) Describe the forces on the system containing only the horse.

Ans.:

(a) Newton's second law deals with the motion of just one object. If we want to know the acceleration of that object we need to consider all the forces on that object. We have to be careful not to include forces that act on any other objects.

Newton's third law deals with pairs of forces, each acting on a different object. When a horse pulls on a wagon, the wagon pulls on the horse. Each of these two forces is an interaction between the horse and the wagon. The two forces have the same magnitude and are in opposite directions.

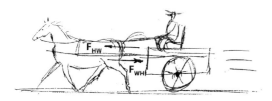

(b) For the system containing both the horse and the wagon the two interaction forces are internal forces. They add up to zero and do not contribute to the sum of the forces on the system or to the system's acceleration.

We can now list the external forces on the system. There are the usual two vertical forces, the weight and the normal force. The horizontal forces are the force of friction and air resistance.

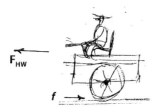

(c) If we want to know the wagon's acceleration, we choose a system containing only the wagon. Just one of the two interaction forces comes into play, the force of the horse on the wagon, F_{HW}. There is also the force of friction on the wagon, and air resistance, in addition to the two vertical forces.

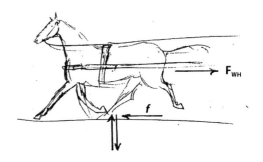

(d) If we want to talk about the horse, we choose a system containing only the horse. As we count the external forces on it, we need to include the force on the horse by the wagon, F_{WH}.

The force of friction acts to push the horse forward. We can see that by looking at what happens at the interaction between the horse's hoofs and the road. The horse pushes back on the road with a force F_{HR}. Paired to this force by Newton's third law is the force of the road on the horse, F_{RH}, forward.

While we dealt with the horse and the wagon we ignored all of their internal structure. In other words we have used a model in which each is a particle. In the real world the horse is a wildly complicated object whose internal motions we can never know completely. Not only that, but we even have trouble deciding what the external forces are. The horse breathes, eats, eliminates, and interacts with its environment in ways that make it quite impossible to delineate its boundaries exactly. The model that we use does not include these forces.

We know that we can ignore its internal forces, the forces between its various parts and between each pair of molecules. Each pair of interaction forces is related by Newton's third law and adds up to zero.

Even the wagon is hardly the rigid and inert body that we might wish for to keep its description simple. It has wheels whose motion we haven't thought about. And on the microscopic scale it is a collection of constantly moving atoms and molecules. Again, a large part of the simplification that allows us to deal with the problem is that we can ignore the internal forces.

(The figure shows the internal forces between the horse's ears and its head.)

EXAMPLE 20

You are pushing two boxes along the floor with a force of 100 N. The box on which you are pushing has a mass of 12 kg and the other one has a mass of 8 kg. What is the force of the first box on the second box, and the force of the second box on the first? (Neglect friction and all other horizontal forces.)

Ans.:
We can find the acceleration most simply by first looking at the system that includes both boxes, with their combined mass of 20 kg. $\Sigma F = F_{pb} = Ma$, $a = \frac{F_{pb}}{M} = \frac{100\,\text{N}}{20\,\text{kg}} = 5\,\text{m/s}^2$.

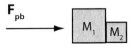

Now we can apply Newton's second law to the lighter of the two boxes.

Since we are neglecting friction, the only horizontal force on the box is F_{12}. We already found the acceleration to be $5\,\text{m/s}^2$. $F_{12} = Ma = (8\,\text{kg})(5\,\text{m/s}^2) = 40\,\text{N}$ in the x direction. By Newton's third law this is also F_{21}.

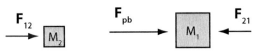

We can check this result by using the force diagram for the heavier box.

$100 - F_{21} = (12\,\text{kg})(5\,\text{m/s}^2)$, or $F_{21} = 100 - 60 = 40\,\text{N}$, this time in the negative x direction.

EXAMPLE 21

An elevator whose mass is 250 kg hangs from a cable. Find the force of the cable on the elevator in the following situations. For each case draw a force diagram, write the relation between the forces using symbols only, and then substitute numbers for the symbols to find the force exerted by the cable.

(a) The elevator is at rest.

(b) The elevator is moving down with a constant velocity of 2 m/s.

(c) The elevator is moving up, and accelerates upward with an acceleration of $1.5\,\text{m/s}^2$.

(d) The elevator is moving up and slowing down with an acceleration whose magnitude is $2\,\text{m/s}^2$.

Ans.:

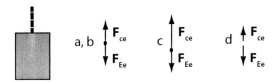

(a) There are two forces on the elevator. Its weight, $F_{ee} = Mg$, and the force of the cable on the elevator, F_{ce}. Let up be positive. The net force is $F_{ce} - Mg$. Since the acceleration is zero, the net force is zero, and $F_{ce} - Mg = 0$. $F_{ce} = Mg = (250)(9.8) = 2450\,\text{N}$.

(b) There is no acceleration. The answers are the same as in part (a).

(c) Let up be positive again. The forces are F_{ce} and $-Mg$. Their sum is $F_{ce} - Mg$, and it is equal to Ma. $F_{ce} - Mg = Ma$. $F_{ce} = Mg + Ma = 2450 + (250)(1.5) = 2450 + 375 = 2.83 \times 10^3\,\text{N}$.

(d) Again the sum of the forces (up) is $F_{ce} - Mg = Ma$.

The acceleration is downward, hence negative, equal to $-2\,\text{m/s}^2$. $F_{ce} = 2450 + (250)(-2) = 2450 - 500 = 1.95 \times 10^3\,\text{N}$.

We can also choose down as positive, so that the acceleration is positive: The sum of the forces (down) is $Mg - F_{ce} = Ma$ and $F_{ce} = Mg - Ma = 2450 - (250)(2) = 1.95 \times 10^3\,\text{N}$, as before.

Later we will spend time with an example that goes beyond mechanics and has the most profound implications. Consider a roomful of air: billions of billions of molecules, most of them moving with speeds far greater than any vehicles on earth or in space, colliding with each other, recoiling, and exerting forces on each other. But if we take the whole amount of gas in the room as our system, all of the forces that the molecules exert on each other as they collide are internal forces, pairs of forces that are related by Newton's third law, and that add up to zero. Only the forces from outside the system, by the walls, as molecules bounce off them, need to be considered in order to deal with the system as a whole. We will see that this example opens the door to the understanding of the relation between mechanics and the phenomena of heat and temperature.

4.4 One more motion that is everywhere: rotation

Uniform circular motion

When a particle does not move in a straight line, even if its speed does not change, it has an acceleration and so there must be a net force on it.

We will look at a particle moving with constant speed in a circle to see what we can say about its acceleration and about the force that must act on it to keep it moving in its circle. This kind of motion is called *uniform circular motion*, where "uniform" refers to the constant, or *uniform* speed.

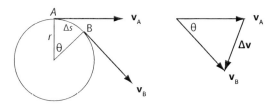

The left-hand part of the figure shows a particle and the circular path along which it moves. As the particle moves from A to B, it moves through a distance Δs along the circumference, and the radius of the circle turns through an angle θ. This angle, in *radians*, is equal to $\frac{\Delta s}{r}$. (For motion all around the circle, through 360°, the number of radians is equal to the circumference of the circle divided by its radius, or $\frac{2\pi r}{r}$, so that 2π radians equal 360°.)

The velocity, as a vector quantity, has both magnitude and direction, and therefore changes as the particle moves. At point A the velocity is $\mathbf{v_A}$ and at point B it is $\mathbf{v_B}$. The change in the velocity between these two points is $\mathbf{v_B} - \mathbf{v_A}$, which we call $\Delta \mathbf{v}$. (Look at it as $-\mathbf{v_A} + \mathbf{v_B}$, i.e., go backward along $\mathbf{v_A}$ and then forward along $\mathbf{v_B}$. The vector from the beginning of this path to the end is $\Delta \mathbf{v}$.)

The right-hand part of the figure is a vector diagram that shows the three vectors $\mathbf{v_A}$, $\mathbf{v_B}$, and $\Delta \mathbf{v}$. The velocity vector remains at right angles to the radius, and as the radius turns through the angle θ, so does the velocity. The length (or magnitude) of the velocity vector is the speed, v, and for the motion that we are considering it remains constant.

The angle ($\theta = \frac{\Delta s}{r}$ radians) between the radii to A and B is the same as the angle between the two velocities. The vector diagram shows that it is also approximately equal to $\frac{\Delta v}{v}$, and becomes closer to it as the angle gets smaller.

Hence $\theta = \frac{\Delta s}{r}$ and $\theta = \frac{\Delta v}{v}$, so that $\frac{\Delta s}{r} = \frac{\Delta v}{v}$. Dividing by Δt (the time interval during which the particle moves from A to B), we get $(\frac{1}{r})(\frac{\Delta s}{\Delta t}) = (\frac{1}{v})(\frac{\Delta v}{\Delta t})$. This is the same as $(\frac{1}{r})(v) = (\frac{1}{v})(a)$. We can now solve for a to get $a = \frac{v^2}{r}$.

It is not surprising that a is proportional to v^2. We know that a is close to $\frac{\Delta v}{\Delta t}$ (and gets closer to that as Δt and Δv get smaller). With larger speed (v), the change in v (called Δv) increases. In addition, as v gets larger, Δt (for the same displacement) decreases. The two changes together cause $\frac{\Delta v}{\Delta t}$, and hence a, to be proportional to v^2.

This acceleration is called the *centripetal* acceleration. (It means "pointing to the center.") When the angle θ gets smaller, $\Delta \mathbf{v}$ becomes perpendicular to $\mathbf{v_A}$ (or $\mathbf{v_B}$), and is directed toward the center of the motion. A net force with magnitude $\frac{mv^2}{r}$ (called the *centripetal force*) is required to keep the particle moving in the circle with radius r with speed v.

We know that a net force is necessary for an object to move in a circle. Without one it would move with constant velocity, i.e., with constant speed in a straight line, in accord with Newton's first law of motion. Now we see that for motion in a circle with constant speed the force has to point toward the center of the motion, and have magnitude $\frac{mv^2}{r}$. There needs to be a centripetal (or center-pointing) force. This is not a new or additional force. Whatever the forces on the object are, if the object is to move in a circle with constant speed (in *uniform circular motion*) their sum has to point to the center and have magnitude $\frac{mv^2}{r}$.

This is true regardless of the nature of the interaction that leads to the circular motion. For the moon going around the earth the centripetal force is the gravitational force. That's also true for the earth going around the sun. For a ball being swung in a circle at the end of a rope it is the force of the tension of the rope. For a particle in a cyclotron it is the magnetic force. In each of these cases there is an object moving in a circle. In each case the magnitude of the net force is equal to $\frac{mv^2}{r}$, regardless of the nature of the force.

except for being displaced with respect to each other along the time axis (or *out of phase*). At $\theta = 0$, $\sin\theta = 0$ and $\cos\theta = 1$. Both components are said to vary *sinusoidally*. (Note that you can change the horizontal and vertical scales of the graphs with the buttons on the right.) Look for the time of a complete revolution on the graph and compare it to the value derived from ω. Use the step function to compare and explore what you see on the platform and on the graph for x and y.

The sinusoidal variation is typical of oscillations in many parts of physics. We will see it for masses on springs, sound, and other waves, and for atomic and molecular vibrations. Motion in which the position varies sinusoidally is called *simple harmonic motion*.

EXAMPLE 22

Go to the PhET website (http://phet.colorado.edu) and open the simulation *Ladybug Revolution*.

  Choose the "Intro" tab. Check "Show velocity vector" and "Show Acceleratin Vector."

(i) Set the angular velocity to zero.

  Drag the ladybug with the mouse. Observe the vectors **v** and **a**.

(ii) Set the disk in motion with the slider at the bottom. Set the angular velocity to about 150 degrees/s. Observe the two vectors.

  Observe the effect on the two vectors of changing the angular velocity and the ladybug position.

(iii) Choose the "Rotation" tab.

  Check "Show Ladybug Graph" "Show Platform Graph," and (θ, ω, v). Set ω to about 4 rad/s. The graphs include the values of ω and v. Calculate r from them.

  Check "ruler" and measure the radius, r, at which the ladybug sits. Compare this to the result of your calculation.

  What are a and α?

  Check the graphs that contain a and α, and compare your results with those on the graphs.

(iv) Check "Show ladybug graph" and check the graph θ, ω, x, and y and "Show X-Position" and "Show Y-Position." Reset all, and set θ to zero. Set the platform in motion again and observe the graphs of x and y. These are the x and y components of the radius vector **r**, i.e., $x = r\cos\theta$ and $y = r\sin\theta$. Observe that they are similar

EXAMPLE 23

Anja swings a ball at the end of a string so that it moves in a horizontal circle. The ball's mass is 0.3 kg. The radius of the circle is 1.2 m. The ball travels around the circle in one second, i.e., it makes one revolution per second. (Neglect the vertical forces.)

(a) What is the magnitude of the force required to make the ball move in the circle?

(b) What force acts on the ball to keep it moving in the circular path?

(c) What pulls on Anja's hand, and how hard does it pull?

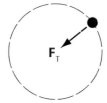

Ans.:

(a) The circumference of the circle is $2\pi r = (2)(\pi)(1.2) = 7.54$ m. The speed of the ball is therefore 7.54 m/s. The force that is required (the centripetal force) is $\frac{Mv^2}{r} = \frac{(0.3)(7.54)^2}{1.2} = 14.2$ N.

(b) The force on the ball is that of the string. The force exerted by the string is its tension of 14.2 N.

(c) The string pulls on the ball toward the center of the motion, i.e., toward Anja's hand. At its other end it pulls on Anja's hand toward the ball. As long as we neglect the mass of the string, the two forces have the same magnitude.

EXAMPLE 24

The astronaut Sally Ride (whose mass is 60 kg) is in her spaceship in a circular orbit 100 km above the surface of the earth.

(a) Use proportional reasoning to find her weight in orbit.

(b) What is the interaction that keeps her in orbit?

(c) How large is the force on her?

(d) What is the time, T, for a complete orbit?

Ans.:

(a) Her weight on earth is $Mg_e = (60)(9.8) = 588$ N, where g_e is the value of g on earth. Since $Mg_o = G\frac{MM_e}{R^2}$, where g_o is the value of g in orbit, g is proportional to $\frac{1}{R^2}$. Hence $\frac{g_e}{g_o} = \left(\frac{R}{R_e}\right)^2 = \left(\frac{6470}{6370}\right)^2 = 1.031$. In other words, her weight on earth is 3.1% greater than it is in this orbit. Her weight in orbit is 570 N.

(b) The gravitational interaction keeps her in orbit.

(c) The force on her is the gravitational force, and it is 569 N.

The spaceship is in orbit with her. There is no force between her and the spaceship. This gives rise to the (incorrect) description of her as being "weightless." Sometimes the interaction between her and the spaceship is called the "apparent weight," which is then zero while she is in orbit. According to our definition the weight is equal to the gravitational force on her and is not zero.

(d) We can find the time T for a complete revolution from the relation for the centripetal force, which here is the gravitational force, so that $G\frac{MM_e}{R^2} = M\frac{v^2}{R}$.

We start by canceling M and dividing each side by R, to get $G\frac{M_e}{R^3} = (\frac{v}{R})^2$. Then we can use the fact that in the time T she and the spaceship go a distance $2\pi R$, so that $2\pi R = vT$, or $\frac{v}{R} = \frac{2\pi}{T}$. We can substitute $\frac{2\pi}{T}$ for $\frac{v}{R}$ to get $\frac{GM_e}{R^3} = (\frac{2\pi}{T})^2$, which we can turn around to get $T^2 = \frac{4\pi^2 R^3}{GM_e}$.

All that's left is to put numbers in: $M_e = 5.98 \times 10^{24}$ kg and $G = 6.67 \times 10^{-11} \frac{\text{Nm}^2}{\text{kg}^2}$. $R_e = 6.37 \times 10^6$ m, so that $R = R_e + 0.1 \times 10^6$ m $= 6.47 \times 10^6$ m. Putting these numbers in leads to $T = 5.18 \times 10^3$ s or 86 min.

If we had used R_e instead of R the result would have been 84 min. We see that while the height above the earth, of 100 km, seems quite large to us, it changes the distance from the center of the earth by only about $1\frac{1}{2}$%. That's why satellites that travel reasonably close to the earth all take about $1\frac{1}{2}$ h for one orbit.

The same considerations can be used for the time of a complete orbit of a planet around the sun. The fact that $T^2 \propto R^3$ for the planets was discovered by Kepler, and is known as Kepler's third law.

The orbits of the planets are actually elliptical, but sufficiently close to being circular that the approximation of considering them to be circular is quite close.

Angular momentum and torque

The momentum of an object or system describes its linear motion. The momentum remains constant unless an external force causes it to change.

There is an analogous quantity called the *angular momentum*, for rotational motion. It also remains constant, unless an external influence (a *torque*) causes it to change. The earth keeps rotating around the sun, and the moon around the earth, and in addition both keep spinning, in accord with the *law of conservation of angular momentum*. Angular momentum is a fundamental property of electrons, protons, and neutrons, and hence of the nuclei, atoms, and molecules of which they are the parts. The angular momentum of an atom depends primarily on its electrons. It plays a major role in determining the atom's properties.

As you can see, rotational motion is in some ways more basic than linear motion. To describe it in detail we will have to introduce some new terms, but each of them will be analogous to the corresponding term for linear motion.

Look at a single particle with mass m moving with constant speed, v, in a circle whose radius is r. Let's start with *angular displacement*. This is the angle θ through which the particle travels. We will measure it in *radians*, so that $\theta = \frac{s}{r}$, where s is the distance along the circumference.

Now we can define the *angular velocity* (ω, Greek *omega*) as the rate of change with time of

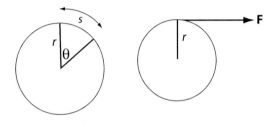

the angular displacement. $\omega = \frac{d\theta}{dt}$, which is also equal to $\frac{1}{r}\frac{ds}{dt}$, or $\frac{v}{r}$.

The *angular acceleration* (α, Greek *alpha*) follows, as the rate of change with time of the angular velocity. $\alpha = \frac{d\omega}{dt}$, which is also equal to $\frac{1}{r}\frac{dv}{dt}$, or $\frac{a}{r}$.

All the relations between x, v, and a have their counterparts in relations between $\theta, \omega,$ and α, and can be used in the same way. For instance, as the relation corresponding to $v = v_0 + at$ we can write $\omega = \omega_0 + \alpha t$ to show how the angular velocity varies with time when there is a constant angular acceleration.

So far we have talked about a single particle, but the same considerations apply to an extended rotating object, as for example, a wheel.

EXAMPLE 25

The drive shaft of a car engine is rotating with an angular velocity of 100 rpm (revolutions per minute). It decelerates with a constant angular acceleration of 2 rad/s² (radians per second squared). How long does it take for the engine to come to rest?

Ans.:
First we have to see that all the quantities that we need to use are in SI units. A *revolution* is 2π *radians*, so that 100 rpm $= (100)(2\pi)$ rad/min or $\frac{100}{60}2\pi$ rad/s. Now we can use $\omega = \omega_0 + \alpha t$, where the final angular velocity, ω, is zero, the initial angular velocity, ω_0, is $\frac{200\pi}{60}$ rad/s, and $\alpha = -2$ rad/s². (α has the opposite sign from ω because the engine is slowing down.) We turn the relation around to read $t = \frac{\omega - \omega_0}{\alpha}$ and substitute the numbers to get $t = \frac{100\pi}{60} = 5.24$ s.

What takes the place of the force? If we apply a force to our particle, tangential to the circle in which it moves, it matters not only how large the force is, but also where it is applied. It will have a greater effect on the rotation if it is applied further from the center. We therefore define the torque (τ, Greek *tau*) as Fr, where F is the tangential force and r is the radius at which it is applied.

We see that $\tau = Fr = mar = m\frac{a}{r}r^2$, or $\tau = mr^2\alpha$. This has the same form as Newton's second law, $F = Ma$, with torque taking the place of force and angular acceleration taking the place of the linear acceleration. The quantity mr^2 takes the place of the mass, and is given the somewhat cumbersome name *moment of inertia* with the symbol I. The torque produces an angular acceleration. The size of the angular acceleration depends on the moment of inertia. $\tau = I\alpha$.

We can extend the description from that of a particle to that of an extended object. The torque and angular acceleration are the same as before. The moment of inertia is the sum of the moments of inertia for all the pieces of the object. We can write it as $I = \Sigma mr^2$, where m is the mass of one piece, at a distance r from the axis about which the object rotates, and the symbol Σ stands for "sum." The moment of inertia is the quantity that for the rotational motion of an extended object is analogous to the mass for linear motion. It depends on how the mass is distributed in the object.

If the mass is distributed in discrete pieces, we can simply add mr^2 for each to find Σmr^2. It may also be distributed continuously, as in a disk or wheel, or sphere. It is then often possible to find a simple expression for the moment of inertia. For example, the moment of inertia of a disk of mass M with radius R, rotating about an axis through its center, is $\frac{1}{2}MR^2$.

The linear momentum is Mv. In analogy we can now define the angular momentum as $I\omega$, the product of the moment of inertia and the angular velocity. For a single particle moving in a circle it is equal to $(mr^2)(\frac{v}{r})$, or mvr.

(There is also rotational kinetic energy, analogous to the linear kinetic energy $\frac{1}{2}Mv^2$. It is equal to $\frac{1}{2}I\omega^2$. More about that in Chapter 6.)

Go to the PhET website and open the simulation *Torque*.

Go to the tab "Moment of Inertia." Explore the relation between torque, moment of inertia, and angular acceleration. Set a torque and observe the acceleration of the platform. Look at the values of these three quantities on the graph and check that their relation is what you expect from the rotational analog of Newton's second law of motion.

With the tab Torque you can explore the relation between the force and the torque, and with the tab Angular Momentum the relation between angular velocity, moment of inertia, and angular momentum.

The moment of inertia of a uniform solid disk is $\frac{1}{2}MR^2$. Explore the relation between M, R, and I by varying these quantities with the sliders.

EXAMPLE 26

A skater has a moment of inertia of 25 kg m² when she stretches out her arms. She starts out with an angular velocity of 3 rpm. She then draws her arms in and her moment of inertia is reduced to 10 kg m². What is her angular velocity now? (Assume that the external torque is zero.)

Ans.:
Angular momentum is conserved, so that $I_1\omega_1 = I_2\omega_2$, or $\frac{\omega_2}{\omega_1} = \frac{I_1}{I_2} = \frac{25}{10} = 2.5$. In the expression $\frac{\omega_2}{\omega_1}$ it doesn't matter what the units are as long as they are the same in the numerator and the denominator. We can therefore leave them as rpm to find that $\omega_2 = 2.5\omega_1 = 7.5$ rpm.

The table shows the various linear quantities and their rotational analogs.

Linear	Rotational
x	$\theta = \frac{s}{r}$
v	$\omega = \frac{v}{r}$
a	$\alpha = \frac{a}{r}$
F	$\tau = Fr$
M	$I = \Sigma mr^2$
Mv	$I\omega$
$\frac{1}{2}Mv^2$	$\frac{1}{2}I\omega^2$
$F = Ma$	$\tau = I\alpha$

The angular momentum of particles

The proton, neutron, and electron each has angular momentum at all times, called their *spin*. This spin angular momentum is also called their *intrinsic* angular momentum, to indicate that it is always there, regardless of any other motion. If an electron is part of an atom it may also have *orbital* angular momentum. Similarly the protons and neutrons move in a nucleus and may have orbital angular momentum in addition to their ever-present intrinsic spin angular momentum. The amounts of the angular momentum of an atom and a nucleus are among their most important properties.

We have to remember that the picture of the particles as little balls spinning about an axis is not correct. This is also true about the picture of an atom, as first envisioned by Bohr, with electrons in orbit, similar to planets about the sun. The picture of a spinning particle, or of an atom with electrons in orbit, may be a simple visualization. But a model in which ordinary (classical) mechanics is used for them does not lead to the correct (observed) results.

We will return to these questions later.

EXAMPLE 27

In its lowest energy state (the ground state) the helium atom has an electronic angular momentum of zero. Explain.

Ans.:
Although each of the two electrons in the helium atom has a spin angular momentum, the spins are in opposite directions. A *mechanical* picture is inappropriate for atoms, but in the ground state the two angular momenta nevertheless add up to zero. Each electron also has an orbital angular momentum. They are also in opposite directions.

As an aside we can mention the nuclear spin angular momentum. Most commonly, the nucleus of helium is an *alpha* particle, which consists of two protons and two neutrons. Again the total spin is zero. In 1.4×10^{-6}% of the atoms, however, the nucleus consists of two protons and only one neutron, to form the isotope ^{3}He, with a net spin angular momentum.

4.5 Summary

Everything we do and everything that happens around us involves forces. The basic feature of forces is that they give rise to acceleration. This is embodied in the relation $\Sigma \mathbf{F} = M\mathbf{a}$, Newton's second law of motion. Here $\Sigma \mathbf{F}$ is the symbol for the *vector sum* of all the forces acting on an object or *system*, or the *net force*, and a is the acceleration. M is the object's mass. It determines the magnitude of the acceleration.

In the SI system of units M is measured in kilograms (kg) and a in $\frac{m}{s^2}$. If these units are used, the unit for force, equal to $\frac{kg\,m}{s^2}$, is the *newton* (N).

Here are some special cases: When ΣF is zero (when the forces on an object add up to zero) the acceleration is zero. The object may be at rest, but it can also be in motion with constant velocity. (There is a nonzero acceleration only when the velocity changes.)

When ΣF is constant, the acceleration is also constant. In that case the relations for constant a apply: $v = v_0 + at$, $x = v_0 t + \frac{1}{2} at^2$, $v^2 = v_0^2 + 2ax$.

When all the forces are along the same line, they can be added by using only positive and negative numbers ("algebraically"). If they are not along one line they have to be added as *vectors*. This can be done in two ways. One is to draw the vectors end to end, one arrow following the other. Their sum is then represented by the single vector from the beginning (the "tail") of the first to the end (the arrowhead) of the last.

The other method of adding vectors is to use a coordinate (x–y) system and to decompose each vector into its x component and its y component. All the x components are then added algebraically to give ΣF_x and the y components are added to give ΣF_y. These two quantities are the components of the single vector ΣF, the vector sum of all the forces.

To find ΣF we draw a *force diagram*. This is a figure that shows vectors representing each of the forces (and nothing else!). The force diagram helps us to visualize the magnitude and direction of all the forces acting on an object.

The *momentum* of an object or system is its mass times its velocity. Since it is proportional to the velocity, it is also a vector quantity. If there is no net force on a system or object, its momentum is constant. This is the *law of conservation of momentum*.

When two objects (A and B) interact, the force of A on B has the same magnitude as the force of B on A and is in the opposite direction. This is *Newton's third law of motion*.

Angular motion is represented by the angular displacement, θ, the angular velocity, ω, equal to $\frac{d\theta}{dt}$, and the angular acceleration, α, equal to $\frac{d\omega}{dt}$. In analogy to linear motion the angular velocity is the rate of change with time of the angular displacement.

"Uniform circular motion" is motion of a particle in a circle with constant speed. It is not constant velocity, because the direction of the velocity keeps changing. The acceleration also has constant magnitude and changing direction. Its direction is toward the center of the motion. It is called the "centripetal" ("toward the center") acceleration. Its magnitude is related to the speed and radius by the relation $a = \frac{v^2}{r}$.

In order for a particle to move in a circle with uniform circular motion there must be a net force toward the center (the "centripetal force"), whose magnitude is equal to $\frac{mv^2}{r}$.

Just as force gives rise to acceleration, *torque* gives rise to angular acceleration. And just as the mass of an object determines how large the acceleration of an object is when a net force acts on it, the *moment of inertia* determines how large the angular acceleration of a rotating object is when a net torque acts on it. The moment of inertia is a measure of how the mass is distributed.

Newton's second law also holds for rotational motion. Analogous to $\Sigma F = Ma$ there is the relation for angular motion $\Sigma \tau = I\alpha$, where $\Sigma \tau$ is the sum of all the torques, I is the moment of inertia, and α is the angular acceleration.

The *angular momentum* (analogous to the linear momentum Mv) is $I\omega$. For a system with no net external torque on it, the angular momentum remains constant. This is the *law of conservation of angular momentum*.

To find the moment of inertia of an object with respect to an axis, we have to imagine cutting it up into pieces of mass m, each a distance r from the axis, and adding the values of mr^2 for each piece to get the sum $I = \Sigma mr^2$.

There is also kinetic energy associated with rotating objects. The *angular kinetic energy* is $\frac{1}{2} I\omega^2$, analogous to the linear kinetic energy, $\frac{1}{2} mv^2$. (See Chapter 6.)

4.6 Review activities and problems

Guided review

1. A book is at rest on a table. What are the forces on it?

2. A car whose mass is 2000 kg accelerates with $a = 2.5$ m/s^2. What is the net force on it?

3. Maja's weight is 120 pounds. What is her mass in kg and her weight in N?

4. A box is being pulled forward horizontally along a table by a force of 10 N. The force of friction is 1.5 N.
   (a) Make a diagram that shows all four forces on the book.
   (b) Write the mathematical statement of Newton's second law with these forces.

5. A wagon whose mass is 12 kg is being pulled by two horizontal forces. The first is 50 N at 30° to the x-axis. The second is 70 N at right angles to the first. What is the magnitude of the wagon's acceleration?

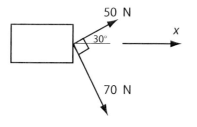

6. F_1 and F_2 are two horizontal forces of 60 N each, acting on a wagon. They are at 45° to the x-axis and at right angles to each other. A third force, F_3, also 60 N, acts along the positive x-axis. What are the magnitude and direction of the net force on the wagon?

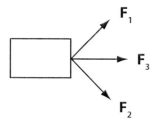

7. For the three forces of the previous question, find $\mathbf{F}_1 + \mathbf{F}_3 - \mathbf{F}_2$.

8. A box whose mass is 23 kg is being pulled horizontally along the floor by a rope with a force of 120 N. The force of friction is 32 N.
   (a) Draw a force diagram, indicating each of the forces on it with a symbol that has the appropriate two subscripts.
   (b) Find the net force on the box, and its acceleration.

9. Go to the PhET website and open the simulation *Forces and Motion*.
   Choose an object other than the crate. Answer the same questions as in the example. (For the first three parts choose a new appropriate applied force.)

10. Repeat Question 8 with the rope at an angle of 20° to the horizontal. Assume that the coefficient of friction (the ratio of the force of friction to the normal force) remains the same.

11. Go to the PhET website and open the simulation *The Ramp: Forces and Motion*.
    (a) Choose the Intro tab.
    Choose the file cabinet. Find the largest angle at which it does not slide down. What are the force of friction and the normal force at this angle? Check that their ratio is the marked coefficient of friction.
    (b) Set the ramp angle at 10°. What is the force of friction now? Why is it not equal to μN?
    (c) Go to *Force Graphs*. Put the file cabinet on the ramp, set the angle at 20°, and apply a force of 700 N. Push to go. Calculate all the forces. Compare your numbers to the ones on the screen.

12. A force in the $x - y$ plane whose magnitude is 12 N makes an angle of 125° with the $(+x)$ direction. What are its x and y components?

13. A car whose mass is 1800 kg is at rest on a road that is inclined at an angle of 8° to the horizontal. Draw a force diagram and find the magnitude and direction of each of the forces on the car.

14. A toboggan slides down a hill inclined at 20° with an acceleration of 1.5 m/s^2. Draw a force diagram showing all the forces. What other information must you know to calculate the magnitude of each of the forces?

15. A stone falls vertically from a roof straight down to the ground. (Air resistance may be

neglected.) Draw a force diagram for the time that it is in the air. What is its acceleration?

16. A rifle bullet is shot out of a gun that points at an angle of 30° to the horizontal. For the moment that the bullet is at its highest point:
    (a) Draw a force diagram.
    (b) What is the direction of its velocity?
    (c) What are the magnitude and direction of its acceleration?

17. Two railroad cars, each with a mass of 10^4 kg, are connected to each other. They are being pulled horizontally by a locomotive with an acceleration of 0.1 m/s^2. (Neglect friction.)
    (a) Draw a force diagram for each of the two cars and for the system containing both.
    (b) Find the magnitude of each of the forces on your force diagrams.

18. A car whose mass is 2000 kg moves with a velocity of 30 m/s. It hits a stationary car that has the same mass. (Ignore all horizontal forces except for the ones that the two cars exert on each other.)
    (a) The two cars stick to each other after the collision. What is their velocity after the collision?
    (b) In a different collision, with the same start, the cars do not stick to each other. After the collision the car that was originally at rest is observed to move with a velocity of 10 m/s at an angle of 35° with the original motion of the other car. Draw vector diagrams that show the momentum vectors before and after the collision. Find the momentum and velocity after the collision of the car that was originally moving.

19. Two railroad cars, each with a mass of 2×10^4 kg, are being pushed by a locomotive with a force of 1.5×10^5 N. (Neglect friction.)
    (a) Draw force diagrams for each of the two cars and for the system containing both.
    (b) Find the magnitudes of each of the forces on your force diagrams.

20. You are pushing three equal boxes (8 kg each) along the floor with a force of 200 N. (Neglect friction.) What are the two horizontal forces on the middle box?

21. A boulder whose mass is 450 kg is being lifted vertically by a chain with an acceleration of 0.2 m/s^2.
    (a) Draw a force diagram for the boulder.
    (b) Find the magnitudes of each of the forces on your diagram.
    (c) The boulder is now moving down and slowing down with an acceleration whose magnitude is 0.8 m/s^2. What is the direction of the acceleration? Draw a force diagram and find each of the forces on it.

22. Go to the PhET website (http://phet.colorado.edu) and open the simulation *Ladybug Revolution*.
    Choose the "Rotation" tab.
    Check "radians," "Show Ladybug Graph," "Show Platform Graph" "Show Acceleration," and uncheck the others.
    Put the ladybug in the green zone and measure its radius with the ruler.
    (a) Set the disk in motion with an angular velocity of about 3 rad/s.
    Calculate the centripetal acceleration.
    Compare your value of the centripetal acceleration with that shown on the graph of $a(t)$.
    (b) Check "X-Acceleration" and "Y-Acceleration." Describe and explain the difference between the graphs of these two quantities.

23. A car travels in a circle whose radius is 20 m, with a speed of 25 m/s.
    (a) What is the magnitude of the car's acceleration?
    (b) What is the nature (origin) of the horizontal force on the car? (What is exerting the force?)

24. (a) At what height above the surface of the earth is an astronaut's weight half of what it is on earth?
    (b) What is the time of an orbit of a satellite around the moon close to the moon's surface? (What quantities do you have to look up to be able to answer this question?)

25. A flywheel speeds up from rest, with an angular acceleration of 1 rad/s^2. What is its angular velocity after one minute in rad/s and in rpm (revolutions per minute)?

26. A disk with a moment of inertia of 3 kg m^2 spins with an angular velocity of 8 rad/s. A second disk, with a moment of inertia of 2 kg m^2 is initially at rest, but is free to rotate on the same shaft. It is now pressed against the first disk, as in a clutch. What is the angular velocity of the

combined system of the two disks in contact with each other?

27. The spin angular momentum of the ground state of the lithium atom ($Z = 3$) and that of the sodium atom ($Z = 11$) is each equal to that of a single electron. Explain why this is so.

Problems and reasoning skill building

1. It is sometimes said that when you travel around a circle in a car there is a "centrifugal" force that pushes you outward. What is it that really happens?

2. A roller coaster has a part with a vertical circular loop where, at its top, the car travels upside down.
    (a) Draw a force diagram of the forces on the car at the top of the loop.
    (b) What must be true about the speed so that the car and passenger do not fall out?

3. A skier comes down a 10° slope.
    (a) Draw a force diagram of the skier as he comes down the slope. Do not neglect friction.
    (b) What measurements could you make to determine if there is any friction between the skier and the slope? If there is friction describe how you could measure the force of friction.

4. Maya pulls three of her children on a sled. You are standing nearby with a meterstick and a stopwatch.
    (a) Draw a force diagram of the sled with the children on it (do not neglect friction)
        (i) when she pulls horizontally
        (ii) when she pulls at an angle of 30° with the horizontal
    (b) What is the relation between the forces when she pulls horizontally
        (i) when the sled moves with constant velocity
        (ii) when the sled accelerates

5. You pull on a rope so that it exerts a 30 N force on a 20 kg sled that moves on a level frictionless icy surface to the right. The force is directed at an angle of 30° above the horizontal.
    (a) Draw a force diagram of the sled.
    (b) A student constructed the equations below to describe the forces and motion of the sled, using SI units. Are these descriptions consistent with the word description? If not, correct the mathematical descriptions. (Let the direction to the right be the positive x direction and let up be the positive y direction.)

$$x : +30 = 20a_x$$
$$y : F_{normal} - (20)(9.8) - 15$$

(c) At time T_1 the sled is moving at a velocity of 1.2 m/s to the right. How fast will it be moving two seconds later?

6. (a) A 4 kg block is pulled from rest along a horizontal surface by a rope with a force of 10 N. It experiences a constant frictional force of 2 N. How far has the block moved in the first three seconds?
    (b) If the rope in part (a) breaks at $t = 3$ s, what will the block do: stop immediately, slow to a stop, or continue moving at a constant speed? Explain your answer.

7. A 15 N thrust is exerted on a 0.5 kg rocket for a time interval of 8 s as it moves straight up. (Neglect all forces except this thrust and the force of gravity.) The rocket then continues to move upward as its speed reduces steadily to zero.
    (a) Draw force diagrams of the rocket (i) for $0 < t < 8$ s and (ii) for $t > 8$ s.
    (b) Describe the motion of the rocket by drawing graphs of v vs. t and a vs. t. You may assume that the rocket does not go so high that g changes significantly.
    (c) What is the maximum height that the rocket attains?

8. The figure shows the force diagram of a 0.9 kg object moving horizontally. $F_1 = 10$ N, $F_3 = 4$ N, and $\theta = 30°$.

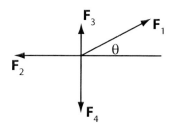

(a) Determine the magnitude of F_2 and F_4 if the object is moving horizontally with constant velocity.

(b) Find the acceleration if F_2 is 4 N.

(c) Describe a physical situation that could be represented by this diagram. What is the source of each of the three forces?

9. Three different motions are described in the diagram: (i) shows a block accelerating down an incline, (ii) is a projectile at the top of its trajectory (neglect air resistance) and (iii) is a car moving to the right and slowing down with constant acceleration.

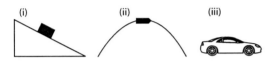

(a) For each of the three motions draw a vector **v** that shows the direction of the velocity and a vector $\mathbf{F}_{net}$ that describes the net force.

(b) For each of the three cases specify a coordinate system and sketch a graph that describes the acceleration as a function of time.

10. For each of the force diagrams, find $\mathbf{F}_{net}$ in terms of the magnitudes F_1, F_2, F_3 and the angle θ.

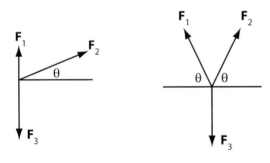

11. For each of the following situations:

(a) Draw a pictorial representation including a symbolic representation of all the information you are given and the assumptions that you make.

(b) Draw a force diagram.

(c) Find ΣF_x and ΣF_y.

(i) A hockey puck is pushed horizontally with a force $F_{stick\ on\ puck}$ across the ice. (Neglect friction.)

(ii) A ball falls through the air.

(iii) A ball has been thrown upward and is now moving up.

12. A hockey puck of mass M is given a horizontal push of magnitude $F_{stick\ on\ puck}$.

(a) At what rate does its speed change?

(b) Use reasonable values to get numerical results. Is the answer reasonable?

13. A crate of unknown mass is pulled upward by a rope. The tension in the rope is 22 N and the crate accelerates upward at a rate of $1.2\ m/s^2$. Determine the mass of the crate.

14. An elevator of unknown mass moves upward with a constant velocity v_0. The tension in the cable pulling the elevator is F_{re} ($F_{rope\ on\ elevator}$).

(a) Draw a pictorial representation including a symbolic representation of all the information you are given and the assumptions that you make.

(b) Draw a force diagram.

(c) Write a mathematical descriptions that allows you to determine the mass of the elevator.

(d) $v_0 = 3$ m/s up and $F_{re} = 5000$ N. Determine the mass. Is your answer reasonable?

15. A child with a mass of 40 kg stands on a spring scale inside an elevator. For each of the scenarios below, draw a force diagram of the forces acting on the child and calculate the reading of the scale in newtons.

(a) The elevator is at rest.

(b) The elevator is accelerating downward at $3\ m/s^2$.

(c) The elevator is moving upward at a constant velocity of 10 m/s.

(d) The elevator is accelerating upward at $2\ m/s^2$.

(e) For which scenario does the child feel the heaviest? The lightest?

16. Two boxes sitting on a frictionless surface are connected to one another by a string of negligible mass. Box 1 has a mass of 5 kg and Box 2 has a mass of 10 kg. The boxes are being pulled to the right with a constant force of 3 N.

(a) Draw force diagrams for each box and for the system consisting of both boxes.

(b) Determine the tension in the string connecting the boxes.

(c) Find the acceleration of the system.

17.

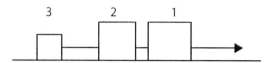

Three boxes sitting on a frictionless surface are connected to one another by strings of negligible mass. Box 1 has a mass of 5 kg and Box 3 has a mass of 2 kg. The boxes are being pulled to the right with a constant force of 22 N and the acceleration of the entire system is 2 m/s^2.

(a) What is the mass of Box 2?
(b) Draw a force diagram for each box and for the system consisting of all three of them.
(c) Determine the tension in each string.

18. Two boxes sitting on a frictionless surface are connected to one another by a string of negligible mass. Box 1 has a mass of 4 kg and Box 2 has a mass of 8 kg. The boxes are being pulled to the right with a constant force of 3 N.

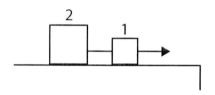

(a) Draw a force diagram for each box and for the system that contains both.
(b) Determine the acceleration of Box 1.
(c) How long does it take Box 1 to reach the end of the table, 2 m away?

19. A rope is causing an object to accelerate upward. Design an experiment to measure the force of the rope. You have a watch, a bathroom scale, and a measuring tape.

20. (a) Can the velocity and the net force ever point in opposite directions? If so, describe a motion where this occurs.
(b) Can acceleration and net force vectors ever point in opposite directions? If so, describe a motion where this occurs.
(c) Is it possible for an object to have zero velocity and also be accelerating? If so, describe a motion where this occurs.

(d) What must be true of the directions of the acceleration and the velocity if an object is speeding up?
(e) What must be true of the directions of the acceleration and the velocity if an object is slowing down?

21. A centrifuge rotates at 40,000 rpm. The bottom of a test tube rotates in it in a circle whose radius is 9 cm. The test chamber contains a sample whose mass is 5 g.
(a) What is the acceleration at the 9 cm radius?
(b) What force must the bottom of the test chamber exert on the sample?

22. A car moves with constant speed in a circular path around a corner on a horizontal road.
(a) What is the force that holds the car in its circular path?
(b) The car's speed is now increased by 50%. By what factor must the horizontal force increase?

23. (a) What is the centripetal acceleration on a 100 kg person at the equator resulting from the earth's rotation?
(b) What percentage of the person's weight is the centripetal force?
(c) What is the interaction that provides the centripetal force?

24. What is the centripetal force on a 100 kg person resulting from the earth's motion around the sun? What percentage of the person's weight is this?

25. A skater has a moment of inertia of 40 kg m^2 when she stretches out her arms. She starts out with an angular velocity of 70 rpm. She then draws her arms in, and her moment of inertia decreases to 10 kg m^2. What is her angular velocity now? (Do you need to convert the units of rpm to SI units?)

26. A beetle whose mass is 2 g sits on a turntable at a distance of 12 cm from the center. The turntable is rotating at 33 rpm.
(a) What are its acceleration and centripetal force?
(b) What is the interaction that provides the centripetal force?

27. A rock at the end of a string is being swung in a circle with higher and higher speed until the

string breaks. Describe the motion after the string breaks.

28. A torque of 30 Nm is applied to an engine shaft that is originally at rest. Its moment of inertia is 8 kg m². What is its angular velocity after 10 s?

29. A car engine is rotating with an angular velocity of 100 rpm. It then decelerates with an angular acceleration of 2 rad/s². How long does it take to bring the engine to rest?

30. Two spaceships are next to each other, both in the same circular orbit around the earth. The mass of the first is three times the mass of the second. Which of the following are the same and which are different for the two spaceships?
  velocity
  acceleration
  gravitational force
  centripetal force

31. A mouse sits on a turntable 20 cm from the center. The coefficient of static friction is 0.2. The angular velocity increases steadily from zero. At what angular velocity does the mouse slide off?

32. Samantha swings a ball at the end of a string in a circle. The ball's mass is 0.2 kg. It is in uniform circular motion with a radius of 1.5 m and an angular velocity of 1.2 rps.
  (a) What are the magnitude and direction of the centripetal force?
  (b) Draw a vector diagram that shows the centripetal force, the weight, the tension in the string, and the relation between these three vectors.
  (c) What is the angle between the string and the horizontal?
  (d) How would the angle change if the ball had twice the mass?

33. A geosynchronous orbit is one where a satellite rotates at the same rate as the earth so that it is always over the same spot with respect to the earth. How far is it from the center of the earth? (Start with the force relation. Then find the relation for the period T.)

34. It takes 5 s for the disk of a record player to accelerate from rest to an angular velocity of 33 rpm.
  (a) What is the angular acceleration?
  (b) How many revolutions does the disk make in this time interval?

35. The wheels on your bicycle have a diameter of 55 cm.
  (a) What is their angular velocity in rps when you ride at 15 mph (=6.7 m/s)?
  (b) What is the angular acceleration if you reach that speed in 5 s, starting from rest?
  (c) How many revolutions does the wheel make in that time?

Multiple choice questions

1. While rock climbing, a 50 kg woman falls off a ledge. She is moving down at a speed of 4 m/s when she lands in a bush that stops her fall in 0.40 m. The magnitude of the average force that the bush exerts on her body as she sinks into the bush is closest to
  (a) 200 N
  (b) 500 N
  (c) 2000 N
  (d) 1500 N
  (e) 1000 N

2. A rope exerts a tension force T on a crate that is initially at rest on a horizontal frictionless surface. When the crate reaches a speed of 6 m/s the tension is abruptly reduced by one-half to $\frac{T}{2}$. Just after the tension is reduced, the speed
  (a) decreases abruptly to 3 m/s
  (b) continues to increase at half the rate
  (c) stops after a short delay
  (d) decreases to 3 m/s after a short delay
  (e) continues to increase at the same rate.

3. At an amusement park a long slide has a sticky portion at the end designed to slow children down before they reach the end. Which of the force diagrams best represents the children while they are on the sticky portion before they come to a stop?

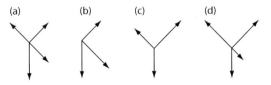

4. Marsha is pulling the front of a sled with a force of 150 N. The sled is on ice and its mass is 45 kg (that includes the mass of her younger

brother). The rope is at an angle of 20° with the ground. Ignore all effects of friction.

A student represented this situation mathematically as follows, using SI units:
x direction: $150 = 45 a x$
y direction: $F_{\text{normal}} - (45)(10) = (45)(0)$

Did the student label anything wrong or forget to include something? Circle all of the corrections that should be made (there may be more than one).

(a) 150 should be 150 sin 20°.
(b) 150 should be 150 cos 20°.
(c) a_y is not zero.
(d) There is a 150 sin 20° term missing in the y direction.
(e) The student did not make any mistakes.

5. Three forces, X, Y, and Z, act on a mass of 4.2 kg. The forces are approximately $X = 2\,\text{N}$ toward the east, $Y = 5\,\text{N}$ acting 45° north of east, and $Z = 3.5\,\text{N}$ acting south.

The direction of the net acceleration is close to being
(a) east
(b) 20° north of east
(c) north
(d) 10° north of east
(e) 44° north of east.

6. When a particle moves in a circle with constant speed, its acceleration is
(a) increasing
(b) constant in direction
(c) zero
(d) constant in magnitude
(e) constant in magnitude and direction.

7. A rock is being swung in a circle at the end of a string, with higher and higher speed, until the string breaks. What is the motion of the rock just after that?
(a) spiraling inward
(b) spiraling outward
(c) tangential
(d) radial
(e) none of the above

Synthesis problems and projects

1. What is the mathematical relation that shows that two spaceships in the same circular orbit are moving with the same speed?

2. You are a ski coach estimating the speed of a skier as she approaches a ski jump. You would like to know whether friction is negligible. You know the angle that the slope makes with the horizontal and you know the length, L, of the ramp. You have a digital video of the skier as she goes from rest down the ramp until she makes her jump. You also have a stopwatch and a measuring tape. What can you do to determine whether friction is negligible? What assumptions are you making?

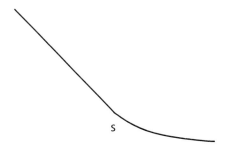

3. A mouse sits on a horizontal turntable that moves with angular velocity ω.
   (a) What is the origin of the horizontal force that keeps the mouse from sliding off?
   (b) What is the maximum value of that force, i.e., the value beyond which the mouse will slide off?
   (c) What is the mouse's acceleration while it is sitting on the turnable?
   (d) Write down the relation between the answers to parts (b) and (c) in terms of ω and R, but not v.
   (e) The angular velocity at which the mouse begins to slide is proportional to R^n. What is the value of n?

CHAPTER 5

Laws and Their Limits: The Organization of Scientific Knowledge

How do we know? Reality and interpretation
> *Theory and model*
> *"Laws"*
> *Constructs: what is real?*
> *From push and pull to operational definition*

The Newtonian model and its limitations
> *Classical physics and the Newtonian world view*
> *Is the universe programmed?*

Mechanics and beyond
> *From elementary particles to composite materials and the edge of the known*

The word *science* comes from the Latin *to know*, and if we follow that origin it should include every kind of knowledge. Instead it is usually used as synonymous with *natural science*, the study of the natural world, and that is the way we will use the word here: a statement of science has to have some relation to the world around us, as we observe it and as we experiment to explore it.

No matter how many symbols we use, no matter how many equations we write down, what we are trying to do is to describe the world. The symbols, units, and the mathematical operations are just parts of the language, as we try to describe, predict, and understand the phenomena that we observe.

There is no science without that contact with the world, without observation and experiment. That's why we won't call mathematics a science. It is wonderful, it can be beautiful, and it is enormously useful. It is the language without which modern science is unthinkable.

Sometimes it seems as if the language is the main part. The technique tends to take over. Which equation should I use? Is this the right unit? How do I get the right answer? But the technique is never the main part of doing or learning science. It may take up most of our time, but it is only the means to the end of greater knowledge.

We need to make time to ask: what does it *mean*? What are we trying to achieve and what have we learned? What do we know and how can we know more?

The essential part is to be able to have a conversation with nature—to let the world speak to us, so as to allow us to see its patterns, to see the relationships between the parts of which it is composed, and to understand its inner workings, its *mechanisms*.

5.1 How do we know? Reality and interpretation

In this chapter we want to look more carefully at how we gain scientific knowledge, how we organize it, and what methods we use. People sometimes talk about a *scientific method*, a regular sequence of steps that leads to scientific knowledge. Although there are common features in the way new science is learned, different questions require different kinds of answers, and different ways of finding answers.

Our knowledge of stars begins with observation, first of their positions, and then of the radiation that they emit. Today we know much more than can be observed directly. We know what stars consist of, how the different chemical elements are created in them, how the stars themselves are created, how they radiate, and how the radiation changes with time, until eventually they cease to be the stars that we know. To gain this knowledge required putting together pieces of knowledge that seemed at first to have nothing to do with stars, such as the interaction between atomic nuclei, which we now know to be responsible for the energy that the stars release.

Our knowledge of atoms began quite differently. It wasn't until long after people were convinced of the existence of atoms that ways were found to observe them directly.

In each case we use the methods that seem to be most appropriate. The most straightforward way to find out is observation. We can create special situations that we can then observe, study, and analyze. That's what we do when we conduct an experiment.

Sometimes the experiment comes first. That was the case for Galileo, when he dropped different objects and found that they all took the same time to reach the ground. At other times it is the analysis that comes first. That's what happened when Einstein thought about the nature of time and space and developed the theories of relativity. In both cases the object was to describe what is observed, and the essential test and requirement for what Einstein and other theoretical physicists achieve is that it must be in accord with what is or can be observed.

We conclude that there is no single method that leads to new knowledge. Rather, we do our best using any method we can.

A list of observations represents only rough and limited knowledge. We look for order in what we observe, and we try to make it quantitative. For example, we see the earth and the sun in motions that repeat, some every day, some every year. We make measurements and we attempt to detect patterns and relationships in the observed quantities. We can then test the scope and validity of the relationships by seeing whether they allow us to predict what will happen in other related situations.

If we describe what happens when we drop a stone, we can look for relationships that we can express in mathematical language, such as those that connect the height from which the stone is dropped to the time it takes to reach the ground and to the speed it has at various points along its path. We can test the relations that we develop by using different heights and different stones. They will be "good" and useful if they also allow us to describe, and so to predict, what will happen in these different circumstances.

Theory and model

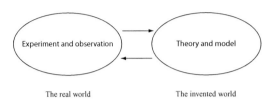

In ordinary language the word "theory" is often used to talk about a speculation or a guess. In science its meaning is different. It usually refers to a set of relationships between observations. In physics a theory is most often expressed in the language of mathematics, and describes relationships between quantities that are observed or are related to observed quantities. Let's look at an example that you are familiar with.

Newton's second law of motion ($\Sigma F = Ma$) tells us what force is: it is what leads to or can lead to acceleration. Newton's law of gravitation ($F = G\frac{M_1 M_2}{r^2}$) describes a particular kind of force, namely the gravitational force of attraction between two objects. Together these two relations can be used to describe the motion of the planets around the sun in great detail. They represent a theory of planetary motion.

If we look closely at what we are doing, we see that we are making some crucial assumptions. We are imagining a place where the sun and the planets have no internal structure; they behave like particles, each with its actual mass. Newton's law of gravitation describes the force between them exactly.

We have created an imagined universe. We call it a *model* of the sun and the planets. In it we leave out many of the sun's and the planets' properties, such as their sizes and shapes, their complex structure, their atmospheres, and their temperature distributions. The model is also different in that it is exact. It is our invention, and we have no doubts about how it is constructed and how every part of it behaves. The construction of the model and the statements about how its parts interact and behave are what we call the *theory*.

Sometimes we will use the word "model" and sometimes the word "theory." They both describe the invented, simplified universe and the relations between quantities in such a universe. If the properties of the model and the relationships in the theory are close to those that are actually observed, then we have succeeded in creating a good model and a good theory.

EXAMPLE 1

Galileo concluded that all objects take the same time to fall to the ground from the top of the leaning tower of Pisa. In light of what we know today, what is the model that is implied by his data, and what are its limitations?

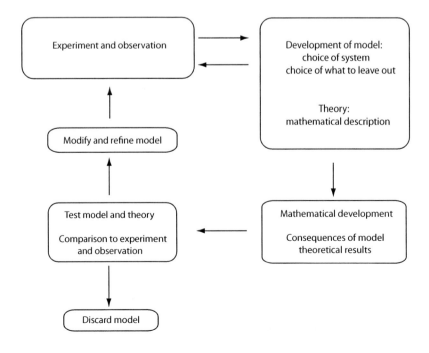

Ans.:
He probably didn't drop a piece of paper or a feather as part of his experiments. He must have known that for them there would be a lot of air resistance, and that this would change the time it takes for these objects to fall.

Galileo's model was that the gravitational interaction was the only one that needed to be considered. A measurable contribution from any other unbalanced forces would have changed his results.

EXAMPLE 2

What assumptions do we make when we use the interaction with the earth, as described by Newton's law of gravitation, with r as the distance to the center of the earth, as the only one that affects the motion of an artificial earth satellite?

Ans.:
The first assumption is that the gravitational interaction is the only one. This will be inappropriate if the satellite's path is so close to the earth that interaction with the atmosphere has a measurable influence. We are also neglecting the presence of all other bodies, such as the moon.

The second assumption is that the earth is spherically symmetrical. (This means that the earth's properties can vary with the distance from its center, but not with its angle, i.e., with the latitude or longitude.) The earth's shape is actually not a sphere. It's an ellipsoid. The distance between the north pole and the south pole is 12,713.6 km, while the diameter at the equator is 12,756.2 km. (The difference between these numbers is more than twice as large as the height of the highest mountain.) The mountains and oceans, as well as the uneven distribution of the elements, also limit the symmetry. In addition, each tree, animal, or person represents a departure from spherical symmetry.

"*Laws*"

Scientists are not very consistent in the way that they use the word "law." We have already used it in situations that are quite different from each other. We have used Newton's second law of motion as the definition of force, so that what it says is exactly true. On the other hand Newton's law of gravitation describes a relation between observed quantities. Its validity depends on whether it is in accord with what is observed.

And we know that although it describes what is observed very well, it is not perfect.

Physical laws that describe the behavior of materials are in a still different category. Hooke's law is the relation between the force (F) on a spring and the amount (x) by which it stretches. It says that these two quantities are proportional. Its validity depends on the particular spring, and we know that this "law" is never exactly true, since with a sufficiently strong force the spring will break. There are many other relations that are called *laws* and that describe the properties of materials. They depend on the interplay of the very large number of atoms in even a small piece of material. This kind of law is often quite approximate.

We see that in physics the word *law* is used in a variety of ways with different meanings. It can be exact, as in a definition. It can be part of a theory that is so well tested that its conclusion is elevated to the status of a *law*. But it is also used for relations that are only approximate, or that are valid only under very restricted circumstances.

Some laws are *empirical*. An empirical law is one that is based on experiment or observation. It describes an observed pattern of results. A law can also be the result of a theory. It then has to be tested by experiment and observation. We would prefer not to call a definition a law, but the term "Newton's laws of motion" is so widely used that we'll stay with it.

If we call a statement a "law" that makes it look as if it were a statement that prescribes what *must* happen. That's not what it means. Scientific theories and models and theoretical laws are attempts to describe natural phenomena. They can never be regarded as final. We have to be ready to find that they are inadequate as better measurements are made or new knowledge is found. If that happens, they may be refined and improved, or they may need to be discarded.

An untested statement or theory is often called a "hypothesis" (plural: hypotheses). Another word that is sometimes used is "postulate." It is a statement or relation that is assumed, usually as the beginning of a theory. A hypothesis or a postulate needs to be tested. Its consequences have to be compared to the results of observations and experiments. Only then can its usefulness and validity be evaluated.

There is some variation in the way that the various terms that we introduce in this chapter are defined and used. In popular usage, and even among physicists, they are not always used in a consistent fashion. We will use the framework that we have described, but in light of these differences it should not be regarded as rigid and absolute.

Most importantly, we have to remember that the models, the theories, the theoretical and empirical laws, the postulates and hypotheses can be considered to be scientific only if they can be tested by observation and experiment. A statement that is not necessarily wrong, if it cannot be tested it is not part of science.

EXAMPLE 3

The *coefficient of sliding friction* is defined as the force of friction on a sliding object divided by the normal force on it. That this coefficient turns out to be constant in certain situations is sometimes referred to as a *law of friction*.

Describe the nature of the term "law" as it is used here and the likely accuracy of the law.

Ans.:

This is an empirical law, based on the observation of objects pulled along a horizontal surface or sliding down inclined planes. The force of friction, and hence the coefficient of friction, are strongly dependent on the nature of the two surfaces that are in contact. Oil, dirt, and irregularities on the surfaces can cause substantial changes. The "law" can be expected to be a rough guide, at best.

EXAMPLE 4

While you are bowling you watch the ball: is it moving with constant speed, slowing down, or speeding up? How can you decide?

Ans.:

You decide to use Newton's second law of motion, $\Sigma F = Ma$. (You are using the *Newtonian model*.) You postulate that you can neglect the rolling motion so that the ball can be treated as a particle. You now have to think of all the possible forces. You make the hypothesis that friction and air resistance are so small that they can be neglected. You assume that the bowling lane is horizontal.

Based on these assumptions you conclude that there is no net horizontal force once the ball leaves your hand. There is therefore no horizontal acceleration, so that the velocity is constant.

To test this theoretical conclusion you measure the position of the ball at equal time intervals. If the ball moves through equal distances along a straight line in equal time intervals, the velocity is constant. If not, the model with its postulates, hypotheses, and assumptions has to be reevaluated and changed.

(Note that the meanings of the terms *postulate*, *hypothesis*, and *assumption* are very similar.)

As another example we describe the interplay of theory and observation in the development of our understanding of the motion of the planets.

Observations of the sun and the planets have been made for at least 5000 years. The Greeks thought that the earth is at the center of the universe, with the stars and the planets moving in circles around it. This was their model.

This model made predictions that were inconsistent with the observations. For example, as seen from the earth the planets sometimes appear to move backward. (This is called *retrograde motion*.) Furthermore, the brightness of the planets changes with time, suggesting that their distance from the earth varies. To account for the discrepancies a different model was introduced. It retained the circles with the earth at their center. The planets, however, did not move along these circles, but rather along smaller circles (called *epicycles*) whose center moved along the larger earth-centered circles.

This model was refined by Ptolemy (85–165 A.D.) to the point where it could account for all of the observations of the time, long before the invention of telescopes. However, the model became very complicated, with many epicycles, and with the earth no longer at the center of the large circles. It lasted to the time when Nicolaus Copernicus (1473–1543), in his last year, published a model in which the sun was at the center of the circular motion of the earth and the other planets.

Copernicus' model was not based on new observations, but rather on the simplicity that resulted from putting the sun at the center. It is interesting to note that all models up to that time, including that of Copernicus, were based on circular motion, to a large extent because that was considered motion that was "perfect" or "divine."

Tycho Brahe (1546–1601), still without telescopes, made observations and measurements that were more accurate than any before him, and showed that none of the models could account for their details.

Johannes Kepler (1571–1630) was able to resolve the difficulties. At first he tried a model in which the planets move around the sun in orbits around the five so-called *perfect solids*. (These are the cube, whose six faces are equal squares, the pyramid with its four equilateral triangles, and three others, also with faces that are equilateral polygons.) This model was inadequate in many ways.

After many years of studying the observations he came to the conclusion that the planets move in elliptical orbits, with the sun at one focus of the ellipse. This statement not only describes accurately what was known in Kepler's time, but also accounts correctly for what was subsequently observed after other planets were discovered. It (and two other statements, that together with this one are known as Kepler's laws) can be shown mathematically to follow from Newton's law of gravitation and Newton's second law of motion, as published by him in his major work in 1687.

EXAMPLE 5

Tycho Brahe measured the orbits of the planets closest to the sun. The following table shows their modern values for the period (T) in seconds and the semimajor axis, which we will use instead of the radius, (R) in meters.

Planet	R	T
Mercury	5.79×10^{10}	7.60×10^{6}
Venus	1.08×10^{11}	1.94×10^{7}
Earth	1.50×10^{11}	3.16×10^{7}
Mars	2.28×10^{11}	5.94×10^{7}

(a) Plot T^2 vs. R^3 for the data in the table.

(b) Describe in words what Kepler could conclude from the data. (This is now called Kepler's third law. Kepler's laws are empirical. He died before Newton showed their origin to be in the law of gravitation.)

(c) How could his conclusion be tested?

(d) Newton showed that Kepler's third law can be derived from his law of gravitation. Show that this is so for circular orbits.

Ans.:
(a)

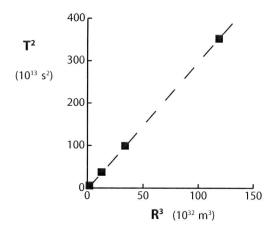

(b) Kepler concluded that for the planets in orbit around the sun the square of the period is proportional to the cube of the radius.

(c) Kepler's conclusion could be tested by using data for other planets to see if for them the values of T^2 and R^3 lie along the same line as for those in the table.

(d) Let M be the mass of the sun, m the mass of the orbiting planet, and R the radius of the orbit. For circular orbits there is a centripetal force equal to $\frac{mv^2}{R}$. If we assume that in this case the centripetal force is the gravitational force, then $\frac{mv^2}{R} = \frac{GMm}{R^2}$. We can cancel m and rewrite this relation as $v^2 = \frac{GM}{R}$.

The planet moves with constant speed, v, a distance $2\pi R$ in the time T, so that $v = \frac{2\pi R}{T}$. We can substitute this expression for v to get $(\frac{2\pi R}{T})^2 = \frac{GM}{R}$. This expression can be written as $R^3 = \frac{GM}{4\pi^2}T^2$, showing that $R^3 \propto T^2$.

Constructs: what is real?

In our description of physical phenomena we have used quantities such as *velocity*, *acceleration*, and *force*. Some are directly observed, such as *distance* and *time*. Others, such as *velocity* and *acceleration*, describe how quantities change. For still others, such as *energy*, the relation to the observations is less direct.

We have invented these quantities. They are sometimes called "constructs," to emphasize that it is we who construct them and decide what we want them to be. They represent choices that we make. Not all are as obvious as *distance* and *speed*. For *energy* and *momentum* the fact that they are invented is more apparent.

Does that mean that momentum is less "real" than velocity? Is the momentum of a baseball more real than that of an electron, just because we can hold a baseball in our hand, while the electron's properties have to be found more indirectly?

Much of our knowledge is obtained indirectly, often through measuring devices that use needles pointing to numbers, or digital electronic readouts. Is the speed of a car more real than that of an atomic nucleus that we can observe only by indirect means? We can see and hear and touch and smell the car. Is that what makes it real?

The question of what is *real* doesn't have a simple answer. Physicists (and others) often disagree about the meaning of the words "real" and "reality." If we can't agree on a definition, we can't decide where reality begins or ends.

When we ask whether something is real we generally think of how the phenomenon or quantity makes contact with our senses. Do we see it or hear it? Can we feel it or smell it? What if we see only the needle pointing to a number on a dial in a complicated piece of equipment? What if we are talking about a construct that appears only in the equations of a theory? In this book we won't attempt a definition, and will use the word *reality* only rarely, in situations where it doesn't seem to be controversial.

If we go back to the question "how do we know?" we see that there are two main parts. On the one hand, there is observation and experiment. On the other there is the building of models and theories with their constructs. Sometimes the observation comes first. It may be accidental, or it may be carefully planned. In other cases the theory and the model building come first. Often progress is made by going back and forth between these two activities. It is the interaction between the two that most often leads to knowledge and understanding.

From push and pull to operational definition

When we first talked about force we weren't very careful about definitions. The common meaning that force is a push or pull was enough to get us going. Similarly, mass is sometimes said to measure the quantity of matter. These statements give only some vague and incomplete idea of the meaning of the words "mass" and "force." They are too indefinite to be good definitions.

We want to use terms like *force* and *mass* in ways that are more precise and also more fruitful than in ordinary speech. A definition is not very useful if it gives just a rough synonym or general notion. What we need is a definition that gives us a recipe for how to find and measure the quantity, so that we will know it unambiguously. That kind is called an *operational* definition.

The recipe may be one that we can actually carry out, or it may be one that we can only *imagine* carrying out, within the rules and laws that we have already agreed on. That's what we did when we defined both force and mass by using $F = Ma$.

EXAMPLE 6

Give the operational definitions of *force* and *mass*.

(a) Start with two objects and describe each step in the procedure that leads to the operational definition of the ratio of the two masses.

(b) What else is necessary to define mass *absolutely*?

(c) Define *force*.

Ans.:

(a) Apply forces with the same magnitude to the two objects ($F_1 = F_2$). This can be done with a spring that applies the same force each time it is compressed by the same amount. Even better is a spring between the two objects. When it is compressed, and then released, it applies forces with the same magnitude to the two objects.

The two objects then move with accelerations that can be measured, a_1 and a_2. The quantity that decides how large each of the accelerations is is the mass: we define the ratio of the masses to be equal to the inverse ratio of the accelerations: $\frac{a_1}{a_2} = \frac{M_2}{M_1}$, or $M_1 a_1 = M_2 a_2$.

(b) To get a definite value we have to decide on a *standard* mass that everyone agrees to, such as a kilogram. The same procedure as in part (a) can be used to compare other masses to it.

(c) The force exerted by the spring in part (a) is defined to be equal to $M_1 a_1$ or $M_2 a_2$, or, in general, $F = Ma$.

These definitions tell us first of all what happens *qualitatively*, namely that *the force on an object gives rise to an acceleration*, and that *the mass determines how large that acceleration is*. But they go further; they tell us *quantitatively* how large the force and the mass are. They provide recipes, using only measured quantities, that we can use to determine the mass and the force. That's what makes them *operational definitions*. They are the connection to the real world.

5.2 The Newtonian model and its limitations

Let's go back to Kepler's and Newton's planetary model. Consider a moment when the particles of the model have the same positions as the centers of the actual sun and planets. We can now calculate the subsequent motions of the particles in the model, using Newton's law of gravitation, and compare them to the observed behavior of the sun and the planets.

It turns out that the model is excellent. The motions of the particles of the model and those of the actual planetary system are very closely the same. The equations of the model and the theory can be used to show that in the model the planets (i.e., the particles that represent the planets) move in elliptical orbits about the sun. This is what is observed in the real world. Past motions can be described, and future motions predicted, in great detail.

The model's success goes even farther. It can be used to represent not just the motion of the planets, but also that of their moons and of the many artificial satellites that have been launched.

You can see that we have gone far beyond speculation. The model is supported by many observations. Its success in describing what we observe gives us the confidence to use it to predict what will happen in the future.

We can now ask a number of questions. Is the model complete? What does it leave out?

Is the model unique, or can we think of other models that can describe the same observations?

There is an even more crucial question: is the model in conflict with any observation? No matter how many observations are in accord with the model, even a single one that is in conflict with it casts doubt on the validity and applicability of the model.

This particular model is so good that it is hard to find discrepancies. But there are limits. When Einstein developed the *general theory of relativity* in 1915, and applied it to the motion of *Mercury*, the planet closest to the sun, his calculations showed small differences from the results of the Newtonian equations, i.e., from the *Newtonian model*.

Einstein used a different theory, a different model. Which one is better? No amount of calculation, no amount of thought or discussion can decide. There is only one way to judge: what are the observations? They were made, and Einstein's results turned out to be better.

Does that mean that Newton's equations and model were abandoned? Not at all. For almost all situations Newton's law of gravitation describes what happens as accurately as we can measure, and does so much more simply. It continues to be used to describe the motions of the planets, their moons, and artificial satellites. But we now know more about its limitations and under what circumstances we may need to use Einstein's more complicated model.

When models and theories lead to conclusions that do not correspond to observations, they must be changed or abandoned. There was, for example, a time when it was thought that heat is a kind of substance that is transferred to a body when it is heated. That idea seemed plausible for a time, but was eventually shown to be in conflict with the observation that heat can be generated in unlimited quantity by friction, and so it was discarded.

Newton's law of gravitation describes the gravitational force. In the Newtonian model the planets are represented by particles, and we use only this single law, together with the definitions given by Newton's second law of motion. The motions within the model describe the motions of the real planetary system over a very long span of time. We have put order into a vast number of separate observations. The model describes and predicts what really happens. That's what we mean when we say that we *understand* the motions of the planetary system.

The success of Newton's law of gravitation goes much further. It was his great insight that it acts between all objects, on earth or elsewhere. That's why it is called the *universal* law of gravitation. This single law describes what happens in a vast variety of situations.

This universality illustrates one of the fundamental aims of science. We try to describe and to explain what happens with as few separate assumptions as possible. If we can describe the motion of the earth around the sun with the same mathematical theory that describes the motion of the falling apple, we have achieved something important, namely an understanding of the relation between two motions that had been thought earlier to be unrelated.

Classical physics and the Newtonian world view

The success of the Newtonian model was so great that the whole world began to be seen as a machine with forces that determine everything that happens. To understand a phenomenon meant to know the forces that were acting and to be able to determine the results of their action, following Newton's laws. The whole universe was looked at as a giant clockwork running according to the laws of Newtonian mechanics, or, as we now call it, "classical" mechanics. Different forces could be incorporated in the model, including the various mechanical pushes and pulls, and later other forces, such as those of electricity and magnetism.

Among the spectacular successes were that the mechanics of wave motion in the air as well as in solid materials could describe the phenomena of sound, and that the mechanics of molecules could describe the phenomena of heat and temperature. Later, as the laws of electricity and magnetism became known, it was recognized that they also lead to a description of the phenomena of the propagation of light and other kinds of "electromagnetic" waves.

Together, the subjects of mechanics, electricity, magnetism, heat, sound, and light were incorporated in the Newtonian model. There seemed to be little reason to doubt that all physical phenomena could be understood and explained within the same framework of what is today called "classical physics." It was tempting to think that the same ideas and methods could be expanded to encompass all of science.

The era of classical physics came to a close near the beginning of the twentieth century, with the development of the theories of relativity, and the realization that classical physics could not account for the existence, the structure, and the behavior of atoms and molecules.

This does not mean that the Newtonian model was then discarded. It continues to be an excellent model. But we now know its limitations. Examples of where we have to go beyond the Newtonian model are at extremely high speeds and very large energies, for which we have to use the special theory of relativity, and at very low temperatures, where quantum effects become important. However, the importance of the theories of relativity and of quantum mechanics goes far beyond these special situations. Our fundamental understanding of structure and interactions, from nuclei to stars, has been profoundly affected by these theories. Still, under many "ordinary" circumstances the Newtonian model and classical physics continue to provide the basic framework and guide to our thinking.

Is the universe programmed?

A peculiar and controversial feature of the Newtonian model is that all motions within it seem to be *determined*: for any particle a knowledge of the initial conditions (the position and velocity at some instant of time) and of the forces (all the forces, for all time) allows the calculation of the position and velocity at any future time. Let's look at that a little further.

The universe is made up of particles. Lots of them. We cannot know the position and velocity of all of them at any given moment. Even if we did they would be different a moment later. But within the framework of classical mechanics each particle *has* a position and a velocity at any moment, whether we know them or not. The forces on a particle depend on its position and velocity with respect to the other particles. It follows that the position and velocity of any particle, at any time, depend on the positions and velocities of all the particles at some moment of time.

The point is not that we might want to know or calculate any or all of the positions and velocities, but rather that they *exist*, and that their values are set and determined. In classical mechanics the set of parameters that describe the universe, i.e., the set of values of the positions and velocities of all of the particles that constitute the universe, is completely determined by the set of these parameters at any given instant of time. Within this framework the future history of the universe develops and unfolds as determined by the state of the universe at the present moment. All of the future is contained in the present.

The urgency and gravity of this question fell away in 1925 when it was realized that classical mechanics does not fully describe the motion of particles (or of any other objects), and has to be extended to include *quantum mechanics*. In quantum mechanics a particle does not have a single path that is determined for it by the forces that act on it. We can only calculate the *probability* of finding the particle at a particular place and moving with a particular velocity at a given time.

There are other questions that first came up in the context of the Newtonian model and its clockwork universe that are still valid today. Do the laws of physics determine everything that we observe? The laws of chemistry are extensions of the laws of physics to systems that are sometimes much more complicated. What about biology? Here the systems are even more complex. Are the laws of physics still applicable? As we have learned more and more about the structures and functions of living matter, we have been able to fit even enormously complex features, such as those of heredity and growth, into the presently known framework of physics.

EXAMPLE 7

(a) A living being, such as a dog, can be thought of as a classical system. Briefly describe the basic elements of the circulatory and digestive systems in terms of this model.

(b) Are there processes that are not well described by this model?

Ans.:
(a) The heart pumps blood through the circulatory system. The lungs take the oxygen from the air, and transport it to different parts of the body by the blood. Food and oxygen undergo chemical changes that result in the action of the muscles and the various organs in the body.

(b) Among the questions that are only partly understood and under active investigation are those of thought, memory, and consciousness. The expectation is that these processes can be described by using the electromagnetic force and the known relations of electromagnetism.

The deterministic character of the Newtonian, classical model seems to require the absence of free will. A more modern model (incorporating quantum mechanics, as developed from 1925 on) does not suffer from this limitation.

It has been speculated that a living organism requires a special "vital" force or ingredient. There is no evidence that this is so. If there are aspects of terms like "soul" and "spirit" that lie outside experimentation and observation, they lie outside the purview of science.

5.3 Mechanics and beyond

So far we have spent most of our time with mechanics, but we have left out quite a lot. There could now be lots of examples and details of techniques. There are many topics in mechanics that we have mentioned, but have not discussed in detail. Projectile motion, circular motion, planetary motion, and oscillating (harmonic) motion come to mind among the more important. We will explore some of these in the problems and projects, and in the next chapters. Electromagnetic forces and the electrical nature of matter will be a major topic later.

With Newton's laws of motion we have the framework for classical mechanics. As long as we stay with this subject, all else is elaboration. There are ramifications and consequences, different quantities, constructs, and relations that illuminate and simplify. We can talk about different kinds of forces. But there will be no other assumptions, no other independent, fundamental physical laws, as long as we stay within the realm of classical mechanics.

With relativity and quantum mechanics we go beyond these limits, with new insights and new and different results. We also go further, in a different direction, and to different forces, with electromagnetism and with atomic and nuclear physics. But we are finished laying the foundation of classical mechanics, all the mechanics that was known before the invention of the special theory of relativity in 1905. That doesn't mean that we are finished using it. We can apply it to different problems. We can invent new combinations of quantities and units. We will do some of that in the next chapters. And we will use the same basic concepts of displacement, velocity, and acceleration, force, mass, and momentum in every succeeding part of physics.

From elementary particles to composite materials and the edge of the known

We have used the word *particle* to mean *elementary particle*, i.e., an object without internal structure. For particles there are only the four fundamental forces.

Normally, however, we deal with macroscopic objects that consist of large numbers of pieces, such as cells, molecules, and atoms. We speak of other forces between them: friction, pushing, pulling, stretching, squeezing. Are these additional forces that we should add to the fundamental four?

As we look at smaller and smaller scales, we see that these forces are the result of electric forces. When objects touch, some of their atoms come so close that they exert electric forces on each other. These are the forces that we feel and observe, on the larger scale, as friction and as pushes and pulls.

There are good reasons why materials are so difficult to describe. First, quantum mechanics is the basic framework for any interaction between atoms or within them. We will get to this subject later. Second, the number of atoms in even a small piece of matter is huge, of the order of 10^{23} per cm^3. Simplification, approximation, and incompleteness are the rule.

That's very different from what we do when we deal with macroscopic objects, using classical, Newtonian mechanics, where we deal with displacement, velocity, acceleration, and time. In that case, once we know the game, we know the straightforward path from question to answer for each problem. Classical mechanics is so clean, so clear: the model, the assumptions, the methods, the results—the completeness within its realm of applicability, the reliability, the evident *truth* of it. We have to face the fact that many subjects in physics aren't like that at all.

There is quite a lot of physics that retains the certainty and clarity of Newtonian mechanics. That includes the other subjects that we consider to be parts of classical physics, particularly the electromagnetic theory. It also includes the special theory of relativity and a portion of quantum mechanics.

Some students get the hang of it, learn how to deal with the symbols and equations, feel comfortable with the apparent certainty of the system, and are then surprised, and even disappointed, when they meet physics that is quite different.

It is primarily the science of materials in its broadest sense that is different: atomic and nuclear phenomena, solid and liquid state physics, down to the smallest constituents and up to the stars and galaxies. In each case the object of attention is so complex that we quickly reach the edge of the known. It is not surprising that we need to simplify and approximate. Rather, the surprise should be how much can be said that is simple and straightforward, and at the same time illuminating and descriptive of the essential features.

We need to learn how to move from the well-marked path, with signs that indicate what is allowed and what is forbidden, to the realms where it is not clear which way to go, and where only portions of the way have so far been explored. That's where we are scientists at the frontier of new knowledge.

5.4 Summary

"It's only a theory!" is said often in the newspapers and in politics. We need to remember that most of our scientific knowledge is embodied in theories and models. They are representations of what we observe, with some features left out and some emphasized.

As we look at the ways in which scientific knowledge is accumulated we see that there is no simple "scientific method." There are *observations*, sometimes of specially created situations that we call *experiments*. There is also the analysis of observations, experiments, and other knowledge, and the creation of *theories* and *models*. A model is an invented representation of a part of the real world. It is simplified and approximate, but attempts to include the essential features of the real system that it is intended to represent. The real system may be so complex that it is difficult or even impossible to analyze. In the model some of the complicating features, such as friction and air resistance, are stripped away.

The word "law" is used for a variety of statements. Most often it is a relation between observed quantities. It may be very general or it may apply only in special situations. It may represent the observations quite precisely or it may be quite approximate. It may be *empirical*, i.e., based on experiments and observation, or it may be *theoretical*, i.e., arrived at by analysis, starting from a *postulate* or *hypothesis*.

In the term "Newton's laws of motion" the word "law" is used differently. We have taken the first two laws to define force and mass. The third law describes the forces between objects that interact with each other.

Velocity, energy, momentum, and others are *constructs* that we have invented to describe observations and quantities derived from observations. We avoid questions of whether one quantity or another is *real* or not, because there is insufficient agreement on the definition of *reality*.

The best definitions are those that provide a definite, quantitative recipe or set of operations for the measurement of a quantity. They are called *operational definitions*.

With Newton's laws of motion and Newton's law of gravitation, a precise description of the solar system, consisting of the sun and its planets, became possible. The success of the procedure and the model led to the notion that other, perhaps all natural phenomena could be described by the forces between the parts of the relevant system and by their motion. A phenomenon would be *explained* by knowing the forces and motions, as in a mechanical clock.

This view was expanded with the knowledge of electric and magnetic forces in the nineteenth century. It formed the basis of what we now call *classical physics*: all motions are determined by the forces, and in fact, so is everything else that happens. Everything about an object or a system can be predicted by knowing the forces, together with the knowledge of the location of the object and its constituents, and their velocities at some one moment.

The *modern* view is quite different. We now know that what happens cannot be exactly predicted. An inherent uncertainty remains, regardless of how precisely we make the measurements. It is built into the best description that we know, that of *quantum mechanics*, which describes (among other things) how materials are built from atoms and their nuclei.

5.5 Review activities and problems

Guided review

1. When we say that the weight of an object is Mg, and that $g = 9.8\,\text{N/kg}$, what assumptions are we making, and what model of the earth are we using?

2. Make a list of what we leave out when we calculate the gravitational force between the earth and the moon by using their masses and the relation $F_g = G\frac{M_e M_m}{R^2}$.

3. State Hooke's law. What kind of law is it? What are its limitations? (See also Chapter 6 on Hooke's law.)

4. "A gas consists of molecules that move randomly with a range of speeds."
   (a) What would make you say that this is a theoretical statement? What would make you say that this is an empirical statement?
   (b) List one or more observations that support it.

5. Explain the circumstances under which T^2 is proportional to R^3 for a planetary system. What are the assumptions that we make when we use this relation?

6. A tree falls in a remote forest, 100 miles from the nearest human being.
   You ask your friends Abigail and Beatrice whether the falling tree makes a sound. A. says *yes*, and B. says *no*. Can they both be right? State a definition that makes A.'s answer right and one that makes B.'s answer right.

7. Describe some aspects of a tree as a classical system. Are there aspects that cannot be described this way?

Problems and reasoning skill building

1. "In winter the sun is farther from the earth than in the summer." Describe an observation that supports this hypothesis and one that contradicts it.

2. State Coulomb's law. What kind of law is it? What is its accuracy? What are its limitations?

3. The evaporation of water causes the remaining liquid to cool. List the characteristics of a model that accounts for this observation.

4. Is a theory ever a fact? Describe what each term means.

5. At one time it was thought that "nature abhors a vacuum." Describe the modern model that has replaced this statement.

6. (i) According to the model called "the ideal gas" the volume of a certain amount of gas held at a constant pressure is linearly related to its temperature. (The graph of volume against temperature is a straight line.)

(ii) When the temperature of water vapor decreases, it can "condense" and form liquid water.

Describe the nature of each of these two statements. Do they contradict each other?

Synthesis problems and projects

1. The *greenhouse effect* is the result of two phenomena. One is that all objects radiate electromagnetic energy and absorb some of the electromagnetic energy that falls on them. The radiated energy is different, depending on the temperature of the radiating object. The sun's temperature is such that some of the energy that it radiates is visible. The energy that is radiated by an object at room temperature is not visible. (It is in the "infrared" part of the spectrum.)

The second phenomenon is that some materials allow visible radiation to pass through, but absorb some of the infrared radiation. This is true, for example, for glass and for some gases, including carbon dioxide.

(a) Describe the model that shows how the greenhouse effect may lead to heating of the earth.

(b) What is the hypothesis that links the model to the burning of coal, oil, and natural gas?

(c) What is the nature of the data that are required to investigate the relevance of the model and the hypothesis?

(d) What are some of the limitations of the model?

2. Evolution describes the progression of life forms as a result of mutations and natural selection. Describe the primary features of the theory of evolution and some relevant observations.

3. Describe observations that provide information on the angle between the earth's axis and the ecliptic (the plane in which the earth moves around the sun).

4. Describe the interplay of theory and observation in the development of our understanding of one or more of the following subjects. In each case describe the system, the model, and the testing of the model.

(a) The shape of the earth, beginning with the hypothesis that the earth is flat.

(b) Gravity.

(c) Astrology, i.e., the belief that the positions of the stars and the planets have a significant influence on human beings.

(d) "A broken mirror brings bad luck."

(e) Earthquakes.

5. Last Saturday you had a dish of rice and beans. The next day you had a stomachache that you think might be a result of eating the rice and bean dish.

(a) What is your hypothesis?

(b) Design an experiment to test your hypothesis. What features, will you want to consider?

CHAPTER 6

Energy and Its Conservation

Work: not always what you think

Energy of motion: kinetic energy

Energy of position: potential energy
> *Gravitational potential energy*
> *The reference level*
> *Mechanical energy and its conservation*
> *Electric potential energy*
> *Springs: elastic potential energy*
> *Hooke's law and the expression for the elastic potential energy*

Friction and the loss of mechanical energy

Internal energy and the law of conservation of energy

Work and energy revisited

Power: not what the power company sells

When you drive a car you have to stop from time to time to add fuel to the tank. In a house or apartment a bill arrives each month from the electric company. And all of us have to keep breathing and eating. In each case we use *energy* that is never created or destroyed, only transformed from one kind to another.

The gasoline in the car makes it possible for the car to move, as some of its internal ("chemical") energy is changed to energy of motion ("kinetic" energy). Electric currents heat the stove or the toaster, make the lamps light up, and the fans and vacuum cleaner run. In each of these cases energy is transformed at the generating station, usually from some kind of stored internal energy, or from the kinetic energy of wind or water. It becomes "electric" energy that can be transmitted through wires, and in turn transformed to the energy that we use as heat or light or motion. The food we eat gives us the energy stored within it, which is then transformed in our bodies to the energy that we need to exist, to move, and to function.

There is hardly a concept more pervasive than energy in all of science. The food we eat is fuel that we count in energy units. We transform it to kinetic energy when we move and to gravitational potential energy when we climb. Changes of material, whether physical, as from ice to water, or chemical, as in the burning of wood, are accompanied by changes of energy. We would not exist without the energy radiated by the sun, which is liberated there by its nuclear reactions. The state of each atom and molecule is characterized by its internal energy, that is, by the motion of its components and by their distribution in space and the resulting electric potential energy.

Our experience with the various kinds of energy brought us to the realization that energy cannot be *made* or created, and that it does not disappear. This is the *law of conservation of energy*. No exceptions to it have been found, and it has become a guiding principle that plays an essential role in every part of science.

6.1 Work: not always what you think

In ordinary language "work" can refer to a variety of activities, from shoveling snow and sawing wood to writing and thinking. In physics its meaning is more limited, but also more precise: you are working when you exert a force on an object and the object moves in the direction of the force (or of a component of the force). You may want to call it work just to hold a book or a rock, even when it doesn't move, but the physicist's definition wouldn't include that. If the rock doesn't move, you're not working.

If the rock does move, and there is a force on it in the direction of the motion, we define the work done on it by the force as equal to the force on the rock times the distance that the rock moves, or $W = Fs$.

EXAMPLE 1

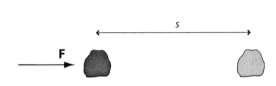

What is the work done by the force $F = 3\,N$?

Ans.:

If the force is 3 N, and the rock moves 4 m, the work done on the rock by the force is $(3\,N)(4\,m)$, or 12 Nm.

The unit Nm is also called the *joule* (J).

The force, F, (3 N) is in the same direction as the distance, s, (4 m) through which the rock moves. The work (W) done by this force is Fs, or 12 J.

This is so even though there are three other forces on the rock. Two of them are vertical, the weight of the rock ($F_{ER} = Mg$) and the upward force on it by the table (F_{TR}). There is no vertical acceleration, and so these two forces add up to zero. Since the rock does not move in the vertical direction, these two forces do not do any work on the rock. There is also the force of friction (f), in the direction opposite to the direction of the motion. However, we are asked about the work done by only one force, the 3 N force, and so we do not need to consider the work done by any of the other forces.

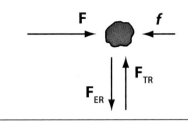

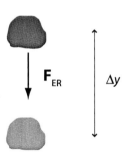

It is not easy to think of a situation where there is only a single force acting on an object unless we make some assumptions that simplify the situation. A freely falling rock is a good example, but only if we assume that we can neglect the air resistance. The weight, $Mg = F_{ER}$, is then the only force that we need to consider.

As the rock falls a distance Δy, the work done on it by the weight is $Mg\Delta y$.

That the work is equal to the force times the distance is the basic definition when the force is constant and in the same direction as the motion. We can expand this statement so that it also works for other cases. If there are several forces on the rock, all in the direction of the motion of the rock, each one does work on it, equal to the magnitude of the force times the distance. If the directions are not the same, we can take care of that too, and will do so later.

Because the force is constant, the acceleration is constant too, and we can use the third relation for constant acceleration from Chapter 3, $v^2 = v_0^2 + 2as$. Since the block starts from rest, $v_0 = 0$, and $v^2 = 2as$, or $as = \frac{1}{2}v^2$. The work done by the force, Fs, or Mas, is therefore equal to $\frac{1}{2}Mv^2$.

We know what happens, but we are going to say it differently. This will open up a whole new way of looking at the same situation. We say that the work done on the rock *changes its energy of motion*, which we call its *kinetic energy*. We define the kinetic energy by saying that it is the work done on the rock, starting from rest, to the point where its speed is v. We see that this is equal to $\frac{1}{2}Mv^2$.

To visualize the energy changes we can use bar charts, in which the height of each bar represents an amount of energy or work done.

6.2 Energy of motion: kinetic energy

The most basic kind of energy is energy of motion, or *kinetic energy*. Let's see how it is related to the quantities that we know, to force and acceleration, and to mass and speed.

Go back to our rock. Suppose that just a single force acts on it. We already know that the force will cause the rock to accelerate. Perhaps it starts from rest, or it may already be moving. The force will cause it to move faster, and after some time it will have a new and higher speed.

If we know the force and the distance, can we find out just how much faster the rock will be moving? Yes, we can. We know Newton's second law of motion, and we can find the acceleration. Once we know the acceleration, we can find how much faster the rock will be moving:

Initial velocity: $v_0 = 0$
Final velocity: v
F is the only force on the rock, and so $F = Ma$
Work done on the rock: $Fs = Mas$

$$W = K_f$$

The two bars have the same height, illustrating the relation that the work done on the rock (W) equals the increase (from zero to K_f) in the kinetic energy.

If the rock starts out with some speed v_0, we can again use Chapter 3's third relation for constant acceleration, $v^2 = v_0^2 + 2as$, but this time we have to include the term in v_0 so that the work done by the force is $Fs = Mas = \frac{1}{2}Mv^2 - \frac{1}{2}Mv_0^2$. We see that doing work on the rock is a mechanism for changing its kinetic energy.

We can write this as $\frac{1}{2}Mv_0^2 + Fs = \frac{1}{2}Mv^2$: the initial kinetic energy of the rock ($K_i = \frac{1}{2}Mv_0^2$) plus the work done on it (W) is equal to the final kinetic energy ($K_f = \frac{1}{2}Mv^2$).

106 / Energy and Its Conservation

W		
	=	K_f
K_i		

That $K_i + W = K_f$ is always true when W is the total work (which we can call W_T) done on an object. W_T is the sum of the amounts of work done by all of the forces on it. It is also the work done by the *net force*, i.e., by the vector sum of all of the forces acting on the object.

The statement that *the total work done on an object is equal to the increase in its kinetic energy* is so universal that it is given a special name. It is called the *work-energy theorem*.

We have to remember that we are not talking about any changes inside the rock or other object. We are using the model in which the object acts as a particle, in other words, as an object without any internal structure. Only then does the work-energy theorem apply.

EXAMPLE 2

A 0.2 kg hockey puck is at rest on horizontal frictionless ice. It is then pushed with a steady horizontal 2.5-N force and moves 1.5 m.

(a) Set up a mathematical description that relates the work to the kinetic energy. Make an energy bar chart.

   Determine how fast the puck is moving after it has gone the 1.5 m.

(b) Repeat part (a) for the same situation, but with the puck initially moving with a speed of 3 m/s in the direction of the force.

Ans.:

(a) The work, W, done by the force, F, when the puck moves a distance, s, is Fs, or $(2.5 \text{ N})(1.5 \text{ m})$, which is equal to 3.75 Nm or 3.75 J. Since $W = K_f$, this is also equal to the final kinetic energy, K_f.

You are, however, asked to find the velocity, v_f, which is related to the kinetic energy by $K_f = \frac{1}{2}Mv_f^2$, or $v_f = \sqrt{\frac{2K_f}{M}}$, which is $\sqrt{\frac{(2)(3.75)}{0.4}}$, or 4.33 m/s.

W		
	=	K_f

(b) This time we first have to find the initial kinetic energy, K_i, which is $(\frac{1}{2})(0.4)(9) = 1.8$ J. W is still 3.75 J, so that $K_i + W$ is $1.8 \text{ J} + 3.75 \text{ J} = 5.55$ J. This is K_f, so that $v_f = \sqrt{\frac{2K_f}{M}} = \sqrt{\frac{(2)(5.55)}{0.4}} = 5.27$ m/s.

W		
	=	K_f
K_i		

We could have kept each unit in the expression for v_f before evaluating it. However, because each of the quantities (energy and mass) was expressed in the appropriate SI units (J and kg), the result for v_f is automatically in the correct SI unit (m/s). If energy and mass were not in SI units, they would first have to be converted. (Some other system of consistent units would work also, but we will stick to the SI system.)

Note also that the mathematical expressions are in terms of work and energy. These are the quantities related by the relevant physical law, namely that the kinetic energy of the puck is increased by the amount of work that is done on it. It is only after we calculate the final kinetic energy that we can find the final velocity. (We cannot add the velocities!)

6.3 Energy of position: potential energy

Gravitational potential energy

When we lift a rock from the floor to the table, we give it energy. This energy depends on where the rock is, so that we can call it *energy of position*. The energy is stored. It has the "potential" to be changed to kinetic energy if the rock falls back down to the floor. That's why it is called *potential energy*.

We are talking about the system containing both the rock and the earth. Take the rock in your hand and lift it. Do it at constant speed so that we don't have to think about a change in its kinetic energy. Your hand exerts a force and does work on the system. The force exerted by your hand, F_{HR} (hand on rock), as the rock moves with constant speed, is equal in magnitude and opposite in direction to the gravitational force of the earth, the weight, F_{ER} (earth on rock), of the rock. The work done by your hand creates *potential energy* (P). It is work done by a force that is in the direction opposite to the direction of the gravitational force, or *against* the gravitational force. We call this potential energy *gravitational potential energy*. We define *the increase in the gravitational potential energy* as the *work done against the gravitational force*.

For the system containing the earth and the rock, the forces exerted by the earth on the rock and by the rock on the earth are internal to the system. The force of your hand is external to this system. It acts to separate the earth and the rock, and increases the gravitational potential energy of the system.

As long as the only other object besides the rock is the earth, there is not much likelihood of confusion, and we may talk about the potential energy "of the rock" while keeping in mind that we are really talking about energy shared by the rock and the earth.

The words *the increase in the gravitational potential energy* and *the work done against the gravitational force* mean the same thing. We can use one or the other. In any one problem we can use either *the increase in the potential energy* or *the work against the gravitational force*, but not both.

Your hand lifts the rock, increases its distance from the earth, and increases the potential energy. Without the presence of the earth, the gravitational potential energy would not change. We see that the potential energy is a property of both the rock and the earth. It is a property of the *system* consisting of both.

How much potential energy did we give the system as the rock moved up? Because the rock moves with constant velocity, the upward force of your hand, F_{HR}, has the same magnitude as the downward gravitational force, F_{ER}. The work (W) done by F_{HR} is the work done against the gravitational force. If the rock goes up a distance y, the force F_{HR} does an amount of work equal to $(F_{HR})(y)$.

The increase in the potential energy (ΔP) of the rock is therefore $F_{HR}y$, and since F_{HR} and $F_{ER} = Mg$ have the same magnitude, it is also equal to Mgy: $W = \Delta P = Mgy$, $\Delta P = P_f - P_i$.

So far we have talked only about a rock so near to the earth that the gravitational force on it can be considered to be constant. Later we will consider the more general situation where objects are so far from each other that the variation of the gravitational force, as given by Newton's law of gravitation, has to be taken into account.

EXAMPLE 3

You do 40 J of work, lifting a rock 1.2 m at constant velocity.

(a) Set up a mathematical description that relates work and energy. Make an energy bar chart.
    How much energy does the rock gain? What kind of energy does it gain?

(b) Determine the mass of the rock.

Ans.:

(a) There is no change in kinetic energy. We will neglect air resistance. Therefore the only energy that is changed is the gravitational potential energy: $W = \Delta P$.

The work done on the rock to lift it is 40 J, and so the rock gains 40 J of potential energy.

(b) The work (W) done against the gravitational force (Mg) is Mgy, where $W = 40$ J, $y = 1.2$ m, and g is, as usual, 9.8 m/s². $M = \frac{W}{gy} = \frac{40}{(9.8)(1.2)} = 3.40$ kg.

Some problems can be solved by using either the concept of force or the concept of energy. But in general there are two great advantages of using energy rather than force. One is that energy is a scalar quantity, while force is a vector quantity. The whole problem of directions and adding vectors falls away when we use the concept of energy. The second advantage is that with energy we need to consider only the initial and final energies. Except for work done on the system, or other changes in its energy, we don't need to think about what happens in between. When we use forces, on the other hand, we have to be able to follow every motion in detail: we have to know the forces at every point and at every moment.

The reference level

The change in the potential energy depends on the height, but we have not said anything about where the height is measured from. As long as we talk only about differences in potential energy, the question doesn't come up. When we want to talk about the actual amount of potential energy, we need to measure it from some *reference level*.

Because only differences in potential energy matter, it makes no difference where we take the reference level to be. It could be the floor, the ceiling, sea level, or any other convenient height. We are free to put it wherever it seems simplest. Regardless of where we choose it to be, the difference in height, $y_1 - y_2$, and the difference in the potential energy, ΔP, will be the same.

If the initial potential energy is P_i, and work W is then done on the rock to lift it, the final energy is given by $P_i + W = P_f$, or $W = P_f - P_i = \Delta P$. If we choose a different reference level, P_i will be different, but so will P_f, and the increase $P_f - P_i = \Delta P$ will remain the same. The bar chart illustrates both of these relations.

Mechanical energy and its conservation

After lifting the rock let's let it drop back, starting from rest at the top. We can use different ways to describe what happens. One is to consider the system of the rock and the earth. Its total energy remains constant. On the way down its potential energy decreases as the rock speeds up and gains kinetic energy.

$K_i = 0, P_f = 0$
$K_i + P_i = K_f + P_f$
$P_i = K_f$

We can also consider the system containing only the rock. The gravitational force (Mg) is now an external force. It does work on the rock so that it speeds up and gains kinetic energy.

How much kinetic energy does it get? Let's go back to the description in terms of the force. The definition of work tells us that on the way down the work done on the rock is equal to the force on the rock times the distance back down. The magnitude of the force is Mg, the distance is again y, and the amount of work is therefore equal to Mgy. This is the amount of kinetic energy that the rock gains on the way down.

Now we have to be careful. Is the force Mg the only force on the rock? Not quite. There is also air resistance, and there may be other forces. For now let's assume that we can neglect them. (We also neglected air resistance when the rock was on the way up.) We will come back to this question later to see the effect of air resistance.

For now, while we neglect air resistance, we see that in the system of the rock and the earth the potential energy that is lost as the rock falls is equal to the kinetic energy that is gained. The sum of the two energies remains constant while the rock falls.

We give this sum a new name, *mechanical energy*, and can then say that the mechanical energy remains constant while the rock falls.

The earth and the rock interact via the gravitational interaction. The gravitational force is *internal* to the system of the earth and the rock. Within this system there is kinetic energy and potential energy. As long as there is no force from *outside* the system, the sum of the two is constant.

As the rock falls, the kinetic energy of the system increases (mostly of the rock, but to a minute extent also of the earth) and the potential energy decreases. When the rock moves away from the earth, the potential energy increases and the kinetic energy decreases. The sum of the two, the mechanical energy, remains constant. Other objects or interactions are *outside* the system. They give rise to forces that *change* the mechanical energy of the system. These *external* forces may be air resistance, or the force of your hand as you separate the earth and the rock, i.e., as you lift the rock. In the absence of any external forces, the mechanical energy of the system remains constant.

This is our first encounter with the law of conservation of energy. It is in a very restricted form, because we are neglecting air resistance and other forces. Only a single force, the gravitational force, the weight, does work.

Under these circumstances, however, it always holds: in the system containing only the earth and another object, the weight of the object is the only force that does work. It is an internal force, and the mechanical energy of the system remains constant.

Remember that the choice of what we call *the system* is up to us. We can also look at the system containing only the rock. For this system, with the earth outside it, the gravitational force is an external force. As long as we continue to use the model where the rock is considered to be a particle, i.e., without internal energy, the only energy that it has is kinetic energy.

We will gradually remove all the restrictions and introduce other kinds of energy. We will then see the law of conservation of energy as one of the most fundamental and general laws in all of science.

We will talk about other kinds of potential energy. We distinguish the one that we started with by calling it gravitational potential energy, but if there is not much chance that we will confuse it with others, we will just call it potential energy.

EXAMPLE 4

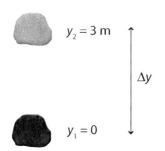

A rock has a mass, M, of 2 kg, and is lifted a distance, y, of 3 m, at constant velocity.

(a) How much potential energy does it gain on the way up? (From its initial amount, P_1, to its final amount, P_2.)

(b) How much kinetic energy does it gain on the way down, if it drops the 3 m, starting from rest?

(From K_2 and P_2 at the top, to K_3 and P_3 when it is back at the bottom.)

(c) What is its speed when it gets back to the starting point?

Ans.:
The system is that of the rock and the earth. As the rock is lifted, work (W) is done on it by your hand. The complete energy relation is $W + P_1 + K_1 = P_2 + K_2$. We can choose the reference level so that $P_1 = 0$. The kinetic energy remains the same ($K_1 = K_2$), so that we are left with $W = P_2$.

$$\boxed{W} = \boxed{P_2}$$

On the way down, $P_2 + K_2 = P_3 + K_3$. It starts from rest ($K_2 = 0$). It goes back to the point where $P_3 = 0$. That leave $P_2 = K_3$.

$$\boxed{P_2} = \boxed{K_3}$$

The rock's weight is Mg, or 19.6 N. On the way up it gains potential energy Mgy, or 58.8 J. On the way down it gains an amount of kinetic energy equal to 58.8 J. Since this is equal to $\frac{1}{2}Mv_3^2$, $v_3 = \sqrt{\frac{2K_3}{M}} = \sqrt{\frac{(2)(58.8)}{2}} = 7.67$ m/s.

EXAMPLE 5

Go to the website phet.colorado.edu and open the simulation *Energy Skate Park*. Check "potential energy reference." Drag the reference level to the lowest point of the skater's path. (The skater's position is indicated by the red dot below the skates.) Slow down the motion as much as possible by using the slider at the bottom. Click on "bar graph."

(a) Observe what happens to the potential energy, (P), the kinetic energy, (K), and the total (= mechanical) energy, (E). How do K and P compare when the skater is half-way down the track? Three quarters of the way down? What is the speed half-way down, compared to its maximum value?

(b) Reset to remove the bar graph. Click on "energy vs. position." Check "potential" and uncheck all others. (All units on the graph are SI units.) Check "measuring tape" and use the tape to measure the maximum height.

Calculate the weight of the skater (in N). Calculate his mass in kg. Are these results reasonable?

(c) Look at K and E and compare to your results in part (a).

Ans.:
(a) E is constant. P and K change form zero to E. Their sum is constant and equal to E. At the top $K = 0$ and P is at its maximum value, which is equal to E. At the bottom $P = 0$ and K is equal to E.

P is proportional to the height, h, above the reference level ($P = Mgh$). Therefore, when the skater is at half the maximum height the potential energy has half of its maximum value, and so has K. When he is one-quarter of the way from the bottom P is at $\frac{1}{4}$ of its maximum and $K = \frac{3}{4}E$.

Half-way down $K = \frac{1}{2}K_{max}$, or $\frac{1}{2}Mv^2 = \left(\frac{1}{2}\right)\left(\frac{1}{2}Mv^2_{max}\right)$. $v^2 = \frac{1}{2}v^2_{max}$, or $v = \frac{1}{\sqrt{2}}v_{max}$.

(b) For a maximum height of 4.1 m and a maximum potential energy ($= Mgh$) equal to about 3000 J, the weight is $\frac{P}{h}$ or 732 N, and the mass is $\frac{P}{gh}$ or 75 kg. This is equivalent to 164 pounds, which seems reasonable.

Electric potential energy

Just as the earth and all other objects attract each other by the gravitational force, positive and negative charges attract each other by the electric force. If we want to separate the charges, we have to pull them apart with a force that acts in a direction opposite to the electric force that attracts them. The force that we exert does work on the charges, and we can talk about electric potential energy in close analogy to the way we talked about gravitational potential energy.

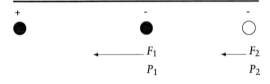

The negative charge is attracted to the positive charge, as described by Coulomb's law. If we push it from point 1 to point 2, doing work, W, it gains potential energy: $P_1 + W = P_2$ or $W = P_2 - P_1 = \Delta P$.

$$W = \Delta P$$

If we let it "fall back" from rest at point 2 to point 1, it will lose potential energy (from P_2 to P_1) and gain kinetic energy (from $K_2 = 0$ to K_1). The sum of the kinetic and potential energies remains constant, just as in the gravitational case, as long as we neglect any forces except for the electric force between the positive and the negative charge. $K_2 + P_2 = K_1 + P_1$ and $K_1 = P_2 - P_1 = \Delta P$.

$$P_2 = \begin{array}{c} K_1 \\ P_1 \end{array}$$

The electric force between two charges is analogous to the gravitational force between two masses. There is the added feature that while all masses are treated equally by the gravitational force, there are two kinds of charge. The electric force can be one of attraction or repulsion, depending on the sign of the charges. Apart from that, the gravitational force and the electric force, as well as the gravitational potential energy and the electric potential energy, are entirely analogous to one another. *The work done against the electric force is equal to the increase in the electric potential energy.*

To find the amount of electric potential energy that the negative charge gains as it is pushed away from the positive charge is not so easy. We can't just say that the work done is the electric force times the distance, because the force becomes smaller as the negative charge moves away. There is no single value for the force! It is still possible to calculate the change in potential energy, and we will do it later.

When we talked about gravitational potential energy we considered only situations at or near the surface of the earth. Because all of the points at the earth's surface are approximately at the same distance from the center of the earth we could take the gravitational force on an object to be constant. If the object moves to an earth satellite, that might no longer be a good approximation. And it certainly won't be if it goes to the moon.

EXAMPLE 6

A single electron ($M_e = 9.11 \times 10^{-31}$ kg) is released from rest near a large positively charged surface. It loses 6×10^{-18} J of electric potential energy after moving 0.2 m.

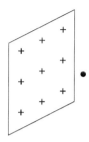

(a) Set up a mathematical description relating the various energies.

(b) Determine the electron's velocity after it has moved the 0.2 m.

Ans.:

(a)

$$P_1 + K_1 = P_2 + K_2$$
$$K_1 = 0, P_2 = 0$$
$$P_1 = K_2$$

(b) The only energies in this problem are the electric potential energy and the kinetic energy. (We will assume that there is no change in the gravitational potential energy, and that there are no forces other than the electric force.)

As the electron loses potential energy, it gains kinetic energy.

$P_1 + K_1 = P_2 + K_2$, where $K_1 = 0$. We can choose our reference level so that $P_2 = 0$. Then $P_1 = 6 \times 10^{-18}$ J, $P_1 = K_2$, and $K_2 = 6 \times 10^{-18}$ J.

$v_2 = \sqrt{\frac{2K_2}{M}} = \sqrt{\frac{(2)(6 \times 10^{-18})}{9 \times 10^{-31}}} = 3.63 \times 10^6$ m/s,

and it moves toward the plane.

Springs: elastic potential energy

Here is another potential energy. Suppose a spring is held stationary at one end and you pull on the other. The spring pulls on your hand in the opposite direction. *The work that you do against the elastic force of the spring is equal to the increase in the spring's elastic potential energy.*

If you let go, the potential energy that was stored in the spring is converted into kinetic energy. If we ignore air resistance, and also any friction in the spring, so that we leave only the elastic force that the spring exerts, the sum of the kinetic and potential energies will again be constant. As the potential energy decreases, the kinetic energy increases.

Like the electric force, the elastic force is not constant as the object on which it acts moves. As the spring is stretched, more and more force is required to stretch it further. We can therefore again not simply say that the work done by the elastic force is equal to the force times the distance.

Here is a spring that starts unstretched. We will use the subscript "1" for the energies there, and take the elastic potential energy P_1 to be zero. Since it is not moving, K_1 is also zero.

It is then stretched by the force, F, until it has a potential energy P_2, with K_2 still zero. When it is released, the potential energy is changed to kinetic energy. When it gets to the original starting point, the potential energy is again zero, and all of the potential energy has changed to kinetic energy (K_3).

$P_1 = 0 \quad K_2 = 0 \quad P_3 = 0$
$P_1 + W = P_2$
$K_2 + P_2 = K_3 + P_3$
$P_2 = K_3$

EXAMPLE 7

A ball whose mass is 0.02 kg is shot horizontally from a spring gun. Initially the spring is compressed and stores 10 J of energy.

(a) Set up a mathematical description that relates the energies.

(b) Determine how fast the ball is moving just after it has been fired from the gun.

Ans.:

Just after the ball leaves the gun, it has not yet moved vertically. Hence there is no change in the gravitational potential energy. We neglect any frictional losses. The only changes are then in the elastic potential energy (P), which is lost by the spring (10 J), and the kinetic energy, which is gained by the ball.

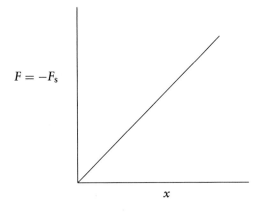

$K_1 + P_1 = K_2 + P_2$, $K_1 = 0$, $P_2 = 0$, and $P_1 = K_2$
$K_2 = 10 \text{ J} = \frac{1}{2}Mv_2^2$
$v_2 = \sqrt{\frac{(2)(10)}{0.02}} = 31.6 \text{ m/s}$

Hooke's law and the expression for the elastic potential energy

Experiments show that for most springs the force that a spring exerts is, to a good approximation, proportional to the displacement from the equilibrium position. This is so until the spring is stretched so much (beyond the "elastic limit") that it distorts permanently and no longer returns to its original shape.

For a force smaller than that of the elastic limit, the force exerted by the spring is $F_s = -kx$, where k is a proportionality constant that is characteristic for the particular spring, and is called the *spring constant*. The *minus* sign shows that the force that the spring exerts is in a direction opposite to the displacement from the equilibrium configuration. The relation $F_s = -kx$ is called *Hooke's law*.

How much work has to be done on the spring to stretch it from its equilibrium position ($x = 0$) to some new position, where the displacement is x?

First we will shift from the force $F_s(=-kx)$ that the spring exerts to the external force with which the spring is being pulled. We will call this force F. It is equal in magnitude and in a direction opposite to F_s, and so is equal to kx. The graph shows the straight line $F(x)$.

If we want to find the work that is done to stretch the spring, we can't just multiply "the force" by the displacement, as we did for constant forces, because now the force changes. In the next figure we have divided the area below the line into a number of vertical strips, six to start with. For each strip the force does not vary as much, and the work done by F is approximately equal to the area of the strip. For example, the area of the crosshatched strip is $(F_3)(x_4 - x_3)$.

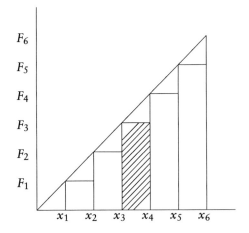

We can repeat this procedure for the other intervals. The work during all six intervals is then the area of all six strips. This is a little less, but close to the triangular area under the line.

If we had used more than six intervals, the area of the strips would be closer to the area under the line, and it becomes equal to it in the limit as the strips become narrower and their number becomes larger. We can see that the work done on the spring is, in fact, equal to the triangular area under the line, which is equal to $(\frac{1}{2})(kx)(x)$ or $\frac{1}{2}kx^2$. This is the work done *against* the spring's force, and is therefore equal to the elastic potential energy that the spring gains when it is stretched from $x = 0$ by the amount x.

The same method can be used for other forces that are not constant. On a graph of F against x, the work done by F is again equal to the area under the curve.

For the special case where F and x are related *linearly*, in other words, when the graph of F

against x is a straight line (as in the case of the spring), the average force is equal to half of the final force. For the stretching of the spring from $x = 0$ to x, this is $\frac{1}{2}kx$, and the work done is $(\frac{1}{2}kx)(x)$, or $\frac{1}{2}kx^2$, as before. But when the graph of F against x is not a straight line, the average force is not half of the final force. The work is then still the area under the curve of F against x, but it is no longer $\frac{1}{2}kx^2$.

Go to the PhET website (http://phet.colorado.edu) and open the simulation *Masses and Springs*.

Move the friction slider to the far left so that there is no friction. Hang one of the masses on one of the springs and watch it oscillate up and down. Click on "Show energy…" and watch the changes in the various energies: kinetic, gravitational potential, and elastic potential. Where is each energy at its maximum and at its minimum? Where is the kinetic energy zero? Note that you can change the reference level for the potential energy.

Explore the changes in the motion with the amount of mass and the softness of the spring. (Note that you can slow the motion down.)

As long as the force follows Hooke's law (and this is programmed into the simulation) the mass moves with simple harmonic motion, i.e., proportional to $x = \sin \omega t$. Here $\omega = \frac{2\pi}{T}$, where T, the *period*, is the time for one complete oscillation. (Add T to t to see that x comes back to the same value after the time T.)

Here is how we can see the relation between Hooke's law and simple harmonic motion. Write Hooke's law as $a = -\frac{k}{m}x$. Remember that a is the slope of the slope (the *second derivative*) of x. Is there a function that has the property that the slope of its slope is the negative of the function itself? Yes! It is the sine or the cosine or some combination of the two.

We can make the relation quantitative. Let $x = A\sin \omega t$, which is the same as $x = A\sin \frac{2\pi t}{T}$. The slope of the slope of this function can be shown to be $-\omega^2 x$. Comparing this relation to Hooke's law shows that $\frac{k}{m} = \omega^2$, which leads to $T = 2\pi \sqrt{\frac{m}{k}}$.

6.4 Friction and the loss of mechanical energy

In each of the three cases (gravitational potential energy, electrical potential energy, and elastic potential energy) the term *potential energy* implies that energy is stored, and that it can be retrieved and converted into an equivalent amount of kinetic energy. Each of the kinds of potential energy changes when the position changes. The gravitational potential energy changes when the rock or other object moves closer or farther from the earth. The electric potential energy changes when the charges move closer to each other or farther apart. The spring's elastic potential energy changes when it is stretched or unstretched. As long as there are no forces other than the gravitational force, the electric force, and the elastic force, the sum of the kinetic energy and the various potential energies is constant, and an object that comes back to a particular position will have the same kinetic energy there that it had at that position at an earlier time. A bouncing ball, for instance, will come back to the same height.

But wait a minute. That doesn't really happen! If I drop a ball, it won't bounce back to the same height. Each time it bounces, it will come back, but not to the same height as before. After some time it will stop bouncing and just sit there.

The ball loses potential energy on the way down, but the kinetic energy that it gains is not the same, because the force of air resistance slows it down. Air resistance causes some of the mechanical energy to be lost. Some mechanical energy is also lost when the ball is distorted when it comes in contact with the floor, and when some sound is produced as the ball bounces.

When we said that the bouncing ball will come back to the same height after it bounces, we assumed that we can neglect the effects of all forces on the ball except for the one that the earth exerts on it, i.e., its weight. For a real ball this is clearly not a good assumption. We can not simply talk about the ball as if it were a particle. Unlike a particle, it takes up space and encounters air resistance. It gets squeezed and deformed as it bounces. It has an internal constitution that we have to consider.

When the ball finally stops, some of its potential energy has been lost but there is no kinetic energy to take its place. The sum of the two, the mechanical energy, is no longer the same. Some of it has been transformed to another kind of energy. It has been *dissipated*. We will use the symbol Q for the mechanical energy that has been dissipated.

114 / Energy and Its Conservation

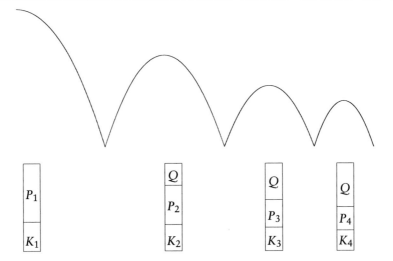

What spoils the earlier story is that there are forces that we have neglected. One of them is air resistance. It prevents us from getting all the potential energy back and keeping the mechanical energy constant.

Air resistance slows the ball down whether it is on the way up or on the way down. This is very different from what the weight does: the weight speeds the ball up on the way down and slows it down on the way up.

Some mechanical energy has been used up to do work against air resistance. You can't get this part of the mechanical energy back. It is gone. It has been *dissipated*. That's why air resistance is called a *dissipative* force.

EXAMPLE 8

A ball whose mass is 0.5 kg is dropped from rest at a height of 2 m above the floor. It bounces and comes back to a height of 1.5 m before dropping again. How much mechanical energy has been dissipated by the time it gets to 1.5 m after the first bounce?

Ans.:
$$P_1 = P_2 + Q$$
$$Q = P_2 - P_1 = Mg(y_2 - y_1) = Mg\Delta y$$

Use the floor as the reference level and measure P from there. At the starting point (point 1) $P_1 = Mgy_1 = (0.5)(9.8)(2) = 9.8$ J.

After the bounce, at 1.5 m (point 2), $P_2 = Mgy_2 = (0.5)(9.8)(1.5) = 7.35$ J.

The kinetic energy is zero at both points. Hence the mechanical energy that has been dissipated is $P_2 - P_1$, or 2.45 J.

The amount of mechanical energy $Mg\Delta y = (0.5)(9.8)(0.5)$ J = 2.45 J is lost to the system. It is *dissipated*.

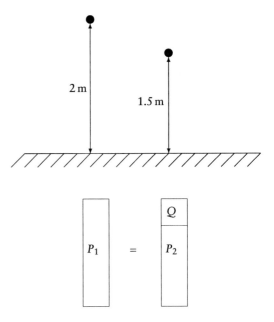

Push a rock along a table so that it moves with constant velocity. You apply a force. The rock moves. You do work. There is another force, and it works against you. That is the force

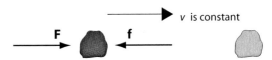

of friction. It works against you regardless of whether you pull or push. If you turn around and push the rock back, the force of friction also turns around, and still opposes the force with which you push the rock.

The work that you do does not increase the rock's energy. It gains neither kinetic energy nor potential energy. Both P and K remain constant.

$$\boxed{\begin{array}{c} W \\ \hline K_1 \end{array}} = \boxed{\begin{array}{c} Q \\ \hline K_2 \end{array}}$$

If there were no friction, the block would accelerate. The work done on it would result in increased kinetic energy. But for a large-scale, macroscopic object, one that you can see, there is always some friction. The increase in the rock's kinetic energy is always less than the work that you do as you push the block.

Go to the PhET website and open the simulation *Masses and Springs*.

Let there be some friction, and explore the various energies, this time including the dissipated energy, here called *thermal energy*.

EXAMPLE 9

You push a block 0.8 m along a horizontal table with a force of 2 N. The block moves with constant velocity.

(a) Set up a mathematical description that relates the energies.

(b) Determine how much mechanical energy is dissipated.

Ans.:
(a) $K_1 = K_2$, $P_1 = P_2$, and $W = Q$.

(b) $W = Fs = (2\,\text{N})(0.8\,\text{m}) = 1.6\,\text{J}$.
This is the amount of mechanical energy that is dissipated.

$$\boxed{\begin{array}{c} W \\ \hline K_1 \end{array}} = \boxed{\begin{array}{c} Q \\ \hline K_2 \end{array}}$$

In the real, large-scale ("macroscopic") world there are always friction, air resistance, or other dissipative forces. The ball does not keep bouncing. The mechanical energy does not stay constant.

It was a revolutionary discovery when it was realized that when the mechanical energy disappears, something else appears, namely an equivalent amount of another kind of energy, which we have not yet considered.

As the block slides along the table, its temperature increases. That's a sign that its *internal energy* increases. Some of the energy is transferred to the table and the air, and their temperature increases, showing that their internal energy also increases. Mechanical energy is lost, but internal energy is gained. The *total energy* remains constant. It is *conserved*.

There are two great insights that led to our understanding of the process that accompanies the decrease of mechanical energy. The first is that it was shown experimentally that a definite amount of internal energy is produced each time a definite amount of mechanical energy is lost.

The second is that on the *microscopic* level, that of atoms and molecules, mechanical energy doesn't disappear at all. What we observe as heating and an increase in internal energy on the macroscopic or large-scale level is, in fact, simple to understand on the microscopic level. The microscopic particles, the atoms and molecules that make up a macroscopic object, are all in motion, randomly, and at all times, each with its own kinetic and potential energy. A change in the total internal energy is actually a change in the kinetic and potential energies of these particles.

In a gas the atoms or molecules move freely, and the internal energy is primarily kinetic energy. In a solid they vibrate about their equilibrium positions, and the internal energy is kinetic and potential in roughly equal parts.

We use the term *internal energy* (U) as the macroscopic quantity that represents the sum of the microscopic kinetic and potential energies of the particles that make up the macroscopic object. One way in which the internal energy can change is when the particles speed up, as they do when the temperature increases. It can also change in other ways, for example, when it is squeezed, or when there is a *phase change* such as that from ice to water. In these two cases the microscopic potential energy changes.

116 / Energy and Its Conservation

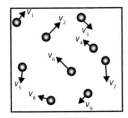 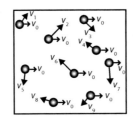

The internal kinetic energy is that of motion of the atoms and molecules in all posssible directions, randomly. This is very different from the motion of an object as a whole, where the atoms move together, all in the same direction. The schematic diagram on the left shows a number of molecules moving randomly. The diagram on the right shows an additional velocity, v_0, for each molecule, representing a movement of the whole piece of material with this velocity. (In reality the velocities of the individual molecules are generally much larger than the velocity of the whole piece.)

We use the word *heat* or *heat energy* only for energy that is transferred from one object or system to another. A material does not *contain* a certain amount of heat, as if heat were a kind of substance. Heating is a *process* by which energy can be transferred from one piece of material to another.

In this sense heat is similar to work. Work is not *contained* in an object. It also represents a process by which energy is transferred from one object to another. There is, however, a profound difference between work and heat. Work can increase mechanical energy, i.e., potential energy and kinetic energy, and it can also change the internal energy.

On the other hand, when an object is heated, its internal energy increases, but there are severe limits on how much of the energy can be changed to mechanical energy. We will come back to this distinction later when we discuss *the second law of thermodynamics*.

EXAMPLE 10

The typical random speed of helium atoms at room temperature is about $v_a = 1300$ m/s.

(a) What is the ratio of the internal Kinetic energy, U, of a mole of helium at room temperature to the kinetic energy, K, of the same amount of gas in a balloon moving up at the rate of $v_b = 1.3$ m/s?

Ans.:
For N atoms $U = (N)(\frac{1}{2}mv_a^2)$ and $K = (N)(\frac{1}{2}mv_b^2)$.
The ratio $\frac{U}{K}$ is equal to $(\frac{v_a}{v_b})^2 = 10^6$.

The internal energy of the gas is about a million times as large as the kinetic energy of the balloon!

6.5 Internal energy and the law of conservation of energy

At first it doesn't seem unreasonable to think of heat as some kind of substance, a fluid that can flow between objects. After all, when a hot body and a cold body are brought together, we say colloquially that heat "flows" from one to the other, until they are at the same temperature. It used to be thought that the heat generated by rubbing two bodies together somehow resulted from the squeezing out of this substance.

The crucial experiments that showed that no such substance exists were done by Benjamin Thompson (1753–1814), one of the strangest figures in the history of science. He was born in Massachusetts, but became a spy for the British, and then fled to England. He was knighted by King George III, and later was in the employ of the Elector of Bavaria, who made him a general and a count of the Holy Roman Empire. He became known as Count Rumford, choosing the name of a village near where he was born.

Benjamin Thompson, Count Rumford.
Courtesy of MIT Press.

In 1798 he was in charge of the making of cannons. The drilling process is accompanied by a very large amount of heating. He first tried to measure the weight of the elusive substance, (termed "caloric") that was supposed to represent heat, and did not find any. Even

6.5 Internal energy and the law of conservation of energy

more interestingly, he showed that there seemed to be no limit to the amount of heating that could be produced. He concluded that heat had to be something quite different from a material substance.

Detailed quantitative experiments were later made by James Prescott Joule (1818–1889), who showed that for a given amount of work done, a definite amount of internal energy (often colloquially called heat) is generated. He stirred water in a container with a paddle wheel. The work was done and measured by a weight attached to the paddle wheel. The internal energy that was produced was determined by measuring the amount of water and the rise of its temperature. Look for "Joule apparatus" on the internet.

James Prescott Joule.

The work done is equal to Mgy, where Mg is the weight and y is the vertical distance through which it moves. In our modern SI units, with M in kg, Mg in newtons, and y in meters, the work is measured in units that we now call *joules*. Heat units were developed independently, at a time when their connection to energy was unknown. The *calorie* is the amount of energy that raises the temperature of 1 gram of water by one degree Celsius. Joule's experiments, in 1847 and later, led to the number of joules equivalent to one calorie, close to the modern value of 4.186 J/cal.

A quite different experiment had already been done in 1807 by Joseph Louis Gay-Lussac. He showed that a gas that was allowed to expand into a container while pushing a piston would cool, while no such temperature change occurred if the gas did not do any work during the expansion.

These experiments were crucial for the realization that mechanical energy and internal energy represent different forms of the same kind of quantity and can be transformed from one form to the other. They led to the establishment and acceptance of the *law of conservation of energy*.

In its most succinct form the law says that *in an isolated system the total energy is constant*. In other words, different kinds of energy can be converted from one to the other, but without any change, by creation or disappearance, of the total amount of energy.

Alternatively, we can count the amount of energy input to a system and say that the system's energy is increased by the net work done on it, and, in addition, by the net amount of heating that is done on it.

Initially the law of conservation of energy applied only to heat and mechanical energy. Other forms of energy were introduced later, especially electromagnetic energy, and the law was generalized to include them. In 1905, as the result of the special theory of relativity, it was seen that the mass of an object could also be transformed to energy, and that the law of conservation of energy needed to be generalized to include this fact. With the inclusion of all forms of energy, as well as mass, it continues to be a cornerstone of modern science.

EXAMPLE 11

Go to the website phet.colorado.edu and open the simulation *Energy Skate Park*. Check "potential energy reference." Drag the reference level to the lowest point of the skater's path. Slow down the motion as much as possible by using the slider at the bottom. Click on "bar graph." Click on "track friction." Put the slider marked "coefficient of friction" at a point two spaces from the left end.

"Pause," put the skater at a point near the top of the track, and "Resume."

What happens to K, P, to the internal (= thermal) energy, U, and to the total energy (E)? How is the mechanical energy related to the total energy at the start? What is it when the skater has come to rest?

Ans.:

The total energy, E, is still constant. The mechanical energy is still $P + K$, but it is no longer equal to the total energy because it is gradually being transformed to thermal energy, i.e., to the internal energy of the skater, the track, and the air. At the start the mechanical energy is the total energy. At the end it is zero and all of it has been transformed to thermal energy.

EXAMPLE 12

How much potential energy does a person whose mass is 75 kg gain as he or she climbs the Empire State Building ($h = 443.2$ m)? How many food calories are equivalent to this amount?

Ans.:
$\Delta P = Mgh = (75)(9.8)(443.2) = 3.26 \times 10^5$ J

1 cal = 4.186 J, and so this is equivalent to 7.79×10^4 cal. However, a food calorie, as listed on a cereal box, is equal to 1000 cal or a kilocalorie. We will write it as a Calorie with a capital C, i.e., 1 kcal = 1 Cal. The answer is therefore 77.9 kcal or 77.9 Cal.

This would be the number of food calories used up in climbing to the top of the Empire State Building if the process were 100% efficient. The actual efficiency of the human body ranges approximately from 3% to 20%, and so you would need at least 400 Cal.

The human body is a complicated engine. The food we eat combines with the oxygen we breathe to liberate the energy that we use. This energy maintains the body's temperature and is, in part, converted to mechanical energy when we move. Excess food leads to stored material and we gain weight. Excess exercise draws on material that contains stored energy.

The quiescent body, as it lies motionless or sleeping, uses energy at a rate called the basal metabolic rate (BMR). The BMR can be measured, for example, by measuring the amount of oxygen that is breathed in and the amount of carbon dioxide that is breathed out in a given time.

You can estimate your BMR crudely by multiplying your body weight in pounds by 10 to get the BMR in kcal/day. A better approximation is the Harris–Benedict equation, which comes in two versions: one for males: $66 + 13.7W + 5H - 6.8A$, and one for females: $655 + 9.6W + 1.85H - 4.7A$, where W is the mass in kg, H is the height in cm, and A is the age in years.

Estimates of the energy used in a variety of activities can be found in many publications. (For example, *Exercise and Weight Control*, The President's Council on Physical Fitness and Sports, http://www.fitness.gov, *Essentials of Exercise Physiology*, W. D. McArdle, F. I. Katch, and V. L. Katch, Lea and Febiger, Philadelphia, 1994). The numbers are often given to three significant figures, but they depend on many factors, both external and internal to the body, and are necessarily quite approximate.

To a reasonable approximation, away from rest and from extremes of stress, the energy used in level walking, running, swimming, and bicycling is constant for a given distance. (At twice the speed you then use the same energy in half the time.) The values are proportional to the weight. (If you weigh 10% less, you use 10% less energy as you move.)

Here are numbers for the energy in kcal used per mile, averaged from several sources: (Remember that these are rough guides, good to perhaps ±15%.)

walking: 85
running: 110
swimming: 400
bicycling: 40

You can see that it takes quite a while to use up the number of Calories represented by a candy bar (150 to 250 and more), or by a glass of orange juice (100).

6.6 Work and energy revisited

Let's look at our definition of work again to see how it can be used when the force and the displacement are not in the same direction. In that case we use only the part of the force that is in the direction of the displacement. This is the only part of the force that does work. We call that part the *component* of the force in the direction of the displacement. If the angle between the force (F) and the displacement (x) is θ, that component

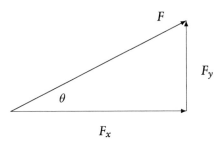

(the x component) is $F \cos \theta$, and the work done by the force is $Fx \cos \theta$.

The figure shows the force, **F**, and its x component, F_x. It also shows the y component, F_y. Since $\cos\theta = F_x/F$ and $\sin\theta = F_y/F$, the x component is $F\cos\theta$ and the y component is $F\sin\theta$.

If F and x are at right angles, so that $\theta = 90°$ and $\cos\theta = 0$, the component of the force in the direction of x is zero, and the force does no work.

EXAMPLE 13

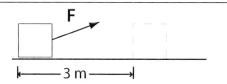

You move a block across a horizontal frictionless surface by pulling with a 4-N force on a rope that is attached to it at an angle of 30° with the horizontal. How much work have you done when the block has moved 3 m?

Ans.:
Work is done only by the component of the force that is in the direction of the motion. This component, F_x, is $(4)(\cos 30°)$ N, or 3.46 N. $W = F_x s = (3.46)(3) = 10.39$ Nm $= 10.39$ J.

EXAMPLE 14

The moon travels around the earth at a steady speed in an orbit that is close to being circular. Use the approximation that the path is exactly circular and neglect the influence of the sun and of any other astronomical objects.

What is the nature of the forces that act on the moon? Give an expression for its acceleration. Is any work done on it? How do its kinetic and potential energies change?

Ans.:
The only force on the moon is the gravitational attraction by the earth, F_{em}, and it is at right angles to the path. No work is done, the kinetic and potential energies each remain constant, and the moon's motion continues. The dissipative forces are so minute that the assumption of constant mechanical energy is extremely close.

If the moon moves in a circle, there must be an acceleration toward the center (the centripetal acceleration) and it has to be equal to $\frac{v^2}{r}$. If it were bigger or smaller, the moon would not continue to move in a circular path. The net force that gives rise to the acceleration is toward the center, and its magnitude must be equal to $\frac{Mv^2}{r}$. This is not an additional force. The gravitational force is the only force.

EXAMPLE 15

A pendulum consists of a string suspended at one end, with a ball swinging back and forth at the other end. It starts from rest at point 1, where it has only potential energy, P_1. Measure the potential energy from the level of its lowest position (point 2), so that $P_2 = 0$, and it has only kinetic energy there.

Describe the forces on the ball, the work that they do, and the energy changes as the pendulum moves back and forth. Make a bar chart at different points along the path of the pendulum.

Ans.:
There are two forces on the ball. One is the force along the string, called the tension. The motion of the ball is in a circle, with the string as its radius. The circular path is at right angles to the radius. Since the tension is along the radius, it is at right angles to the path, and so does no work.

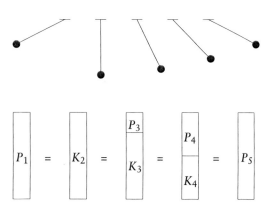

The second force is the weight. It acts straight down. It is not at right angles to the path (except at one point—do you see where?).

Consider the system of the earth and the pendulum. As the pendulum swings downward, the kinetic energy increases and the potential energy decreases. The pendulum then swings upward, the potential energy increases, and the kinetic energy decreases until it becomes zero. The pendulum comes back, and continues to swing back and forth.

There is also air resistance, and friction at the place where the string is suspended. These are the dissipative forces. If we neglect them there are no dissipative forces and the mechanical energy (the sum of the kinetic and potential energies) remains constant. The pendulum comes back to the same height, with the same potential energy at the top of each swing, and back to the same speed, and the same kinetic energy at the bottom of each swing. It continues forever.

Of course this is not what happens for any real pendulum. The dissipative forces of air resistance and friction cause the mechanical energy gradually to diminish, and the pendulum eventually comes to rest.

If we neglect friction and air resistance, the mechanical energy remains the same. It is *conserved*.

Go to the PhET website (http://phet.colorado.edu) and open the simulation *Pendulum Lab*.

Put the friction slider to the left so that there is no friction. Check to show velocity and acceleration. Set the pendulum into oscillation and watch the two vectors. Slow the motion down to see what happens more clearly. Where is the velocity zero? What is the angle between it and the string of the pendulum? Where is the acceleration tangential to the motion? Where is it perpendicular to the motion and along the pendulum string? Follow what happens to the component of the acceleration parallel to the motion and perpendicular to it.

Click on "Show energy." Follow the kinetic energy, the potential energy, and the total energy. Where is each at its maximum and where are they zero? (Don't let the energy get so large that the bar graph no longer follows it.)

Explore the effect of changing the length and the mass of the pendulum, adding friction, and going to the moon and to other planets.

EXAMPLE 16

A pendulum starts from rest at the height y_1, 30 cm above its lowest point, where the height is y_2. What is the maximum speed of the pendulum bob as it swings?

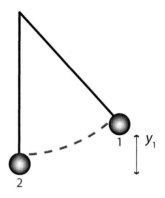

Ans.:

The maximum speed is reached at the lowest point. We will assume that the only forces that we need to consider are the tension of the string and the weight of the pendulum bob. Since the tension is at all times perpendicular to the path of the pendulum bob, it does no work and does not change the mechanical energy. In the system that contains the pendulum bob and the earth, the mechanical energy is in part the kinetic energy and in part the potential energy of the pendulum bob.

Take the reference level at the lowest point, so that $y_2 = 0$. The potential energy at that point is then $P_2 = 0$ and $P_1 = Mgy_1$, where $y_1 = 0.3$ m. Since the pendulum is at rest at point 1, $K_1 = 0$.

Mechanical energy is conserved, so that $P_1 + K_1 = P_2 + K_2$, which here reduces to $P_1 = K_2$, or $Mgy_1 = \frac{1}{2}Mv_2^2$ or $v_2 = \sqrt{2gy_1} = 2.42$ m/s.

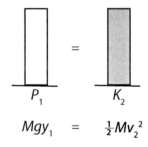

EXAMPLE 17

A baseball moves freely through the air, in *projectile motion*. Describe the motion, the forces on the ball, and the changes in energy as it flies through the air from the time just after it leaves the bat to the time just before it returns to the ground. Make a bar chart.

Neglect air resistance. Draw graphs that show the horizontal and vertical components of the velocity at various points along the path of the ball. Also draw a graph that shows the potential, kinetic, and mechanical energies as a function of the horizontal distance along the path.

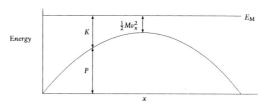

Ans.:

Once the ball has left the bat the only force on it is the force exerted on it by the earth, i.e., its weight, vertically downward. There is also air resistance, but we will neglect it. This may or may not be a good approximation, but if it is, the only force that does work is the weight, and mechanical energy is conserved.

As long as we neglect all dissipative forces, and the only force is the weight, there is no horizontal force and no horizontal acceleration, and so the horizontal velocity, v_x, remains constant.

Taking "up" as positive, the vertical acceleration is $-g$, and $v_y^2 = v_{y0}^2 - 2gy$.

Take the reference level at the starting point (point 1), so that $P_1 = 0$. $K_1 = \frac{1}{2}Mv_0^2 = \frac{1}{2}M(v_x^2 + v_{y0}^2)$. These are also the values when the projectile returns to the same height at point 4.

At the highest point (point 3) $v_y = 0$ and $v = v_x$, $K_3 = \frac{1}{2}Mv_x^2$, and $P_2 = Mgy_2$. At intermediate points there are intermediate values for v_y, K, and P.

The figure shows graphs of P, K, and the mechanical energy, E_M, which is equal to $P + K$:

Since $P = Mgy$, a graph of P against x will look just like the graph of y against x. As long as we neglect all dissipative forces, E_M remains constant. K is the difference between E_M and P.

EXAMPLE 18

A stone with mass 0.2 kg is thrown at an angle to the horizontal, with an initial velocity whose horizontal component is 5 m/s and whose vertical component is 3 m/s.

Neglect all dissipative forces. Use the starting point as a reference level for the height and the potential energy.

(a) What is the initial kinetic energy, K_1?

(b) What is the kinetic energy, K_4, at the point where the rock comes back to the height where it started?

(c) What is the kinetic energy, K_3, at the highest point?

(d) What is the potential energy, P_3, at the highest point?

(e) At point 2 the rock has reached half of its maximum height. What are the kinetic and potential energies at this point?

(f) What are the height and the horizontal and vertical components of the velocity at the half-way point?

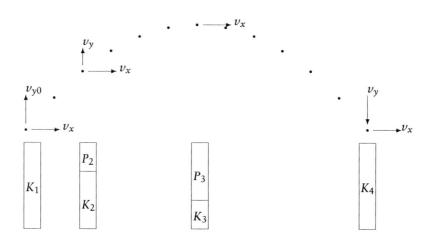

122 / Energy and Its Conservation

Ans.:

(a) The initial velocity v_0, has an x component v_x and a y component v_{y0}. The magnitude v_0 is $\sqrt{v_x^2 + v_{y0}^2}$ and the initial kinetic energy is $\frac{1}{2}Mv_0^2$ or $\frac{1}{2}M(v_x^2 + v_{y0}^2)$. This is equal to $(\frac{1}{2})(0.2)(25 + 9) = 3.4$ J.

(b) Since we are neglecting all dissipative forces, the mechanical energy is conserved. The height at the starting and ending points is the same, and so is the potential energy. The kinetic energy is therefore again 3.4 J.

(c) At the highest point the vertical component of the velocity is zero. In the absence of any horizontal forces the acceleration has no horizontal component, and the horizontal component of the velocity remains constant, at 5 m/s. The kinetic energy at that point is therefore $(\frac{1}{2})(0.2)(25) = 2.5$ J.

(d) Let the potential energy be zero at the starting point. $P_1 = 0, K_1 = 3.4$ J, so that the mechanical energy is 3.4 J. In the absence of dissipative forces the mechanical energy is conserved and remains constant at this value. The potential energy at the highest point is therefore $3.4 - 2.5$, or 0.9 J.

(e) At a point half-way to the highest point, the height and the potential energy are half of what they are at the highest point. $P_2 = 0.45$ J. The mechanical energy is still 3.4 J, so that the kinetic energy at this point is 2.95 J.

(f) $P_2 = Mgy_2$, so that $y_2 = \frac{P_2}{Mg} = \frac{0.45}{(0.2)(9.8)} = 0.23$ m.
v_x remains constant at 5 m/s.
$K_2 = 2.95$ J $= (\frac{1}{2})(0.2)(5^2 + v_y^2)$, so that $5^2 + v_y^2 = 29.5$. Hence $v_y^2 = 4.5$, and $v_y = 2.12$ m/s.

EXAMPLE 19

A cart rolls along a track as in the figure, starting from rest at point 1 at the top. Neglect friction and other dissipative forces and neglect any internal motion in the cart. There is no engine, and we will neglect any complication from the fact that the wheels are turning and therefore have *rotational kinetic energy*.

Draw energy bar charts and write mathematical descriptions for the total energy at point 1, point 2 in the middle, and point 3 at the end.

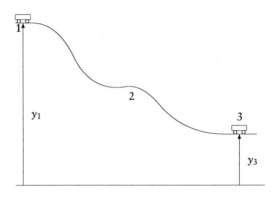

Ans.:
We can use the system containing the cart and the earth. The force of the track on the cart consists of two components: one is the component perpendicular (or normal) to the track, which does no work. The other is the component parallel to the track. This is the force of friction. We are assuming it to be zero.

With our assumptions the sum of the potential energy (P) and the kinetic energy (K) (of the system containing the cart, the track, and the earth) remains constant. We can look at any two points along the path of the cart, characterized by subscripts 1 and 2 at a difference in height $y_1 - y_2$, and write $P_1 + K_1 = P_2 + K_2$, just as in the previous examples.

$$\boxed{\begin{array}{c} P_1 \\ K_1 \end{array}} = \boxed{\begin{array}{c} P_2 \\ K_2 \end{array}}$$

Because $K_1 = \frac{1}{2}Mv_1^2, K_2 = \frac{1}{2}Mv_2^2$, and $P_1 - P_2 = Mg(y_1 - y_2)$, we can also write

$$Mgy_1 + \frac{1}{2}Mv_1^2 = Mgy_2 + \frac{1}{2}Mv_2^2$$

There are several interesting aspects of this relation. First, the mass, M, appears in each term, and can be cancelled. This shows that the relation between the heights and the speeds is the same, regardless of the value of the mass, just as was true for the pendulum.

Second, the vertical distance appears only as $y_1 - y_2$, the difference in the height of the two points. As before, changing the reference level changes y_1 and y_2, P_1 and P_2, as well as the mechanical energy at both points, but the differences, $y_2 - y_1$ and $P_2 - P_1$, remain the same. The mechanical energy is different, but the kinetic energies and the speeds at any point remain the same. That's why we say that it is only differences in the potential energies that matter, and not their values.

Finally, the path of the car as it moves between the two points does not affect the result. As long as we stay within our self-imposed limitation that the only force that does work is the gravitational force, we don't have to know anything about what happens between points 1 and 2. Whatever the path, the sum of the kinetic and potential energies is the same at the two points. Whenever the cart gets back to the same height (y_1, y_2, or any other), its potential energy and therefore also its kinetic energy will be the same as when it was at that height earlier.

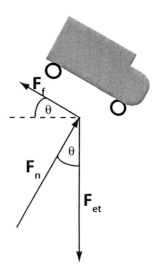

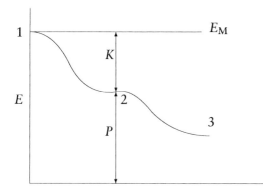

The total force on the cart includes the force that the road exerts on the car. Its component parallel to the road is the force of friction, which we are neglecting. Its component perpendicular to the road does no work and therefore we don't have to know it. All we need to know is that the mechanical energy at point 1 is the same as the mechanical energy at point 2.

A graph of P against x again looks like the graph of y against x. The mechanical energy, E_M, is constant, and $K = E_M - P$.

The figure shows the force diagram of the cart, with the friction that we have been neglecting, but will include in Example 21.

EXAMPLE 20

The cart, on the same hill as in the previous example, starts from rest at a height of 20 m above the ground. (Take the ground as the reference level.) What will be its speed and kinetic energy when it is at a height of 5 m? What is the total mechanical energy? For the kinetic energy we need to know its mass: it is 50 kg.

Ans.:

We have $y_1 = 20$ m, $y_2 = 5$ m, and $v_1 = 0$. We are trying to find v_2. We start with $P_1 + K_1 = P_2 + K_2$, where $P_1 = Mgy_1$ and $K_1 = \frac{1}{2}mv_1^2$ at point 1, with similar relations at point 2. M is in each term and can be cancelled. The rest can be rearranged to give $v_2^2 = v_1^2 + 2g(y_1 - y_2)$, or, since $v_1 = 0$, $v_2^2 = 2g(y_1 - y_2)$. All the quantities are already in SI units, and we know that if we put all of them in these units the result will also be in the same system of units, so that $v_2^2 = (2)(9.8)(15) = 294$ and $v_2 = 17.15$ m/s. If we assume that we know all quantities to three significant figures, the result should also have three significant figure, and we round it to 17.2 m/s.

The corresponding kinetic energy is $(\frac{1}{2})(50)(17.2)^2 = (\frac{1}{2})(50)(296) = 7400$ J.

[There is a small problem. We saw earlier that $v_2^2 = 294$ SI units. By rounding 17.15 to 17.2 we have changed its square from 294 to 296! As you can see, it is best to keep an extra digit, and decide only at the end how many figures are significant and should be kept. Here the better value for the kinetic energy is $(\frac{1}{2})(50)(294)$, or 7350 J.]

What is the total mechanical energy? We don't have to specify when, since it is constant. At point 1 there is only potential energy. As long as we continue to assume the reference level to be at the bottom of the track, it is Mgy_1, or $(50)(9.8)(20)$, or $9800\,\mathrm{J}$. At point 2 the potential energy is $2450\,\mathrm{J}$, and the kinetic energy is $7350\,\mathrm{J}$, adding up to the same $9800\,\mathrm{J}$.

EXAMPLE 21

Now let's change the problem so that it is more realistic. Let's say that the speed at point 2, at a height of 5 m, is only 10 m/s, because some mechanical energy has been lost as a result of dissipative forces, like friction and air resistance. How much mechanical energy has been lost, and what happened to it?

Ans.:
There is no doubt about what happened to it. It has been transferred as heat, and changed to internal energy. This internal energy is partly in the road, partly in the car, and partly in the air. This time we include all of these as part of the system, and also the earth. Let's see how much the internal energy has increased in these various places. K_2 is now $(\frac{1}{2})(50)(100) = 2500\,\mathrm{J}$. P_2 is, as before, Mgy_2 or $2450\,\mathrm{J}$ for a mechanical energy of $4950\,\mathrm{J}$. Since the mechanical energy started out at point 1 ($h = 20\,\mathrm{m}$) as $9800\,\mathrm{J}$, $4850\,\mathrm{J}$ of mechanical energy has been dissipated and changed to internal energy.

EXAMPLE 22

You are an engineer investigating the rollercoaster shown on the figure, and wish to determine the spring constant of the launching mechanism shown there. You may assume that the track is frictionless except for the rough patch, where the force of friction on a cart is 0.2 times the weight of the cart. (The coefficient of sliding friction is 0.2.)

You have a scale and a tape measure.
Describe a procedure to determine the spring constant, k. List the quantities that you need to measure and describe how you will measure them.

Ans.:
Choose the system that includes the track, cart, spring, air, and the earth.

Energy is conserved: $K + P_g + P_s + U$ is constant, where P_g is the gravitational potential energy, P_s is the elastic potential energy of the spring, and U is the internal energy. Let $P_g = 0$ for the initial position of the cart, let $P_s = 0$ when the spring is not compressed, and let the height of the cart on the rough patch be h_1 above the beginning position.

Measure the mass, M, of the cart with the scale. Push the cart against the spring and measure the distance, x, by which the spring is compressed with the tape measure. Let go. The cart moves either (a) up the track to a maximum height, h, and returns, or (b) it moves up the track, over the hump, and comes to rest after moving a distance L along the rough patch. Measure h for case (a) or L for case (b).

(a) At the initial position $P_{g1} = 0, P_{s1} = \frac{1}{2}kx^2, K_1 = 0$; At the final position $P_{g2} = Mgh, P_{s2} = 0, K_2 = 0, U_1 = U_2$.

Hence $P_{s1} = P_{g2}$, so that $\frac{1}{2}kx^2 = Mgh$. Do this for several values of x and h, and solve for k in terms of the measured quantities.

$$\boxed{P_{s1}} = \boxed{P_{g2}}$$

(b) The initial position is same as in part (a). At the final position is $P_{g2} = Mgh, P_{s2} = 0$. The energy transformed to internal energy is $2MgL$.

Hence $P_{s1} = P_{g2} + 0.2MgL$, or $\frac{1}{2}kx^2 = Mgh_1 + 0.2MgL$. Solve for k in terms of the measured quantities.

$$\boxed{P_{s1}} = \boxed{\begin{array}{c} P_{g2} \\ \Delta U \end{array}}$$

6.7 Power: not what the power company sells

Power is another word that is loosely used in everyday language but has a precise meaning in science, and not necessarily what you might have expected. We talk about the electric company as the *power company*, and many of them have the word *power* in their name. But what they sell is energy.

We use the word *power* to refer to something different, namely the rate at which energy is transferred, i.e., transformed from one kind to another. It is the quantity that tells us how much energy is transformed per second from electric potential energy in a lightbulb, by a household per month, or by the whole country per year. It is the energy divided by the time during which the transfer takes place.

The SI unit for energy is the joule. The unit for power is the *joule/second*. It is given its own name, the watt (W): 1 J/s = 1 W.

While it is turned on, a 40-W bulb uses energy at the rate of approximately 40 W, or 40 J in each second. In an hour it uses (40 J/s)(3600 s) or 144,000 J. A device that uses energy at the rate of 1000 W, or one kilowatt (kW), will consume (1 kW)(1 h) or one kilowatt-hour of energy in one hour. That's equal to (1000 W)(3600 s) or 3.6×10^6 J, and that's what you pay the electric company for.

EXAMPLE 23

As an example, look at a person whose mass is 75 kg, running up a flight of stairs, through a vertical height of 15 m in 12 s. What is the change in the potential energy, and what is the corresponding power?

Ans.:
The change in P (ΔP), is $Mg\Delta y$, or $(75)(9.8)(15)$ J, or 11,025 J. The power is $\Delta P/t$, or $(\frac{11,025}{12})$ W, or 919 W.

6.8 Summary

The law of conservation of energy is very simple: *energy is conserved*, i.e., the energy of a closed system remains constant. If the system is not closed, i.e., if energy enters the system or leaves it, you have to count the amount of the change of energy. The energy of the system increases by the net amount that is transferred to the system. You have to know what the boundaries of your system are. You have to make sure that you count all the energy that is transferred in or out. Apart from obvious points that's it for one of the most important principles of science.

There are various kinds of energy:

Kinetic energy or energy of motion: $\frac{1}{2}Mv^2$; this is the difference in energy that an object or system has compared to what it has when it is at rest.

Potential energy is energy of position. There are several kinds. The *gravitational potential energy* changes by the amount of work that is done against the gravitational force.

The *electric potential energy* changes by the amount of work that is done against the electric force. The *elastic potential energy* changes by the amount of work that is done against the elastic force (like the force of a spring). A spring that follows Hooke's law ($F_s = -kx$) has an elastic potential energy $\frac{1}{2}kx^2$ when it is stretched by a distance x from its equilibrium position.

The gravitational potential energy increases by Mgy when a mass M is moved in the direction opposite to the gravitational force through a distance y. The electric potential energy increases when a charge Q is moved in the direction opposite to the electric force.

An object has various kinds of internal energy. It is the kinetic and potential energy of the object's constituents. The atoms and molecules are in motion and have electric potential energies with respect to each other. The sum of these microscopic energies is called *thermal energy*. In addition, the electrons and nuclei of each atom have kinetic and mutual potential energies. So do the nucleons in the nucleus.

Energy can be transferred by doing work and by heating.

Power is the rate at which work is done or energy changes. The SI unit of energy is the joule (J); the SI unit of power is the watt (W): 1 W = 1 J/s.

6.9 Review activities and problems

Guided review

1. A box is pushed horizontally by a force of 4 N. A frictional force of 1 N acts in the opposite direction. The box moves forward through 2 m.
   (a) What is the work done by the 4 N force?
   (b) What is the total work done on the box?
   (c) What is the work done by the frictional force?

2. A box whose mass is 2 kg starts from rest, is pushed to the right by a force of 4 N, and moves through 3 m. There are no other horizontal forces.
   (a) Set up a mathematical description that relates work and energy.
   (b) What is the speed of the box after it has moved through the 3 m? (Use work and energy relations.)
   (c) Repeat parts (a) and (b), but with the box initially moving with a speed of 3 m/s in the direction opposite to that of the force.

3. You do 60 J of work lifting a box 2 m. The box is at rest initially and again after it has been lifted.
   (a) Set up a mathematical description that relates work and energy.
   (b) How much energy does the box gain? What kind of energy does the box gain?
   (c) Determine the mass of the box.

4. A stone has a mass of 0.5 kg and is lifted through 1.5 m without gaining any kinetic energy. Answer the following questions:
   (a) How much potential energy does it gain?
   (b) It is then dropped from rest. How much kinetic energy does it gain on the way down?
   (c) What is its speed when it gets back to the starting point?

5. Go to the website phet.colorado.edu and open the simulation *Energy Skate Park*. Put the reference level at the lowest point on the path of the skater. Slow down the motion as much as possible. Open the bar graph.
   (a) Click on "energy vs. position." Check "kinetic" and uncheck the others. Use the results of the example to calculate the maximum speed of the skater.
   (b) At what height, as a fraction of the maximum, is K equal to half the total energy? At what height, as a fraction of the maximum, is the speed half of the maximum speed?
   (c) Pause. Change the track by dragging on the blue circles. Experiment with the following:
   (i) Make the track much steeper on the left than on the right. What remains constant about the points where K is zero?
   (ii) Lower the right end of the track below the height of the starting point. What happens? Why?
   (iii) Make a hump in the middle of the track. What does it take for the skater to overcome this obstacle?

6. A single proton ($M_p = 1.66 \times 10^{-27}$ kg) is released from rest near a large positively charged surface. It loses 4×10^{-18} J of electric potential energy after moving 0.1 m.
   (a) Set up the mathematical description that relates the energies.
   (b) Determine the proton's velocity after it has moved the 0.1 m.

7. A rock whose mass is 0.5 kg is shot horizontally from a spring gun. The spring of the spring gun is initially compressed with an elastic energy of 11 J.
   Use energy bar charts to set up the problem, and determine how fast the rock is moving just after it leaves the spring gun.

8. A ball whose mass is 0.2 kg is dropped from rest at a height of 1.5 m above the floor. It bounces and comes back to a height of 1.2 m. How much mechanical energy is lost? What happened to it?

9. A force of 10 N acts on a box, which moves horizontally with constant velocity through 2 m. What is the total work done on the box? Make a force diagram, showing all forces on the box, an energy bar chart description, and a mathematical description that relates work and energy.

10. For two moles of oxygen at room temperature the internal kinetic energy is about 7500 J. Each mole consists of 6×10^{23} molecules, each of which has a mass of 2.7×10^{-26} kg.

(a) What is the typical value of the random speed of an oxygen molecule?

(b) How long would it take the molecules to move 20 m across a room with this speed?

(c) Why is this not likely to be the actual time that it takes a molecule to get from one side of the room to the other?

11. Go to the website phet.colorado.edu and open the simulation *Energy Skate Park*. Check "potential energy reference" and drag the reference level to the lowest point on the path of the skater.

(a) Click on "track friction." Put the slider marked "coefficient of friction" at the point two spaces from the left end. "Pause," and put the skater back up on the track. "Resume." How many complete cycles does it take for the skater to stop?

Increase the friction by a factor of two and repeat this.

(b) Click on "Bar graph." Observe the changes in K, P, and U. What are the beginning and end values of K, P, U, and of the mechanical energy and the total energy (E) as a fraction of the initial potential energy?

Lower the reference level. What are the beginning and end values of K, P, U, E, and the mechanical energy now in terms of the initial potential energy?

12. A person whose mass is 70 kg runs up a flight of stairs through a vertical height of 6 m. If she uses her food with an efficiency of 10%, how many food calories does she have to consume to do this work?

13. A force of 12 N acts at an angle of 28° to the horizontal. What are its horizontal and vertical components? If it moves an object vertically through 2 m, how much work does it do?

14. A proton moves at constant speed in a circular path in a cyclotron. Explain how you know what the total work done on the proton must be during this motion.

15–16. A pendulum bob whose mass is 0.25 kg is attached to a string whose mass is so small that it may be neglected. As the pendulum swings, its bob moves from a starting height of 18 cm to a height of 12 cm, measured from the lowest pont of the motion.

(a) What is the work done by the tension in the string?

(b) What is the speed at the height of 12 cm?

(c) Use energy bar charts to show the changes in the energy of the pendulum during this segment of its motion.

17–18. A ball whose mass is 0.15 kg has an initial velocity whose horizontal component is 3 m/s and whose vertical component is 2.5 m/s. Neglect dissipative forces.

(a) What is its kinetic energy at the starting point and at the ending point when it returns to the same height?

(b) What are its kinetic and potential energies at the highest point?

(c) At a point $\frac{1}{3}$ of the way up, what are the potential and kinetic energies, the height, the speed, and the components of the velocity?

19–21. A car whose mass is 200 kg moves on a rollercoaster, starting from rest at a height of 20 m above ground. If we neglect dissipative forces, to what maximum height can it return? At what height would it be moving with a speed of 5 m/s?

However, when we measure its actual speed we find that it moves with a speed of only 4 m/s at that height. Describe all energy changes, and calculate their amount, from the starting point.

22. You are part of a team of amusement park engineers. You want to know precisely how much the spring in the figure should be compressed so that the cart moves up the ramp, down on the other side, and comes to rest at the exit, which is to be placed half-way along the rough patch. You may assume that the track is frictionless except for the rough patch.

(a) Describe the energy relationships. Bar charts may be helpful to set up the mathematical relationship.

(b) What is the minimum distance through which the spring must be compressed to reach the rough patch?

(c) Calculate how much the spring needs to be compressed so that the cart comes to rest halfway along the rough patch.

List the quantities that need to be known, the quantities that need to be measured, and how they are to be measured.

23. An 80 kg person runs up a flight of stairs whose vertical height is 12 m. In how much time does she have to do that to use energy at the rate of one horsepower? Is this a realistic possibility?

Problems and reasoning skill building

1. Two forces with equal magnitude pull on a box: F_1 to the right and F_2 to the left. The box moves to the right with constant velocity, through a distance x. There are no other horizontal forces or forces with horizontal components.

(a) How much work is done by F_1?
(b) How much work is done by F_2?
(c) What is the total (or net) work, W_T, done on the box?

2. Two forces pull on a box, F_1 to the right and F_2 to the left. The magnitude of F_1 is twice the magnitude of F_2. There are no other horizontal forces or forces with horizontal components. The box moves to the right through a distance x.

In terms of the magnitude F_2:

(a) How much work is done by F_2?
(b) How much work is done by F_1?
(c) What is the total work, W_T, done on the box?
(d) Does the kinetic energy change as the box moves through the distance x, and if so, by how much?

3. An alpha particle (a positively charged helium nucleus) is emitted from a radium nucleus with a kinetic energy of 3.6 MeV. It moves straight toward a gold nucleus (also positively charged). Assume that the gold nucleus remains at rest.

Describe the motion and energy changes of the alpha particle with the help of energy bar charts. (This is the experiment in Rutherford's laboratory that established the existence of the atomic nucleus.)

(a) Sketch a graph of the speed of the alpha particle as a function of its distance, r, from the center of the gold nucleus.

(b) Sketch a graph of the velocity of the alpha particle as a function of r.

4. A ball is released from rest, bounces four times, and then stops.

Sketch graphs of the following quantities as a function of the time from the moment when the ball is dropped:

(a) potential energy,
(b) kinetic energy,
(c) mechanical energy,
(d) internal energy.
(e) What happens to the mechanical energy?

5. A vertical spring is held fixed at one end, and a weight is attached at the other end. The weight is pulled down, then released, and the spring oscillates up and down as the weight moves from its minimum height to the equilibrium position and then to its maximum height and back.

At what positions of the weight are the following quantities at their maximum, at zero, and at their minimum:

(a) gravitational potential energy,
(b) elastic potential energy,
(c) kinetic energy.

6. A baseball is batted, hits the ground, and is then caught.

(a) Describe the sequence of energy transformations of the ball that takes place from the time it leaves the bat.

(b) Start at some point in the past (before the player's breakfast) and describe the energy transformations leading up to the moment when the ball leaves the bat.

7. Describe the energy transformations of a pendulum from the time when it is released, neglecting dissipative forces.

8. Describe the energy transformations of a pendulum from the time when it is released (not neglecting dissipative forces) until it stops and remains at rest.

9. Describe each of the lines on the energy diagram (P, E_M, and K) of Example 17.

10. A ton of coal is hoisted up at constant speed through a height of 15 m by a crane.
    (a) What are the forces on the load of coal?
    (b) What is the total (net) force on the load?
    (c) What is the work done by each of the forces on the load?
    (d) What is the total work done on the load?

11. An elevator whose mass is 200 kg moves up at a constant speed through a distance of 20 m.
    (a) What is the gravitational force on it?
    (b) What other forces act on it?
    (c) What is the work done on the elevator against the gravitational force?
    (d) What is the work done on the elevator by the gravitational force?
    (e) What is the total work done on the elevator?
    (f) What are the energy changes of the elevator?

12. A stone falls from rest and hits the ground with speed v. Neglect air resistance.
    (a) What is its speed as it passes a point halfway to the ground?
    (b) What is its speed as it passes a point three quarters of the way to the ground?

13. A ball whose mass is 0.25 kg is thrown straight up. It reaches a height of 16 m before falling back to where it started. Neglect air resistance, and use the starting point as the reference level where $y = 0$ and the potential energy, P, is zero.
    (a) Draw a graph of P against y.
    (b) On the same coordinate system, draw graphs of the mechanical energy, E_M, and the kinetic energy, K.
    (c) Use the graphs to set up mathematical expressions for P, E_M, and K in terms of y.

14. A block whose mass is 2.4 kg slides down an inclined plane, starting from rest. It slides to the bottom through a height, y, of 1.2 m, while the distance along the plane, s, is 2 m.
    (a) For this part neglect friction and air resistance, and take the reference level to be at the bottom. Draw graphs of the potential energy (P), the kinetic energy (K), and the mechanical energy (E_M) in terms of the distance (s). (Take $s = 0$ at the starting point.)
    (b) Now assume that there is a constant frictional force of 3 N and no other dissipative forces. What is the mechanical energy when the block reaches the bottom?
    (c) Again draw graphs of P, E_M, and K, and also the dissipated energy, Q, in terms of s.

15. A rock whose mass is 0.2 kg is dropped from a height of 10 m.
    (a) First assume that there is no air resistance, and take the reference level to be at the bottom. Draw graphs of P, K, and E_M as a function of the height until the rock hits the bottom.
    (b) Use the graphs to set up mathematical expressions for the three energies.
    (c) Now assume that there is some air resistance. Again draw graphs of P, K, E_M, and the dissipated energy, Q.
    Use the graphs to write the mathematical expressions for each of the energies.
    (d) Repeat parts (a) and (b), but this time use the starting point as the reference level.
    (e) Still using the starting point as the reference level, write down the relation between y and the time, t. Use it to get an expression for K in terms of t, and draw a graph of it.

16. A spring hangs vertically, with one end fixed. At the other end a weight whose mass is 0.15 kg hangs from it. The spring constant is 12 N/m. Neglect all dissipative forces, and also the mass of the spring. Assume that the spring obeys Hooke's law.
    (a) The system is at rest. What is the elastic potential energy?
    (b) You pull the weight down with a force of 1.2 N until it is again at rest. Which energies change? By how much?
    (Sketch a figure with the spring and weight. Where have you chosen the reference level? Mark it on your figure. You can only choose one! In this problem there is really only one good choice.)
    What is the mechanical energy now?
    (c) You then release the 0.15 kg weight. The spring returns to the equilibrium point of part (a). What are the various energies now?
    (d) The weight continues upward. What is the criterion for how far it goes? As an extra and harder part, find the value of y at the highest

point, measured from the reference level that you chose in part (b).

17. A ball is thrown straight up from ground level, and eventually returns to the same height h. It has an initial velocity v_{0y}. Neglect all dissipative forces.

Write expressions in terms of these two quantities and m for the following:

(a) The largest kinetic energy. Where does it occur?

(b) The smallest kinetic energy. Where does it occur?

(c) The maximum potential energy. (Be sure to state what reference level you have chosen.)

(d) The maximum height.

(e) The mechanical energy.

(f) What is the amount of work that had to be done on the ball as it was thrown?

18. Repeat the previous problem, but with an additional component of the initial velocity in the horizontal direction, v_{0x}.

19. For each of the listed situations, neglecting all dissipative forces, do the following:

(i) Identify the system that you wish to consider; describe the initial and final states.

(ii) Describe the energy transformation processes in words and with bar charts.

(iv) Draw a force diagram showing all the forces on the system.

(a) A pendulum bob swings down, starting from rest.

(b) An elevator, moving up, slows down until it stops.

(c) A block is launched up an inclined plane by a compressed spring.

(d) Another system of your choice.

20. A bucket hangs from a rope. The other end of the rope is held by a woman. The tension in the rope is 50 N. All movements take place with constant velocity. What is the work done by the rope in each of the following cases?

(a) She moves 5 m horizontally without changing the height of the bucket.

(b) She lifts the bucket 0.5 m.

(c) She lowers the bucket 0.5 m.

(d) She pulls the bucket up an incline, moving it 1 m vertically and 0.5 m horizontally.

(e) She lowers the bucket down the same incline by the same distance as in part (d).

21. (a) Corresponding to each of the following mathematical relations, describe a possible process, in words, with a sketch, and with bar charts:
$K_1 + W = P_2$ and
$P_1 + W = K_2$.

(b) Write a problem with numerical values for each of these situations.

22. A tennis ball has a mass of 0.057 kg. It falls 18 m vertically, from rest, and then has a speed of 12 m/s.

(a) Identify the system that you wish to consider.

(b) What are the initial and final states?

(c) What are the energy transformations?

(d) Draw the corresponding energy bar chart and draw a force diagram.

(e) Write a mathematical statement that describes the energy transformations.

(f) What is the magnitude of the average resistive (dissipative) force?

23. A spring has a spring constant of 1.2×10^4 N/m. A 10 kg block is placed against it, and the spring is compressed 6 cm. The spring is then released, shooting the block forward along a horizontal surface against a force of friction of 15 N.

Follow steps (a) to (e) of the previous problem.

(f) How far does the block travel before coming to rest?

24. A block is pulled up an inclined plane with constant speed against a constant frictional force.

(a) Identify the system that you wish to consider.

(b) Draw a force diagram, showing all forces on the system.

(c) Describe all energy transformations.

(d) What quantities do you need to know to calculate the dissipated energy.

25. A 900 kg car starts from rest, with its engine off, and rolls a distance of 50 m down a hill. A 400 N frictional force opposes the motion. When the car reaches the bottom it is moving with a speed of 8 m/s.

(a) Describe the energy transformations using energy bar charts.

(b) Use your charts to write a mathematical statement that describes the energy transformation processes.

(c) What is the vertical height through which the car moves?

26. A slide is 42 m long and has a vertical drop of 12 m. A 60 kg man starts down the slide with a speed of 3 m/s. A frictional force opposes the motion, causing 20% of the kinetic energy to be dissipated.
   (a) Identify the system that you wish to consider and its initial and final states.
   (b) Describe the energy transformation processes using energy bar charts.
   (c) Use your charts to write a mathematical statement that describes the energy transformations.
   (d) What is his speed at the bottom?
   (e) What is the magnitude of the average frictional force?

27. Two children ride a sled down a hill, starting from rest. Neglect all dissipative forces.
   (a) Identify the system that you wish to consider, and its initial and final states.
   (b) Use the following representations to describe what happens: a sketch of the hill, sled, and children; a force diagram; energy bar charts; a mathematical statement of the relation between the various initial and final energies.
   (c) The mass of the sled with the children in it is 150 kg. Their speed at the bottom is 13 m/s. What is the vertical height through which they move?

28. Repeat parts (a) and (b) of the previous problem, but with an initial velocity at the top of the hill, and some dissipated energy.
   (c) Make up a problem for this situation by giving some numerical quantities and asking for two others. Show the solution of your problem.

29. A ball ($M = 0.60$ kg) is dropped off a cliff, starting from rest. What is its kinetic energy just before it hits the ground 30 m below? (Neglect air resistance.) What is its speed?

30. A ball ($M = 0.50$ kg) is thrown vertically, with an initial kinetic energy of 50 J. (Neglect air resistance.)
   (a) What is its kinetic energy when it comes back to the starting height?

31. (a) When a ball ($M = 0.50$ kg) is thrown at an angle of 24.6° with an initial energy of 50 J it is observed that its kinetic energy as it hits the ground is only 90% as high as when air resistance is neglected. How much mechanical energy has been lost during the flight?
   (b) When the ball is thrown at a different time, it hits the ground with a speed that is 90% of that calculated when air resistance is neglected. How much mechanical energy is lost during the flight this time?

32. A ball ($M = 0.40$ kg) drops from a height of 5 m. After it bounces it rises only to a height of 4.0 m. How much mechanical energy has been lost?

33. A block ($M = 1.5$ kg) slides down an incline at a constant speed. Its vertical height decreases at a rate of 0.50 m/s. How much internal energy is produced in eight seconds? What is the rate of energy dissipation in watts?

34. A toaster is marked 500 W, indicating that this is its power consumption when it is turned on. It takes three minutes to toast two pieces of bread. How many joules are used during that time? What is the cost if the electric energy costs 20 cents per kwh?

35. A ball ($M = 0.50$ kg) is thrown straight upward with an initial kinetic energy of 75 J. Neglect dissipative forces, assume that the potential energy, P, is zero at the starting point, and find the following:
   (a) K at the highest point,
   (b) P at the highest point,
   (c) the distance between the lowest and the highest point,
   (d) K when the ball falls back to the starting point.

36. A block whose mass is 2 kg slides along a table, starting with a speed of 4 m/s. It gradually comes to a halt as a result of friction. How much internal energy (in J and in cal) is produced during this process?

37. A child is on a swing that starts from rest at a point 1.8 m above the lowest point. The mass of the child together with the swing is 30 kg. If you neglect all dissipative forces, what are the maximum kinetic energy and speed of the swing with the child?

38. Consider the child and swing of the previous problem, but this time do not neglect dissipative forces. Without a push the swing does not

come back to the same height, but stops at a point 60 cm lower. How much energy does the father have to supply with each push if the swing is to come back to its original height of 1.8 m above the lowest (central) position?

39. The unit called the horsepower is equal to 550 $\frac{\text{ft-lb}}{\text{s}}$. Calculate how many W this corresponds to.

40. A proton ($M_p = 1.66 \times 10^{-27}$ kg) is 10 cm from a large, positively charged plane, as in the figure accompanying Example 6. The charges on the plane give rise to a constant force of 3×10^{-10} N on the proton, away from the plane. Use energy bar charts to set up the problems and answer the following questions.

(a) The proton is initially at rest, and then moves through a distance of 1 cm. What is its kinetic energy after it has moved through the 1 cm? What is its speed?

(b) The proton is initially moving toward the plane with a kinetic energy of 5×10^{-12} J. Describe its subsequent motion. What is the distance of its closest approach to the plane? How fast is it moving when it is 11 cm from the plane?

41. In the ground state of the hydrogen atom the electron has a kinetic energy of 13.6 eV. (1 eV = 1 electron volt = 1.6×10^{-19} J.) The electric potential energy is less by 27.2 eV than when the electron is far removed from the proton.

Use as reference level for the potential energy the state where the electron and the proton are infinitely far from each other. (This is then the state where the potential energy is zero.) With this reference level, what are the potential energy and the total energy of the atom in its ground state?

How is this amount of energy related to the amount that has to be supplied to the atom to separate the proton and the electron to a point where each is at rest far from the other? (This amount is called the ionization energy of the electron, or its binding energy.)

42. A satellite moves around the earth at a height that is small compared to the radius of the earth. An object or person in the satellite is often described as "weightless."

(a) What has happened to the weight of the object or person, defined, as we do in this book, as the gravitational attraction of the earth?

(b) Which force on the object or person, sometimes defined as the weight, is zero?

43. How much energy does a diet of 2000 Calories per day supply to the human body in one day? (These are food calories, or kilocalories, each equal to 4186 J.) Use the number of seconds in a day to find the rate of energy use in watts corresponding to this diet.

44. You are told that a car has 10^6 J of kinetic energy and 2×10^6 J of potential energy. In what respect is this an incomplete statement?

45. When you go up two floors, approximately how much does your potential energy increase? If you run up in 10 seconds, what is the rate at which you expend energy (in W)?

46. A pendulum consists of a mass (M) and a string of length L. How fast must the mass be moving at the lowest point to be able to just move in a full circle?

Multiple choice questions

1. A rock falls from rest and hits the ground with speed v. (Neglect air resistance.) Halfway along its path to the ground its speed is
   (a) $2v$
   (b) $\sqrt{2}v$
   (c) $\frac{1}{2}v$
   (d) $\frac{1}{\sqrt{2}}v$

2. A ball loses a third of its kinetic energy as it bounces off the floor. What is the fraction of its original height, H, to which it rises before coming to rest? (Neglect air resistance.)
   (a) $\frac{1}{3}$
   (b) $\frac{2}{3}$
   (c) $\frac{1}{\sqrt{2}}$
   (d) $\sqrt{\frac{2}{3}}$

3. A ball loses a third of its speed as it bounces off the floor. What is the fraction of its original height, H, to which it rises before coming to rest? (Neglect air resistance.)
   (a) $\frac{4}{9}$
   (b) $\frac{2}{3}$

(c) $\sqrt{\frac{2}{3}}$

(d) $\frac{1}{3}$

4. Two astronauts sit in a spaceship that orbits the earth at a height of 20 km. Which of the following forces is zero?

(a) The gravitational force of one astronaut on the other.

(b) The gravitational force of the earth on one astronaut.

(c) The force of the floor of the spaceship on one astronaut.

(d) The centripetal force on one astronaut.

5. Assume the approximation that the moon (mass m) travels around the earth in a circular orbit of radius R with speed v. The work done on the moon in one complete revolution is

(a) mgR

(b) $\frac{1}{2}mv^2$

(c) $\frac{mv^2}{R}$

(d) zero

Synthesis problems and projects

1. Fill in the missing steps at the end of Section 6.3 leading to $T = \frac{2\pi}{\omega}$ and $T = 2\pi\sqrt{\frac{m}{k}}$.

2. Go to the PhET website and open the simulation Masses and Springs.

(a) Hang one of the masses on one of the springs and set it into oscillation without friction. Measure the period, T. Calculate the spring constant, k from T and m.

(b) Put different known masses on the same spring and measure the equilibrium position for each. (It is easiest to do this after putting on a lot of friction to stop the motion.) Draw a graph of F against x and find the spring constant from your measurements.

(c) Find the masses of the unmarked weights. Calculate T for one of these weights. Check it with a direct measurement of T.

3. You have the following equipment:

(a) a spring whose spring constant you know,

(b) a track that can be inclined and whose friction is negligibly small,

(c) a glider whose mass you know,

(d) a motion detector,

(e) a ruler.

Describe an experiment that you can do with this equipment to test the law of conservation of energy.

4. On a pool table the "cue" ball is given a velocity of magnitude v in the y direction. It hits another equal ball. After the collision the two balls move with velocities whose magnitudes are v_1 and v_2 at an angle of 90° with respect to each other.

(a) Draw a vector diagram of the momenta before and after the collision.

(b) After the collision, what is the relation between the components in the x and y direction of the velocity of each ball?

(c) What is the relation between v, v_1, and v_2?

5. A block with mass M hangs at rest from a string whose upper end is fixed. A pistol shoots a bullet (mass m and velocity v) into the block. The block with the bullet embedded in it then begins to move with a horizontal velocity V.

(a) What law of physics governs the collision of the bullet and the block and determines the velocity V?

(b) What is the relation between v and V?

(c) Following the collision the mass rises. What law of physics determines the height, H, to which the block (with the bullet in it) rises?

(d) Derive the relation that gives v in terms of the quantities V, m, M, and H.

(This arrangement is called a ballistic pendulum. It can be used to measure the speed of a bullet.)

6. The amount of energy used each day by primitive people was roughly what they ate, of the order of 2000 kcal. Look up the total energy used per year in the United States and use this number to calculate the yearly consumption per person, and the consumption per person, per second, in watts. By what factor is this larger than the amount used by primitive people?

7. What, approximately, is the average speed (in m/s) corresponding to the world record of running 100 m? One mile? How much kinetic energy does each of these correspond to? Answer the same questions for your walking fairly fast for one mile.

What, roughly, are the corresponding kinetic energies? Make an assumption of how long it takes the 100 m sprinter to reach a kinetic energy equal to his average kinetic energy, and calculate the required power in watts. How many horsepower is that equivalent to?

8. You are sliding quarters down a ramp. You want to know how large the effect of friction is, and whether it can be ignored.

You have a motion detector and a meterstick. Show how you can use them to answer these questions.

9. A spring, sitting vertically on a table, is compressed. It is released, and jumps up in the air. You want to check whether mechanical energy is conserved as this happens. You have a scale, a ruler, and a set of weights. Describe a possible procedure.

10. Describe an experiment that you can use to convince a classmate that $\cos \theta$ is important in the definition of work.

CHAPTER 7

Materials and Models

Ideal systems and models: the ideal gas
Bouncing molecules: the microscopic point of view
The macroscopic point of view
Pressure
Archimedes' principle
Bernoulli's equation
Absolute temperature
The surprising bridge to temperature
Internal energy and heat capacity
Heating
Work
The first law of thermodynamics

Other systems: adding pieces from reality
Molecules
Phase changes
Condensed matter
Metals
Chemical energy
Quantum theory
Back to heat capacities
Diatomic gases

We have applied Newton's laws of motion to the objects they were meant for: objects that are large enough to see or feel. In this chapter we use them to describe the mechanics of atoms and molecules. As we do that we have to keep in mind that Newton's laws and classical mechanics apply in the microscopic realm only to a limited extent, and that the ideas and methods of quantum mechanics may be necessary.

So far we have applied Newton's laws to "objects" without considering the material that they are made of or their internal structure. To treat them in terms of their atoms and molecules takes us from a few simple particles and unchanging objects to composite objects that consist of very large numbers of particles. The real systems are so complex that we approximate them with *models*, invented, imagined systems that, however, retain the essential properties of the systems that we seek to understand.

We begin this chapter by introducing the model called the *ideal gas*. It is an enormously fruitful model, which describes some of the most important properties of real gases, and, to some extent, also those of other materials. It also provides a bridge to a seemingly entirely different subject, that of heat and temperature.

In addition, it shows the correspondence between two very different aspects of the world around us. On the one hand there are the large-scale objects of our direct sensory experience. Our instruments show us that underlying this *macroscopic* world is a *microscopic* world that we are normally not aware of, the world of moving, vibrating, rotating, colliding atoms and molecules, absorbing, emitting, and exchanging energy.

7.1 Ideal systems and models: the ideal gas

How can we get started talking about the vastly complex real world? We know that we need to talk about one that is simpler, a *model* system, sometimes called an *ideal system*. Let's review the rules of the game. We decide what the model is and how it works. It is an invented system that shares some of the characteristics of the corresponding system in the real world.

The words *ideal* and *model* describe a system with properties that we make up and laws that we prescribe. We set the rules that we need to calculate the properties and behavior of our model system.

We can then compare the model and the real material to see to what extent their properties overlap. We can also ask what the model *predicts* for behavior under conditions that were not originally considered, and to see how the ideal and the actual systems then compare.

Here is the model that we will look at in some detail: it's a gas whose constituents (its atoms or molecules) are particles, i.e., they have no size or internal structure. They exert forces and experience forces only when they touch each other or the walls of their container. They move in accord with the laws of Newtonian mechanics. This model is called *the ideal gas*.

A real gas is very different. Its smallest units may be molecules, composed of two or more atoms, or single atoms, consisting of nuclei and electrons. They are not zero-size particles, and they exert forces on each other even when they don't touch. Nevertheless, the ideal-gas model describes and predicts the behavior of real gases very well under many circumstances.

Bouncing molecules: the microscopic point of view

The basic feature of the model that we call the ideal gas is that its components are always in motion and have only kinetic energy. They have no internal structure, no internal motion, and no internal energy. They exert no forces on each other except when they touch, so that there is no mutual potential energy. Since the model requires that these elementary components are point particles with no internal structure, it doesn't matter whether we call them atoms or molecules. (We'll call them molecules.) The figure shows a portion of a gas with just four molecules.

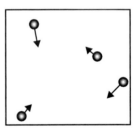

When the molecules collide with each other they can interchange energy and momentum in accord with the laws of conservation of energy and momentum. The total momentum of the system of which they are parts does not change. Neither does the total kinetic energy, since we have decided that in this model there is no other kind of energy.

When we put the gas in a box or other container the molecules bounce off the walls. At each bounce the wall exerts a force on a molecule. From Newton's third law we know that there is at the same time a force *by* the molecule *on* the wall.

Our aim is to relate what the molecules are doing, as described by their *microscopic* variables, namely their velocities and kinetic energies, to the *macroscopic* variables that we observe and measure directly, namely the pressure (equal to the average force on the walls divided by the area of the walls) and the volume of the container.

We start with a single molecule moving back and forth in a cubical box whose sides have length L and area $A = L^2$, with volume $V = L^3$. The molecule moves with an initial velocity $\mathbf{v_i}$ in the x direction toward a wall that is at right angles to the x-direction. We assume that we can neglect the effect of the gravitational force on its path.

Like a ball hitting a solid wall at right angles to it, the molecule will bounce back along the same line, with its kinetic energy unchanged, and the direction of its momentum reversed. The kinetic energy is a scalar quantity that does not depend on direction, but the velocity and the momentum are vector quantities. The initial momentum vector $m\mathbf{v_i}$ (where m is the mass of a molecule) and the momentum vector after the collision, $m\mathbf{v_f}$, differ by the vector $m\mathbf{\Delta v}$, where $\mathbf{v_i} + \mathbf{\Delta v} = \mathbf{v_f}$.

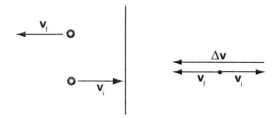

Since all the velocities and momenta are along the x direction, we can call the initial momentum mv_x and the final momentum $-mv_x$. The change in momentum is $-2mv_x$, with a magnitude of $2mv_x$. Between a collision with one wall and a collision with the opposite wall the molecule moves a distance L in a time that we'll call Δt, so that $v_x \Delta t = L$, or $\Delta t = \frac{L}{v_x}$.

To find the force exerted by the molecule on the wall we use Newton's second law of motion. Instead of $F_{net} = ma$ we use the momentum form. Since $a = \frac{\Delta v}{\Delta t}$ (the change in the velocity divided by the time it takes for the velocity to change), ma is equal to $\frac{\Delta(mv)}{\Delta t}$, i.e., the force is equal to the change in momentum divided by the time during which the momentum changes. For our single molecule the magnitude of the momentum change, $\Delta(mv)$, is $2mv_x$ at each collision, and this happens once in every time interval Δt, on one wall and then on the other, so that the average force is $\frac{\Delta(mv)}{\Delta t} = \frac{2mv_x}{L/v_x} = \frac{2mv_x^2}{L}$.

The molecule travels back and forth and contributes to the pressure on both walls. To find the pressure we need to divide the force by the area of both walls, $2L^2$, to get $\frac{2mv_x^2}{2L^3}$ or $\frac{mv_x^2}{V}$. This is the contribution to the pressure of a single molecule bouncing back and forth. For a gas of N molecules it is $\frac{Nmv_x^2}{V}$. We can also take into account that the velocities are not all the same, and use the average of mv_x^2, which we write as $\overline{mv_x^2}$, so that $P = \frac{N\overline{mv_x^2}}{V}$, which can be rewritten as $PV = N\overline{mv_x^2}$.

This is really the end of the derivation. But we can also look at what happens if the molecules do not just move along the x direction. In that case there are the additional velocity components v_y and v_z. The velocity v is related to them by $v^2 = v_x^2 + v_y^2 + v_z^2$. With a large number of molecules, randomly moving in all possible directions, all components will be equally represented, and their averages $\overline{v_x^2}, \overline{v_y^2}$, and $\overline{v_z^2}$ will be equal. $\overline{v^2}$ is then equal to $3\overline{v_x^2}$. This allows us to rewrite our result as $PV = \frac{1}{3}N\overline{mv^2}$, which we can also write as $PV = \left(\frac{2}{3}N\right)\left(\frac{1}{2}\overline{mv^2}\right)$.

We have achieved our goal. We now have a relation between the macroscopic variables, P and V, quantities that apply to the whole container with the gas in it, on the left, and the microscopic variables that refer to the molecules, i.e., their number and their average kinetic energy, on the right.

EXAMPLE 1

Joe and Jane are practicing tennis by hitting balls against a wall. The mass of a ball is 0.15 kg. The speed of each ball as it hits the wall is 20 m/s. There are 40 hits in one minute. Assume that the balls hit the wall at right angles and do not lose any kinetic energy when they bounce back.

(a) What average force does the wall experience in one minute?

(b) What is the average pressure on the section of the wall, whose area is $3\,\text{m}^2$, where the balls hit?

Ans.:

(a) $mv_x = 3\,\text{kg m/s}$.

At one hit of a ball the momentum changes from mv_x to $-mv_x$. This is a change of $2mv_x = 6\,\text{kg m/s}$. For 40 hits it is $240\,\text{kg m/s}$.

In a time of 60 s this is a rate of $\frac{240\,\text{kg m/s}}{60\,\text{s}} = F = 4\,\text{kg m/s}^2$, or 4 N.

(b) The pressure is $\frac{F}{A} = \frac{4}{3} = 1.33\,\text{N/m}^2$.

EXAMPLE 2

At atmospheric pressure ($1.01 \times 10^5\,\text{N/m}^2$) and room temperature the number of molecules in $1\,\text{m}^3$ of air is about 2.7×10^{25}. The average mass of an air molecule is about $5 \times 10^{-26}\,\text{kg}$. What is the average kinetic energy of a molecule?

Ans.:

$$PV = (\tfrac{2}{3}N)(\tfrac{1}{2}\overline{mv^2})$$
$P = 1.01 \times 10^5$, $V = 1\,\text{m}^3$, $N = 2.7 \times 10^{25}$
$\tfrac{1}{2}\overline{mv^2} = 1.5\,\text{PV/N} = 5.6 \times 10^{-21}\,\text{J}$

Atomic and molecular energies are most often expressed in *electron volts* (eV) where $1\,\text{eV} = 1.6 \times 10^{-19}\,\text{J}$. Here the energy is .035 eV.

We can find the corresponding speed: $\sqrt{\overline{v^2}} = \sqrt{\frac{2K}{m}} = \sqrt{\frac{(2)(5.6 \times 10^{-21})}{5 \times 10^{-26}}} = \sqrt{2.24 \times 10^5} = 473\,\text{m/s}$.

This is the square root of the average of the square of the speed. It is called the *root mean square speed*, v_{rms}.

You might think that it would be simpler just to say "average." But since all directions are equally represented, the average velocity is zero! (For two velocities of 10 m/s in opposite directions, the average velocity is zero, but v_{rms} is $\sqrt{\tfrac{1}{2}(100 + 100)} = 10\,\text{m/s}$.)

The macroscopic point of view

A number of pioneering scientists in the seventeenth and eighteenth centuries performed experiments on gases at different pressures and temperatures and with different volumes. Their results can be summarized by the relation that came to be known as *the ideal gas law*, $PV = nRT$. Here P is the pressure, V the volume, T the absolute temperature (measured in *kelvins*), and n is the amount of gas measured in *moles*. R is a proportionality constant, called the *universal gas constant*. In this section we review the work that led to the ideal gas law and examine the quantities P, T, and n.

Pressure

Water in a glass exerts a force on the bottom of the glass equal to the water's weight, Mg. This is in addition to the weight of the air and the resulting *atmospheric pressure*. We can express the force of the water in terms of the height, h, and the density, ρ (Greek *rho*) of the water. The density is the mass per unit volume, and the volume is the height times the cross-sectional area. $\rho = \frac{M}{V}$ and $V = hA$, so that $Mg = \rho hAg$. The pressure on the bottom is $\frac{Mg}{A}$, which is equal to $\rho g h$. The SI unit of pressure is N/m^2, which is also called the *pascal*, Pa, after Blaise Pascal (1623–1662), who, among other contributions, suggested correctly that atmospheric pressure should be smaller on top of a mountain. ($1000\,\text{Pa} = 1\,\text{kPa}$.)

The pressure, $\rho g h$, is not only there at the bottom of a column of fluid. It is the pressure at the depth h regardless of the direction. If this were not so, a drop of water at that depth would not be in equilibrium, would experience a net force, and would accelerate.

EXAMPLE 3

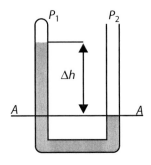

The figure shows a U-tube partly filled with mercury. (Mercury, because it is a very dense liquid, with a density 13.6 times that of water.) The tube is open on the right side and closed on the left side. On the left side the mercury column is higher by $\Delta h = 10\,\text{cm}$. What is the difference in pressure between the top of the mercury column on the left and the top of the mercury column on the right?

Ans.:
At the level marked A the pressures on the two sides are the same. Call the pressure above the mercury on the right side P_2 and on the left side P_1. P_2 is equal to P_1 plus the pressure of a column of mercury 10 cm high, $P_2 = P_1 + \rho g \Delta h$, where ρ is the density of mercury, 13.6×10^3 kg/m³, and $\Delta h = 0.1$ m. The difference between P_2 and P_1 is $\rho g \Delta h = (13.6 \times 10^3)(9.8)(0.10) = 1.33 \times 10^4$ Pa.

EXAMPLE 4

What is the pressure on the bottom of a glass of water?

Ans.:

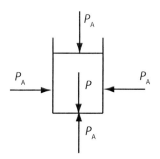

The pressure on the top of the water is the atmospheric pressure P_A. On the bottom there is the additional pressure $\rho g h$, where h is the height of the water. The total pressure on the bottom is $P = P_A + \rho g h$.

Atmospheric pressure also acts on the outside of the glass of water. It acts everywhere, including on the bottom of a glass sitting on a table. (There is some air between the glass and the table unless special efforts are made to exclude it.)

All of us live at the bottom of an ocean, not of liquid, but of air. The atmosphere is held to the earth by its weight. We are so used to the pressure that it exerts that we are normally not aware of it.

The first to understand the existence of atmospheric pressure and to measure it was Evangelista Torricelli (1608–1647). In 1643 he performed the following experiment. He took a glass tube closed at one end, with a length of

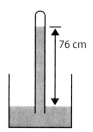

about 1 m, and filled it with mercury. He put his finger over the open end, turned it upside down, and submerged the open end in a pool of mercury. Some of the mercury ran out of the tube into the pool, until the difference in height between the surface in the tube and the surface of the pool was about 0.76 m.

The pressure on the surface of the mercury pool (outside the U-tube) is just the atmospheric pressure. Inside the U-tube, above the mercury, there is no air and no atmospheric pressure. Inside the same tube, at the level of the mercury surface outside the tube, there is only the pressure of the mercury column.

In effect Torricelli produced a U-tube with the pressure of the column of mercury on one side and atmospheric pressure on the other. The two pressures are equal, so that what he showed is that the pressure of the air in the atmosphere is the same as that of a column of mercury whose height is 0.76 m.

It was known at that time that a suction pump can lift water only to a height of about 10.3 m. Torricelli's experiment showed why this is so. The situation is just as in his experiment, but with water instead of mercury. The pump removes the air (and its pressure) above the water. The atmospheric pressure forces water up.

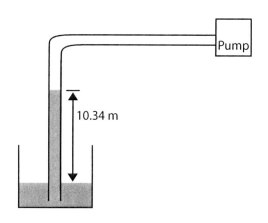

With no pressure above the water, the pressure of the water column is equal to atmospheric pressure. Since the density of mercury is 13.6 times as great as that of water, the water column is 13.6 times as high as the mercury column. (13.6×0.76 m $= 10.34$ m.)

EXAMPLE 5

How large is atmospheric pressure in pascals when it is said to be "760 mm"? (The height of a column of mercury, here 760 mm, is often used as a measure of the pressure at its base.)

Ans.:
The density of mercury is 13.6×10^3 kg/m^3. At the bottom of a column of mercury whose height is 760 mm or 0.76 m the pressure is $\rho g h = (13.6 \times 10^3)(9.8)(0.76) = 1.01 \times 10^5$ Pa or 101 kPa.

Atmospheric pressure at the surface of the earth varies, but is generally close to 101 kPa or 760 mm of mercury.

Archimedes' principle

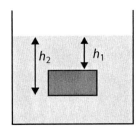

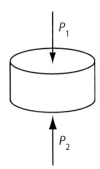

At the depth h_1 the pressure is atmospheric pressure (p_A) plus $\rho g h_1$. At the depth h_2 it is greater and equal to $p_A + \rho g h_2$. A disk with cross-sectional area A experiences the pressure P_1 downward at its top surface and the pressure P_2 upward at its bottom surface. There is a net pressure upward, $P_2 - P_1$, equal to $\rho g h_2 - \rho g h_1 = \rho g(h_2 - h_1)$ and hence a net upward force A times as large, equal to $\rho g(h_2 - h_1)A$. This force is called the *buoyant force*.

The volume of the disk is $(h_2 - h_1)A$. ρ is the density of the liquid, so that $\rho g(h_2 - h_1)A$ is the weight of the liquid that is displaced by the disk. That the magnitude of the buoyant force is equal to that of the weight of the displaced liquid is called *Archimedes' principle*.

EXAMPLE 6

A block of wood whose height is 10 cm, and whose density is 900 kg/m^3 floats in water. How much of the block is under water?

Ans.:
The block sinks until its weight is equal to the buoyant force. The net force on the block is then zero, and the block is in equilibrium. The block's weight is $Mg = \rho_{\text{wood}} A h g$, where $h = 10$ cm. Since $\rho_{\text{water}} = 1000$ kg/m^3 it is also $.9 \rho_{\text{water}} A h g$.

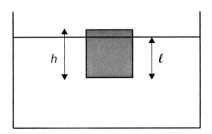

If the wooden block is submerged to a distance ℓ, the buoyant force is $\rho_{\text{water}} A \ell g$. For it to be equal to the weight of the wooden block, $\ell = 0.9h$, and 9 cm of the block is submerged. The ratio $\frac{\ell}{h}$ is equal to the ratio of the densities $\frac{\rho_{\text{wood}}}{\rho_{\text{water}}}$.

Bernoulli's equation

So far we have considered only liquids at rest. We now look briefly at pressure in a flowing liquid.

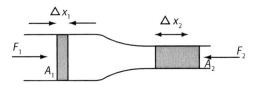

The figure shows a liquid in a tube whose cross-sectional area changes from A_1 to A_2. Look at the part of the liquid that is shaded. It starts on the left, where the area is A_1. It is pushed to the right with a force $F_1 = P_1 A_1$. Later the same volume is at the right where the area is A_2. The force in the opposite direction is $F_2 = P_2 A_2$. As this amount of liquid moves to the right the distance through which it moves is different in the wide part and in the narrow part. It moves forward by Δx_1 in the wide part and by Δx_2 in the narrow part. The volumes $A_1 \Delta x_1$ and $A_2 \Delta x_2$ are the same.

The work done on this amount of liquid is $F_1 \Delta x_1 - F_2 \Delta x_2$, which is equal to $P_1 A_1 \Delta x_1 - P_2 A_2 \Delta x_2$. This amount of work is equal to the gain in kinetic energy, $\frac{1}{2} m v_2^2 - \frac{1}{2} m v_1^2$, where m is the mass of the liquid whose volume is $A_1 \Delta x_1$ (or $A_2 \Delta x_2$).

Since the density is equal to the mass divided by the volume, $\rho = \frac{m}{V}$, we can write $m = \rho V = \rho A_1 \Delta x_1$ (or $\rho A_2 \Delta x_2$), so that the change in kinetic energy is $(\frac{1}{2})(\rho A_2 \Delta x_2)(v_2^2) - (\frac{1}{2})(\rho A_1 \Delta x_1)(v_1^2)$. This is equal to $P_1 A_1 \Delta x_1 - P_2 A_2 \Delta x_2$.

The volume $A_1 \Delta x_1$ (or $A_2 \Delta x_2$) appears in each term and we can divide by it to get $P_1 - P_2 = \frac{1}{2} \rho v_2^2 - \frac{1}{2} \rho v_1^2$ or $P_1 + \rho v_1^2 = P_2 + \rho v_2^2$.

This relation, based on the law of conservation of energy, is called *Bernoulli's equation*, after Daniel Bernoulli (1700–1782). We see that in the wide part, where the speed of the flowing liquid is smaller, the pressure is larger. This is an important insight, with many applications.

There are, however, some major limitations. Bernoulli's equation assumes that the liquid flows smoothly, every part of it in the same direction. This is called *laminar flow*, in contrast to the much more common *turbulent flow*, where the motion is more complicated. The equation also ignores any loss of energy from frictional effects that result from the *viscosity* (the "stickiness") of the liquid.

With these limitations you may question the usefulness of the equation. But although its exact quantitative application is limited to special situations, *Bernoulli's principle*, namely the decrease in pressure where the liquid flows faster, is of great importance.

EXAMPLE 7

In a hurricane the wind blows horizontally across the 300 m² roof of a house with a speed of 30 m/s (67 mph). What is the resulting lifting force on the roof?

Ans.:
We are looking for the pressure difference that results from the change in speed of the air, from zero to 30 m/s. The density of air is 1.29 kg/m³. $\Delta P = \frac{1}{2} \rho v^2 = (0.5)(1.29)(30^2) = 581$ Pa. $P = \frac{F}{A}$. $F = (581)(300) = 1.74 \times 10^5$ N, which is equivalent to the weight of 1.77×10^4 kg, or about 18 metric tons. (1 metric ton is 1000 kg.)

We came to Bernoulli's equation by using the fact that the work done on the liquid is equal to the increase in its kinetic energy, and then dividing each term by the volume. We can also include the potential energy. Since we are using the symbol P for pressure in this section, we will use $\mathcal{P}$ for potential energy.

We can use the system consisting of the fluid and the earth, with its kinetic and potential energies, and external work done on it: $K_1 + \mathcal{P}_1 + W = K_2 + \mathcal{P}_2$, where the subscripts 1 and 2 indicate two different positions in the fluid. As before, we divide each term by the volume, so that $\frac{1}{2} m v^2$ becomes $\frac{1}{2} \rho v^2$, and the work done on the liquid becomes $P_1 - P_2$. Similarly the difference in the potential energies, $Mgy_1 - Mgy_2$, becomes $\rho g y_1 - \rho g y_2$. The expanded Bernoulli equation is then $\frac{1}{2} \rho v_1^2 + \rho g y_1 + P_1 = \frac{1}{2} \rho v_2^2 + \rho g y_2 + P_2$.

EXAMPLE 8

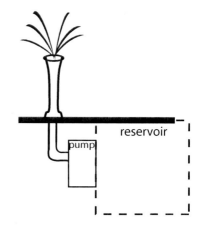

A pump is 1.2 m below the surface of a reservoir. It provides the difference in pressure for a fountain 3 m above the surface of the reservoir, from which water shoots out with a speed of 2 m/s. What is the pressure difference that must be supplied by the pump?

Ans.:
Assume that v_1, the velocity at the intake to the pump, is zero. Take the reference level to be at y_1, i.e., $y_1 = 0$. Bernoulli's equation then reduces to $P_1 - P_2 = \frac{1}{2}\rho v_2^2 + \rho g y_2$.

The density of water is 10^3 kg/m^3, so that $P_1 - P_2 = (\frac{1}{2})(10^3)(2^2) + (10^3)(9.8)(4.2) = 43.2 \times 10^3$ Pa.

Absolute temperature

After Torricelli's work a number of other discoveries were made in the seventeenth and eighteenth centuries in the study of gases. One is that the product of the pressure and the volume of a given amount of gas is constant if the temperature remains constant. This relation is usually called Boyle's law, after Robert Boyle (1627–1691). It is an empirical law. Like the other relations between pressure, volume, and temperature that we discuss next, it is approximate and subject to various limitations.

Another discovery was that at a given pressure the volume and the temperature are linearly related, in other words, the graph of volume against temperature is then a straight line. The graph of pressure against temperature is also a straight line. A particularly significant feature of this relation can be seen in the figure.

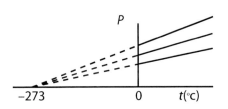

The straight lines of pressure against temperature are different for different volumes, but when they are extrapolated to lower pressures they converge at the point where the pressure is zero. At this point the temperature is $-273°C$. Since no lower pressure is possible, this is also the lowest possible temperature. It is called the *absolute zero* of temperature. Temperatures measured from absolute zero are called *absolute temperatures*, and the symbol T is used for them. (t is used for other temperature scales, such as the Celsius scale. On this scale $0°C$ is the temperature of melting ice and $100°C$ is the temperature of boiling water.) T (in K) $= t$ (in °C) $+ 273$.

The scale that starts at the absolute zero of temperature, and whose units are the same as for the Celsius scale, is called the Kelvin scale (after William Thomson, Lord Kelvin, 1824–1907). Its unit is called the *kelvin* (K). If T is the temperature on the Kelvin scale and t is the temperature on the Celsius scale, $T = t + 273$. On the Kelvin scale volume and temperature are proportional.

The two relations for pressure, volume, and temperature can be combined to say that $\frac{PV}{T}$ is constant for a given amount of gas. If we compare the pressure, volume, and absolute temperature at two different times, i.e., in two different *states* of a certain amount of gas, we can write $\frac{P_1 V_1}{T_1} = \frac{P_2 V_2}{T_2}$. Another way to write the same relation is that the product PV is proportional to T ($PV \propto T$), i.e., it is equal to a proportionality constant times T.

EXAMPLE 9

By what fraction does the pressure in a tire increase when the temperature goes from 20°C to 30°C?

Ans.:
Assume that the volume remains constant. The absolute temperature goes from $273 + 20$K or 293K to $273 + 30$K or 303K. The ratio is $\frac{303}{293} = 1.034$. The tire pressure increases by .034 of 3.4%.

So far we have considered a definite quantity or mass of gas. We can now incorporate the mass in our relation. If P and T are held constant, what can we say about V? For a given amount of gas it is constant, but what if the amount changes? Then the volume is proportional to the mass.

Experiments show that if the mass is measured in *moles* the relation between P, V, T and the mass is *universal*, i.e., independent of the kind of gas.

What Is a *Mole*?

In Chapter 1, we talked briefly about *atomic weight*. In the meantime we have also talked about mass, and you will not be surprised that we prefer talking about *atomic mass* rather than atomic weight. After all, your weight depends on where you are and is different at the top of a mountain from what it is at the bottom. And it is quite different on the moon. The mass remains the same, no matter where you are.

At first only relative masses were known, and one unit was assigned to hydrogen, which led to about four for helium, about 12 for carbon, and so on. Eventually it was realized that these numbers corresponded approximately to the number of particles in the nuclei of the various atoms.

To get exact numbers a standard was adopted, namely exactly 12 units (now called "atomic mass units," with the symbol u) for the most abundant kind of carbon, which is called "carbon 12" (^{12}C). It is the kind, or isotope, with six protons and six neutrons in its nucleus. It was then possible to assign atomic masses (in atomic mass units) to each element on the basis of measurements.

The quantity 12 grams of carbon 12 is called a mole of ^{12}C. (12 kg is a kilogram-mole.) The number of atoms in a mole is called *Avogadro's number*, given by the symbol N_A. A mole of hydrogen atoms consists of N_A atoms. A mole of hydrogen molecules consists of N_A molecules. Experiments show that N_A is 6.023×10^{23} molecules/mole or 6.023×10^{26} molecules/kg-mole.

If we measure the mass of gas by the number of moles, n, we can write the relation between P, V, T and the mass as $PV = nRT$, where R is the proportionality constant. Its value is determined from experiments and is about 8.31 J/mol K, but remember that the number depends on the units that are used.

The relation $PV = nRT$ summarizes the results of many experiments. This single statement describes how the macroscopic quantities P, V, T, and n are related, regardless of which kind of gas is being measured. It is therefore sometimes called the *universal gas law*. More often it is called the *ideal gas law* to emphasize that real gases follow it only approximately, as an "ideal" to be approached under special circumstances.

$PV = nRT$ is an empirical law, i.e., it is based on experiment. It represents a model. Real gases follow it more closely as the density decreases i.e., when the molecules are far apart and so have little interaction with each other.

EXAMPLE 10

What is the volume of 1 mole of gas at atmospheric pressure and 0°C?

Ans.:
$V = \frac{nRT}{P} = \frac{(1)(8.31)(273)}{1.01 \times 10^5} = 0.0225 \, \text{m}^3$ or 22.5 liters.

We have to be careful with the units. We have not been sticking to SI units, since we have defined the mole as the number of molecules in 12 g of ^{12}C. R is then 8.31 J/mol K. If we were to use SI units consistently we would have to use the kg-mole, which is 10^3 as large. R is then also 10^3 as large, and equal to 8.31×10^3 J/kg-mol K. The volume of 1 kg-mole at the same pressure and temperature is $\frac{(1)(8.31 \times 10^3)(273)}{1.01 \times 10^5} = 22.5 \, \text{m}^3$.

The surprising bridge to temperature

The number of molecules in one mole is Avogadro's number, N_A. The number of molecules in n moles, N, is therefore nN_A, so that $PV = \frac{N}{N_A}RT$. The constant $\frac{R}{N_A}$ is called Boltzmann's constant, k, so that $PV = NkT$. ($k = \frac{8.31}{6.023 \times 10^{23}} = 1.38 \times 10^{-23} \frac{\text{J}}{\text{molecule-K}}$.)

We now have two relations for the same quantity, PV, on the left-hand side. One is for the microscopic model of molecules flying freely except when they bounce into each other or off the walls. It is $PV = \left(\frac{2}{3}N\right)\left(\frac{1}{2}mv^2\right)$. The other is for the approximate observed macroscopic behavior $PV = nRT$ or $PV = NkT$.

If both are to describe the same situation, the right-hand sides of the two relations must be the same, i.e., $\left(\frac{2}{3}N\right)\left(\frac{1}{2}mv^2\right) = NkT$ or

$$\overline{\frac{1}{2}mv^2} = \frac{3}{2}kT.$$

This is one of those very special relations that provide a link between two different realms. We have here two different ways of looking at the same set of phenomena using two quite different languages. On the left-hand side is the average kinetic energy of a molecule. It provides a look at the microscopic constituents. On the right-hand side is the *temperature*. It is a macroscopic quantity, which characterizes the whole container of gas.

We see that the quantity that we observe on the macroscopic level as temperature is a manifestation of the motion, on the microscopic level, of the molecules: *the average kinetic energy of a* molecule in a gas is proportional to the absolute temperature. Moreover, this is so regardless of what kind of molecules we are talking about.

We first defined the ideal gas as a theoretical microscopic model with small, rigid molecules that have only kinetic energy. Then we used the relation $PV = nRT$ to describe the approximate experimental behavior of real gases. It is the confluence of the two approaches, the assumption that both describe the same system, that leads to the result $\overline{\frac{1}{2}mv^2} = \frac{3}{2}kT$.

This relation is only for the ideal gas of particles whose internal structure plays no role. But the conclusion that temperature is a manifestation of energy on the microscopic scale turns out to be universally valid, and is one of the great insights of science.

EXAMPLE 11

What are the average energy and the root-mean-square speed, v_{rms}, of oxygen molecules at a room temperature of 300 K (=27°C)?

Ans.:

It is important to stay with SI units for all of the calculations. Only if you use a consistent system of units (such as the SI system) do you know that the answer will also come out in the same system of units. (Remember that "constants" may have units also.) You may occasionally want to use other units, such as grams, moles, or electron volts, but the units on the two sides of an equation need to be the same. If there is any doubt it is best to include units in every term.

The energy is $\frac{3}{2}kT = (1.5)(1.38 \times 10^{-23})(300) = 6.21 \times 10^{-21}$ J.

We can express the energy in electron volts (eV), where 1 eV = 1.6×10^{-19} J. We can convert the energy from joules to electron volts by multiplying by $\frac{1 eV}{1.6 \times 10^{-19} J}$ to get 0.039 eV.

The energy is kinetic energy, $\frac{1}{2}mv^2$, so that $\overline{v^2} = \frac{2E_K}{m}$.

One mole of oxygen molecules has 6.023×10^{23} molecules and a mass of 32 g. The mass of one molecule is therefore $\frac{32}{6.023 \times 10^{23}}$ g = 5.31×10^{-23} g = 5.31×10^{-26} kg.

$v_{rms}\sqrt{\overline{v^2}} = \sqrt{\frac{(2)(6.21 \times 10^{-21})}{5.31 \times 10^{-26}}} = \sqrt{2.34 \times 10^5} = $ 484 m/s. Note that this is quite a high speed, larger than the speed of sound.

Internal energy and heat capacity

The only energy of a molecule of an ideal gas is its kinetic energy. For an amount of gas consisting of N molecules the energy is $(N)(\frac{1}{2}mv^2)$, which we now see to be equal to $\frac{3}{2}nRT$. This is the *internal energy* of the gas, which we have previously called U.

Can we observe and measure the internal energy? We can measure how much energy it takes to *change* it, as, for example, by heating. The amount of added energy for each unit of temperature change is called the *heat capacity*, C. The heat capacity per unit mass is called the *specific heat capacity*, c.

We can see what the heat capacity is for our model. If we increase the internal energy U to $U+\Delta U$, the temperature will increase to $T+\Delta T$, and $U+\Delta U = \frac{3}{2}nR(T+\Delta T)$. We can subtract U from the left side and the equal quantity $\frac{3}{2}nRT$ from the right side, to leave $\Delta U = \frac{3}{2}nR\Delta T$. We see that the heat capacity, i.e., the amount of energy to change ΔT by one kelvin, is $\frac{3}{2}nR$.

For the ideal gas model the heat capacity is $\frac{3}{2}nR$. We can say that the model *predicts* a heat capacity of $\frac{3}{2}nR$. This is a remarkable result. It says that the specific heat capacity per mole of an ideal gas (its *molar heat capacity*) is $\frac{3}{2}R (=12.5 \frac{J}{mol\,K})$, no matter what the gas is. Here are the experimentally observed molar heat capacities per mole for some gases. (These are the "noble" gases. Because of their complete electron shells they interact with each other only extremely weakly, and do not form molecules.)

Gas	C (J/mol K)
He	12.5
Ne	12.5
Ar	12.5
Kr	12.3

We see that the measured quantities are close to each other and to $\frac{3}{2}R$ for each of the gases. This is a major triumph for the model.

You notice that the list of gases does not include the common ones, oxygen, nitrogen, and hydrogen. The reason is that for these gases each molecule consists of two atoms. Each of these molecules has both kinetic and potential energy, and our model, which treats each as a particle, is inadequate.

EXAMPLE 12

6 J of energy is transferred by heating 1 g of neon (atomic mass 22.5 g/mol). What is the rise in temperature?

Ans.:
$$Q = \Delta U$$
$$\Delta U = \tfrac{3}{2} nR\Delta T$$
$$Q = C\Delta T, \text{ where } C = \tfrac{3}{2} nR$$
$$n = \tfrac{1}{22.5}$$
$$C = \tfrac{3}{2} \tfrac{1}{22.5}(8.31) = 0.554 \text{ J/K}$$
$$\Delta T = \tfrac{Q}{C} = \tfrac{6}{0.554} = 10.8 \text{ K}$$

EXAMPLE 13

0.1 kg of water at 80°C is mixed with 0.2 kg of water at 20°C. What is the final temperature?

Ans.:
Call the final temperature t. (Here there is no need to use absolute temperature, T. We use °C and the symbol t.) The 0.1 kg of water changes from 80°C to t, and gives up an amount of energy of $mc\Delta t = (0.1)(c)(80-t)$. This amount of energy goes to the 0.2 kg of water, which gains $0.2(c)(t-20)$. Since the two amounts of energy are the same, $(0.1)(c)(80-t) = (0.2)(c)(t-20)$, which leads to $t = 40°C$.

Heating

Let's look more closely at what happens when we increase the internal energy by heating. We'll talk about ideal gases, but the process is similar in all materials. We start with a container of gas at some temperature T and internal energy $\tfrac{3}{2} nRT$. We take a second container of gas, at a higher temperature, and bring the two gases into contact.

From the macroscopic point of view energy "flows" from the hotter to the cooler gas. The microscopic point of view helps us to understand how this happens. The molecules of the hotter gas move, on average, with greater speeds. When they collide with the molecules of the cooler gas they transfer some of their kinetic energy to them. The internal energy of the hotter gas decreases and it becomes cooler, and the internal energy of the cooler gas increases and it becomes hotter, until their temperatures are the same. There is then no further net transfer of energy, and the two gases are in *thermal equilibrium*.

Note that we use the words *heat* and *heating* only for the *transfer* of energy. (The gas does not *have* heat energy.) The transfer of energy changes the *internal* energy, which in an ideal gas is the kinetic energy of the molecules.

Work

A second way in which the internal energy can be changed is by doing work. (We will again talk about an ideal gas, but the considerations are similar for other materials.) We know that work is done on an object when a force acts on it and it moves. We can do work on a gas by compressing it with a piston that moves in a cylinder. If the force, F, is applied with a piston, and the piston moves a distance s, the work done on the gas is Fs. We can write this as $(\tfrac{F}{A})(sA)$, where A is the cross-sectional area of the cylinder. $\tfrac{F}{A}$ is the pressure (P) applied to the gas and sA is the change in the volume (ΔV) of the gas in the cylinder. The work done on the gas can therefore be written as $P\Delta V$.

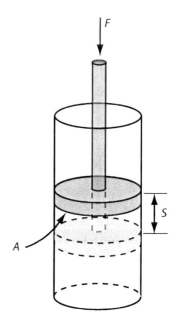

We can describe the process on a graph of P against V (a *PV diagram.*) In the figure the pressure is constant and the volume changes from V_1 to V_2. The magnitude of the work done on the gas is equal to that of the rectangle $P\Delta V$ under the curve. If the pressure is not constant, the work is still equal to the area under the curve.

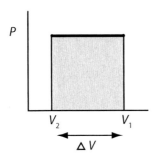

EXAMPLE 14

An ideal gas is compressed to $\frac{1}{3}$ of its initial volume by a piston in a cylinder.

(a) What is the change in the pressure if the temperature does not change?

(b) What is the change in the temperature if the pressure does not change?

(c) What is the change in the temperature if at the same time the pressure increases by a factor of 4?

Ans.:

(a) $PV = nRT$. If T does not change, PV remains constant, and $P_1 V_1 = P_2 V_2$, or $\frac{V_2}{V_1} = \frac{P_1}{P_2}$. Since $\frac{V_2}{V_1} = \frac{1}{3}$, $\frac{P_1}{P_2} = \frac{1}{3}$ and $P_2 = 3P_1$. The pressure increases by a factor of 3.

(b) Since P, n, and R are constant, $V \propto T$, and both the volume and the absolute temperature change by the same factor.

(c) $PV = nRT$. Since n and R are constant, $\frac{PV}{T}$ is constant, and $\frac{P_1 V_1}{T_1} = \frac{P_2 V_2}{T_2}$. $\frac{V_2}{V_1} = \frac{1}{3}$, $\frac{P_2}{P_1} = 4$, so that $\frac{T_2}{T_1} = (\frac{P_2}{P_1})(\frac{V_2}{V_1}) = \frac{4}{3}$.

The first law of thermodynamics

Both heating and doing work on a gas are processes that increase the internal energy of the gas. The initial internal energy of the gas (U_1), plus the energy transferred to it by heating it (Q) and doing work on it (W), is equal to its final internal energy (U_2). In this form and context the law of conservation of energy is called *the first law of thermodynamics*: $U_1 + Q + W = U_2$.

EXAMPLE 15

The pressure on a container with 18 moles of an ideal gas increases from 1 atmosphere to 2.5 atmospheres while the volume decreases from $0.4\,\text{m}^3$ to $0.2\,\text{m}^3$ along a straight line on a PV diagram.

(a) Draw a PV diagram.

(b) How much work is done on the gas?

(c) What are the initial and final temperatures?

(d) What are the initial and final internal energies?

(e) How much heat is transferred to the gas during this process?

Ans.:

(a)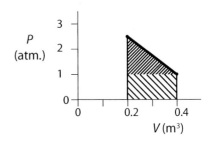

(b) Area under the line on the diagram: $(1)(.2) + (\frac{1}{2})(.2)(1.5) = .35$. This is in (atmospheres)(m³) and has to be changed to SI units. 1 atmosphere $= 10^5$ Pa. The area in joules is therefore $.35 \times 10^5$ J. This is the work done on the gas.

(c) $T_1 = \frac{P_1 V_1}{nR} = \frac{(10^5)(0.4)}{(18)(8.31)} = 267$ K or $-6°$C.
$T_2 = \frac{(2.5 \times 10^5)(0.2)}{(18)(8.31)} = 334$ K or $61°$C.

(d) $U_1 = nRT_1 = 4 \times 10^4$ J.
$U_2 = nRT_2 = 5 \times 10^4$ J.

(e) $U_1 + W + Q = U_2$.
$Q = 5 \times 10^4 - 4 \times 10^4 - 3.5 \times 10^4 = -2.5 \times 10^4 = -2.5 \times 10^4$ J.
2.5×10^4 J is transferred out of the gas.

Go to the PhET website (http://phet.colorado.edu) and open the simulation *Gas Properties*.

Explore the relationship between volume, pressure, and temperature, and the effect of work done or heat applied to a system.

7.2 Other systems: adding pieces from reality

The ideal-gas model assumes that the constituents have no internal structure and no forces between them except on contact. In this section we explore some other systems and the models that help us to understand their properties and behavior.

In contrast to what we have done so far, we will need to consider forces, within the molecules as well as those between them. These forces are electric forces, following Coulomb's law.

The gravitational forces are still there, but they are smaller by a huge factor, so that we can neglect them. The nuclear forces act only over distances much smaller than the size of atoms and play their role almost completely within the nuclei. On the other hand, in each atom the electrons are held to the nuclei by electric forces. The forces between atoms in molecules and solids, and those between molecules, are also entirely electric.

EXAMPLE 16

In a hydrogen atom the average distance between the proton and the electron is 0.53×10^{-10} m. The mass of the proton is 1.67×10^{-27} kg and the mass of the electron is 0.91×10^{-30} kg. The charge on each is $\pm 1.6 \times 10^{-19}$ C.

(a) What is the magnitude of the gravitational interaction between the two particles?

(b) What is the magnitude of the electric interaction between them?

(c) What is the ratio of the electric force to the gravitational force that they exert on each other?

Ans.:

(a) $F_g = G\frac{M_1 M_2}{r^2} = 1.67 \times 10^{-11} \frac{(1.67 \times 10^{-27})(0.91 \times 10^{-30})}{(0.53 \times 10^{-10})^2}$
$= 9.04 \times 10^{-48}$ N.

(b) $F_e = K\frac{Q_1 Q_2}{r^2} = 9 \times 10^9 \frac{(1.6 \times 10^{-19})^2}{(0.53 \times 10^{-10})^2} = 8.5 \times 10^{-8}$ N.

(c) $\frac{F_e}{F_g} = 0.9 \times 10^{40}$.

We will, for the time being, not consider what happens inside the nuclei. They have internal structure and internal energy, but their energy levels are separated by energies of the order of 10^5 times as large as those of atoms. The amount of energy that is required to change their energy is so large that we are justified in continuing to consider each nucleus to be a structureless particle.

Molecules

In molecules the interatomic electrical forces hold two or more atoms, the same kind or different, to each other. The molecules have internal motions of vibration and rotation, and as a result have additional kinetic and potential energies.

Diatomic molecules, such as those of hydrogen, nitrogen, and oxygen, consist of two atoms. They can be very successfully described by a model in which the force between the two atoms is represented by a spring. In this model the two atoms vibrate along the line that joins them.

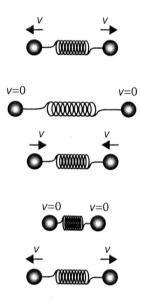

Let's look at the vibrating motion in more detail. We'll assume that the spring follows Hooke's law, $F = -kx$. The vibrational motion of an object connected to a spring that follows Hooke's law is called *simple harmonic motion*. As each atom goes back and forth, its energy changes from kinetic to potential and back. At the two ends of the motion the atoms are stopped and all of the energy is stored in the spring as potential energy. In between, in the middle, no energy is stored in the spring and all of the energy is kinetic. On average, as the atoms vibrates, it turns out that there is just as much kinetic energy as potential energy. The diagram shows the model of a diatomic molecule in five different positions of its vibration.

Vibration is not the only possible motion. The molecules can also rotate about an axis perpendicular to the line that joins them. The diagram shows the model in three different positions as it rotates about an axis at right angles to the paper. In addition, it can rotate about a second axis that is also perpendicular to the line that joins the atoms, but lies in the plane of the paper. In this rotation one of the atoms starts by coming out of the paper, while the other one moves into the paper. (There is a third axis, joining the two atoms, but rotation around it contributes so little to the internal energy that we can neglect it.)

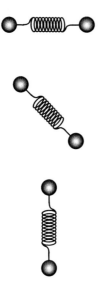

Just as for monatomic gases, we can find out about the internal energy of the gas by measuring the heat capacity. If we heat a monatomic gas, all the energy that we transfer to it becomes kinetic energy of the motion of the individual atoms. For the diatomic gas some of it becomes energy internal to the molecules, namely energy of vibration and energy of rotation. Because any added energy is divided between these various kinds of motion and their kinetic and potential energies, it now takes more energy to raise the temperature, and so the heat capacity is larger.

If we know how the energy is divided we can calculate the heat capacity. It turns out that at sufficiently high temperatures each part of the internal energy takes up an equal amount of energy. The vibrational energy has two such parts, the elastic potential energy of the spring and the kinetic energy, back and forth, of the vibrating motion. The rotational energy has to be counted as two parts, because the molecule can rotate about two different axes at right angles to each other, both of them perpendicular to the line joining the two atoms. (There is no potential energy associated with the rotational motion.) For the translational energy of the molecule as a whole we also have to count more than one part,

because there are three different perpendicular directions, $x, y,$ and z.

What we have called "parts of the energy" are called *degrees of freedom*. The rule that the energy is divided equally among them is called the *principle of equipartition of energy*. It holds as long as each kind of motion, i.e., each of the degrees of freedom, is equally accessible.

The energies are not all equally accessible when the energy is *quantized*, i.e., when only certain values of the energy are allowed. The energies between the allowed energy levels are then "forbidden" and are not accessible. At high temperatures the allowed energies are very close to each other, so that the forbidden energies do not play a determining role. The principle of equipartition therefore holds most closely at high temperatures. It fails at temperatures that are so low that the spacing between the allowed energies is of a size comparable to the *thermal energy*, ie., to about kT.

EXAMPLE 17

What is the internal energy of a mole of an ideal diatomic gas under the assumption that the principle of equipartition holds?

Ans.:

The internal energy of a diatomic molecule is $\frac{1}{2}kT$ for each degree of freedom (if the temperature is sufficiently high that the quantization of energy has no measurable effect). For one mole the internal energy is N_A times as large, or $\frac{1}{2}RT$. At room temperature (about 300 K), $\frac{1}{2}RT = (.5)(8.31)(300) = 1.25 \times 10^3$ J/mol for each degree of freedom.

There are three degrees of freedom for the translational (linear) motion, two degrees of freedom for the vibrational motion, and two more degrees of freedom for the rotational motion, for a total of seven. The internal energy is 8.73×10^3 J/mol.

Phase changes

In a gas the molecules fly around freely. If the temperature goes down, the molecules slow down. In the ideal gas, where $PV = nRT$, the molecules can slow down until they don't move at all when the absolute temperature reaches zero. A real gas behaves differently. At some temperature it *condenses* and becomes a liquid. At some still lower temperature it freezes and becomes a solid.

When water freezes to become ice, boils to become steam, or evaporates to become water vapor, its molecules don't change. Ice, water, and steam are different *phases* of the same material.

The phases have different properties. Most obviously, their densities are very different. In the gas phase the molecules are far apart. In the liquid phase they are close to each other, but are still able to move. In the solid phase each atom or molecule has a fixed place about which it can vibrate.

We know that a phase change takes place at a fixed temperature. Water freezes and ice melts at 0°C. Water boils and steam condenses at 100°C. At each phase change the internal energy changes by a definite amount. When ice changes to water its internal energy increases by 80×10^3 cal/kg, or 333.5 kJ/kg, and this amount of energy, called the *heat of fusion*, needs to be supplied to it. When it boils, water requires the *heat of vaporization* of 540×10^3 cal/kg or 2.257×10^3 kJ/kg.

When a solid is heated, its temperature rises until it begins to melt. Further heat causes more and more of it to melt while the temperature remains the same. When all of it has melted the temperature rises again, until it reaches the boiling point. The phase change from liquid to gas then continues until all of the material has changed to the gas phase.

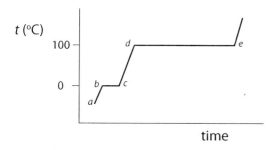

Here is a schematic graph to show the progression of the phases of water when it is heated at a steady rate, starting with ice. From *a* to *b* it is in the solid phase (ice), and warms until the melting point is reached at 0°C. The temperature then remains constant until all of the ice has melted at point *c*. The liquid water now warms to the boiling point, 100°C, point *d*. The temperature again remains constant, this time at 100°C, until all of the water has turned to steam at point *e*. With continued heating the temperature of the steam then continues to rise.

EXAMPLE 18

How much energy needs to be transferred to 0.2 kg of ice at $-10°C$ to raise its temperature so that it is water at $+10°C$?

Ans.:
We will need to know the specific heat capacity of water, which is $c_w = 10^3 \frac{cal}{kg\,K}$ or 4.18×10^3 J/kg K, and the specific heat capacity of ice, which is $c_i = 2.09 \times 10^3$ J/kg K. If the heat of fusion h_{iw} is equal to 333.5×10^3 J/kg then the required energy is $10mc_i + mh_{iw} + 10mc_w = (0.2)(10)(2.09 \times 10^3) + (0.2)(333.5 \times 10^3) + (0.2)(10)(4180) = 4.18 \times 10^3 + 66.7 \times 10^3 + 8.36 \times 10^3 = 75.1 \times 10^3$ J.

What happens from the microscopic point of view? What does our gas model of flying molecules suggest? As the temperature goes down and the molecules slow down, they spend more time near each other. The forces between them then have a greater chance to affect their motion. Eventually the disruptive effect of the ceaseless motion decreases sufficiently for the forces to bring about the *phase transition* from gas to liquid.

In the liquid phase the atoms or molecules continue to move, but their motion is much more confined. At the freezing (or melting) temperature there is again a phase transition as they become locked into place. The motion continues, but it is now a vibration around a fixed position.

In each of the phases or *states* of a material, the atoms or molecules are in motion, but not all with the same energy. There is a range or *distribution* of energies. In the liquid state, for example, some of the molecules move so rapidly that they are able to escape. This is the process of *evaporation*.

We can look at the progression from gas to solid as a path to greater order. The gas has the least amount of order as the molecules race around. In the solid the order is greatest as each atom assumes its position.

Both liquids and solids are called *condensed matter*. We will look at models for solids in the next section. In liquids the atoms or molecules are neither completely free to move nor locked into place. That makes them more difficult to describe, and there is no simple model. That's why we will not consider the microscopic aspects of liquids.

Condensed matter

Some solids consist of molecules held in place by the forces between them. This is so, for example, when large organic molecules form solids. In other solids the individual atoms are more likely to be the building blocks. When we talked about gases we called their constituents molecules. For solids it is usually more appropriate to think of the atoms.

In solids the interatomic forces are so strong that the atoms remain close to each other. The atoms vibrate about some equilibrium position, and so have both kinetic and potential energy. This vibration, the continuous shaking of the solid, is called its *thermal motion*. The sum of the kinetic and potential energies is the internal energy.

A model that we can call the *ideal solid* accounts for some of the thermal properties of solid materials. In this model the atoms are again without internal structure. They are connected to their neighbors by springs that represent the chemical bonds. The springs follow Hooke's law, $F = -kx$. The atoms vibrate about their equilibrium position. The internal energy is the energy of vibration, which is partly the kinetic energy of the motion and partly the elastic potential energy stored in the springs. Both of these energies increase when energy is transferred to the solid.

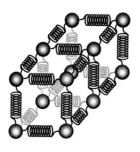

In this model the vibrational motion of each atom is *simple harmonic motion*, just as for the diatomic molecules that we discussed earlier. Again, as each atom vibrates, its energy changes from kinetic to potential and back. Again, on average, there is just as much kinetic energy as potential energy. This time, however, the vibration can be in any of the three directions, x, y, and z. We have to count the different parts of the energy, the different *degrees of freedom*, separately for each of these three directions. There is kinetic energy and potential energy along each axis, so that there are six degrees of freedom.

We can compare the average internal energy of an atom in the ideal solid to that of an atom in a monatomic ideal gas. The internal energy of the ideal gas is only kinetic. In the ideal solid, with the addition of an equal amount of potential energy, it is twice as large. In the model therefore, for a given number of atoms, the heat capacity is twice as large as for the ideal gas, namely $3R$ for one mole of material.

That this is in fact the case for a number of solids at and above room temperature was discovered experimentally by Pierre Dulong and Alexis Petit in 1819 and is known as the law of Dulong and Petit.

The model can be extended: as the vibration increases, eventually the springs no longer follow Hooke's law. They "break," and the atoms are cast adrift. The solid *melts*.

Metals

The model of atoms connected by springs cannot account for the phenomena that characterize metals, the most important of which is that they are good conductors of electricity. What characterizes metals is that some of the electrons in the atoms are separated from the nuclei to which the others are bound.

The simplest model that includes this possibility is the *free electron model*. In it one or more of the electrons from each atom becomes detached and "free." The free electrons can move throughout the material, leaving behind the remaining parts of the atoms, which are now positively charged ions. (In the figure the electrons are shown as the small circles.)

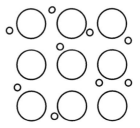

This model neglects the forces between the free electrons. They fly freely through the solid, and their only energy is kinetic energy, just as for the molecules of an ideal gas.

It would seem to be straightforward to see what the internal energy and hence the heat capacity are for this model. We expect the internal energy to be larger by the amount of the kinetic energy of the electrons. Since they are free to move, they act just like the molecules of an ideal gas. For a metal in which one electron from each atom is free, the model therefore leads to an additional energy of $\frac{3}{2}nRT$ and an additional heat capacity of $\frac{3}{2}R$ per mole.

This, however, is not observed! Experiments show that the heat capacity of metals is close to that of nonmetals. This was well known by the early part of the twentieth century. That a metal has free electrons was also well known. It was therefore a mystery that the free electrons did not seem to participate in the internal energy.

This is an example of a model that did not predict what was observed. The model seemed reasonable in terms of the knowledge at that time. Nevertheless its conclusions did not correspond to the observed facts. The discrepancy made it clear that in some fundamental way the nature of metals was not understood. Eventually it was learned that in metals electrons were indeed detached from the atoms, but that their behavior could not be described by classical mechanics.

There were other indications that the structure of atoms and the interactions between them were not understood. The turning point came in 1925 when quantum mechanics was first developed, and led to today's understanding of atomic phenomena. We discuss some of the principles of quantum mechanics later.

Chemical energy

The forces between atoms are stronger, and the atoms are held more tightly, in some molecules, and in some solids, than in others. Chemical transformations can take place, with atoms of one kind changing places with those of another. The positions and the momenta of the atoms, and of the electrons within them, change, as well as their potential and kinetic energies.

On the macroscopic scale we call the energy that is then released or absorbed *chemical* energy. On the microscopic scale it represents the changes in the kinetic and in the electric potential energies of the electrons and atoms.

Quantum theory

Except for our brief mention of the detachment of electrons in metals, we have not considered changes within atoms in this chapter. Such changes generally require more energy than changes that are external to the atoms.

Each system, whether it is a solid, a molecule, an atom, or a nucleus, can exist only with certain definite amounts of energy. They are said to be in particular *energy states*. The lowest state is called the *ground state*. The others are called *excited states*. It therefore takes a definite minimum amount of energy to change the state of a system. This energy gets larger as we go down in size to the atom and to the nucleus. It takes a minimum of 10.2 eV to change the energy of a hydrogen atom from its ground state to the first excited state. It takes 2.3 MeV to change a nucleus of nitrogen (^{14}N) from its ground state. In each of these cases a smaller amount of energy cannot be accepted by the system.

Our consideration of the ideal gas showed that the average kinetic energy per molecule is $\frac{3}{2}kT$. Let's see how much that is. Boltzmann's constant, k, is equal to 1.38×10^{-23} J/molecule-deg. At room temperature, with T about 300 K, $\frac{3}{2}kT$ is about 0.04 eV. This is the average kinetic energy per molecule in a gas. The amount of energy that can be exchanged between molecules when they collide is therefore also about 0.04 eV. It takes about 250 times as much energy to raise the energy of a hydrogen atom from that of its lowest state. (For other atoms the amount of energy is different, but it is of the same order of magnitude.) We see that in most collisions between hydrogen atoms the amount of energy that can be transferred is not enough to change the internal energy of the atoms.

This is a fact of great importance. After all, it was the starting point of our discussion of the ideal gas that its constituents behave as particles without internal structure or internal energy. If the kinetic energy (of about $\frac{3}{2}kT$) were large enough to change the internal energy of an atom, the assumption of a rigid atom, one that is unchanged by collisions, would not have been appropriate or fruitful.

Suppose for a moment that there were no such minimum energy. Even a small amount of energy given to an atom, either in collisions or in any other way, could then become internal kinetic and potential energy of the atom. The assumption that we made in the ideal gas model that the atoms have no internal energy, and are like rigid marbles without internal structure, could not be made. The real gas would behave quite differently, and the ideal gas model that we have discussed here would not represent its properties.

The definite, finite energy required to change the energy of an atom, as well as that of a molecule or solid, plays a crucial role in the stability of these structures. If they were able to accept any amount, no matter how small, they would have properties quite different from those of a model that treats the atoms or molecules as particles.

If this had been understood when the ideal gas model was developed, the *quantization* of energy and the *quantum theory* might have been discovered much earlier than they were, by a much more circuitous route, by Planck in 1900.

Back to heat capacities

The heat capacities of all solids decrease as the temperature is lowered. The molar heat capacities are then no longer $3R$, as expected from the law of Dulong and Petit. The graph shows the specific heat capacity of silver as a function of temperature.

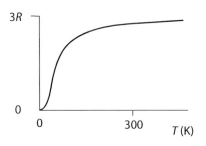

Again the model's predictions do not correspond to what is observed. It was Einstein, in 1907, who first realized why the old model was inadequate: it did not take into account the quantization of energy, i.e., the fact that individual atoms, as well as solid materials, can exist only with definite energies separated from each other.

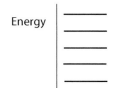

The figure shows an *energy level diagram*. Each horizontal line represents one of the allowed energies. For a particle in simple harmonic motion (i.e., vibrating as if attached to a spring following Hooke's law) the energy levels are equally spaced.

Just as in the ideal gas, the internal energy is higher at higher temperatures. At temperatures near room temperature and higher, the spacing between the possible energies is small compared to the internal energy and the fact that the energy is quantized does not play a significant role. As the temperature decreases, on the other hand, the internal energy becomes smaller, and the spacing between the energy levels plays a more important role.

We saw earlier that when there is potential energy in addition to kinetic energy, the heat capacity is larger. In general, the heat capacity is larger when there are more kinds of energy and more ways for the energy to be distributed. This time we have the opposite situation. Because the energy is quantized there are gaps between the allowed energies. There are fewer possible values of the energy and the heat capacity is smaller.

This is a good example of how the limits of one model led to the invention of a different model. In this case it led to important new knowledge, and the confirmation of the quantization of energy, which had been discovered a short time earlier.

Diatomic gases

The heat capacity of the monatomic gases is $\frac{3}{2}nR$, i.e., $\frac{1}{2}nR$ for each of the three degrees of freedom, represented by mv_x^2, mv_y^2, and mv_z^2. For the diatomic gases there are two more degrees of freedom for the vibrational motion's additional kinetic and potential energies, and another two for the rotations about the two axes. With contributions of $\frac{1}{2}nR$ for each, you expect heat capacities of $\frac{7}{2}nR$.

Gas	C (J/mol K)
H_2	20.4
N_2	20.8
O_2	21.1
CO	21.0

The table shows heat capacities per mole for some of the diatomic gases at room temperature, and they are seen to be only about $\frac{5}{2}R (= 20.8$ J/mol K$)$. Where did the other two degrees of freedom go?

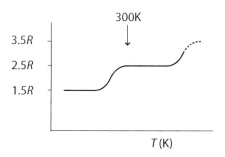

Here is a schematic graph of the molar heat capacity of hydrogen as a function of temperature. We see that at sufficiently high temperatures the heat capacity really does go up toward $\frac{7}{2}R$. To see what happens to the heat capacity at lower temperatures we have to look at the spacing of the energy levels of the hydrogen molecule. It turns out that for the rotational motion the spacing is much smaller than for the vibration. So much so that room temperature is "high" for the rotational motion and "low" for the vibrational motion. At room temperature the molecules don't vibrate, but they do rotate, leading to the $\frac{5}{2}R$ molar heat capacity. At much lower temperatures the spacing between the rotational energy levels also becomes significant (compared to kT), and the molar heat capacity decreases to $\frac{3}{2}R$.

The spacing between the vibrational levels is 0.53 eV, so that the energy available, on average, at room temperature, equal to kT, or about 0.025 eV, is insufficient to raise the energy of the molecules from that of the ground state. In other words, the molecules can vibrate only at much higher temperature. On the other hand, the energy difference between the ground state and the first excited rotational state is about 7×10^{-3} eV. Since this is much smaller than kT at room temperature, there is enough energy at room temperature to populate many of the rotational energy levels. The energy of the molecules then includes the rotational energy. This can be seen in the heat capacity, which is $\frac{5}{2}R$, representing both the translational and the rotational degrees of freedom.

7.3 Summary

Whenever we talk about materials, whether they are gases, liquids, or solids, we have to remember that we are dealing with enormous numbers of particles—the atoms and molecules of which the material is made. It is remarkable how much we can say about these complex systems by using simple models. We look in detail at one of the most successful models, the *ideal gas*. What we observe with our senses are the large-scale or *macroscopic* objects. Each is made of small particles (atoms, molecules) on the *microscopic* scale. The microscopic features are those of the particles and their motion. The discussion of the ideal gas shows how the microscopic features are related to the properties that we observe on the macroscopic scale, such as pressure, volume, and temperature.

On the microscopic level we think in terms of the number of molecules, N. On the macroscopic level we think in terms of the number of moles, n. The two are linked by Avogadro's number, N_A, the number of molecules in a mole: $N = nN_A$.

A real gas consists of particles that move freely. They have both kinetic energy and mutual potential energy as they move toward and away from each other. Together, the sum of these two kinds of energy for all of the molecules is the internal energy (often called *thermal* energy) of the gas. Each atom and molecule also has internal energy, but since these energies do not change so easily they can often be ignored when we talk about the properties of gases.

For the model that we call the ideal gas we make several assumptions. It consists of particles that have no internal energy or internal structure and no size. There are no forces between them except when they collide, so that there is no mutual potential energy. The only energy of an ideal gas is the kinetic energy of its particles. The sum of the kinetic energies of the particles is the internal energy of the gas.

From our knowledge of mechanics we showed that $PV = \frac{1}{3}N\overline{mv^2}$ or $PV = \frac{2}{3}N\overline{\frac{1}{2}mv^2}$. Here P is the pressure ($P = \frac{F}{A}$), N is the number of particles, and $\overline{\frac{1}{2}mv^2}$ is the average energy per particle.

Empirically, i.e., from observation and experiment, a relation followed approximately by many gases is $PV = nRT$. Here n is the number of moles and T is the absolute temperature, i.e., the temperature measured from absolute zero, the lowest possible temperature, equal to zero kelvins (0 K) and to $-273°$C. R is the *universal gas constant*. By comparing the relation from the application of mechanics to the motion of the particles to the empirical macroscopic relation we see that $\overline{\frac{1}{2}mv^2} = \frac{3}{2}kT$, where k (Boltzmann's constant) is equal to $\frac{R}{N_A}$. The relation shows that the average energy per particle is proportional to the absolute temperature. The relation is for the *ideal gas*, but the conclusion that the particles of a gas have greater average energy at higher temperature is always true.

That shows what temperature is: when energy is transferred to an object (as by heating), it becomes energy of its constituents, i.e., internal energy. We detect and perceive the higher internal energy as higher temperature. A change in the internal energy can also accompany a change of structure, i.e., a *phase change* such as that from solid to liquid or from liquid to gas.

The pressure at the bottom of a column of liquid of height h is $\rho g h$, where ρ is the density of the liquid. Normal atmospheric pressure is about the same as the pressure at the bottom of a column of mercury whose height is 76 cm.

In a flowing liquid, energy considerations lead to *Bernoulli's equation*, $\frac{1}{2}\rho v_1^2 + \rho g y_1 + P_1 = \frac{1}{2}\rho v_2^2 + \rho g y_2 + P_2$.

The work done on a gas when its volume changes is $P\Delta V$.

7.4 Review activities and problems

Guided review

1. A cubical box, 1 cm to the side, holds 1000 molecules, each with a mass of 10^{-25} kg. The molecules move back and forth in the x direction with speeds of 1000 m/s.
   (a) What is the magnitude of the momentum of each molecule?
   (b) What is the change of momentum when a molecule hits a wall?
   (c) What is the time between collisions of a molecule with one wall?
   (d) What is the average force on one of the walls perpendicular to the x direction?
   (e) What is the pressure?

2. A cubical box, 1 cm to the side, holds 10^{19} molecules of an ideal gas, each with a mass of 10^{-25} kg. The pressure is 8×10^4 N/m².
   (a) What is the average kinetic energy per molecule?
   (b) What is v_{rms}?

3. A U-tube contains oil whose density is 0.92 times that of water. The height of the column of oil on one side is 15 cm greater than that on the other side. What is the difference in the pressures above the oil on the two sides?

4. What is the total pressure at the bottom of a lake at a depth of 30 m?

5. (a) A hole in the ground is filled with water to a depth of 30 m. A pump is put next to the top of the hole to pump out the water through a tube to the bottom of the hole. What happens when the pump is turned on?
   (b) What would be a better arrangement to pump out the water?

6. An iceberg whose density is 900 kg/m³ floats in a lake, 90% submerged. When the iceberg melts, will the level of the lake rise, fall, or remain the same?

7. To separate two pieces of paper whose area is 10 cm by 10 cm, you blow across them with an air speed of 2 m/s. What is the resulting force on each?

8. A pump that produces a pressure difference of 70 kPa is 0.8 m below ground level. A faucet

connected to it is on the second floor, 5 m above ground level. What is the speed of the water coming out of the faucet?

9. As a balloon rises, its volume goes from $10\,\text{m}^3$ to $12\,\text{m}^3$. The temperature goes from $20°\text{C}$ to $-20°\text{C}$. The initial pressure in the balloon is 1.1 atmospheres. What is the final pressure in atmospheres?

10. (a) How many moles of air are in the balloon of the previous question?
    (b) What is the mass of air in the balloon? (The average molecular mass is about 29 g/mol.)

11. The main constituents of air are oxygen and nitrogen. What is the ratio of v_{rms} for oxygen to that of nitrogen?

12. How much energy does it take to heat 1 mole of a monatomic ideal gas through $18°\text{C}$?

13. 1 kg of water ($C = 4.18\,\frac{\text{J}}{\text{g K}}$) at $50°\text{C}$ is mixed with 0.2 kg of ethanol ($C = 2.46\,\frac{\text{J}}{\text{g K}}$) at $0°\text{C}$. What is the final temperature?

14. The pressure on five liters of an ideal gas in a container with a movable wall is raised by 50%. The temperature goes from $20°\text{C}$ to $80°\text{C}$. What is the final volume?

15. Two moles of an ideal gas go through a process during which 10^4 J of work is done on it without any change in temperature. Is there a heat transfer, and if there is, what are its magnitude and direction?

16. The distance between the two protons in a hydrogen molecule is 1.5×10^{-10} m. What is the net force between them?

17. At room temperature hydrogen molecules rotate freely, but do not vibrate. (The difference between the vibrational energy levels is much larger than kT.) What is the internal energy of one mole of hydrogen at $0°\text{C}$?

18. An ice block whose mass is 2 kg is at $0°\text{C}$. It is put into a container with 10 kg of water at $50°\text{C}$. What is the final temperature?

Problems and reasoning skill building

1. In the relation $v_{\text{rms}} \propto m^x$, where m is the molecular mass, what is x?

2. About how many molecules are in $1\,\text{mm}^3$ of air?

3. The internal energy of an ideal gas is $\frac{3}{2}nRT$. Why does the internal nuclear energy not get counted?

4. A scuba diver has 2.5 liters ($= 2.5 \times 10^{-3}\,\text{m}^3$) of air in his lungs when he is 10 m below the surface. He holds his breath as he rises. What will the volume (in liters) be when he reaches the surface?

5. A balloon is filled with air at $20°\text{C}$ and has a volume of $4 \times 10^{-3}\,\text{m}^3$ at atmospheric pressure. It is then submerged and tied to a coral reef, where the pressure is 250 kPa and the balloon's volume is $1.58 \times 10^{-3}\,\text{m}^3$.
   (a) What is the temperature (in $°\text{C}$) at the reef?
   (b) What are the internal energies of the gas at the surface and at the reef?
   (c) Is work being done? By what? Can you tell how much?
   (d) Is heat being transferred? By what and to what? How can you tell?

6. 0.3 moles of an ideal gas at atmospheric pressure is enclosed in a container whose volume, V, is fixed. It is first submerged in ice water. Later it is submerged in boiling water.
   (a) Describe the changes in the macroscopic quantities, T and P, in words. Calculate the final values of T and P.
   (b) Describe the changes in the microscopic quantity v_{rms} in words. Calculate v_{rms} at both temperatures.
   (c) Draw a graph of P against T for this process.
   (d) Calculate the change in the internal energy.
   (e) Is there a heat transfer? In what direction? How large is it?

7. A copper coin whose mass is 5 g is dropped 400 m from a building. Air resistance causes it to slow down so that it has a velocity of 45 m/s just before it hits the ground. What is the change in the temperature of the coin as it falls? ($C = 0.39$ kJ/kg.) Assume that all of the dissipated energy goes to the coin.

8. What is the amount of energy that must be removed from 2 kg of water at $20°\text{C}$ to change it to ice at $-10°\text{C}$? (The specific heat capacity

is 4.18 $\frac{kJ}{kg\,K}$ for water and 2.09 $\frac{kJ}{kg\,K}$ for ice. The heat of fusion is 334 kJ/kg and the heat of vaporization is 2.26×10^3 kJ/kg.)

Multiple choice questions

1. 4 mol of an ideal gas is compressed at a constant pressure of 110 kPa from 0.25 m³ to 0.12 m³.

   The increase in internal energy is about
   (a) 21 kJ
   (b) 36 kJ
   (c) −21 kJ
   (d) −14 kJ

2. An ideal gas is in a container whose volume is V at a pressure P. The rms speed of the molecules is v_{rms}. When V increases by a factor of 4, and P decreases by a factor of 2, v_{rms} is multiplied by a factor of
   (a) $\frac{1}{2}$
   (b) 4
   (c) 2
   (d) $\sqrt{2}$
   (e) $\frac{1}{4}$

3.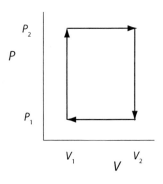

   An ideal gas is initially at P_1 and V_1. It changes along the path shown in the PV diagram. In one cycle (from the starting point all around and back to the same point on the diagram) the net work done by the gas is
   (a) $V_2(P_2 - P_1)$
   (b) $P_2 V_2 - P_1 V_1$
   (c) $(P_2 - P_1)(V_2 - V_1)$
   (d) $P_2(V_2 - V_1)$
   (e) $P_1 V_1$

4. 7.4×10^{-3} m³ of neon is at a pressure of 470 kPa and a temperature of 87°C. (The atomic mass of neon is 20.2 g/mol.) The mass of the gas (in kg) is about

   (a) 0.014
   (b) 0.023
   (c) 0.096
   (d) 0.059
   (e) 0.0014

The pressure of one mole of an ideal gas is reduced from 100 kPa to 50 kPa without changing the volume of 2×10^{-2} m³. How much heat, in kJ, is released by the gas during this process?
   (a) 2.5
   (b) 1.0
   (c) 3.0
   (d) 6.0
   (e) 1.5

5. The graph shows P plotted against T for a process that takes one mole of an ideal gas from point A to point B. Which of the following is correct?

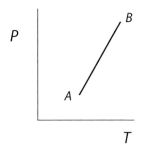

   (a) The volume increases.
   (b) The volume decreases.
   (c) The volume remains the same.
   (d) Any one of the above is possible.
   (e) none of the above is possible.

6. The diagram describes a cyclic process for an ideal gas. How much work (in J) is done in one cycle if $P_0 = 8.1 \times 10^5$ Pa and $V_0 = 7.0 \times 10^{-3}$ m³?

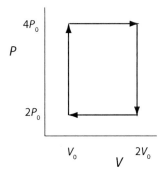

(a) 11,300
(b) 4320
(c) 21,200
(d) 22,600
(e) 43,200

7. A compound has a solid phase, a liquid phase, and a gas phase. A sample of this material is heated at a constant rate. The graph shows the temperature as a function of time. Which of the following conclusions can be drawn from this graph?

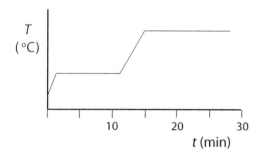

(a) The sample never boiled.
(b) The heat of fusion is greater than the heat of vaporization.
(c) After 5 min the sample is partly solid and partly liquid.
(d) The heat capacity of the solid phase is greater than that of the liquid.
(e) After 20 min the sample is totally liquid.

8. The figure shows a process from point A to point B for an ideal gas. Select the correct answers from the following:

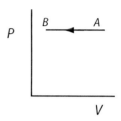

(a) Work is done on the gas (W is positive).
Work is done by the gas (W is negative).
(b) The internal energy increases.
The internal energy decreases.
(c) Heat is transferred to the gas.
Heat is transferred from the gas.

9. When P and V of a certain amount of gas are both raised by 50%, v_{rms} is multiplied by a factor of
(a) 0.5
(b) 1.5
(c) 2.25
(d) 3
(e) 5

CHAPTER 8

Electricity: It Is Everywhere

The electric force
> *A world full of charges*
> *Electricity and gravitation*
> *Separating charges: polarization*

The electric field
> *What is a field?*
> *Coulomb's law revisited: force and field*

Field lines and flux
> *Lines to represent the field*
> *Electric flux*
> *Charge, field, and flux: Gauss's law*
> *The gravitational field: solving Newton's problem*
> *Gauss's law and symmetry*

We think of the phenomena of mechanics as being everywhere around us, while electricity is more remote and unusual. It turns out instead that it is the electric force that dominates almost everything that impinges on our lives, although in ways that we are not so immediately aware of.

The first thing that comes to mind is the gadgetry of civilization. But more fundamentally, it is the electric forces between atoms that give rise to the existence of molecules, liquids, and solids with their endlessly varied properties, and it is the electric forces within atoms that are responsible for the existence of atoms and for the structure of the elements.

In this chapter we examine the properties of the electric force, and introduce the concept of the electric field, which profoundly changed the way we think about forces.

8.1 The electric force

A world full of charges

We rarely associate everyday experiences with electric forces. We know that electricity causes the shock that we sometimes feel after walking across a carpet, and that it is responsible for lightning. But its sweep goes enormously farther. It is at the root of almost everything that we are aware of, our very existence in all its vast variety.

On the one hand, our civilization depends, in some places almost entirely, on electricity. We use it for light, for heat and cold, and for motors in factories, farms, trains, and households. Without it there would be no communication by telephone, radio, or television, and none of the information technology based on computers.

It is, however, in the microscopic realm that the electric force reigns supreme. If we look at the size scale beyond the nucleus, but below the size of planets and stars, the electrical force is just about the only one that matters. It holds the electrons to the nuclei to form atoms and the atoms to each other to form molecules and solids.

All contact forces, i.e., all the pushing and pulling that we do, is electrical at the point where atom meets atom. All chemical changes and transformations are the result of changes in the position and motion of electrons. This is true as well of all biological processes.

The electrical nature of our civilization is apparent all around us. Each time we switch on the light, or the toaster, or the vacuum cleaner, pick up the telephone or turn on the radio or the television set, we affect the motions of electrons. Each one is so light and small that it is far below any possibility of having an impact on our senses. But they cooperate, and flow through wires to bring about the large-scale, macroscopic effects that we experience.

The electrical nature of matter is far less apparent. All atoms contain protons with their positive charge in the nucleus, surrounded by electrons with their negative charge. Yes, there are also neutrons and neutrinos, but except in the nuclear realm or in the astronomical realm that is so closely related to it, these neutral particles play, at best, a subtle role.

We already saw a drastic example in Chapter 2 that showed how strong the electric force is. We imagined all the protons from just one gram of hydrogen at the north pole, and all the electrons at the south pole, 4000 miles away. The force of attraction between the two turned out to be the enormous force equal to the weight of about 50 tons.

The force between protons and electrons is so large that we rarely encounter a situation where their numbers are not equal. Only under special circumstances can they be separated from one another, at least here on earth, and then only a minute fraction of them.

What we can do is to shift them around with respect to each other. We can move the electrons a little further away from the protons. We can cause them to move differently, so that, on average, the distance between the electrons and protons in a piece of material is just a little different. This is what happens each time a chemical reaction occurs.

The energy that changes when atoms and molecules combine or dissociate is commonly called *chemical energy*. Each such change, whether it is as subtle as in biological processes or as violent as in burning and explosion, is, on the tiny scale of atoms, a change in electric energy, i.e., in the electric potential energy and the kinetic energy of the electrons and atoms.

Electricity and gravitation

The force between two point charges is described by Coulomb's law, $F_e = k\frac{Q_1 Q_2}{r^2}$. If we compare it to the force between two masses, as given by Newton's law of gravitation, $F_g = G\frac{M_1 M_2}{r^2}$, we see that the two kinds of forces depend on the distance, r, in the same way. In both cases the force between the two interacting objects (charges or masses) varies inversely with the square of the distance between them. That gives us the comfort of familiarity. It also allows us to take over some of the concepts and calculations from the earlier discussion.

But the differences between the two forces are great. Masses always attract, but charges can attract or repel each other. The magnitudes are widely different. The electric force between a proton and an electron is larger by a very large factor (about 10^{40}) than the gravitational force. Hence, where the electric force plays any role at all, its effect is likely to be far more important than that of the gravitational force.

EXAMPLE 1

A proton is attracted to a stationary electron and is accelerated toward it.

(a) What is the acceleration of the proton when they are separated by a distance of 1 m?

(b) As you answered part (a), did you include all of the forces of interaction between the proton and the electron? Why or why not?

Ans.:

(a) The electric force on the proton is $k\frac{e^2}{r^2}$. The proton's acceleration is $k\frac{e^2}{m_p r^2} = (9 \times 10^9) \frac{(1.6 \times 10^{-19})^2}{(1.67 \times 10^{-27})(1^2)} = 0.14 \text{ m/s}^2$.

(b) There is also the gravitational force, $F_g = G\frac{M_p M_e}{r^2}$, and the proton's gravitational acceleration, $a_g = G\frac{M_e}{r^2} = 6.67 \times 10^{-11} \frac{0.91 \times 10^{-30}}{1^2} = 6.1 \times 10^{-41}$ m/s^2, which is smaller by a factor of about 10^{40} and can be neglected.

Separating charges: polarization

When we talk about mechanical forces on a rock or other object we don't usually think about motion that might result *inside* the object. It is different for electrical forces. Electrons, and to a lesser extent nuclei, can move within atoms and molecules. In metals some of the electrons can move quite freely. The way charges are distributed in objects can therefore change in response to forces on them.

Consider this example: put a positive charge near an uncharged block. The block may not have any *net* charge, but like everything else it is full of charged particles—its electrons and protons. The positive charge outside will repel the positive charges inside, and they will move a little away from it. The negative charges will be attracted and will move closer. The positions of the charges will now be different for the positive charges and for the negative charges. The object has become *polarized*.

Not only that, but the attractive force on the negative charges, which are now closer, is larger than the repulsive force on the positive charges, which have moved away. Hence there is a net attractive force between the uncharged block and the charge outside it!

Charged objects occur in nature as ions, nuclei, and particles such as electrons. Macroscopic objects can be charged by rubbing two dissimilar objects against each other. Electrons can then transfer from one to the other, leaving both objects charged. Some materials, such as hair, lose electrons easily. Others capture electrons easily.

Charges are rarely fixed in space, and actual situations can be quite complicated. Electrons in atoms and solids, for example, are never at rest. As we study the effects that charges have on one another we may have to consider examples and problems that are simplified and not completely realistic.

EXAMPLE 2

Rub a balloon with your hair. Some electrons transfer from the hair to the balloon. Rub a second balloon with a sheet of plastic. Electrons will transfer from the balloon to the plastic.

(a) What happens when the two balloons are brought close to each other?

(b) What happens to your hair?

Ans.:

(a) The first balloon carries an extra negative charge. The second balloon carries an extra positive charge. The balloons attract.

(b) The hair is left positively charged. The individual hairs repel each other and will tend to "stand up."

8.2 The electric field

What is a field?

The idea of the electric field, and of fields more generally, is one of the most powerful that physics has given us. It has changed the way we think of interactions and forces, whether they are electric, gravitational, or of any other kind.

We introduce it in a way that gives barely a hint of its possibilities. One charge exerts a force on a second charge. Let's say the same thing differently: the first charge creates an electric field; the second charge is in that field, and as a result it is acted on by a force.

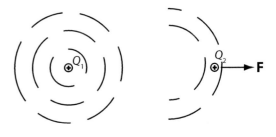

Nothing is changed *operationally*. Only the force is observable. The field is an artifice, an invention. Yet our perception has changed.

Can we think of another example of an "immaterial" field? Here is one that is more elusive and fanciful: think of a person, first alone. By outlook and appearance, manner and gesture, that person affects his or her environment. If another person enters the "field" of the first one, this second person is affected differently, and as a result behaves differently, depending on the perception of "friendliness," or otherwise, of the first one.

The interaction requires two people, but you can imagine a "friendliness field" created by just one. Similarly, the electric field is there even if you have just one charge. But for a force to exist it takes two.

The first charge, just by its existence, creates a field. The field is everywhere. It is unseen and undetected, until another charge is placed somewhere, and experiences a force as a result of being in the field.

We can use the second charge as a probe, or *test charge*. If it is acted on by a force, it shows that there is a field. We can calculate what the force on the test charge will be once we put it there. After all, we know Coulomb's law. All we need to know is how large the charges are and how far they are from each other.

We define the direction of the field at a point to be the same as the direction of the force there on a positive charge. We define the magnitude of the field as the magnitude of the force divided by the magnitude of the test charge.

EXAMPLE 3

Go to the PhET website (http://phet.colorado.edu) and open the simulation *Charges and Fields*. Select "Run now." Select "grid."

With the mouse, pull a positive charge from its "box" on the right to the middle of the screen. Select "Show E-field." The arrows show the direction of the electric field. They are fainter where the field is smaller. Pull an "E-sensor" from its box to a place near the charge. Its arrow shows the direction and the magnitude of the field. Compare its arrow to the other field arrows. Deselect "show E-field" to see the E-sensor arrow more clearly. Move it around the charge at different distances from the charge to explore the field.

Ans.:
The field is away from the charge. It becomes smaller as you move away from the charge.

In one sense we have done nothing. There isn't anything, so far, that we can describe or calculate by using the concept of the electric field that can't also be done without it. Yet the way we look at what is happening, and the way we think about it, has profoundly changed.

We no longer need to think of the two charges exerting forces on each other across the empty space between them. We now have a way for them to communicate directly. The field is created by a charge and spreads out from it. It is the field that then exerts a force on any other charge in it, right there where the charge is.

The story is similar for the gravitational field. The earth, just by its existence, creates a gravitational field. The field changes the space around the earth. A ball flying through the field is affected by it, and experiences a force. It is no longer a force of the earth, somehow reaching out through empty space. The field provides the communication and interaction between the earth and the ball.

EXAMPLE 4

On a sketch of the earth, draw vectors for the force on an object outside it, at eight evenly-spaced places.

(a) Where do they point?

(b) What approximation did you have to make to reach the answer to part (a)?

(c) We define the gravitational field similarly to the way we define the electric field: it is the force on a test mass divided by the test mass. What is the direction of the gravitational field near the surface of the earth?

Ans.:

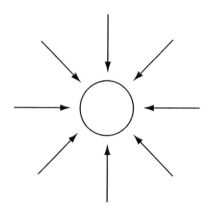

(a) They point toward the center of the earth.

(b) The force vectors point to the center only if the earth is spherically symmetrical, i.e., if we assume that the earth's density at any point depends only on the distance from the point to the center, and not on the angle, i.e., the density, ρ, is a function only of the radial distance, r: $\rho = f(r)$. In addition, we are assuming that the influence of all other astronomical bodies is negligible.

(c) The gravitational force on the test mass and the gravitational field are in the same direction. Since the force is down, toward the center of the earth, the gravitational field is also.

From what we have said so far, we see the electric field as an aid to the calculation of electric forces. That alone would be useful. But there is much more. It turns out that the field has its own independent existence. It is created by a charge, but it can move off, through empty space, in ways that are no longer tied to its origin.

This is the real triumph of the field: electric and magnetic fields can propagate through space. Light waves, radio waves, *electromagnetic waves* of all kinds move through otherwise empty space. They are created by charges, but float off, disembodied, on their own.

The electric field is still what we said it was at the beginning: it tells us that at any point where there is an electric field, there will be an electric force on a charge, if we put one there. Since force is a quantity that has both magnitude and direction, i.e., is a vector quantity, the same is true of the electric field. Quantitatively, if we know the magnitude and direction of the electric field at some point and the sign and magnitude of the charge that we put at that point, we can calculate the force on the charge, exerted by the field.

EXAMPLE 5

Here is a different kind of field. The cheese creates the field. The mice feel the effect of the field. The field is stronger closer to the cheese, as is shown on the diagram by the greater enthusiasm of the mice as they get closer.

In this case the field consists of molecules released by the cheese and diffusing in the air. This is very different from the electric field, where no atoms or molecules of any material are involved.

(a) Which of the following graphs could describe the field strength as a function of the distance from the smelly cheese?

(b) Which graph could describe the electric field as a function of distance from a point charge, q?

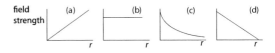

Ans.:
The "cheese field" is very different from an electric field. What it is likely to have in common with the electric field of a point charge is that it decreases as the distance from the "source" increases. The decrease is, however, very unlikely to be linear, with a definite end, as in (d). We reject the increasing and the constant field. This leaves (c), where the field decreases strongly close to the cheese, and then more gradually as the distance from it increases.

The same is true for the electric field of a point charge. In this case we know the way in which the magnitude of the field changes with the distance from the source. We know that the force decreases as $\frac{1}{r^2}$, so that $E \propto \frac{1}{r^2}$.

Coulomb's law revisited: force and field

Let's look at Coulomb's law again. We can think of one of the charges, Q_1, as the source of the electric field. The other, Q_2, at the point P, *experiences* the field created by the first charge. It is acted on by the electric force. The force on Q_2 is described by Coulomb's law, and has the magnitude $F_{12} = k\frac{Q_1 Q_2}{r_{12}^2}$, where r_{12} is the distance between Q_1 and Q_2. The direction of the force is such that the two charges repel if they have the same sign and attract if their signs are opposite.

Now separate the force equation into two parts:

$$F_{12} = \left[\frac{kQ_1}{r_{12}^2}\right][Q_2]$$

We call the first part on the right-hand side of this relation E_P, the electric field created by Q_1 at the point P, a distance r_{12} from it.

We can now write the force relation as $F_{12} = E_P Q_2$.

This is a relation between the *magnitudes* of the force and the field. But both are vector quantities that also have *direction*. We can incorporate the direction by making it a vector equation and letting Q_2 be positive or negative:

$$\mathbf{F}_{12} = \mathbf{E}_P\, Q_2$$

$\mathbf{F}_{12}$ and $\mathbf{E}_P$ are in the same direction if Q_2 is positive and in opposite directions if Q_2 is negative. The direction of the electric field is defined so that a positive charge experiences a force in the same direction as the field and a negative charge experiences a force opposite to the field direction.

EXAMPLE 6

(a) What are the units of E?

(b) What is the electric field of a $-5\,\text{nC}$ (-5×10^{-9} C) charge at a point 1.2 m to the left of the charge?

Ans.:

(a) The relation $E_P = \frac{F_{12}}{Q_2}$ shows that the units are those of force divided by charge, or N/C.

We can also look at the source relation for E: $E = \frac{kQ_1}{r^2}$. We know from Coulomb's law that the units of k are $\frac{\text{N m}^2}{\text{C}^2}$. The units of E are then $(\frac{\text{N m}^2}{\text{C}^2})(\text{C/m}^2)$ or N/C.

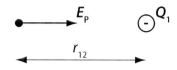

(b) The magnitude is $E_P = \frac{kQ_1}{r_{12}^2} = \frac{(9\times10^9)(5\times10^{-9})}{1.2^2} = 31.25\,\text{N/C}$. Since Q_1 is negative, the electric field vector is in this case toward the charge, to the right.

EXAMPLE 7

A charge of $+4\,\text{nC}$ is at a point P where the electric field is 20 N/C to the left.

(a) What is the force on the charge?

(b) What do we know about the source of the field?

Ans.:

(a) The charge experiencing the force is 4 nC. Hence the magnitude of the force is $F_{12} = Q_2 E_P = 80 \times 10^{-9}$ N. Since Q_2 is positive, F_{12} and E_P are in the same direction, to the left.

(b) The force on the charge tells us the magnitude and direction of the electric field, but provides no information about the sources of the field.

We have taken the perfectly symmetric Coulomb's law, where each charge plays the same role, and separated it into two parts: the source (Q_1) and the probe (Q_2). The first charge, Q_1, is the source of the electric field, the second charge, the test charge Q_2, probes the effect of the field.

As long as there is no charge anywhere except for Q_1, the field is a figment of our imagination. Only when a second charge (Q_2) is put at some point does the field become observable. There is then a force on Q_2, equal to the field at that point, multiplied by Q_2.

We see that there are two ways to find out what the electric field is. We can either start with the source charge (Q_1) or with the test charge (Q_2).

To use the test charge, we have to know or determine what force, $\mathbf{F}$, it experiences. Since $\mathbf{F} = Q_2\mathbf{E}$, the electric field at a point, $\mathbf{E}$, is equal to the force on the test charge at that point, ($\mathbf{F}$), divided by the amount of the charge Q_2, i.e., $\mathbf{E} = \frac{\mathbf{F}}{Q_2}$. In this case we need to know nothing about the sources of the field.

The second way to find **E** is to start with the sources of the field. If there is only one point-like source charge, Q_1, the magnitude of the field at a distance r_1 from it, at a point P, is $\frac{kQ_1}{r_1^2}$. If Q_1 is positive, the direction of the field is away from the charge, and it is toward Q_1 if the charge is negative.

If there is more than one source charge, each one, Q_n, contributes a field vector at P, a distance r_n away, whose magnitude is $\frac{kQ_n}{r_n^2}$, and whose direction is away from Q_n if Q_n is positive and toward Q_n if the charge is negative. To find the total electric field **E** at P, all of these vectors have to be added up.

(b) The field is upward above the line and downward below the line.

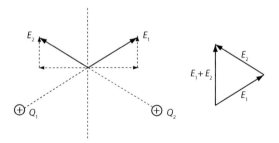

EXAMPLE 8

(a) Go to the PhET website (http://phet.colorado.edu) and open the simulation *Charges and Fields*. Select "grid." Put two equal positive charges on a horizontal line 20 divisions apart. Move an E-sensor along that line from the far left to the far right and describe the variation of the field qualitatively. Could you have predicted the direction of the field to the left of both charges, to the right of both charges, and between the charges?

(b) Do the same along the vertical line half-way between the two charges.

Ans.:

(a) While the sensor is to the left of both charges, the field is to the left. This is as you expect; the force on a positive test charge in this region is the vector sum of two forces, both to the left, one from each of the two charges on the line.

The field becomes very large as you get close to the left charge. It reverses direction and is to the right as you move the sensor to the region between the two charges. As you move the sensor further to the right the field becomes smaller, until it reaches zero in the middle. Again this is what you expect. The force on a test charge is to the right from the left charge and to the left from the right charge. (In both cases it is away from a positive charge.) The two contributions are in opposite directions, and in the middle they cancel. Moving closer to the charge on the right the field becomes larger, as the contribution from the right charge predominates.

In the region to the right of both charges the field is away from both, to the right, and becomes smaller as the sensor is moved further to the right.

We already know that on the line, half-way between the charges the field is zero because the contributions to the field from the two charges add up to zero. At other points along the vertical line each contribution to the field has two components: the horizontal components have the same magnitude and are in opposite directions. They cancel. The vertical components are in the same direction, and they add. Alternatively, we can add the vectors representing the fields E_1 and E_2 from the two charges end to end, as on the right side of the figure.

The field is zero half-way between the charges. It also goes to zero far from the charges. Its magnitude therefore has a maximum at some point above the line joining the two charges and at a point symmetrically located below.

EXAMPLE 9

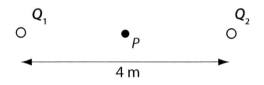

Two equal charges, Q_1 and Q_2, each $8\,\mu C$, are 4 m from each other. The point P is midway between the two charges.

(a) Both charges are positive. What is the electric field at P?

(b) Q_1 is negative and Q_2 is positive. What is the electric field at P now? What is the force on a $+5\,\mu C$ charge at P?

Ans.:

(a) The fields at P of Q_1 (E_1) and Q_2 (E_2) have the same magnitude and are in opposite directions. The two vectors add up to zero, so that the electric field at P is zero.

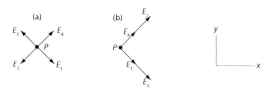

(b) In this part the fields E_1 and E_2 are in the same direction, to the left. (E_1 is toward Q_1 and E_2 is away from Q_2.) Each has magnitude $\frac{(9 \times 10^9)(8 \times 10^{-6})}{2^2} = 1.8 \times 10^4$ N/C, so that the field from both is 3.6×10^4 N/C. The force on a $+5$-μC charge is $(5 \times 10^{-6})(3.6 \times 10^4)$ or 0.18 N to the left.

EXAMPLE 10

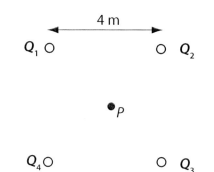

Four equal charges, Q_1, Q_2, Q_3, and Q_4, each 5 μC, are at the corner of a square whose sides are 4 m.

(a) All charges are positive. What is the electric field at the point P in the center of the square?

(b) For this part Q_1 and Q_4 are positive and Q_2 and Q_3 are negative. What is the direction of the electric field at P?

(c) What is the magnitude of the field at P in part (b)?

Ans.:

(a) The four vectors E_1, E_2, E_3, and E_4 point away from the four charges. They add up to zero.

(b) This time the two vectors E_1 and E_3 point in the same direction (away from Q_1 and toward Q_3), and so do E_2 and E_4. Their y components cancel and their x components add, so that the total field at P is in the x direction, to the right.

(c) The distance from any corner to the center of the square is $2\sqrt{2}m$. The field from Q_1 at P, E_1, is $\frac{kQ_1}{(2\sqrt{2})^2} = \frac{(9 \times 10^9)(5 \times 10^{-6})}{8} = 5625$ N/C. Its x component is $E_1 \cos 45° = (5625)(0.707) = 3977$ N/C. The x components of E_1, E_2, E_3, and E_4 are the same. The total field at P is therefore $4E_1 \cos 45°$, or 1.59×10^4 N/C.

We could use the same procedure to find the field at some other point, not at the center of the square. There would again be four vectors to add, but each would have a different magnitude and point in a different direction. The addition procedure would be much more time consuming. We see that the *symmetry* of the problem that we have solved makes it much simpler.

8.3 Field lines and flux

Lines to represent the field

There is a wonderful and productive geometric representation that allows us to get an intuitive but rigorous and detailed feeling for the electric field.

We associate with each positive charge, Q, a number of *electric field lines* emerging from it. The direction of an electric field line, at any point on it, is the same as the direction of the electric field. If the charge is negative, the lines go toward it, so that the lines begin at positive charges and end at negative charges.

So far a drawing of the field lines is a representation that shows the direction of the electric field. We can go further and also incorporate its magnitude. We do this by letting the number of lines divided by the area that they cross (at right angles to the lines) represent the magnitude of the electric field. This means that the lines are closer together where the field is stronger.

Here is why the field concept works: think of an imaginary sphere with a charge at its center. Coulomb's law tells us that the field is proportional to $\frac{1}{r^2}$. The area, A, of the sphere is proportional to r^2. The two factors cancel in the product EA so that the result is independent of r. E is the number of lines through the surface of the sphere, per unit area. EA is therefore just the number of lines through the surface of the sphere. If this number does not depend on r, the number of lines is the same through all possible spheres with this charge at the center, regardless of their radius.

The cancellation of r^2 is the crucial fact that leads to the importance and usefulness of the electric field lines. It is the result of the confluence of two quite separate features, one that is characteristic of the electric field and one that is purely geometric. The first is the power "2" of r in Coulomb's law, so that $E = k\frac{Q}{r^2}$. The second is the power "2" of r in the surface area of a sphere, $A = 4\pi r^2$. It is the fact that these two numbers are precisely the same, so that they cancel, that makes the description of the field by field lines possible. If these two powers were not equal, a set of continuous lines could not describe the field.

We can look at the same story more precisely: what is the actual number of lines emerging from a single positive charge, Q? Draw an imaginary spherical surface around Q, with radius r. The electric field at the surface of the sphere is perpendicular to the sphere, and its magnitude is $k\frac{Q}{r^2}$, where k is the proportionality constant in Coulomb's law. The surface area, A, of the sphere is $4\pi r^2$. The number of lines through the spherical surface is EA, or $(k\frac{Q}{r^2})(4\pi r^2)$, which is equal to $4\pi kQ$.

k is also often written as $\frac{1}{4\pi\epsilon_0}$, where ϵ_0 is called *the permittivity of free space*. (We will generally just use the symbol, rather than this unwieldy terminology and its numerical value.) The number of lines through the sphere can then be written as $\frac{Q}{\epsilon_0}$.

Electric flux

So far we have just one charge, Q, with $\frac{Q}{\epsilon_0}$ lines through the surface of our imaginary sphere, coming from the charge at its center if it is positive and going toward it if it is negative. The number of lines crossing the surface per unit area represents the magnitude of the electric field, and the direction of the lines is the same as the direction of the electric field.

The total number of the electric field lines crossing the surface area of the sphere is EA, where E is the magnitude of the electric field and

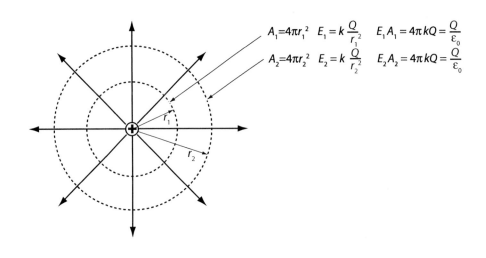

A is the surface area of the sphere. The quantity EA is called *the flux of the electric field*, or just the *electric flux*, through the surface of the sphere. We see that the flux of E through the imaginary surface is equal to $\frac{Q}{\epsilon_0}$. The symbol Φ (Greek capital phi) is usually used for it.

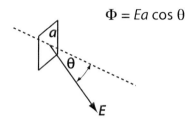

$$\Phi = Ea \cos\theta$$

For the sphere with the charge at its center the electric field lines are perpendicular to the surface. In cases where they cross at other angles we define the flux as $E_\perp A$, where $E_\perp$ is the component of E perpendicular to the area.

Again, we haven't done anything. We have introduced the terms *electric field* and *flux*, but except for these new words we still have only Coulomb's law, expressed just a little differently. With the introduction of the electric field lines, however, we have a new representation.

EXAMPLE 11

A charge Q is at the center of a cube. What is the flux through the top surface of the cube?

Ans.:
With Q in the center, the sides are located symmetrically with respect to Q. In other words, it doesn't matter which side we call the top surface. The flux through each is the same and is $\frac{1}{6}$ of the total flux of $\frac{Q}{\epsilon_0}$. The flux through any one side is therefore $\frac{Q}{6\epsilon_0}$.

Note that we can answer the question without knowing how large the cube is, or finding E at any point on the surface of the cube. To find the field at every point on one of the square sides, and then to add them up to find the flux, would be a very difficult problem. It is the *symmetry* of the configuration that makes the problem easy. This kind of simplification is characteristic of questions that can be answered by symmetry considerations.

Charge, field, and flux: Gauss's law

The electric field lines are much more than a tool for the visualization of the field. They give us a description that can be vastly generalized. It doesn't even matter whether Q is at the center of the imaginary sphere. The number of lines through the sphere, i.e., the flux of E through it, is still equal to $\frac{Q}{\epsilon_0}$. And if there is more than one charge, this result remains, as long as we let Q represent the total net charge within the sphere. It doesn't matter how many charges there are, where within the sphere they are, or how much of the charge is positive and how much is negative.

In fact, why stick to a sphere? All closed surfaces will have the same property: the flux of E through any closed surface is equal to $\frac{1}{\epsilon_0}$ times the total net charge within the surface. This statement is called *Gauss's law*.

All we have done is to follow Coulomb's law, and let it lead us in a new direction. No additional physical law has entered our development. Gauss's law is entirely equivalent to Coulomb's law. Each can be shown to follow from the other. But Gauss's law gives us an important additional tool for looking at electric fields.

The two laws look very different. Gauss's law allows new calculations and leads to new insights. It, rather than Coulomb's law, is generally considered to be one of the fundamental laws of electromagnetism, known as *Maxwell's equations*.

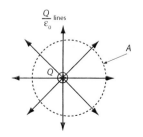
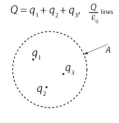
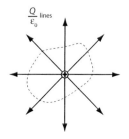

EXAMPLE 12

A sphere whose radius is R carries a net positive charge Q, uniformly distributed. (This means that for each piece of the sphere with volume ΔV, with a charge ΔQ in it, the *charge density* $\frac{\Delta Q}{\Delta V}$ is the same throughout the sphere. Here the symbol "Δ" denotes a small amount of charge or volume, not, as before, a change in these quantities.)

(a) What is the flux through the surface of the sphere?

(b) What is the electric field at the surface of the sphere?

(c) What is the flux through an imaginary spherical surface outside the charged sphere, a distance r from the center?

(d) What is the electric field at the imaginary surface?

(e) How do the answers to parts (c) and (d) differ from what they would be if all of Q were concentrated at the center?

Ans.:

(a) From Gauss's law the flux through any closed surface is equal to $\frac{Q}{\epsilon_0}$, where Q is the net charge inside the surface. Hence the flux is $\frac{Q}{\epsilon_0}$.

(b) The electric field has the same magnitude at each point on the surface of the sphere and is directed outward. The flux through the surface of the sphere is EA, so that the field is $E = \frac{Q}{\epsilon_0 A}$, where $A = 4\pi R^2$, so that $E = \frac{Q}{4\pi\epsilon_0 R^2}$ or $\frac{kQ}{R^2}$.

(c) $\frac{Q}{\epsilon_0}$.

(d) $\frac{kQ}{r^2}$.

(e) The field lines outside the charged sphere look exactly as they would if all of the charge were at the sphere's center. Therefore the field a distance r from the center (and outside the charged sphere) is the same as if Q were at the center.

The example shows that the number of lines from a single point charge is the same as for a sphere in which this same amount of charge is distributed uniformly throughout the sphere. We can make the result more general: *if the charge is distributed with spherical symmetry (i.e., if the charge distribution looks the same from any angle), the electric field outside the sphere is the same as for the same charge concentrated at the center of the sphere.* The example illustrates how Gauss's law can use symmetry to simplify a problem.

The gravitational field: solving Newton's problem

Use mass instead of charge. Change the proportionality constant. That's all that distinguishes Newton's law of gravitation from Coulomb's law. The essential part is the same: the power of r is still -2. The lines of the gravitational field have all the properties of the lines of the electric field. We know of no negative mass, so that the story is simpler, and the field always points toward a mass. Gauss's law holds for the gravitational field just as it does for the electric field.

We can now use Gauss's law, just as for the electric field, to describe the gravitational field outside a sphere whose mass is uniformly distributed, or more generally, distributed with spherical symmetry. It is the same as if all the mass were at its center.

That the gravitational field outside a spherically symmetric mass distribution is the same as it would be if all the mass were concentrated at the center is the single important result of the application of Gauss's law to gravitation. Of course the earth is not a perfect sphere. Each person walking across the street, each ant, for that matter, destroys that symmetry.

Even the highest mountain in America is smaller than $\frac{1}{1000}$ of the earth's diameter. A greater departure from sphericity is that the diameter from pole to pole is about 43 km less than the diameter of the equator. That's about $\frac{1}{300}$ of the earth's diameter, so the difference is still not great. The density of the earth changes as we go closer to the center, but we ask only for spherical symmetry, not for homogeneity. A homogeneous object has the same properties everywhere. For a spherically symmetric one they may change, but only with the distance from the center. A spherically symmetric object looks the same from any angle. Although the earth (as well as the other planets and moons, and the stars) is not a perfect sphere, the deviations from spherical symmetry are so small that the simple result of Gauss's law for a sphere is sufficient and appropriate most of the time.

Gauss's law and symmetry

We have seen that if there is some symmetry in a physical situation we can often draw some very general conclusions quite simply. Most of us don't have much experience with this kind of thinking, so it may be helpful to look at some more examples. Problems with symmetric charge distributions are just what Gauss's law is *good* for.

We have just seen the most important one: for a charge Q at a point, or uniformly distributed throughout a sphere or on the surface of a sphere, or, most generally, if the charge density depends only on r, the distance from the center, the electric field outside the charge distribution is directed along the radius, and its magnitude is $\frac{kQ}{r^2}$ or $\frac{Q}{4\pi\epsilon_0 r^2}$. In each of these cases the charge is distributed with spherical symmetry.

There are two other situations where Gauss's law leads to a simple relation for the field, that of cylindrical symmetry, and that for an infinite plane.

The cylindrical case is that of a long charged line, like a metal rod, or a cylinder, such as a tube or thick cable. The linear charge density is the total charge, Q, divided by the length, L, and is usually called λ (Greek *lambda*), equal to $\frac{Q}{L}$. The electric field is at right angles to the line or to the curved part of the cylinder. (This is exactly true only if the line is infinitely long, but it may be a good approximation for a finite line far enough from its ends.)

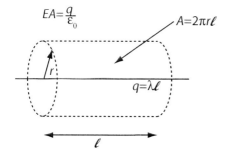

Look at an imaginary cylindrical surface of radius r and length ℓ surrounding the line or cylinder concentrically, with an amount of charge Q inside the surface. The surface area of its curved part is $A = 2\pi r\ell$. Because the electric field is at right angles to the curved surface, there is no flux through the flat ends if the cylinder.

The total flux through the cylinder is equal to the flux through the curved part of its surface, A, and is equal to EA. From Gauss's law it is equal to $\frac{Q}{\epsilon_0}$, so that $E = \frac{Q}{\epsilon_0 A}$, which is equal to $\frac{Q}{\epsilon_0 2\pi r\ell}$ or $\frac{\lambda}{2\pi\epsilon_0 r}$.

EXAMPLE 13

A wire, 10 m long, is charged uniformly with a charge of 5 μC. A charge of $Q = 3\,\mu C$ is at a point on a line perpendicular to the wire at its center, 1.5 m from the wire. What is the force on the charge Q?

Ans.:

The question can be answered with the relation that we have derived from Gauss's law if we can assume that the wire is infinitely long. Since the wire is much longer than the distance from the wire to the charge, this is probably a good approximation.

The electric field of the wire, at the point where the 3 μC charge is located, is

$$E = \frac{\lambda}{2\pi\epsilon_0 r}, \text{ or } \frac{2\lambda}{4\pi\epsilon_0 r} = \frac{2k\lambda}{r}$$

$$\lambda = \frac{5 \times 10^{-6}\,C}{10\,m} = 0.5 \times 10^{-6}\,C/m$$

$$E = \frac{(2)(9 \times 10^9)(0.5 \times 10^{-6})}{1.5} = 6 \times 10^3\,N/C$$

The force on the charge is $F = EQ = (6 \times 10^3)(3 \times 10^{-6}) = 0.018\,N$.

Because the wire is not infinitely long there are field components parallel to the wire at all points except on the central plane, perpendicular to the wire, on which Q is located. At Q the field is perpendicular to the wire and away from it, and this is also so for the force on Q.

Because the wire is not infinitely long, E will actually be somewhat smaller than the result that we have calculated.

The third case that we will look at is that of an infinitely large plane, charged uniformly with a surface charge density (charge per unit area) of σ. (There are, of course, no infinitely large planes. Our main application will be to the field between two planes, separated by a small distance.) We can tell from the symmetry that E must be at right angles to the surface. Pick a part of the surface, with area a, and surround it on both sides of the plane with an imaginary cylinder. The charge inside the cylinder is then σa.

Because the field is at right angles to the plane there is no flux through the curved part of the cylinder. But lines go out from the plane on both sides of it through the top and the bottom of the cylinder, each with area a, for a total area $2a$. With an electric field magnitude E outward from each surface, the flux out of the cylinder is $(E)(2a)$.

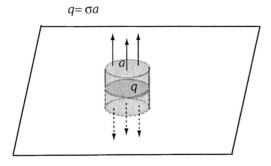

From Gauss's law we know that the flux is equal to $\frac{Q}{\epsilon_0}$, where Q is the amount of charge inside the cylinder, which here is σa, so that the flux is $\frac{\sigma a}{\epsilon_0}$. Putting the two relations for the flux together, we see that $(E)(2a) = \frac{\sigma a}{\epsilon_0}$, or $E = \frac{\sigma}{2\epsilon_0}$.

That's it for Gauss' law. It is always true, but these are among the few situations where it leads to simple results.

EXAMPLE 14

A negative charge of -5 nC is at a point P near a large (assume infinite) single plane that carries a charge with a constant surface charge density of $+2$ nC/m^2.

(a) Is it possible to find the electric field at a point near the plane without knowing how far the point is from the plane?

(b) What are the magnitude and direction of the field at P?

(c) What is the force on the charge?

Ans.:
(a) The electric field near an infinite plane is constant and equal to $\frac{\sigma}{2\epsilon_0}$, which is independent of the distance from the plane.

The field lines are perpendicular to the plane. Their density (the number crossing a unit area), which is equal to E, does not change.

Qualitatively, the further away from the plane a point is, the greater the contributions to the field from distant points on the plane. Since the plane is assumed to be infinitely large, there will be distant regions that will contribute, no matter how far away from the plane a point is.

(b) $E = \frac{\sigma}{2\epsilon_0}$, with lines away from the plane and perpendicular to it. $\frac{1}{4\pi\epsilon_0} = 9 \times 10^9$, so that $\frac{1}{\epsilon_0} = 36\pi \times 10^9$ SI units.

$$E = (2 \times 10^{-9})(\frac{1}{2})(36\pi \times 10^9)$$
$$= 36\pi \text{ or } 113 \text{ N/C}$$

(c) $F = EQ = (113)(5 \times 10^{-9}) = 5.5 \times 10^{-7}$ N.

The direction of the force is opposite to the direction of the field because the charge is negative. It is toward the plane.

EXAMPLE 15

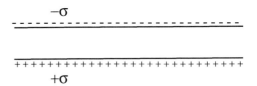

Two large metal planes are separated by a distance of 2 cm. The upper one is charged negatively and the lower one positively, each with a surface charge density of 5 nC/m^2.

(a) Describe the electric field lines.

(b) What is the electric field between the planes?

(c) What is the electric field away from both planes?

Ans.:
(a) Electric field lines start at positive charges and end at negative charges. (Although we have talked about isolated charges, for each positive charge there is an equivalent negative charge somewhere.) In this case the lines go from the positively charged plane to the negatively charged plane.

Alternatively, we can add the field contributions from the two planes. Between them there is a field $\frac{\sigma}{2\epsilon_0}$ upward from the positive plane and a field equally large toward the negative plane. Together the two fields add up to a field $\frac{\sigma}{\epsilon_0}$ upward.

(b) $E = \frac{\sigma}{\epsilon_0} = (5 \times 10^{-9})(36\pi \times 10^9) = 180\pi = 5.65 \times 10^2$ N/C.

(c) In the region above both planes there is a field down toward the negative plane and a field up away from the positive plane. They have the same magnitude, and add up to zero. Similarly, the field is zero below both planes.

The configuration described in this example is called a *capacitor*. It stores charge on each of the two plates. It also stores the uniform electric field in the region between the plates. Since the problem states that the plates are "large," we have tacitly assumed that we can use the approximation that they are infinitely large. This is a good approximation as long as the distance between the plates is small compared to their length and width. The approximation is best in the central region and least appropriate near the edges, where the field lines curve outward. The field near the edges, where it is no longer uniform, is called the *fringing* field.

8.4 Summary

The electric force is the second of the fundamental forces of nature. While Newton's law describes the interaction between bodies that have mass, Coulomb's law describes the interaction between charges. The gravitational force is always attractive, but the electric force can be one of attraction or repulsion. The electric force depends on the magnitude and the sign of the charges and on how far apart they are from each other: $F_e = k\frac{Q_1 Q_2}{r^2}$.

The attractive force between the negatively charged electrons in the atoms and the positively charged protons in the nucleus of the atom holds the atom together and is responsible for the structure of all atoms, molecules, and their combinations.

An object is *polarized* when the average positions of the positive and negative charges in it are not the same.

We can describe the electric force by using the concept of the *electric field*: each charge creates an electric field that surrounds it. Another charge that is in the field experiences a force. The electric field is a *vector* quantity. Its calculation can quickly become complicated as the number of charges grows. We therefore looked at some of the simplest charge distributions, particularly those where we can make use of *symmetry* to simplify the calculations.

A charge Q_1 has an electric field around it with magnitude $E_1 = k\frac{Q_1}{r^2}$. The direction of the field is away from a positive charge and toward a negative charge.

If there are several charges, each contributes to the field. The total field at a point is equal to the vector sum of the field vectors from each charge.

A mass m has a gravitational field around it, of magnitude g, toward the mass.

In an electric field E_1 a charge Q_2 experiences a force $E_1 Q_2$. The force is in the direction of the field if Q_2 is positive and in the direction opposite to the field if the charge is negative.

As an aid to visualizing the field we use *electric field lines*. The lines begin at positive charges and end at negative charges. The direction of the lines is the direction of the electric field.

When a field, E, is perpendicular to an area, a, the electric flux through the area is Ea. If the normal to the area (the line perpendicular to a) makes an angle θ with the field, the flux through the area is $Ea \cos \theta$.

The flux (Ea) from a point charge, or the number of field lines coming from a point charge, is $\frac{Q}{4\pi\epsilon_0 r^2} 4\pi r^2$, or $\frac{Q}{\epsilon_0}$. Here $k = \frac{1}{4\pi\epsilon_0} = 9 \times 10^9 \frac{\text{Nm}^2}{\text{C}^2}$. This is also the flux (or the number of lines) emerging from a closed surface that contains a net charge Q. (This statement is called *Gauss's law*.)

The field outside a spherically symmetric charge distribution is the same as that of a point charge (with the same total charge) at its center.

The field on each side of a positively charged, infinitely large plane is $\frac{\sigma}{2\epsilon_0}$. If the field is only on one side, its magnitude is $\frac{\sigma}{\epsilon_0}$. This is also the field inside a (infinitely large) capacitor.

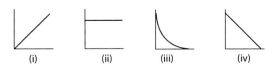

(i) (ii) (iii) (iv)

8.5 Review activities and problems

Guided review

1. A proton is released from rest at a distance of 0.5 m from an electron.
    (a) What is its initial acceleration?
    (b) What is the ratio of this acceleration to the gravitational acceleration that results from the gravitational force between the two particles?

2. Go to the PhET website and open the simulation *Balloons and Static Electricity*. The following should be checked as you start: "show all charges," "ignore initial balloon charge," and "wall."
    (a) Drag the balloon to the sweater and rub it against it. Move it back to where it started and let go. What happens? Why?
    (b) Move the charged balloon toward the wall. What happens to the charges in the wall? What happens when you let go of the balloon near the wall? Why?
    (c) Click on "two balloons." Charge both of them. Get one of them to move slowly in the middle. Can you push it toward the wall with the other balloon? What is meant by *polarization*? Describe the polarization of the wall.

3. Go to the PhET website and open the simulation *Charges and Fields*. Pull out a charge. Use the E-field numbers and the tape measure to find the magnitude of the charge. (The unit V/m is the same as the unit N/C.)

4. (a) On the sketch of a positively–charged sphere, draw vectors for the force on another positive charge outside the sphere at eight places, each at the same distance from the sphere.
    (b) Repeat for a negative charge in the vicinity of the same positively–charged sphere.

5. (a) Which of the graphs best describes the magnitude of the gravitational field as a function of distance, going from the surface of the earth to a point half-way to the moon?
    (b) Which one best describes the magnitude of the gravitational field as you go from the earth's surface to a height of 20 m?

6. (a) What are the magnitude and direction of the electric field 0.6 m to the right of a +5 nC charge?
    (b) Compare the magnitude and direction of the field to the result of Example 6, and explain the difference.

7.

⊖   •   ⊕
A   P   B

There is a negative charge at the point A and a positive charge at the point B.
A charge of -3 nC is at the point P, where the electric field is 15 N/C to the left.
    (a) What are the magnitude and direction of the force on the charge at P?
    (b) Which of the following statements is correct?
    (i) The net charge at A is greater than the one at B.
    (ii) The net charge at B is greater than the one at A.
    (iii) It is not possible to conclude anything about the relative magnitudes of the charges.

8. Two charges are 3 m apart. $Q_1 = 1\,\mu C$ is on the left and $Q_2 = 4\,\mu C$ is on the right. Where is the point P at which the electric field is zero?

9.

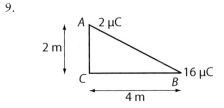

The figure shows a right-angled triangle with positive charges of $2\,\mu C$ and $16\,\mu C$ at the corners A and B.
    (a) What is the electric field at C?
    (b) What is the force on a 10 μC charge at C?

10. Four charges, each of 5 µC, are at the corners of a square with 4 m sides, as in the example.

(a) All charges are positive. What are the magnitude and direction of the electric field halfway between Q_2 and Q_3?

(b) The configuration is changed so that the magnitudes of all charges remain the same, but Q_1 and Q_4 are negative and Q_2 and Q_3 are positive. How does the field halfway between Q_3 and Q_4 compare with that found in part (a)?

11. A charge Q is at the center of a regular dodecahedron. (This is a polygon with 12 equal faces, each of which is a regular pentagon.) What is the flux through one of the faces?

12. A sphere with radius R carries a net positive charge, Q, uniformly spread over its surface.

(a) What is the electric field at the sphere's surface?

(b) What is the flux through an imaginary spherical surface just outside the sphere?

(c) What is the flux through a spherical surface surrounding the sphere at $r = 2R$?

(d) What is the electric field at this spherical surface?

(e) How are the answers to parts *a* to *d* different if the same charge, Q, is spread uniformly throughout the sphere's volume?

13. A charge of $Q = 3\,\mu C$ is 2 m from an 18 m-long wire. It experiences an electric force of 0.012 N. What is the total charge on the wire?

14. A charge of +4 nC and mass 2×10^{-15} kg is held at rest at a point P near a large (assume infinite) plane with a constant charge density of $\frac{3\,nC}{m^2}$.

(a) If the charge is let go, in which direction will it move?

(b) What is its acceleration?

15. Each of the two parallel plates of a capacitor has a surface charge density whose magnitude is $\frac{8\,nC}{m^2}$.

(a) Draw a sketch of the plates to show the charge distribution and the electric field lines.

(b) What is the electric field between the plates?

(c) Describe the path of an electron that is shot into the capacitor along a path parallel to the plates half-way between them.

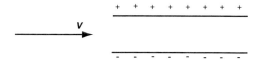

Problems and reasoning skill building

1. When you run a plastic comb through your dry hair it becomes electrically charged as electrons move to it from your hair. Small pieces of paper are then attracted to the comb. Describe the sequence of movements of charge that leads to this attraction.

2. (a) How many electrons are there in a (neutral) molecule of water?

(b) How many electrons are there in a liter ($= 1000\,cm^3 = 10^{-3}\,m^3$) of water? (The density of water is $1\,g/cm^3$. The molar mass is 18 g, i.e., the mass of N_A (Avogadro's number) molecules is 18 g.)

(c) What is the charge (in C) of all the electrons in a liter of water?

3. A sphere has a charge of $+4\,\mu C$. An additional 6×10^{13} electrons are placed on the sphere. What is the net charge now?

4. A charge $Q_1 = 6\,\mu C$ is at the origin of an x–y coordinate system. A charge $Q_2 = 4\,\mu C$ is at the point $(x = 1.5\,m, y = 0.6\,m)$ and a charge $Q_3 = -4\,\mu C$ is at the point $(x = 1.5\,m, y = -0.6\,m)$.

(a) What are the magnitude and direction of the force of Q_2 on Q_1?

(b) What are the magnitude and direction of the force of Q_3 on Q_1?

(c) Draw a vector diagram of the forces in parts (a) and (b) and their sum.

(d) What are the magnitude and direction of the net force of both Q_2 and Q_3 on Q_1?

5. A charge is located on the x-axis at $x = 0$. At $x = 1\,m$ the magnitude of the electric field, **E**, is 1 N/C. Make a graph of E as a function of x from $x = 0.4$ to $x = 5$.

6. (a) Use Coulomb's law to determine the units of k.

(b) What is the magnitude of the electric field at a distance of $10^{-9}\,m$ from a proton?

7. A charge of –3 nC experiences an electric force of $10^{-7}\,N$ to the right. What are the magnitude

and direction of the electric field at the point at which this charge is located?

8. At every point on the surface of a sphere of radius 0.4 m the electric field is radially outward, with a magnitude of 20 N/C. What is the flux through the spherical surface?

9. The flux through an imaginary spherical surface is $12\, \text{Nm}^2/\text{C}$.

(a) What can you conclude about the charge within this surface?

(b) What can you conclude about the charge within the spherical surface if, in addition, the electric field is radial, with the same magnitude at every point on the surface?

10. (a) In Example 10, part (a) there are four $+5\,\mu\text{C}$ charges at the corners of a square, and the field is zero at the center. What changes can you make in the number, the magnitudes, and the positions of the four charges that will leave this result ($E = 0$) unchanged? What is the most general change that you can think of?

(b) For the charges as in part (b) of Example 10, what are the magnitude and direction of the field midway between Q_2 and Q_3?

11. There is a uniform electric field on both sides of a large plane. Its magnitude is 80 N/C and its direction is toward the plane, on each side of it. What is the surface charge density on the plane?

12. A 100 kg sphere and a 100 g sphere outside it are held fixed with the distance between their centers equal to 0.5 m. Each carries a charge of $+2.5\,\mu\text{C}$. Assume that the charge on each sphere is and remains uniformly distributed.

(a) What are the magnitude and direction of the force on each of the spheres?

(b) What is the acceleration of each when they are released?

(c) What will you observe when they are released?

13. Two equal positive charges, Q_1 and Q_2, each $5\,\mu\text{C}$, are 4 m from each other. The point P is between the two charges on the line joining them, 1 m from Q_1. What are the magnitude and direction of the field at P?

14. There is a uniform electric field of 2×10^4 N/C in the x direction. A point charge of $4\,\mu\text{C}$ is placed at the origin. What are the coordinates of the point or points where the total electric field from both the point charge and the uniform field is zero?

15. Three equal charges, each $+4\,\mu\text{C}$, are at three corners of a square whose sides are 2 m long. Add up the vectors representing the contributions from the three charges to find the magnitude and direction of the field at the fourth corner of the square.

16. The two opposite surfaces of a cell membrane act like the plates of an empty parallel plate capacitor. Assume that the charge densities on the two surfaces are $\pm 6.5 \times 10^{-6}$ C/m^2.

(a) What do you need to assume before you can calculate the electric field within the membrane?

(b) With these assumptions, what are the magnitude and direction of the electric field within the membrane?

(c) What are the magnitude and direction of the force on a singly-charged ($q = +e$) ion within the membrane?

17.

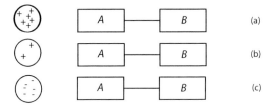

Two metal blocks, A and B, are attached to each other with a removable metal wire. A charged object is brought near block A. The wire is then removed. Rank the four situations shown in the figure in order, with the largest positive charge remaining on B ranked first, to the most negative last.

Multiple choice questions

1. If the distance between two positive ions is doubled, and the charge on each is also doubled, what will happen to the force on each?

It will

(a) stay the same,
(b) increase by a factor of 2,
(c) decrease by a factor of 2,
(d) decrease by a a factor of 4.

2. The point P is a distance L from a uniformly-charged sphere. The electric field at P is E. What is the value of the electric field for each of the following changes.

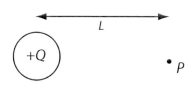

(i) Q is doubled.
(ii) L is doubled.
(iii) Both Q and L are doubled.
(a) $2E$
(b) $\frac{1}{2}E$
(c) $4E$
(d) $\frac{1}{4}E$

3. An electron is brought to a point 1 cm from a positively-charged Na$^+$ ion. A proton is brought to a point 2 cm from an SO$_4^{2-}$ ion.
   (a) The force on the electron is half as large as that on the proton.
   (b) The force on the electron is four times as large as that on the proton.
   (c) The force on the proton is half as large as that on the electron.
   (d) The force on the proton is four times as large as that on the electron.

4. Which of the following best describes why the sublimation (transformation to the gaseous phase) of "dry ice" (solid CO_2) is not increased in a microwave oven.
   (a) Dry ice is too cold for microwaving.
   (b) There are no water molecules in dry ice.
   (c) Microwaves do not work with frozen substances.
   (d) Adding thermal energy does not affect the rate of sublimation.

5.

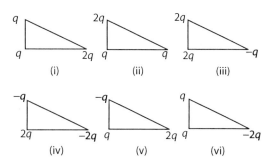

The triangles in the figure all have the same size, 1 unit for the vertical side and 2 units for the horizontal side. The charges are shown on the figure.

Rank the triangles in order of the magnitudes of the net force on the charge in the lower left corner of each triangle.

6.

The figure shows three fixed charges, two of them positive and one negative. The net force on #3 is zero. Which of the following must be true? (There may be more than one.)
   (a) #1 carries more net charge than #3.
   (b) #3 carries more net charge than #2.
   (c) #1 carries more net charge than #2.
   (d) #2 carries more net charge than #1.
   (e) #3 carries more net charge than #1.

CHAPTER 9

More on Electricity: From Force to Energy, from Field to Potential

Electric potential energy and electric potential
> *The electric potential*
> *Equipotentials*
> *Simple and ideal: the uniform field*
> *The uniform gravitational field*
> *Motion in a uniform field*
> *Closer to reality: the field of a point charge*
> *Approximate method, right result*
> *The point charge, properly*
> *Finding the potential energy and the potential*
> *Starting with the sources*
> *Moving charges*

Energy transformations and electric circuits
> *Battery and resistor*
> *What happens in the wire?*
> *Resistivity: separating out the property of the material*

To describe the forces between charges at rest there is only Coulomb's law (or its equivalent, Gauss's law). But by using the concept of energy instead of that of force we gain two great advantages, just as we do when we go from gravitational force to gravitational potential energy. One is a mathematical advantage, the other is a physical one.

The mathematical advantage is that force is a vector quantity and energy is a scalar. Adding scalars is easy. It's just like adding numbers. Adding vectors is more involved and takes more effort. The physical advantage depends on a physical law, the law of conservation of energy.

If we use energy we don't need to know what the forces are at every point and at every moment. We can look at just two points along the path of a particle, say *A* and *B*. If no energy has been given to the particle between these two points, or taken from it, then its energy at *A* must equal its energy at *B*.

More generally, if we use the concept of energy, it doesn't matter what happens to a system between start and end. It doesn't matter what the paths of its constituents are. All we need to know is how much energy, if any, has been added to the system or taken away from it. Moreover, the law of conservation of energy is not limited to physics. It is a great unifying principle that plays an important role in every part of science.

9.1 Electric potential energy and electric potential

The electric potential

We have seen that the concept of the electric field takes us beyond the knowledge of the force between two charges. It allows us to describe the effect on a charge by many charges. It tells us how the presence of one or more charges affects the space around them. If we know the magnitude and direction of the electric field at a point, it allows us to *predict* the force on any charge that we might put there.

We would now like to make use of the advantages that we get if we use energy rather than force. Let's remind ourselves of the definition of potential energy: the work done on a system against the gravitational force is equal to the increase in its gravitational potential energy, and the work done against the electrical force is equal to the increase of its electrical potential energy. (The definition tells us only about changes in energy. Just as for gravitational potential energy, if we want a definite, *absolute* value, we need to choose a reference level where the potential energy is zero.)

We can define a new construct, one that is related to the electric potential energy in the same way that the electric field is related to the electric force. It is the *electric potential* and it is equal to the electric potential energy that a charge has at a point, divided by the value of the charge.

That gives us two quantities that describe the space around a charge: the electric field and the electric potential. They provide information about two properties of the charge, namely the electric force on it and its electric potential energy. One is a vector quantity, the other is a scalar quantity. Together they give a more complete description than either one alone.

Our new quantity, the electric potential, is a property at a point in space. It is defined as the electric potential energy that a charge Q has (or would have) at that point, divided by the

amount of that charge, Q. If the potential energy of a charge of Q coulombs is P joules, then the electric potential, V, is $\frac{P}{Q} \frac{\text{joules}}{\text{coulomb}}$. If Q is positive, P and V have the same sign, but if Q is negative, they will have opposite signs.

The unit J/C is called a *volt* (V). A difference in electric potential is also called a *voltage difference*. When we buy a nine–volt battery, for instance, it means that there is a potential difference, or a voltage difference, of nine volts between the two terminals of the battery. (This is the "nominal" value, i.e., it is approximate, and decreases with time as the battery becomes exhausted and the chemical processes within it cease.)

A volt is a joule per coulomb. For each coulomb of charge that moves through a 1-volt battery, a joule of energy is transferred from the battery (where it was stored as internal energy) to the electrons as electric potential energy.

One more reminder: we often talk about the gravitational potential energy of an *object* in the vicinity of the earth, even though we know that it is really the potential energy of the *system* containing the object and the earth. Similarly we say that a charged object has electric potential energy. Again it is the system that we are talking about, containing the charge on which we are focusing, and perhaps other charges that have an effect on it.

EXAMPLE 1

A rock whose mass is 0.5 kg is lifted through a distance of 3 m.

(a) What is the force on the rock exerted by the gravitational field?

(b) What is the magnitude of the work done against the field when it is lifted?

(c) What is the increase of the gravitational potential energy when the rock is lifted?

A charge of 0.5 nC is in a uniform electric field of 12 N/C. It is moved a distance of 3 m in the direction opposite to the field direction.

178 / More on Electricity: From Force to Energy, from Field to Potential

(d) What is the force on the charge by the electric field?

(e) What is the magnitude of the work done against the field as the charge is moved?

(f) What is the increase in the electric potential energy when the charge is moved?

(g) What is the increase in the electric potential between the beginning point and the endpoint of this path?

Ans.:

(a) $F_g = Mg = (0.5)(9.8) = 4.9$ N.

(b) The force, F, moving the rock is in the direction opposite to the force of the gravitational field, but they have the same magnitude, 4.9 N. The work done is $Fh = (4.9)(3) = 14$ J.

(c) From the definition it is the same as the answer to part (b), namely 14.7 J.

(d) $F_e = QE = (0.5 \times 10^{-9})(12) = 6 \times 10^{-9}$ N.

(e) The force moving the charge is in the direction opposite to the force of the electric field, but they have the same magnitude, 6×10^{-9} N. The work done against the electric field is $(6 \times 10^{-9})(3) = 18 \times 10^{-9}$ J.

(f) From the definition of electric potential energy the answers to parts (e) and (f) are the same, namely, 1.8×10^{-8} J.

(g) The difference in electric potential is equal to the difference in the electric potential energy of a charge divided by the charge, or $\frac{18 \times 10^{-9}}{0.5 \times 10^{-9}} = 36$ V.

Equipotentials

If we want to show how high points on a map are, we draw *contour lines*. These are lines that connect points that are at the same height, i.e., points that have the same gravitational potential energy. Similarly, on a map of charges, we can draw *equipotentials*, lines that connect points that are at the same electric potential.

What happens if we move a charge along an equipotential? The potential energy remains the same, and no work is done against the electric force. That must mean that there is no component of the electric field along the equipotential. Since the field has no component along the equipotential, it must be perpendicular to it. Along the field lines, perpendicular to the equipotentials, the potential energy and the potential change more quickly than along any other direction.

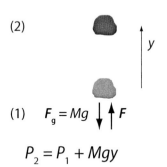

$P_2 = P_1 + Mgy$

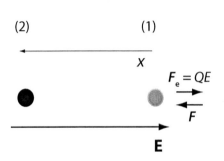

$P_2 = P_1 + QEx$

EXAMPLE 2

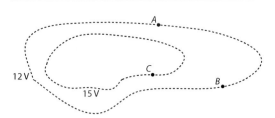

The figure shows two equipotentials, one at 12 V and the other at 15 V.

What is the work that needs to be done to move a 9 nC charge from A to B? From A to C?

Ans.:

A and B are on the same equipotential. They are at the same electric potential and it therefore takes no work against the electric force to move them from A to B.

C is at a higher potential, and it therefore takes work to move the charge there: it is $Q\Delta V = (9 \text{ nC})(3 \text{ V}) = 27 \times 10^{-9}$ J.

EXAMPLE 3

Go to the PhET website (http://phet.colorado.edu) and open the simulation *Charges and Fields*. Select "grid" and "show numbers."

(a) Pull one positive charge and one negative charge with the mouse to points along a horizontal 3 m apart, equidistant from each side. Select "show E-field." Drag the blue voltage sensor so that its crosshairs are at points 0.5 m apart along a vertical line first through one of the charges and then through the other. At each point (except right at the charges) click "plot" to draw the equipotential that includes this point. Explore other points. If you select "direction only" it is easier to see the field vectors, but the magnitude is no longer indicated by the intensity of the arrow color.

Turn off "Show E-field" and pull out an "E-field sensor." Drag it around the screen and observe the field arrow. What is its orientation with respect to the equipotentials?

(b) Clear all and repeat, this time with both charges positive and 1 m apart. Describe the equipotentials close to one charge, and then far from both.

(c) Explore other charge configurations.

Ans.:
(a) The electric field is a vector field. The red arrows on the screen show the electric field. Both the magnitude and the direction of the field can be seen with the "E-field sensor." The field is strongest close to a charge, pointing away from a positive charge and toward a negative charge. The equipotential lines are perpendicular to the field lines.

(b) Close to one charge the equipotentials surround it. Far from both the equipotentials surround both charges.

Simple and ideal: the uniform field

An electric field, **E**, is said to be *uniform* in a region where it has the same magnitude and direction at every point. No matter where in this region we put a charge of Q coulombs, the electric force on it is the same: $\mathbf{F} = Q\mathbf{E}$ newtons, with the force in the same direction as the field if Q is positive and in the opposite direction if Q is negative. A good example of a nearly uniform field is that between the plates of a capacitor, as in the last example of Chapter 8.

If we move a positive charge along a field line in the direction opposite to the field, we have to do work against the electric force (against the field) and both the electric potential energy of the charge and the electric potential increase.

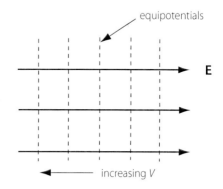

EXAMPLE 4

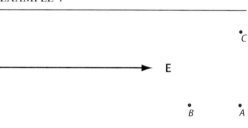

There is a uniform field of 5 N/C in a region. A charge of 2×10^{-4} C is moved from a point A to a point B along an electric field line in the direction opposite to the field, a distance of 3 m.

(a) What is the electric force on the charge at A?

(b) What is the electric force on the charge at B?

(c) What is the work done on the charge, against the electric field, as it is moved from A to B?

(d) Which is larger, the electric potential energy at A (P_A) or at B (P_B)?

(e) What is the difference in the potential energies, $P_B - P_A$?

(f) What is the potential difference, $V_B - V_A$?

The point C is 4 m from A, in a direction at right angles to the field. Let the potential at A be zero.

(g) What is the potential at B and at C?

(h) What is the work that is done on a 2×10^{-4} C charge to move it from C to A?

(i) What is the work done on this charge to move it from C to B?

Ans.:

(a) $F_e = QE = (2 \times 10^{-4} \text{ C})(5 \text{ N/C}) = 10^{-3}$ N. The direction of the force on this positive charge is in the same direction as the field, E.

(b) Because the field is uniform, $E_A = E_B$, and the forces are the same.

(c) The work done is $F_e \Delta x = 10^{-3}$ N $\times$ 3 m = 3×10^{-3} J or 3 mJ.

(d) As work is done against the field, the potential energy increases. It is therefore larger at B than at A.

(e) The difference in potential energy between points A and B is equal to the work done against the field as the charge is moved from A to B. It is equal to 3 mJ.

(f) $V_B = \frac{P_B}{Q}$, $V_A = \frac{P_A}{Q}$. $V_B - V_A = \frac{P_B - P_A}{Q} = \frac{3 \times 10^{-3} \text{ J}}{2 \times 10^{-4} \text{ C}} = 15$ V.

(g) The line from A to C is perpendicular to the electric field. Therefore there is no change in potential between A and C. The points A and C are on an equipotential.

$$W \quad = \quad \Delta P \quad = \quad P_B - P_A$$

Now find the potential at B. The work done on a charge is equal to the force on the charge times the distance that the charge is moved. On a unit charge (a charge of 1 C) the force is equal to the electric field, E, and the work done on it is the increase in the electric potential. Here $V_B - V_A = E \Delta x$.

The potential at B is higher than at A by $\Delta V = E \Delta x = (5 \text{ N/C})(3m) = 15$ J/C or 15 V. To determine the actual value of the potentials at B and C we have to decide on a reference level. Here $V = 0$ at A, so that $V_C = 0$ and $V_B = 15$ V.

(h) It takes no work to move a charge along an equipotential.

(i) Since the work to move the charge from C to A is zero, the work to move it from C to B is the same as to move it from A to B, i.e. (15 J/C)(2×10^{-4} C) = 3×10^{-3} J. This is so regardless of the particular path from C to B.

(We have assumed that the force of the electric field is the only one that has to be overcome as the charge is moved. If there are other forces, such as gravitation or friction, they need to be taken into account. The values of the potential and the potential differences will still be the same.)

The uniform gravitational field

Near the earth the gravitational force on an object of mass M is downward and equal to Mg. The gravitational field is the force divided by the mass, also downward, and it is equal to g. We can make the approximation that the field is uniform if we ignore the curvature of the earth's surface. In other words, we are confining ourselves to distances so small compared to the radius of the earth that we can treat the earth as though it were flat. Within this approximation the gravitational field can then be considered to be uniform, with lines of the gravitational field that are parallel, equally spaced, and downward.

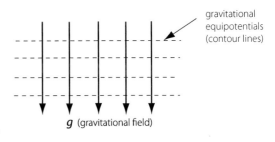

When we move an object upward, we have to do work against the gravitational force. Both the gravitational potential energy of the object and the gravitational potential increase. Since we know of no negative masses the complication of having two possible signs (as for positive and negative charges) does not arise in the gravitational case.

Motion in a uniform field

We know what happens when we release an object from rest in a gravitational field: it experiences a force (F_g) in the direction of the field (downward), whose magnitude is Mg. Its acceleration is also in the direction of the field, and its magnitude is g. We can write down the expressions for the velocity (v) and the displacement (y). If we use the field direction (down) as positive, they are $v = gt$ and $y = \frac{1}{2}gt^2$.

We can do the same to describe the motion of a charge in a uniform electric field: the force (F_e) is equal to QE. This time it is in the direction of the field only if Q is positive, and it is in the direction opposite to the field if the charge in negative. The acceleration is in the same direction as the force and has the magnitude $\frac{QE}{M}$. (In the gravitational case there is a term M both in the numerator and the denominator, and they cancel.) The velocity is $v = at$ and the displacement is $\frac{1}{2}at^2$.

What if the start is not from rest but with an initial velocity v_0? The force and the acceleration remain the same. But just as in the gravitational case the nature of the motion of a charge in an electric field is different depending on the direction of v_0. If v_0 is along a field line it's easy: $v = v_0 + at$, $x = v_0 t + \frac{1}{2}at^2$, and the motion is along a field line. As usual, we have to choose a direction to be positive and keep track of the plus and minus signs.

You also remember what happens to a component of the initial velocity that is perpendicular to the field and the force: nothing! Since there is no component of the acceleration in the direction perpendicular to the field, a velocity component in that direction does not change.

In the gravitational case this is what happens when a ball or other projectile is thrown and we neglect all other forces, such as air resistance. The force and the acceleration are straight down. The horizontal component of the initial velocity does not change. The vertical component changes in accord with the general relation $v = v_0 + at$. This is what we call *projectile motion*. The path of the projectile is a parabola.

The motion of a charge in a uniform electric field is similar. It is in a straight line along the field if there is no initial velocity, or if v_0 is also along the field direction. The path is a parabola if there is a component of the initial velocity that is perpendicular to the field.

If you turn the page through 90°, so that the electric field vector points toward the bottom of the page, the three paths in the figure are just like paths of an object in a gravitational field: the first one for an object thrown straight down, the second one for an object thrown down at an angle, and the third for an object thrown upward at an angle. In each case there is a force and an acceleration in the direction of the field. In the second and third cases there is also a component of the initial velocity (v_{0y} on the figure), which does not change because it is at right angles to the acceleration.

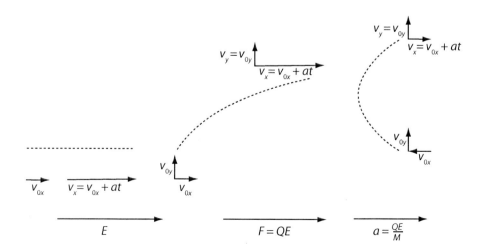

182 / More on Electricity: From Force to Energy, from Field to Potential

EXAMPLE 5

In the same 5 N/C uniform field as in the previous examples, the $+2 \times 10^{-4}$ C charge is released from rest at B.

(a) Describe its subsequent motion.

(b) What is its kinetic energy when it reaches A?

(c) When the charge is released, as in part (a), it moves toward A. What change would be necessary in the initial motion so that the charge will, instead, move to C? What additional information would be required for a detailed calculation?

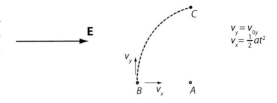

Ans.:

(a) Since the charge is positive, it experiences a force and an acceleration in the direction of the electric field. It starts from rest and moves along the field toward A in accord with the relations that describe the motion of a particle with constant acceleration.

(b) Let the potential and the potential energy of the charge at A be zero. Its potential at B is larger by 15 V. This is the difference in potential energy per unit charge, so that the difference in the potential energy of the charge at the two points is $Q\Delta V = (2 \times 10^{-4} \text{ C})(15 \text{ J/C}) = 3 \times 10^{-3}$ J.

$K_B + P_B = K_A + P_A$, where $K_B = 0$ and $P_A = 0$, so that $P_B = K_A$, i.e., the potential energy at B is transformed to kinetic energy at A, equal to 3 mJ.

(in the x direction) from B to A. The presence of a velocity component in the y direction will not change this time. It will take just as long to get to B from C. We can then see how large the y component of the velocity has to be for the charge to go the required distance in the y direction.

To find the time we can use $x = v_{0x}t + \frac{1}{2}at^2$, with $v_{0x} = 0$. Although we know the force, we do not know the acceleration. To go further we need to know the mass of the charged object. Let's say the mass is 5 g, or 5×10^{-3} kg. Then $a = \frac{F}{M} = \frac{QE}{M} = \frac{10^{-3} \text{ N}}{5 \times 10^{-3} \text{ kg}} = 0.2$ m/s^2.

$x = \frac{1}{2}at^2$, so that $t^2 = \frac{2x}{a} = \frac{(2)(3)}{0.2} = 30$, and $t = \sqrt{30} = 5.48$ s.

During that time v_y has to have a magnitude such that the charged object moves the distance from A to C of 4 m. This velocity component is constant (since there is no acceleration in the y direction), so that $y = v_y t$. Hence $v_y = \frac{y}{t} = \frac{4}{5.48} = 0.73$ m/s.

The path from B to C is a parabola. (If you turn the page one-quarter turn clockwise, you see that it looks like the path of a ball moving in a gravitational field.)

$$P_B + K_B = P_A + K_A$$

$$P_B + K_B = P_C + K_C$$
$$K_B = \tfrac{1}{2}Mv_y^2 \qquad K_C = \tfrac{1}{2}Mv^2 = \tfrac{1}{2}M(v_x^2 + v_y^2)$$

(c) If the charge is released from rest in this uniform field, it moves along a field line toward A. To move, instead, to C it has to have an initial velocity component perpendicular to the field. Since there is no field component and no force component in this direction, this velocity component does not change.

To find this component (v_y) we need to know how long the charge will take to go along the field

Closer to reality: the field of a point charge

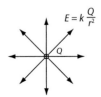

$$E = k\frac{Q}{r^2}$$

The electric field lines of an individual electron or proton are *radial*. They are straight lines, equally spaced, coming from or going to the charge. This is also true for a uniformly charged sphere, or for any other spherically symmetric charge distribution, as we saw from our discussion of Gauss's law. In the gravitational case we rarely deal with a *point mass*, but stars, planets, and moons are close to being spherically symmetrical.

A radial field is far from uniform. Its magnitude gets smaller as r increases, proportional to $\frac{1}{r^2}$. On the diagram that shows the field lines, the lines are close to each other near Q, and become farther apart as r increases.

We know how the electric potential energy of a system of two positive charges (Q_1 and Q_2) changes when we move Q_2 closer to Q_1: the definition of electric potential energy tells us that the increase in potential energy is equal to the work done *against* the electric force on Q_2. The closer we move Q_2 to Q_1 the more work we have to do against the force of repulsion, and the greater is the resulting potential energy.

When we set out to calculate the work to move Q_2 closer to Q_1 we encounter an obstacle: the field and the force are not constant. We remember that work equals force times displacement, but if the magnitude of the force changes, what are we to do?

Q_1

r_1

r_2

$E_1 = k\frac{Q_1}{r_1^2}$ $E_2 = k\frac{Q_1}{r_2^2}$

$F_{12} = k\frac{Q_1 Q_2}{r_1^2}$ $F_{12} = k\frac{Q_1 Q_2}{r_2^2}$

The magnitude of the field created by Q_1 is $k\frac{Q_1}{r^2}$, and the magnitude of the force on Q_2 is $k\frac{Q_1 Q_2}{r^2}$. We want to calculate the work done as we move the charge Q_2 closer to Q_1, from a distance r_2 to a smaller distance, r_1.

The force changes as r changes and the charge moves through the distance $r_2 - r_1$. We can't multiply the expressions for the force and the distance because we don't know what value to use for F.

This is the kind of problem for which the calculus was invented. It allows us to deal with variables that change, so that we can calculate force times distance even when the force changes.

Approximate method, right result

There have been many attempts to calculate the potential energy of two charges with only algebraic methods. We will look at one of the simplest.

Since r varies from r_2 to r_1, r^2 will vary from r_2^2 to r_1^2. The force, F_1, on Q_2 varies from $k\frac{Q_1 Q_2}{r_2^2}$ to $k\frac{Q_1 Q_2}{r_1^2}$. To get a value between the two we might take one value from each end, using $r_1 r_2$ instead of r^2 in the denominator. This looks like a rough approximation, but it turns out, surprisingly, to give the correct answer. The force is then $k\frac{Q_1 Q_2}{r_1 r_2}$ and the distance is $r_2 - r_1$. The work done against the repulsion is therefore $(k\frac{Q_1 Q_2}{r_1 r_2})(r_2 - r_1)$, which is equal to $kQ_1 Q_2 \frac{r_2 - r_1}{r_1 r_2}$ or $kQ_1 Q_2(\frac{1}{r_1} - \frac{1}{r_2})$.

Let's look more closely at the sign of this result for the work done. If r_1 is less than r_2 then $\frac{1}{r_1}$ is greater than $\frac{1}{r_2}$, and the amount of work is positive. This is as we expect: we have to do (positive) work on the system to move the two positive charges closer to each other.

The point charge, properly

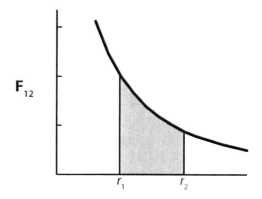

A little calculus, and the use of the expression for the integral of r^n that we cited near the end of Chapter 3, allows us to get away from guesses and approximations. For a constant force the work is equal to the force times the displacement. This is equal to the area under a curve of force against displacement. It is still the area under that curve if the force changes. For a force equal to $k\frac{Q_1Q_2}{r^2}$, and a displacement from r_2 to r_1, the area is equal to the integral $\int_{r_2}^{r_1} k\frac{Q_1Q_2}{r^2}dr$. In the expression for the force, $k\frac{Q_1Q_2}{r^2}$, k and the charges are constant, so that the integral is equal to $kQ_1Q_2 \int_{r_2}^{r_1} \frac{1}{r^2}dr$. We only need to evaluate the integral of $\frac{1}{r^2}(=r^{-2})$, i.e., $\int_{r_2}^{r_1} r^{-2}dr$. Since the integral of r^n is $\frac{r^{n+1}}{(n+1)}$, the integral of r^{-2} is $\frac{r^{-1}}{-1}$ or $\frac{-1}{r}$.

If the force moves the charge from r_2 to r_1, we need to evaluate the integral at these two values and calculate the difference, i.e., $\left(\frac{-1}{r_2} - \frac{-1}{r_1}\right)$, or $\left(\frac{1}{r_1} - \frac{1}{r_2}\right)$. We still have to multiply this result by kQ_1Q_2 to get $kQ_1Q_2\left(\frac{1}{r_1} - \frac{1}{r_2}\right)$, the same result as before.

EXAMPLE 6

A fixed charge, Q_1, of 4 µC and a second charge, Q_2, of -3 µC are initially at rest, separated by a distance of 5 cm. The second charge moves toward the first as a result of the Coulomb attraction, until the distance between them is 2 cm. What is then the kinetic energy of the negative charge?

Ans.:

As Q_2 moves toward Q_1, potential energy is lost, and Q_2 gains an equal amount of kinetic energy.

First, calculate the magnitude of the difference in the potential energies. It is $kQ_1Q_2\left(\frac{1}{r_1} - \frac{1}{r_2}\right)$, where $r_1 = 2$ cm and $r_2 = 5$ cm, i.e. $(9 \times 10^9)(4 \times 10^{-6})(3 \times 10^{-6})\left(\frac{1}{2 \times 10^{-2}} - \frac{1}{5 \times 10^{-2}}\right) = (0.108)(10^2)(0.5 - 0.2) = 3.24$ J.

As the charges move toward each other, they lose 3.24 J of electric potential energy and gain the same amount of kinetic energy.

Finding the potential energy and the potential

In the previous section we used the fact that the electric potential energy is equal to the work done against the electric force. Can we also talk about the actual amount of the potential energy, the *absolute* potential energy? Yes, but just as for the gravitational potential energy we need first to choose a reference level where the potential energy is zero.

We know that we can put it anywhere we want, but there is one place that is used most often when we talk about point charges: we let the potential energy, P, be zero when the charges are very far (infinitely far) from each other.

In the calculation of the previous section the two charges are initially separated by the distance r_2. The work done as Q_2 is pushed closer to Q_1, until they are separated by a distance r_1, is $kQ_1Q_2\left(\frac{1}{r_1} - \frac{1}{r_2}\right)$. If we let r_2 become very large, $\frac{1}{r_2}$ becomes very small. When r_2 goes to infinity, $\frac{1}{r_2}$ goes to zero. The work done on Q_2 as it moves from infinitely far away to a distance r_1 from Q_1 becomes $kQ_1Q_2(\frac{1}{r_1})$.

Let's look at the definition again. The increase in potential energy is equal to the work done against the electric force. If we start from the reference level, where $P = 0$, this will also be the potential energy. To bring Q_2 closer to Q_1 means that we have to overcome the electric repulsion between the two positive charges. To bring it from far away to r_1 requires an amount of work equal to $k\frac{Q_1Q_2}{r_1}$, and this is, therefore, their potential energy when they are at the distance r_1 from each other.

We have to be careful with the plus and minus signs. If one charge is positive and the other is negative, for example, the charges attract, and the signs are reversed. There are schemes to keep track of the signs in the formulas, but it is easier first to calculate the magnitude of the work or the change in the potential energy, and to think separately about the signs. Is the work to move a charge against the electrical force positive or negative, i.e., do we have to force them together or hold them apart? Is the value of the potential energy increased or decreased?

Now let's look at the potential. We can again call Q_2 a *test charge*. Take it to be positive. If at a certain point its potential energy is P, the electric potential at that point is $\frac{P}{Q_2}$.

If there are just two charges, Q_1 and Q_2, a distance r_1 apart, then $P = k\frac{Q_1Q_2}{r_1}$, and the potential, a distance r_1 from Q_1, is $k\frac{Q_1}{r_1}$.

We can put our test charge at any point. If its magnitude is Q_2, and the electric force on it is F, then the electric field at that point is $E = \frac{F}{Q_2}$. If its electric potential energy at that point is P, then the electric potential at that point is $\frac{P}{Q_2}$.

Starting with the sources

If we know the sources we can start with them. For a point charge, Q, we know that at a distance r from it the electric field is radial, with magnitude $k\frac{Q}{r^2}$, outward if Q is positive and inward if Q is negative. The potential, V, is $k\frac{Q}{r}$, with the sign of Q, positive or negative, depending on which kind of charge Q is. (Here we have taken the reference level $P = 0$ when r is infinitely large and $\frac{1}{r} = 0$.) If Q is not a point charge, but consists of charges distributed with spherical symmetry, these results will still be correct outside the charge distribution. If there are several point charges, we have to take the fields and potentials from each, and add them up. The electric fields are vectors. The potentials are scalars, so that we need only add them up like numbers (which can be positive or negative).

EXAMPLE 7

A charge of $Q_1 = 4\,\mu C$ is at a distance of $L = 50$ cm from a charge of $Q_2 = -3\,\mu C$.

(a) Is there a point on the line between the two charges where the electric potential is equal to zero, and, if so, how far is it from Q_1?

(b) Are there any other points along the line on which the two charges are located where the potential is zero? (To the left of Q_1 or to the right of Q_2.)

Ans.:

Between the two charges, at a distance x from Q_1, the potential is $k\frac{Q_1}{x} + k\frac{Q_2}{L-x}$. It is equal to zero when $\frac{Q_1}{x} = \frac{-Q_2}{L-x}$ or $Q_1(L-x) = -Q_2 x$, i.e., when $(4 \times 10^{-6})(0.5 - x) = (3 \times 10^{-6})(x)$ or $4(0.5 - x) = 3x$. $2 - 4x = 3x$, $7x = 2$, and $x = 0.286$.

There is a second point where $P = 0$: outside both charges, and nearer to the smaller charge, at a point a distance x from Q_2, where $\frac{Q_1}{L+x} + \frac{Q_2}{x} = 0$. Here $\frac{4}{0.5+x} - \frac{3}{x} = 0$, $4x - 3(0.5 + x) = 0$, or $4x - 1.5 - 3x = 0$, so that $x = 1.5$ m.

(Note that there is no corresponding point on the other side of the two charges.)

EXAMPLE 8

Four charges are at the corners of a square. What has to be true for them so that the electric potential at the center of the square is zero?

Ans.:

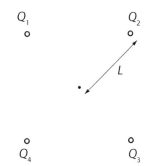

Let the distance from each of the charges to the center be L, and the charges Q_1, Q_2, Q_3, and Q_4. The potential in the center is $k\frac{Q_1}{L} + k\frac{Q_2}{L} + k\frac{Q_3}{L} + k\frac{Q_4}{L}$ or $\frac{k}{L}(Q_1 + Q_2 + Q_3 + Q_4)$. If this is to be zero, $(Q_1 + Q_2 + Q_3 + Q_4)$ must be zero, i.e., there must be just as much positive as negative charge at the corners, but otherwise it doesn't matter what the charges are. (Of course the electric field will be different for each charge configuration. Under what conditions will it be zero at the center?)

Moving charges

We know the advantages of using energy concepts over those of force whenever we can. Let's see how we can use energy concepts when we talk about charges moving in an electric field. Look at a charge, Q, at a point where the potential is V_1, so that its electrical potential energy is $P_1 = QV_1$. It moves to a point where the potential is V_2 and its potential energy is QV_2. Its kinetic energy at the first point is K_1 and at the second point it is K_2. If there are no other energies that we need to consider, then

$$P_1 + K_1 = P_2 + K_2, \text{ or}$$

$$QV_1 + \frac{1}{2}mv_1^2 = QV_2 + \frac{1}{2}mv_2^2$$

To use these relations we don't need to know anything about where there are other charges or what the electric fields look like!

EXAMPLE 9

Bombardment with alpha particles (discovery of the nucleus)

The bombardment of a gold foil with alpha particles from a radioactive source was a groundbreaking experiment by Ernest Rutherford in 1911. The result that the alpha particles rebounded showed that the positive charge of the gold atom was concentrated in a very small region. In fact, the experiment could not distinguish it from a point charge. This was the first demonstration of the existence of the atomic nucleus.

An alpha particle consists of two protons and two neutrons. It has a charge of $+2e$ or 3.2×10^{-19} C and a mass of 6.65×10^{-27} kg.

An alpha particle has an initial kinetic energy of 4 MeV. (1 MeV = 1.6×10^{-13} J.) It is shot toward the center of a gold nucleus ($Z = 79$, i.e., a charge of $79e$).

(a) What is the subsequent motion of the alpha particle?

(b) Draw a graph of the potential energy as a function of the distance between the alpha particle and the gold nucleus. On the same graph draw a line that represents the total energy of the alpha particle. Show the distance of closest approach, D, on the graph.

(c) Calculate the distance of closest approach.

Ans.:
(a) Let the system of the alpha particle and the nucleus have zero potential energy when they are very far from each other and the kinetic energy is 4 MeV. The total energy is then 4 MeV.

As the alpha particle moves toward the gold nucleus, it loses kinetic energy and gains potential energy because of the repulsion between the two positively charged objects. It continues to gain potential energy until all of its kinetic energy has been transformed into potential energy. It is then momentarily at rest at the distance of closest approach (D) before rebounding and moving away from the gold nucleus. (Note that when the alpha particle gets close to the nucleus the electrons of the gold atom are far away, and we are ignoring them.)

(b)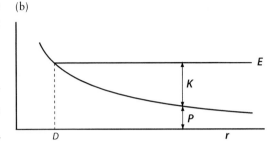

(c) If we call the initial point (far away) A and the point of closest approach B, $P_A = 0$, $K_B = 0$, and $K_A = 4$ MeV. Since $P_A + K_A = P_B + K_B$, the potential energy at the point of closest approach is $P_B = K_A = 4$ MeV. We know that $P_B = \frac{kQ_1Q_2}{D}$, and we can now find $D = \frac{kQ_1Q_2}{P_B} = \frac{(9 \times 10^9)(79 \times 1.6 \times 10^{-19})(2 \times 1.6 \times 10^{-19})}{4 \times 1.6 \times 10^{-13}}$, which is equal to 5.69×10^{-8} m or 56.9 nm.

EXAMPLE 10

Here is a similar example, but in a gravitational field.

A rocket is launched, straight up, from the surface of the earth, with a velocity v_e. Assume that it has this velocity right after launch, and that the engines then shut off so that there is no further energy input. It moves away from the earth and eventually is so far from the earth that the gravitational potential energy of the earth-rocket system is zero. (Using the usual reference level $P = 0$ when $\frac{1}{r} = 0$.) Neglect any forces other than the gravitational force of the earth.

(a) The minimum velocity that the rocket must have at the beginning to get that far is called the escape velocity. How large is that velocity, v_e?

(b) Draw a graph of the gravitational potential energy P as a function of the distance r. Mark on it the total energy, E, the initial kinetic energy, $\frac{1}{2}mv_e^2$, and the radius of the earth, R_e, and on it show the kinetic energy K of the rocket.

Ans.:
(a) The rocket starts out with a potential energy $P = -G\frac{mM_e}{R_e}$, where m is the mass of the rocket and M_e and R_e are the mass and the radius of the earth. Its initial kinetic energy is $\frac{1}{2}mv_e^2$.

At its destination $P_f = 0$. If it just gets there, without energy to spare, its kinetic energy, K_f, is also zero. Since both P_f and K_f are zero, so is the total energy.

If there is no energy input, this is also the energy at the start, i.e., $-G\frac{mM_e}{R_e} + \frac{1}{2}mv_e^2 = 0$ or

$\frac{1}{2}mv_e^2 = G\frac{mM_e}{R_e}$, so that $v_e = \sqrt{\frac{2GM_e}{R_e}}$, which we see to be independent of the mass of the rocket and equal to $\sqrt{\frac{(2)(6.67\times 10^{-11})(5.97\times 10^{24})}{6.38\times 10^6}}$, which is equal to 1.12×10^4 m/s, or close to 7 miles per second.

(b)

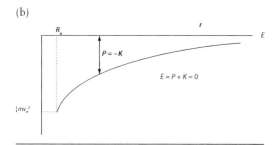

High-speed electrons are used in many devices. In an *X–ray tube*, for instance, they hit a "target" from which the X–rays are emitted.

Another application is the vacuum tube rectifier or *diode*. It contains two metal *electrodes*. One of them (the *cathode*) is heated, so that some of its electrons are given enough energy to leave it. The other one (the *anode*) is at a higher potential, so that the electrons liberated at the cathode are accelerated toward it. Since the anode is not heated, electrons cannot leave it, and there is current in only one direction. If the electrons start with negligible kinetic energy at the cathode, they will arrive at the anode with a kinetic energy equal to $Q\Delta V$ joules, where ΔV is the potential difference between the anode and the cathode.

If the anode has a hole in it, or is in the shape of a ring, some of the electrons, instead of slamming into it, will continue to the region on the other side of the anode where the electric field is zero, or at least quite small. Such an arrangement is called an *electron gun*.

These applications are so important that they led to the introduction of the electron volt as a unit of energy. We have used it before, and it is used universally in the discussion of atomic, molecular, and nuclear energies. One electron volt is equal to 1.6×10^{-19} J. Hence electrons accelerated by a potential difference of ΔV volts will gain a kinetic energy of ΔV electron volts.

EXAMPLE 11

In a vacuum tube diode the anode voltage is 60 V. The cathode voltage is zero, and the electrons leave the cathode with negligible energy.

(a) What is the kinetic energy of the electrons when they arrive at the anode? Give the answer in joules and in electron volts.

(b) What is the speed of the electrons at the anode?

Ans.:

(a) At the cathode the potential energy of the electrons, P_C, is zero, and so is their kinetic energy, K_C. At the anode the potential energy is $P_A = QV = (-1.6\times 10^{-19})(60) = -9.6\times 10^{-18}$ J. Since $P_C + K_C = P_A + K_A$, $0 = K_A + P_A$, and $K_A = -P_A = 9.6 \times 10^{-18}$ J or 60 eV.

(b) $K = \frac{1}{2}mv^2$ so that $v = \sqrt{\frac{2K}{m}} = \sqrt{\frac{(2)(9.6\times 10^{-18})}{0.91\times 10^{-30}}} = 4.6 \times 10^6$ m/s.

9.2 Energy transformations and electric circuits

Battery and resistor

An electric battery is a device that transforms stored internal ("chemical") energy to electric potential energy. The chemical reactions inside it produce a difference of electric potential between its two terminals. We can use this energy by changing it into other kinds of energy when we make the battery part of an electric circuit.

Here is the simplest circuit: we attach a wire to the battery between its positive and negative terminals. As a result of the potential difference electrons move in the wire. There is now a *current* in the wire.

Here is a "schematic" diagram to represent this circuit.

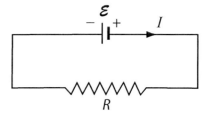

The symbol in the top part with the + and − signs represents the battery. The symbol with the lines going back and forth in the bottom part represents the wire. Its shape is meant to indicate that there is quite a bit of it, as in a coil of wire. It is called a *resistor* and is marked R. The straight lines between the symbols for the battery and the coil of wire are there only to connect them.

Since the electrons are negatively charged, they move through the wire from the battery's negative terminal toward the positive one. However, as an accident of history, the electric *current* is defined as the amount of positive charge passing a place in the wire in a second. If the charges that move are actually negative (as is most often the case, since it is the electrons that move in a wire), the direction of what we call the current is opposite to the direction of motion of the charges. In other words, a current to the right can consist of positive charges going to the right or of negative charges going to the left. The symbol I is used for the current, and its SI unit is the coulomb/second, which is also called the *ampere*.

What happens in the wire?

You might be surprised that the current is the same in every part of the circuit shown in the figure. In fact, you should be. There is an electric field in the metallic wire, from the positive terminal toward the negative terminal of the battery, just as there is from any positive charge toward a negative charge. Aren't the electrons accelerated? Wouldn't that make the current larger and larger along the direction of motion?

That's what would happen if the wire were an empty tube. But the wire is made of ions (the parts of the atoms that are left behind when the moving "free" electrons are detached). The electrons bounce along among them, making collisions along the way that slow them down.

Let's see what happens to the electrons. After each collision an electron is scattered and goes in some new direction. This new direction can be at any angle, and all directions are equally probable. For all of the scattered electrons, going in all of the equally probable directions, the average velocity is zero. We can look at a single electron that represents the average behavior of all of them. It is as if this electron were stopped by the collision. It is again accelerated, but has to start over after each collision. The result is that the average velocity of the electrons is constant, and the current is the same in each part of the wire.

The situation is somewhat analogous to that of falling raindrops. If they fell through empty space they would be accelerated by the gravitational force. (It would be very dangerous to be outside and to be bombarded by them!) Instead they collide with air molecules and move much more slowly than they would in empty space and with constant velocity.

When the electrons collide with the ions they slow down and give up some of their kinetic energy. The energy is shared by the ions and is now internal energy of the wire. This increased internal energy is perceived by us as a higher temperature.

Look at a coil of wire. There is a difference between the values of the electric potential at the two ends of the coil. This potential difference across the coil, $\Delta V = V_2 - V_1$, divided by the current, I, through the coil, is called the coil's *resistance*, R, equal to $\frac{\Delta V}{I}$. The SI unit for the resistance, equal to the volt/ampere, is called the ohm, with the symbol Ω (Greek capital *omega*).

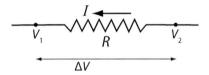

The electrons carry electric potential energy from one part of the circuit to another. In the battery they get energy from the internal energy of the battery materials. In the resistor they give up energy to the material of the wire through collisions with the metal ions. This energy becomes internal energy of the wire.

The potential difference, or *voltage difference*, ΔV, represents a difference in potential energy per coulomb. A volt is a joule/coulomb. The current is I amperes, or I coulombs/second. $I\Delta V$ is therefore the number of joules per second, or the power in watts, transformed in the wire from electric potential energy to internal energy of the wire.

The resistance of a wire varies with temperature, but is otherwise constant. It is a property of the wire. This is so for metals, but not for all other materials. When the resistance is constant (at a given temperature), so that it does not depend on the current, the material is said to follow *Ohm's law*.

The resistance of metal wires used for connections in circuits is usually so small that it can be neglected. On the other hand, the energy

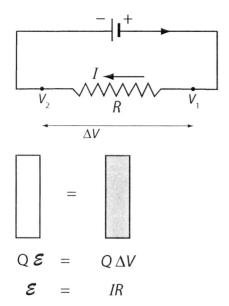

dissipated in the wire ("filament") of a light bulb is so large that the wire gets white hot, and part of the transformed energy is radiated away as visible light.

One more new term: the amount of energy that is transformed in the battery (or other device) from some other kind of energy to electric potential energy divided by the charge that passes through it is called its *emf*. A 9 volt battery has an emf of 9 volts. For each coulomb that moves through it, 9 joules of energy are transformed from stored chemical energy to electric potential energy.

"Emf" stands for *electromotive force*, but physicists tend to be reluctant to write this out, because it is not a force at all, but an energy divided by a charge.

EXAMPLE 12

Go to the PhET website (http://phet.colorado.edu) and open the simulation *Ohm's Law*. Set R at 140 Ω by dragging the resistance button up and down.

(a) Change the voltage while R remains constant. How do V and I change with respect to each other?

(b) Change R while V remains constant. How do I and R change with respect to each other?

Set V at 4 V. Set R at 400 and 600 Ω. For each case calculate I and compare the result to the value on the screen.

Go to the PhET website and open the simulation *Battery–resistor Circuit*. Uncheck the three buttons on the right. Set $V = 12.00$ V and $R = 0.4\,\Omega$.

Calculate the current and compare it to the value on the ammeter.

Click on "show inside battery" to see the electrons being pushed (by little elves?).

Click on "show cores" to see the ions. Reduce the voltage and observe the change in the motion of the ions and in the temperature.

EXAMPLE 13

In a flashlight a 1.5-V battery is connected to a light bulb.

(a) How much energy is given to each electron as it passes through the battery? Where does this energy come from?

(b) What happens to the electric potential energy of the electrons as they pass through (i) the wire and (ii) the bulb?

Ans.:

(a) Each electron gets an amount of energy, $Q\Delta V$, whose magnitude is $(1.6 \times 10^{-19})(1.5) = 2.4 \times 10^{-19}$ J. This energy is transformed from the battery's internal energy to electric potential energy.

(b) (i) If you assume that the wire has no resistance there is no difference of potential across the wire and no energy transformation. (ii) The full potential difference is then across the bulb, and in the bulb some of this electric potential energy is transformed to internal energy (thermal energy) of the filament and some is radiated away as visible light.

EXAMPLE 14

A 25 Ω resistor is connected across the terminals of a battery whose emf is 6 V.

(a) Draw a schematic diagram of the circuit.

(b) What is the potential difference across the resistor?

(c) What is the current through the resistor?

(d) What is the current through the battery?

(e) What is the power generated by the battery?

(f) What is the power dissipated in the resistor?

(g) What is the energy (in joules) transformed in the resistor in one minute?

(h) What is the amount of energy in kilowatt-hours (*kwh*) generated by the battery in one day?

Ans.:
(a)

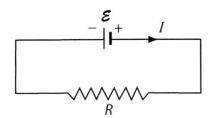

(b) The potential difference across the resistor has the same magnitude as the potential difference across the battery. The connecting lines in the schematic diagram are assumed to have no resistance, so that there is no potential difference across them. ($\Delta V = IR = 0$ for them.) In an actual circuit the wires have resistance, but it can usually be neglected because it is much less than the resistance of the resistors.

(c) $I = \frac{\Delta V}{R} = \frac{6}{25} = 0.24\,\text{A}$.

(d) The current is the same in every part of this circuit, 0.24 A.

(e) Power = $I\Delta V$. For the battery it is equal to $\mathcal{E}I = (0.24)(6) = 1.44\,\text{W}$.

(f) $I\Delta V = 1.44\,\text{W}$.

(g) The power is the energy transformed per second. In 60 s the energy transformed is therefore $(60\,\text{s})(1.44\,J/s) = 8.64\,\text{J}$.

(h) In one day there are (24)(3600) seconds. The energy generated in a day is therefore $(1.44)(24)(3600)\,\text{J} = 1.24 \times 10^5\,\text{J}$.

A *kilowatt-hour* is the energy transformed in an hour if the rate is 1000 W. It is $(3600)(1000)\,\text{J}$ or $3.6 \times 10^6\,\text{J}$. $1.24 \times 10^5\,\text{J} = \frac{1.24 \times 10^5}{3.6 \times 10^6} = 0.035\,\text{kwh}$.

EXAMPLE 15

Two buzzers (which can be treated as resistors) are connected, end to end, to a battery. When resistors are connected so that there is the same current in each, they are said to be connected *in series*. The emf of the battery is 12 volts. The resistors are $R_1 = 6\,\Omega$ and $R_2 = 9\,\Omega$.

(a) Draw a schematic diagram for this circuit.

(b) What is the resistance of a single resistor that could replace the two resistors without changing the current through the battery? This is called the "equivalent resistance," R_{eq}.

(c) What are the potential differences across each of the two resistors in the original circuit?

(d) What is the power dissipated in each resistor?

(e) What is the power dissipated in R_{eq}?

(f) What is the power generated in the battery?

Ans.:
(a)

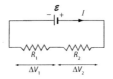 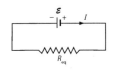

(b) The voltages ΔV_1 and ΔV_2 add up to the voltage ΔV across the equivalent resistance. $\Delta V = IR_1 + IR_2 = I(R_1 + R_2)$. $R_{eq} = \frac{\Delta V}{I} = R_1 + R_2$.
$R_{eq} = R_1 + R_2 = 6 + 9 = 15\,\Omega$.

(c) $\Delta V_1 = IR_1$.
$I = \frac{\mathcal{E}}{R_{eq}} = \frac{12}{15} = 0.8\,\text{A}$.
$\Delta V_1 = (0.8)(6) = 4.8\,\text{V}$.
$\Delta V_2 = (0.8)(9) = 7.2\,\text{V}$.
Note that $\Delta V_1 + \Delta V_2 = 12\,\text{V} = \mathcal{E}$.

(d) $P_1 = I\Delta V = I^2 R_1 = (0.8)(4.8) = 3.84\,\text{W}$.
$P_2 = I^2 R_2 = (.64)(9) = 5.76\,\text{W}$.
$P_1 + P_2 = 9.60\,\text{W}$.

(e) $P = I^2 R_{eq} = (.64)(15) = 9.60\,\text{W}$.

(f) $P = \mathcal{E}I = (12)(.8) = 9.60\,\text{W}$.

EXAMPLE 16

Two resistors are connected separately across the battery. When resistors are connected so that the potential difference is the same through each, they are said to be connected *in parallel*. The emf of the battery is 12 volts. The resistances of the two resistors are $R_1 = 6\,\Omega$ and $R_2 = 9\,\Omega$.

(a) Draw a schematic diagram for this circuit.

(b) What is the current through each resistor?

(c) What is the current through the battery?

(d) What is the resistance of a single resistor that could replace the two resistors without changing the current through the battery? This is the equivalent resistance for the combination.

(e) What is the power dissipated in each resistor?

(f) What is the power dissipated in the equivalent resistance?

(g) What is the power generated by the battery?

Ans.:

(a)

(b) $I_1 = \frac{\Delta V}{R_1} = \frac{12}{6} = 2$ A.
$I_2 = \frac{\Delta V}{R_2} = \frac{12}{9} = 1.33$ A.

(c) The currents I_1 and I_2 come together to form the current I through the battery: $I = I_1 + I_2 = 2 + 1.33 = 3.33$ A.

(d) If there were just a single resistor, R_{eq}, across the battery, its resistance would be $\frac{\mathcal{E}}{I}$ or $\frac{\Delta V}{I}$. Here the potential difference across each resistor is the same, but the current through each is different. (For the series connection of the previous example the current through each resistor is the same, but the two voltages are different.)

$$R_{eq} = \frac{\mathcal{E}}{I} = \frac{12}{3.33} = 3.60 \, \Omega.$$

We could also do the calculation more generally: $I = \frac{\Delta V}{R_{eq}} = I_1 + I_2 = \frac{\Delta V}{R_1} + \frac{\Delta V}{R_2}$ or $\frac{1}{R_{eq}} = \frac{1}{R_1} + \frac{1}{R_2}$.

(e) $I_1 \Delta V = (2)(12) = 24$ W, $I_2 \Delta V = (1.33)(12) = 16$ W.

(f) $I^2 R_{eq} = (3.33)^2(3.60) = 40$ W. [This is the sum of the two values calculated in part (e).]

(g) $\mathcal{E} I = (12)(3.33) = 40$ W.

When lights, toasters, vacuum cleaners, and other appliances are "plugged in" at home, they are connected in parallel. That allows each of them to be connected independently. The "electric meter" measures the amount of energy that is used, in kilowatt–hours, and this is what shows up in the monthly bill.

Resistivity: separating out the property of the material

The resistance of a wire depends on its shape, i.e., its length and cross-sectional area, and on the material of which it is composed. Can we separate the effect of the material's properties from those of the wire's size?

Just think of the wire as a number of pieces in series. Suppose there are five pieces, each with a length L and a resistance R. The resistance of the whole wire (its equivalent resistance) is $5R$, and the length is $5L$. We see that the resistance is proportional to the length.

Similarly we can think of the wire as made up of five pieces in parallel, each with resistance R and cross-sectional area A. The resistance of the whole wire (its equivalent resistance) is then $\frac{R}{5}$. The total area is $5A$. We see that the resistance is inversely proportional to the area.

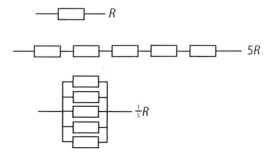

In general then we can write $R = \rho \frac{L}{A}$, showing the length and area dependences. The remaining factor, ρ (Greek *rho*), depends only on the material of the wire and its temperature. It is called the *resistivity*.

EXAMPLE 17

Go to the PhET website and open the simulation "Resistance in wire." Vary ρ, L, A in turn to see the effect on R. Choose values of $\rho, L,$ and A and compare them to the values on the screen.

EXAMPLE 18

What is the diameter of a 2-cm-long tungsten filament in a "60 W" incandescent light bulb? (At the operating temperature the resistivity is about $10^{-8}\,\Omega\,\text{m}$.)

Ans.:
Since the power is $P = IV$, or $\frac{V^2}{R}$, we can write $R = \frac{V^2}{P} = \frac{110^2}{60} = 201.7\,\Omega$. Since $R = \rho\frac{L}{A}$, we can write $A = \frac{\rho L}{R} = \frac{(10^{-8})(0.02)}{201.7} = .99 \times 10^{-12}\,\text{m}^2$. $A = \frac{\pi}{4}D^2$, so that $D^2 = 1.26 \times 10^{-12}$ and $D = 1.1 \times 10^{-6}\,\text{m} = 1.1\,\mu\text{m}$.

9.3 Summary

Like the gravitational potential energy the electric potential energy has two great advantages. One is that it is a scalar quantity. That makes it much simpler to use than the gravitational and electric forces, which are vector quantities. The other is that when a particle moves from one point to another, the difference in potential energy depends only on the potential energies at the starting point and at the end point, and not on the particular path between the two.

When a charge moves in an electric field the increase in electric potential energy is equal to the work done against the electric field.

The change in the potential energy of a charge, divided by the magnitude of the charge, is called the change in the electric potential.

No work is done when a charge moves along an equipotential.

In an electric field the acceleration of a charge is along the electric field line. In a uniform field, if the initial velocity of the charge is zero or along the field, the charge moves along the field line. If the charge has an initial velocity that is not along the field it moves in a parabola. The velocity component parallel to the field changes, but the component perpendicular to the field remains constant.

The electric potential energy of a system of two point charges is $k\frac{Q_1Q_2}{r}$, where the reference level is taken to be at $r = \infty$ ($\frac{1}{r} = 0$). With this reference level the potential at a distance r from a point charge is $k\frac{Q}{r}$.

In our homes and in our other surroundings, electric circuits are everywhere. They can have many purposes, but their basic property is that wires allow the transport of charges (*currents*), and with them the transport of energy. The simplest circuits consist of batteries and resistors.

The current in a wire is the charge passing a cross section per second.

The resistance of a wire is the potential difference across the wire divided by the current through the wire. $R = \frac{\Delta V}{I}$.

If R is constant when ΔV and I vary, the wire is said to follow Ohm's law. (R can and usually does vary with temperature.)

When a current passes through a resistor some electric potential energy is changed to internal energy. The amount of power (= energy per second) that is transformed is $I\Delta V [= I^2 R = \frac{(\Delta V)^2}{R}]$ watts.

When two resistors are connected so that the same current passes through both, they are said to be connected in series. The equivalent resistance of a combination of resistors in series is their sum. $R_{eq} = R_1 + R_2 + R_3 + \cdots$.

When two resistors are connected so that the potential difference across each is the same, they are said to be connected in parallel. The reciprocal of the equivalent resistance of a combination of resistors in parallel is the sum of their reciprocals. $\frac{1}{R_{eq}} = \frac{1}{R_1} + \frac{1}{R_2} + \frac{1}{R_3} + \cdots$.

The resistivity, ρ, is a property of the material of which a wire is made. A wire of length L and cross-sectional area A has a resistance of $\rho\frac{L}{A}$.

9.4 Review activities and problems

Guided review

1. The electric field in a region is uniform, in the y direction, with magnitude 12 N/C. Use the point A (0,0) as the reference point where $V = 0$. For each of the points $B(3,0), C(0,3)$, and $D(3,3)$ (where all distances are in meters):

(a) What is the force on a $+3 \times 10^{-5}$ C charge at B, C, and D?

(b) How much work needs to be done to move this charge from the origin to B? to C? to D?

(c) What is the electric potential energy of the system when the charge is at B? at C? at D?

(d) What is the electric potential at B? at C? at D?

(e) How do the answers to the previous parts change if the charge is negative instead of positive?

2. Go to the PhET website and open the simulation *Charges and Fields*. Select "grid." Select "show numbers" to see the length scale. Put four positive charges at the corners of a square with sides 1 m. Deselect "show numbers."

(a) With the blue voltage sensor explore the equipotentials (i) close to the charges and (ii) outside the square and in the intermediate regions.

What is the shape of the equipotentials as you come close to one of the charges? Why?

(b) Use an "E-field sensor" to check what you know to be the magnitude of the field in the center. Explore the various regions with this sensor.

(c) Plot the equipotentials outside the square at regular intervals from one of the charges. What happens to the shape of the equipotentials as you move away from the square? Why?

(d) "Clear All" and replace the two charges on the right with negative charges. Plot equipotentials near to and far from the charges. What is the direction of the field at the center this time? What is it along the horizontal and vertical lines through the center?

3. A positive charge of $3\,\mu C$ is at the origin. A negative charge with the same magnitude is 8 m away along the x direction.

(a) What are the magnitude and direction of the electric field at the points between the two charges, 2, 4, and 6 m from the positive charge?

(b) What is the electric potential at each of these three points? (The reference point, where $V = 0$, is infinitely far away.)

4. In the field of problem 1, how much work is necessary to move an electron from C to D, from C to A, and from C to B?

5. (a) In the field of problem 1, what is the path of an electron released from rest at point C?

(b) How long will it take the electron to reach point A?

(c) The electron has an additional velocity v_{0x} so that its path will take it to B. How long will it take for the electron to reach B?

6. A proton is fixed at a point. A second proton is released from rest at a point 2 nm from the first proton. What is its kinetic energy when it is 4 nm from the first proton?

7. Four charges, $+5\,\mu C$ each, are at the corners of a square whose sides are 2 m. What is the potential at the center of the square?

8. Five identical charges are at the points of a symmetrical five-pointed star, so that the distances from each charge to its neighbor are the same and the distances from each to the center are the same. Is it possible for the electric potential at the center to be zero? If yes, what must be true? If not, why not?

9. A proton with initial kinetic energy 2 MeV is shot straight at an alpha particle.

(a) Describe the motion of the proton.

(b) Draw a graph of the electric potential energy of the proton as a function of its distance from the α particle. Show the distance of closest approach on the graph.

(c) Calculate the distance of closest approach.

10. A rocket is launched straight up at twice the escape velocity. How fast is it moving when the gravitational potential energy is zero?

11. An electron, a proton, and an alpha particle are accelerated through the same potential difference of 50 V. What are the resulting kinetic energy and speed of each?

12. A battery with an emf of 20 V is connected to a lightbulb whose filament has a resistance of $400\,\Omega$.

(a) What is the current in the circuit?

(b) The resistance is doubled by adding a second, equal, bulb to the circuit (in series). What is the current now?

(c) What assumptions did you have to make to get the results for parts (a) and (b)?

(d) An ammeter is added to the circuit to measure the current. It reads 0.020 A. This is not the current that you calculated in part (b). What can you conclude from this reading about the ammeter? What should you look for in a different ammeter to get a result closer to that of your calculation in part (b)?

13. A 9-V battery supplies a lamp with a current of 5 mA. How much energy does the lamp use in 10 s?

14.

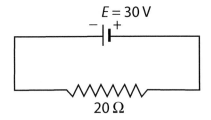

$E = 30$ V

$20\,\Omega$

(a) What is the voltage across the resistor?
(b) What is the current through the resistor?
(c) In what direction does the current flow (clockwise or counterclockwise)?
(d) In what direction do the electrons move (clockwise or counterclockwise)?
(e) How much energy is supplied by the battery in each second?

15. Three resistors, $2\,\Omega, 4\,\Omega$, and $6\,\Omega$, and a battery whose emf is 24 V are connected in series.
(a) What is the potential difference across each of these four elements?
(b) What fraction of the total power is dissipated in each of the resistors?
(c) What energy transformation takes place in each of the four circuit elements?

16. Two electronic devices are separately plugged in at home. Each has a resistance of 60 kΩ. (Do the problem as if the source of emf was 110 V DC.)
(a) Are they connected in series or in parallel?
(b) What is the current in each?
(c) How much energy is used by each in one second?
(d) Find the equivalent resistance of the two devices.
(e) What is the power supplied by the source of the emf?
(f) How long would both have to be connected to use 1 kw-h of energy?

17. A wire has a resistance of $R\,\Omega$. A second wire (of the same material) has half the length of the first and twice its diameter. What (in terms of R) is the resistance of the second wire?

18. How long must a copper wire ($\rho = 1.7 \times 10^{-8}\,\Omega\,\text{cm}$) of diameter 0.2 mm be to have a resistance of 1 Ω?

Problems and reasoning skill building

1.

There is a uniform electric field of 3 N/C in the $(-y)$ direction, as in the figure. Take the point A as a reference, with $V = 0$. What is the potential at points B, C, and E?

2. An electron is released from rest at point A in the field of the previous question.
(a) What is its path?
(b) What is its kinetic energy (in J and in eV) after it has traveled 4 m?

3. Can you put charges at the corner of a square so that both the electric field and the electric potential at the center (with the usual reference at infinity) are equal to zero?

4. An electron passes a point at which the potential is 5 V, with a kinetic energy of 7 eV. Some time later it passes a point where the potential is 8 V. What is its kinetic energy there, in eV and in J?

5. A radio draws a current of 150 mA when plugged into a wall socket with a voltage of 120 V. How much power does the radio use?

6. (a) A circuit consists of two similar lamps in series across a 110 V source of emf. The current through one lamp is 1 A. What is the current through the other lamp?
(b) A circuit consists of two similar lamps in parallel across a 110 V source of emf. The current through one lamp is 1 A. What is the current through the other lamp?
(c) What is different about the lamps in parts (a) and in (b)? (Give a quantitative answer.)

7. The resistance between two points in a cell is 1.6×10^{10} Ω.

(a) There is a potential difference of 60 mV between the points. What is the current?

(b) The current consists of singly charged Na$^+$ ions ($Q = +e$). How many ions flow past each point in 0.3 s?

8. A toaster has a resistance of 22 Ω. It takes 2 min for one slice of toast. What is the energy that is used by the toaster in that time?

9. You have a battery and some bulbs, and you make three circuits. Rank the circuits in order of the largest current to the smallest, and in order of the brightest to the dimmest.

(a) one bulb only;
(b) two bulbs in series;
(c) two bulbs in parallel.

10. A proton is shot directly at a nucleus of sulfur ($Z = 16$). If it gets to within about 7×10^{-15} m of this nucleus it will be so close as to be within the range of the nuclear force, and can initiate a nuclear reaction. What initial kinetic energy (in MeV) must the proton have to achieve this?

11. In a uniform electric field of 5 N/C in the y direction, an electron is given an initial kinetic energy of 12 eV in the x direction.

(a) Sketch the path of the electron on a diagram, on which you also show E and the initial velocity, v_0.

(b) Later the electron passes a point where the value of y has changed by 5 cm from where it started. What is the potential at that point? What is the electron's kinetic energy there? Take the potential to be zero at the electron's initial position.

12.

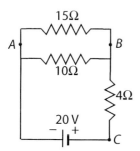

(a) What is the equivalent resistance of the 15 Ω and the 10 Ω resistors between points A and B?

(b) What is the equivalent resistance of the three resistors between points A and C?

(c) What is the current through the battery?

(d) A wire (with zero resistance) is connected between points A and B. What is the current through the battery now?

(e) Why would the analysis that you used for part (d) fail if the question is instead to find the current when points A and C are connected with a wire? (In other words, which of your assumptions would no longer be tenable?)

13.

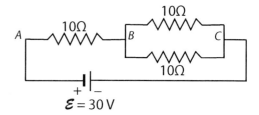

(a) What is the equivalent resistance between points B and C? Between A and C?

(b) What is the current in each part of the circuit?

(c) What is the potential difference between points A and B? Between points B and C? A and C?

(d) What is the power dissipated in each resistor? What is the power generated in the battery?

14. A coil whose resistance is 36 Ω is connected to a source of 120 V and immersed in $\frac{1}{2}\ell$ (=500 cm^3) of water at 20°. Assume that all of the energy produced in the coil is used to heat the water. How long does it take to bring the water to the boiling point?

15. A 1 cm^3 copper block is used to make a wire whose resistance is 10 Ω. What are the length and diameter of the wire? ($\rho = 1.7 \times 10^{-8}$ Ω m.)

Multiple choice questions

1. Two equal light bulbs (assume that they have the same constant resistance) are connected across a constant 110-V source of emf, first in parallel and then in series. The ratio of the power developed in the parallel connection to that in the series connection is

(a) 4
(b) 2

(c) 1
(d) $\frac{1}{2}$
(e) $\frac{1}{4}$

2. The resistance of a light bulb with a metal filament increases as the temperature of its filament changes. If we repeat the previous question, the answer is now
   (a) larger
   (b) smaller
   (c) the same
   (d) unknown

3. Adding impurities to a metal increases its resistivity. A certain alloy of copper has twice the resistivity of copper. A length of wire made of this alloy has a resistance R_a and a diameter d_a. A copper wire with the same length and resistance has a diameter d_{Cu}. The ratio $\frac{d_{Cu}}{d_a}$ is
   (a) 4
   (b) 2
   (c) 1
   (d) $\frac{1}{2}$
   (e) $\frac{1}{4}$

4. A resistor is connected across a battery with constant emf. A second resistor whose resistance is half as large is now connected across the battery in parallel with the first. That changes the total power dissipated by a factor of
   (a) $\frac{1}{3}$
   (b) $\frac{1}{2}$
   (c) 3
   (d) 2
   (e) 1

5. The resistivity of a metal increases when the temperature increases because of which of the following:
   (a) The electrons move faster.
   (b) The metal ions vibrate with greater energy.
   (c) The electrons collide with each other more often.
   (d) More of the electrons become free to move.
   (e) More of the metal atoms become ionized.

6. How much current flows in a 6 W clock radio plugged into a household receptacle?
   (a) 50 A
   (b) 660 A
   (c) 50 mA
   (d) 18 A
   (e) 180 mA

7. A computer is connected across a 120 V source for 10 hours. It requires an average of 0.20 A. The energy that is used is
   (a) 240 J
   (b) 860 J
   (c) 860 kJ
   (d) 14.4 kJ
   (e) 24 J

Synthesis problems and projects

1. Go to the PhET website and open the simulation *Charges and Fields*. Select "grid" and "show E-field."

   (a) Make a horizontal line of eight negative charges, spaced five divisions apart, near the middle of the screen. What happens to the field direction below the line, near its middle, as you add more and more charges?

   (b) Make a line of the same number of equally spaced positive charges, parallel to the first, and 10 divisions away. What happens to the field between the lines as you add more and more charges?

   (c) Plot a series of roughly equally spaced equipotentials, about half a division apart along a vertical line near the middle of your capacitor. What do they tell you about what the potential and the field are in the capacitor?

   Check "clear" on the blue sensor. Check "show numbers" and use the equipotential plotter to measure the potential difference between the two lines of charge. (Stay along a line between the charges.) Calculate the electric field in the middle of the capacitor from the potential difference. Now use an E-field sensor to see what the field is. Are the two at least approximately the same? This is, of course, not a perfect capacitor. What changes would be needed to make it closer to perfect?

   (d) Describe the variation of the electric potential and the electric field as you move in directions perpendicular and parallel to the lines of charge between the charged lines.

   (e) Compare your observations to your expectations based on the last example in Chapter 8 and Section 9.1.

2. Go to the PhET website and open the simulation *Circuit Construction Kit* (DC only). The

components can be moved with the mouse and connected by placing the red circles on top of each other. The length of the connecting wires can be changed and they can be turned by "pulling" on them. A component can be removed by selecting it and pressing the backspace key on your keyboard. A connection can be broken by clicking on it (a yellow ring appears) and pressing the backspace key.

Construct a circuit consisting of a battery and a light bulb. You can tell whether you have made the connections by seeing whether the electrons move.

Look at "lifelike" and "schematic." You can also hide the electrons (select "Advanced") and make the schematic diagram look more like the usual schematics. Select "show values."

(a) Select "voltmeter." It and its probes can be moved with the mouse. Measure the voltage across the battery, the bulb, and a wire.

(b) Select "ammeter." It can then be moved with the mouse and connected. Connect it in the circuit so that the current goes through it (it replaces a wire). How does the current differ in different parts of this series circuit?

(c) Initially the connecting wires do not have resistance. Select "advanced" to change their resistivity. Set the resistivity slider near the middle and measure the voltages across the battery, the light bulb, and the wires. What is the resistance of a wire that is about two inches long?

3. Go to the PhET website and open the simulation *Circuit Construction Kit*. (See the more detailed instructions in the previous question.)

(a) Construct a circuit with two light bulbs in series with a battery and an ammeter. Measure the voltage across each bulb and across the battery. How are they related? What is the resistance of each bulb?

(b) Construct a circuit with two light bulbs in parallel, connected to a battery. How many ammeters do you need to measure all of the currents at the same time? Measure the voltage across each bulb and across the battery and all of the currents. How are the currents related?

(c) Calculate the equivalent resistance of each circuit. Which circuit has the greatest equivalent resistance?

(d) Calculate the power dissipated in each circuit. Which circuit uses the largest power?

(e) What happens in each of the circuits when you unscrew a bulb? You can incorporate a switch next to each bulb to disconnect it from the battery.

(f) Are the lights in a house connected in series or in parallel? Explain.

CHAPTER 10

Magnetism: Electricity's Traveling Companion

Again—force, field, and motion
> *Poles and currents*
> *The force between two parallel currents*
> *The magnitude of the magnetic field of a current*
> *The direction of the magnetic field of a current*
> *Motion of a charged object in a magnetic field*
> *Solenoids*
> *The earth's magnetic field*
> *Ampere's law*

The electron: an old friend turns out to be the elemental magnet
> *Spin and magnetic moment*
> *Magnetic materials*

Generating electricity: motional emf and Faraday's law
> *The motional emf*
> *The emf in terms of the change in the flux: Faraday's law.*

Magnetic forces have been known since iron-containing rocks were found in antiquity. The magnetic compass was known in China in the second century, and later made possible Columbus's visit to America, but it was not until 1600 that Gilbert suggested that the earth was itself a giant magnet. Today current-carrying coils can produce magnetic fields that are much stronger than those of iron magnets. Most of the electric energy that we use comes from the motion of wires in magnetic fields. Magnetic forces drive the motors in our fans and vacuum cleaners. And perhaps the most startling impact on our civilization comes from the interplay of electric and magnetic fields that we call electromagnetic waves.

We play with magnets from the time when we are children. We pick up pins and nails with them and use them to stick notes to the refrigerator. The forces that they exert are more familiar to us than electric forces, and it comes as a surprise to learn that they are more complex.

The magnetic effects of electric currents were discovered in the early part of the nineteenth century by Oersted, and Ampere soon suggested that the magnetism of iron is the result of internal currents. They were called *amperian* currents, but their detection proved elusive.

There was no way to understand the situation in more detail at that time, since electrons were not identified until near the end of the nineteenth century. An additional obstacle was that it is an unforeseen property of electrons, their *spin*, not discovered until a century after Ampere's suggestion, that is responsible for the strong magnetic effects of iron that we call ferromagnetism. Ampere had the right general idea, but the facts showed themselves to be more complicated and also more interesting than could have been envisioned.

Each electron, and to a lesser extent each proton, is a little magnet. In addition, electrons are in orbit around the nuclei. As a result all materials are magnetic to some extent. As with electric properties, the question becomes "Why are we not more often aware of this?" Electric properties tend to cancel because there are just as many positive as negative charges. A similar cancellation happens with magnetic properties when there are many electrons. Surprisingly, the cancellation is sometimes less complete than in the electric case, especially in the materials that we call *magnetic*, of which iron is the most important.

The origin of the magnetism of materials is now well understood in terms of the properties of atoms and electrons. Today magnetism is a vital subject, under intense investigation for its intrinsic interest and for its important applications, such as magnetic recording and magnetic memories.

10.1 Again—force, field, and motion

Poles and currents

Electric charges are everywhere, but we rarely think about them or the electric forces between them. Magnetic forces are more familiar. We know them from toys and compasses, and we are aware that the earth is a giant magnet. But what gives rise to the magnetic forces? Is there something like the electric charges that are responsible for electric forces?

Our first experience with this question may be with a compass needle and its north and south poles. If we have two of them, we see that a north pole attracts a south pole and repels another north pole, just as a positive charge attracts a negative charge and repels another positive charge. But as we look further, the situation turns out to be a good deal less simple. If we cut a compass needle in half, we get two new magnets, each with its north and south pole. In fact, all efforts to get an isolated pole have failed.

In some ways the poles act like "magnetic charges," but, as the cutting of a magnet shows, the two kinds cannot be separated. The clue to a more fruitful approach came when Oersted discovered in 1819 that magnets exert forces on electric currents, that is, on moving charges. Soon (in 1825) Ampere suggested that currents play the same role for magnetism that charges play for electricity. That turned out to be the fundamental origin of magnetic effects: magnetic forces are forces between moving charges, over and above the electrostatic (*Coulomb*) forces, and arise when charges are moving with respect to each other.

The force between two parallel currents

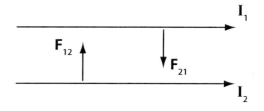

If we take two wires, parallel to each other, each with a current in the same direction, we observe that they attract each other. Is that in some ways analogous to the attraction between electric charges with opposite signs and to the gravitational attraction between two objects? Can we use this experiment as a basis for our knowledge of magnetism, just as we used the force law between charges as the basis for our understanding of electricity? The answer is a qualified "yes." There are some similarities, but also important differences.

Currents are not at points, as charges are. We can think of parallel currents as "like" currents and currents in opposite directions as "unlike." Experiments show that like currents attract and unlike currents repel. This is opposite to the force between charges. There are other differences. If we change the separation between the wires with the currents in them, and measure the force of one on the other, we observe that the force is proportional to $\frac{1}{r}$, and not to $\frac{1}{r^2}$, as in Coulomb's law.

We can't really think of a wire with a current as just ending somewhere in the middle of space. In general, there can be a continuous current only in a closed loop, i.e., in a complete circuit. That means that we can't have two straight wires, isolated as in the diagram. As an approximation we can imagine two very long wires, with the rest of the circuit so far away that we don't need to consider it. This becomes the model that we use when we talk of two infinitely long current-carrying wires.

That raises another problem. The longer the wires, with a given current, the greater is the force on each. To separate the dependence on length we need to talk about the force per unit length, $\frac{F_m}{L}$.

We can now combine the observations. The force (F_m) changes with the length of the wires (L), the distance between them (r), and the magnitudes of the currents (I_1 and I_2), so that the magnitude of the force is given by

$$\frac{F_m}{L} = k' \frac{I_1 I_2}{r}$$

k' is the proportionality constant, which in the SI system is 2×10^{-7} N/A^2. We see that it is very small, while the constant in Coulomb's law, $k = 9 \times 10^9 \frac{Nm^2}{C^2}$, is very large. In part this reflects the fact that the *Coulomb* is a very large unit, but also that magnetic forces tend to be much smaller than electric forces. Just as the constant in Coulomb's law can be written as $\frac{1}{4\pi\epsilon_0}$, k' is often written as $\frac{\mu_0}{2\pi}$, where μ is the Greek *mu*.

EXAMPLE 1

Two parallel wires, each 4 m long, with currents of 4 A in opposite directions, are separated by 5 cm. What is the force on each?

Ans.:
We will assume that the wires are sufficiently long, and the distance between them sufficiently small, that the relation for the force between infinitely long wires gives an adequate approximation.

$$F = (4)(2 \times 10^{-7})\frac{(4)(4)}{.05} = 2.56 \times 10^{-4} \text{ N}.$$

The direction of the force is such that the wires repel each other. We see that although four amperes is a substantial current, the force is quite small.

The magnitude of the magnetic field of a current

The relation for the magnetic force between parallel currents is not as universally useful as Coulomb's law. It describes only what happens in a special situation, and not even one that can actually be realized. Nevertheless we can use it to define the magnetic field.

As in the electric case, we separate the force relation into two parts,

$$\frac{F_m}{L} = \left[k' \frac{I_1}{r}\right][I_2]$$

We define the magnitude of the magnetic field by saying that the infinitely long current, I_1, is the source of the magnetic field, $B = k'\frac{I_1}{r}$. The second (finite) current, I_2, with length L, by being in this field, experiences a force, given by $\frac{F_m}{L} = BI_2$, or $F_m = BLI_2$. Just as in the electric case, the magnetic field is introduced only as an aid to calculation, but is later seen to have an independent existence of its own.

The direction of the magnetic field of a current

We still have to define the direction of the magnetic field. This is not as simple as for the electric field. In the electric case there is only a single direction that we have to think about in addition to the field direction, and that is the direction of the force. We let them both lie along the same line, so that there is just one orientation for both field and force.

The magnetic force is an interaction between moving charges, so that there is one more direction that needs to be considered, namely the direction of the current (or moving charge) on which the force acts. That means that we have three directions to keep track of: those of the field, the force, and the current.

Experiments show that a current in the vicinity of another current experiences a force. That's true for almost all angles between them. There is just one direction of a current for which there is no force on it. This single direction is the one that we use to define the direction of the magnetic field. We do that by defining the field direction so that a current parallel to the field experiences no magnetic force. A current at right angles to the field experiences the maximum force. We'll go on to develop the relation between the size and orientation of the force and the sizes and directions of the magnetic field and the current.

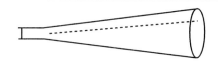

We can do the experiment by actually using a test current whose orientation we can change. To visualize the deflection we can use a small *cathode–ray tube*. This is a vacuum tube with a beam of electrons. (Before their nature was understood, the electrons in such a beam were called cathode rays.) A fluorescent material on the face of the tube lights up (usually green) when hit by the electrons. (Does that sound like a TV tube? Yes. Before the development of flat-panel monitors, all computer and TV screens were in cathode-ray tubes.)

We can move the tube around in the region of a magnetic field and see the deflection of the electron beam. The one direction along which the beam is not deflected is the direction of the magnetic field.

Another way to show the direction of the magnetic field is with a compass needle. A small needle made of iron, free to rotate, will line up along the magnetic field.

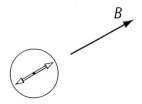

Here, once more, is the current I_1. Surrounding it is its magnetic field B. With the help of observations, using a small cathode–ray tube or a compass needle, we can show that the direction of B is perpendicular to I_1 and tangent to circles whose center is on the wire, as in the head-on view.

The diagram also shows a side view. At the two points that are shown, the magnetic field is perpendicular to the plane of the paper. We represent a vector perpendicular to the paper and into the paper by a circle with a cross (like a feathered arrow from the back) and a vector out of the

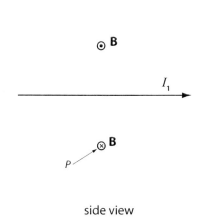

side view

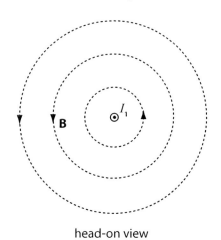

head-on view

paper by a circle with a dot (like an arrow seen head-on). That still leaves us with two possibilities. Is the field direction into the paper or out of it? We choose it to be into the paper at the point P.

Is that up to us? Doesn't nature tell us what to do? Isn't this supposed to be an operational definition?

The experiment tells us only about the force and its direction. The field is a *construct* that we have invented to help us talk about the force. We can decide its direction, as long as the relation between the field and the force is in accord with the experimental observations.

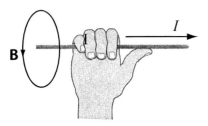

This is what has been agreed on: grasp the current with your right hand, so that your thumb points in the direction of the current. Your fingers then curl in the direction of the field lines. This gives the directions that we show in both views. (We'll call this the first right-hand rule.)

Now let's look at the force on a second current, I_2, in the magnetic field. We already know that when I_2 is in the direction of the magnetic field, it experiences no force, because that's how we defined the direction of the field, and that the maximum force occurs when I_2 and **B** are at right angles to each other. What happens when the magnetic field and the current are at some other angle? In that case we can separate the field into two components. The component parallel to the current does not lead to a force. Only the perpendicular component contributes to the force.

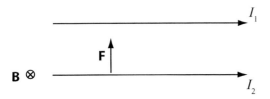

The direction of the field B of I_1 at the location of I_2 is *into* the plane of the paper. We can now look at the force on I_2 in that field. From the experimental fact that the two currents attract we know that the force on I_2 is toward I_1.

$$\mathbf{F} = \mathbf{I}L \times \mathbf{B}$$

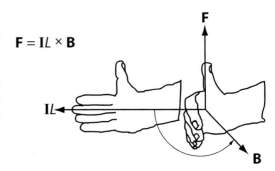

Here is one of the various rules that have been invented to remind us of the relation between the three directions. Point the fingers of your right hand along the direction of the current. Now curl the fingers so that they point in the direction of the field, letting the fingers of the right hand go from pointing along the current toward pointing along the magnetic field. The thumb then points in the direction of the magnetic force on the current. (We'll call this the second right-hand rule. It relates the directions of the magnetic field, the current, and the magnetic force on the current. The first right-hand rule relates the directions of the current and of the magnetic field that it gives rise to.)

A shorthand way to write a relation that describes both the magnitude and the direction of the force is $\mathbf{F} = I\mathbf{L} \times \mathbf{B}$, where the vector **L** points in the direction of the current.

$\mathbf{C} = \mathbf{A} \times \mathbf{B}$ is called the *cross-product* or *vector product* of the two vectors **A** and **B**. Its magnitude is $C = AB\sin\theta$, where θ is the angle between **A** and **B** and the direction of **C** is given by the second right-hand rule.

The cross-product succinctly describes the experimental result: the magnitude $ILB\sin\theta$ is zero when $\sin\theta$ is zero, i.e., when the current and the field are in the same direction. It is at its maximum when $\theta = 90°$, i.e., when the current and the magnetic field are perpendicular.

EXAMPLE 2

(a) What are the units for the magnetic field?

(b) What is the magnetic field of a long wire with a current of 4 A at a distance of 5 cm from it?

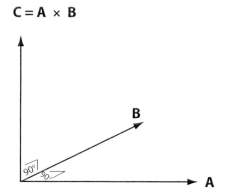

Ans.:

(a) From $F = BLI$, the units of B are the same as those of $\frac{F}{LI}$, i.e., $\frac{N}{A\,m}$.

This unit is also called the *tesla* (T). Another unit for B is the *gauss*, equal to 10^{-4} T.

(b) $B = k'\frac{I}{r} = (2 \times 10^{-7})\frac{4}{.05} = 1.6 \times 10^{-5}$ T.

EXAMPLE 3

A wire 30 cm long, carrying a current of 2 A in the y direction, is in a uniform magnetic field of 0.8 T in the x direction. What are the magnitude and direction of the magnetic force on the wire?

Ans.:

The current and the magnetic field are at right angles, so that $F = ILB = (2)(0.3)(0.8) = 0.48$ N.

The direction is given by the second right-hand rule (and by the vector relation $\mathbf{F} = I\mathbf{L} \times \mathbf{B}$). The magnetic force is at right angles both to the current and to the magnetic field, into the plane of the paper.

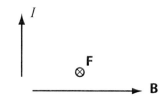

EXAMPLE 4

For realistic descriptions we have to go beyond infinitely long wires. Most often wires are in closed loops.

A rectangular loop of wire has a width of 20 cm and a length of 30 cm and carries a current of 4 A. It is horizontal and suspended in a uniform magnetic field (0.7 T in the x direction) in such a way that it can turn about an axis through its center, parallel to the long sides.

(a) What are the magnetic forces on each of the four sides of the loop?

(b) What is the total magnetic force on the loop?

(c) What is the torque on the loop?

(d) What will be the subsequent motion of the loop?

Ans.:

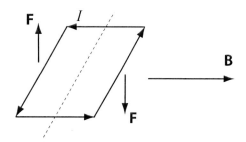

(a) For a wire perpendicular to the field the magnitude of the force is ILB. Here this is so for the two forces on the longer sides. The magnitude of the force on each is $(4)(0.3)(0.7) = 0.84$ N. The direction of each force is perpendicular both to the field and to the current, so that one is up and the other is down, as shown on the figure. The two short sides are parallel to the field, and there is no force on them.

(b) The two forces are in opposite directions so that the total force on the loop is zero.

(c) Each of the two forces is 0.1 m from the axis of rotation. This is the perpendicular distance from the line of action of the force to the axis. The magnitude of the torque of each is equal to the product of this distance and the force, 0.1 m × 0.84 N or 0.084 N. Both torques act to turn the loop clockwise as seen in the diagram, and add for a total torque of 0.168 Nm.

(d) The loop will start out with an angular acceleration clockwise about the axis (the dotted line in the diagram).

EXAMPLE 5

The same loop as in the previous example is rotated through 90° so that it is at rest perpendicular to the field. Answer the same questions as before.

The orientation of the loop can be described by specifying the direction of the line perpendicular or *normal* to the loop. It is often just called the *normal*. In this example the normal is parallel to the field. Of the two possible directions of the normal we use the one given by a variation of the first right-hand rule: let the fingers of the right hand curl so as to follow the current. The thumb then points in the direction of the normal. Here the normal is in the same direction as the field. In the previous example the normal is perpendicular to the loop and up.

Ans.:
(a) The two long sides are perpendicular to the field. The magnitude of the two forces on them is the same as before. The short sides are also perpendicular to the field. There are now also forces on them, equal to $(4)(0.2)(0.7) = 0.56$ N. The directions are shown on the figure. They are such as to tend to stretch the loop.

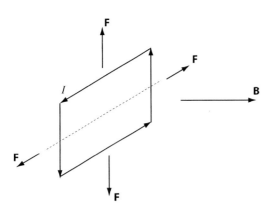

(b) Both pairs of forces add up to zero, so that the total force is zero.

(c) This time the forces on the long sides act along the same line, so that there is no torque. This is also true about the two forces that act on the short sides.

(d) There is no torque and no angular acceleration. The loop is at rest to begin with and remains at rest.

EXAMPLE 6

(a) Write a relation that describes the torque on the loop of the previous examples as a function of the angle that the normal to the loop makes with the magnetic field.

(b) In the position of the previous example there is no torque, and the loop will not continue to turn. What change would cause it to continue to have an angular acceleration and to continue to rotate through another half turn? What would keep it rotating, as in an electric motor?

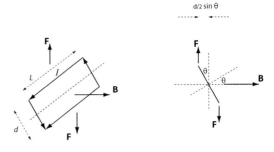

Ans.:
(a) Let the angle between the normal to the loop and the magnetic field be θ. The long sides (of length L) are still perpendicular to the magnetic field. The magnitude of the force on each of them is ILB.

Let the short sides have length d. The forces on them are in opposite directions, and they do not contribute to the torque. The contribution to the torque of each of the forces on the long sides is $(ILB)(\frac{d}{2} \sin \theta)$. The two torques add, to give $ILBd \sin \theta$. Since $Ld = A$, the area of the loop, the magnitude of the torque, τ, is $IAB \sin \theta$. (Both the magnitude and the direction of the torque are given by the vector relation $\tau = I\mathbf{A} \times \mathbf{B}$, where the vector $\mathbf{A}$ is in the direction of the normal to the loop. The vector τ gives the direction of the torque. It is related to the sense of rotation by a rule analogous to the first right-hand rule. The magnitude of the vector product is $IAB \sin \theta$, where θ is the angle between the vectors $\mathbf{A}$ and $\mathbf{B}$.)

(b) The current would have to be reversed. (It could also be the magnetic field that is reversed, but this is usually more difficult.) The torque would then also reverse, and the loop would turn through another 180°. For continuous rotation, the current, and hence the torque, would have to be reversed every half turn. A built-in switch that makes this happen is called a *commutator*. The diagram shows how it works. A split ring is attached to the axis on which the loop rotates and turns with it. The stationary brushes make contact with the half rings, changing from one to the other each half turn. The current in the loop changes direction every half turn, but the current in the external circuit (connected to the brushes) stays in the same direction. This is the principle of the *DC motor*, i.e., a motor whose current comes from a battery or other source whose emf remains in the same direction.

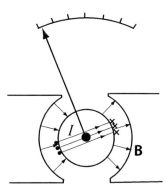

With a current in it, the loop rotates as a result of the magnetic torque. The spring opposes this motion. The loop turns until the magnitude of the magnetic torque is equal to the opposing torque of the spring. The larger the current, the larger the torque, and the more the coil will rotate. The scale is marked to indicate the magnitude of the current.

For the loops of the previous examples the torque varies as sin θ, where θ is the angle between the normal to the loop and the magnetic field. In a galvanometer we want the magnetic torque to depend only on the current and not on the position of the loop, in other words, not on sin θ. This can be accomplished with a magnet that has curved pole pieces, as on the diagram. The field is then radial, and remains at right angles to the loop as it turns.

(a) Mark the direction of the forces of the magnetic field on the coil.

(b) What is the relation between the torque on the coil and the current through it?

(c) If the torque on the spring follows Hooke's law, what is the relation between the current and the angle through which the coil turns?

Ans.:

(b) $\tau = IAB$.

(c) Hooke's law for the spring is $\tau = k\theta$. Hence $k\theta = IAB$. k, A, and B are fixed, so that the angle θ is proportional to the current, I.

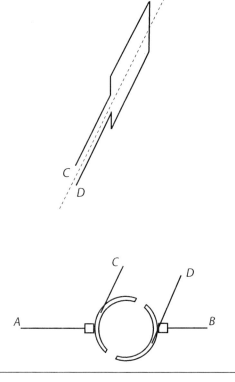

EXAMPLE 7

The three previous examples show what happens to a loop of wire carrying a current when it is in a magnetic field. The interaction between the current and the field results in a torque, and the loop turns. The magnitude of the torque is proportional to the current. The *galvanometer* is a device that uses the torque on a loop to measure the current. The figure shows its essential features.

A wire coil is wound around the cylinder in the middle. The cylinder and coil can turn in the field of a permanent magnet. A needle pointing to a scale is attached to the cylinder. When there is no current in the coil a spring (not shown) fixes the orientation so that the needle points to the left end of the scale.

Motion of a charged object in a magnetic field

We have talked about electrons moving in wires as an electric current. Can they also move when there is no wire, for example in air? It's not so easy. First they have to be liberated from their "home" atoms. And then they face the obstacle of all the atoms that are in the way and prevent them from moving freely.

Electrons travel through air in sparks and lightning, and after they are emitted from radioactive materials (when they are called beta rays). But they move much more easily when the air is removed, such as in x-ray tubes, vacuum tube rectifiers, cyclotron chambers, and electron microscopes.

One or more charges, moving through empty space, represent a current. Let's see how we can adapt the relation that we have used for the magnetic force on a current for this situation.

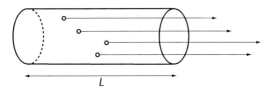

Look at a tube of length L in which charges move at a speed $v = \frac{L}{t}$, so that each takes a time t to traverse the tube. In that time all the charges that were in the tube (a total charge Q) will have left it. The ones that started out at the left-hand end will just make it to the other side. All the others will get further and pass the right-hand end. The current is the rate at which charge passes any cross section. The direction of the current is the same as the direction of the velocity of a positive charge and opposite to the direction of the velocity of a negative charge. It is therefore $I = \frac{Q}{t}$. Hence $Q = It$ and $Qv = Ivt = IL$. We see that we can replace IL in the force law by Qv: the force on a current of length L in a field B perpendicular to it is $F = ILB$, and the force on a charge moving with velocity v is QvB.

These relations apply if the current (or the velocity of the moving charge) is perpendicular to the magnetic field. If a particle moves parallel to the field there will be no force on it and it will continue with constant velocity. The relation that gives the force on the moving charge for all angles between the field and the velocity of the charge is $\mathbf{F} = Q\mathbf{v} \times \mathbf{B}$. The magnitude of the force is $QvB \sin \theta$, and its direction is given by the second right-hand rule.

The figure shows what happens. The magnetic field is uniform, i.e., its magnitude and direction are the same everywhere. When the velocity is v_1, at right angles to the field, there is a force, and hence an acceleration, at right angles to v_1. The direction of the motion changes. With the velocity changed (in direction but not in magnitude) to v_2, the magnetic force is still at right angles to the field and also to the new velocity. As the charge moves, the velocity vector continues to change direction, and so does the force. The result is that the charge moves in a circle, in a plane perpendicular to the field.

In a magnetic field the magnetic force on a charge is always perpendicular to the direction

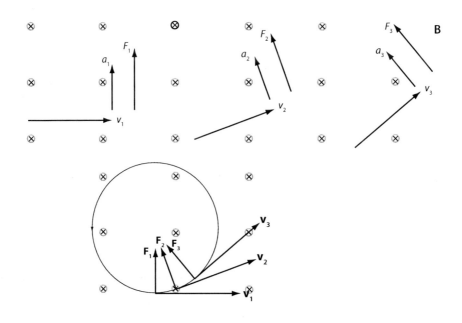

of motion of the charge. Hence it can change the direction of the motion, but not the magnitude of the velocity or the particle's kinetic energy.

If the velocity is at an angle to the field, then the component parallel to the field will continue unchanged. The perpendicular component, on the other hand, will cause circular motion in the plane perpendicular to the field. The two together will lead to spiral motion.

We got fairly far, starting with the force between two infinitely long wires, even though we knew from the start that this was a special case. Can we do better? Yes, but at considerable cost in complexity. It is possible to write a relation that describes the contribution to the magnetic field of a tiny ("infinitesimal") piece of current. The contributions from all the pieces can then be added. We won't do that here.

EXAMPLE 8

Circular motion of a charge in a magnetic field

There is a uniform vertical magnetic field of 0.1 T in a region. A charge of 0.1 C enters the field region traveling horizontally with a velocity of 20 m/s.

(a) What is the force on the charge?

(b) What is the path of the charge? Describe it quantitatively.

Ans.:

(a) For a charge moving at right angles to a magnetic field $F = qvB = (0.1)(20)(0.1) = 0.2$ N.

(b) The path is a circle in the plane perpendicular to the magnetic field. We can find the radius of the circular path by noting that the centripetal force is provided by the magnetic force, i.e., $qvB = \frac{mv^2}{r}$. To calculate the size of the radius we need to know the mass. Let's say it is 10^{-3} kg: $r = \frac{mv}{qB} = \frac{(10^{-3})(20)}{(0.1)(0.1)} = 2$ m.

The direction of the force, and hence of the path, can be found from the right-hand rule.

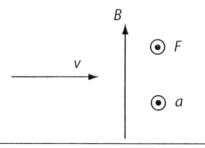

EXAMPLE 9

A proton whose energy is 1 MeV moves in a circle perpendicular to a magnetic field of 1.2 T.

(a) What is the time, T, for one complete revolution?

(b) What are the frequency, f, (in revolutions per second) and the angular velocity, ω?

(c) How do the quantities T, f, ω, v, and r change when the proton is accelerated to 2 MeV?

Ans.:

(a) The force on a charge in a magnetic field perpendicular to its motion is qvB. Here this is the force that causes the charge to move in a circle, the centripetal force, equal to $\frac{mv^2}{r}$.

We can solve the relation $qvB = \frac{mv^2}{r}$ for the velocity, v, to give $v = \frac{qBr}{m}$.

The velocity times the time for one revolution is equal to the distance that the particle travels in one revolution, i.e., $vT = 2\pi r$, so that $T = \frac{2\pi r}{v}$.

If we now substitute the relation for v in terms of the magnetic field, we see that $T = 2\pi r \frac{m}{qBr}$, which is equal to $\frac{2\pi m}{qB}$. For a proton $m = 1.67 \times 10^{-27}$ kg and $q = e = 1.6 \times 10^{-19}$ C, so that $T = \frac{(2\pi)(1.67 \times 10^{-27})}{(1.6 \times 10^{-19})(1.2)} = 5.5 \times 10^{-8}$ s.

(b) The frequency is $\frac{1}{T} = \frac{qB}{2\pi m} = 1.83 \times 10^7$ revolutions per second, or 1.83×10^7 Hz.

The number of radians per second is 2π times the number of revolutions per second, so that the angular velocity is $(2\pi)(1.83 \times 10^7)$ or 1.15×10^8 radians per second.

(c) We see that we did not use the fact that the energy is 1 MeV. The time T and the frequency, as well as ω, are independent of the energy! Of course v depends on the particle's energy ($\frac{1}{2}mv^2$) and r is proportional to v.

What are the speed and the radius? Here $v = \sqrt{\frac{2K}{m}} = \sqrt{\frac{(2)(1)(1.6 \times 10^{-13})}{1.67 \times 10^{-27}}} = 1.38 \times 10^7$ m/s.

$r = \frac{mv}{qB} = \frac{(1.67 \times 10^{-27})(1.38 \times 10^7)}{(1.6 \times 10^{-19})(1.2)} = 3.3 \times 10^{-2}$ m or 3.3 cm.

The proton moves with the same angular velocity in the magnetic field, regardless of its energy. If it could be accelerated each time it crosses a diameter, first to the right and then to the left, with a frequency equal to the frequency of its motion, it could gain energy each time it goes through 180°. This is what happens in a *cyclotron*. The frequency $\frac{qB}{2\pi m}$ is called the *cyclotron frequency*.

The way this is done is that the particles are made to move in the magnetic field in two semicircular metal enclosures (called "dees", because they have the shape of the letter D). If an alternating voltage is now applied between the dees, at the cyclotron frequency, the particles will get a kick, i.e., a force and an acceleration along their motion, each time they go through 180°. After each kick from one dee to the other the velocity and hence the radius will increase, so that the particles will spiral outward until they reach the boundary of the magnetic field.

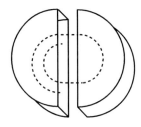

The figure shows the dees and a particle path with an exaggerated acceleration. Inside the dees the path is circular. The voltage between the dees accelerates the particles, so that they move with increased velocity, energy, and radius after they move from one dee to the other. By the time they have moved through a half circle and are about to return to the first dee, the voltage has reversed, and they are again accelerated. (Can you tell what the direction of the magnetic field is?)

Solenoids

A current in a coil gives rise to a magnetic field inside it that is stronger than that of a single loop because the fields of the different loops are roughly in the same direction and add up. A coil used to produce a magnetic field is also called a *solenoid*, meaning "pipeshaped." You can see that the field lines seem to flow through the coil like a liquid through a pipe.

The figure shows the magnetic field of a coil. We see that near the center, some distance from the ends, the field is uniform. We can use the first right-hand rule to relate the direction of the current to that of the magnetic field: grasp the coil with the right hand, with the fingers curling in the direction of the current. The thumb points in the direction of the magnetic field.

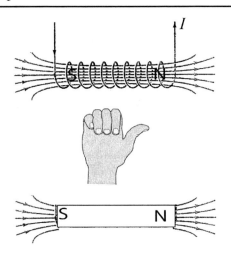

The figure also shows a magnet whose size is the same as that of the coil. Its field distribution is the same! Both the coil and the magnet have a north pole from which the magnetic field lines emerge and a south pole where they enter.

The earth's magnetic field

The earth is a giant magnet. It acts as if it had a magnet or a current-carrying coil inside it. In fact, that's just what seems to be the case. The inside of the earth is so hot that part of it is molten, and its rotation gives rise to the earth's magnetic field. Our knowledge of the origin of the earth's field is quite limited. Surprisingly perhaps, the earth's interior is much less accessible to us than the surfaces of the moon, the planets, and even the stars. Most of our knowledge comes from mechanical (sound-like "seismic") waves, generated by natural phenomena, such as volcanic activity and earthquakes.

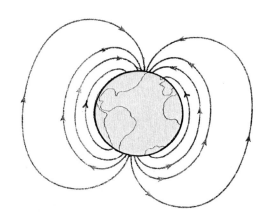

The magnetic field of a magnet points away from its north pole, which is attracted to the south pole of another magnet. That's where we run into an inconsistency: early on, the earth's pole to which the north pole of a compass needle is attracted was called the north pole. We still call it that, but if we think of the earth as a giant coil or magnet, it is the other way around. What we call the earth's north pole is the south pole of its internal magnet!

Charged particles are part of the "cosmic rays" coming toward the earth. They are deflected by the magnetic field of the earth into spiral paths surrounding the magnetic field lines and form belts of radiation around the earth named after their discoverer, the *Van Allen* belts.

Only when the charged particles come toward the earth parallel to its magnetic field are they undeflected. This happens when they come toward the poles parallel to the earth's axis. Therefore more particles come through the atmosphere there. They ionize the air molecules. When the ions recombine with electrons, radiation is emitted, some of it visible, called the *aurora* (the *aurora borealis* or Northern Lights in the north and the *aurora australis* or Southern Lights in the south).

Ampere's law

Gauss's law is a general law about charges and electric fields. Is there also a general law for currents and magnetic fields? Yes, there is, but it looks quite different. It is called *Ampere's law*, and we will show what it is in this section. In the next chapter we will see that in this form it is incomplete, and how Maxwell, by extending it, was able to develop the concept of *electromagnetic waves*.

Look again at the magnetic field surrounding a long straight wire and follow one of the circular field lines with radius r. The magnitude of the field is $k'\frac{I}{r}$ or $\frac{\mu_0 I}{2\pi r}$. Now calculate B times the length of the path $(2\pi r)$ as we go around. We get $(B)(2\pi r)$ or $(\frac{\mu_0 I}{2\pi r})(2\pi r)$, which is equal to $\mu_0 I$.

We can go around the wire along other paths, again multiplying B for each piece of path times the length Δs of the piece of path. We can approximate the path by a series of segments parallel and perpendicular to the magnetic field. There is no field radially out from

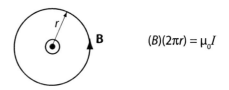

the wire, only the field tangential to the circular field lines, as before. The result of multiplying each piece of path, Δs, by the magnetic field component parallel to it is therefore again $\mu_0 I$.

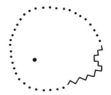

We have taken the paths around a long straight wire, but the result is the same for wires with other shapes. The path times the magnetic field component parallel to it, around a current I, is always $\mu_0 I$. This is *Ampere's law*.

EXAMPLE 10

The field inside a long solenoid (or coil) is uniform, i.e., it has the same magnitude and direction everywhere inside the solenoid. What is its magnitude?

Ans.:

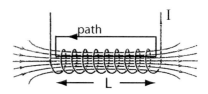

Use Ampere's law. Take a path of length L inside the solenoid and return outside. Inside the path length is L in the field B. For N turns within the path, the enclosed current is NI. The outside part of the path does not contribute, since the field is zero there. The total contribution is therefore BL, and it is equal to $\mu_0 NI$. Hence $BL = \mu_0 NI$ and $B = \mu_0 \frac{N}{L} I$.

10.2 The electron: an old friend turns out to be the elemental magnet

Spin and magnetic moment

We are familiar with the fact that the electron's charge is the fundamental unit of charge. The electron, with its *spin*, is also the fundamental magnet. The quantity that describes its magnetic properties is called its magnetic moment.

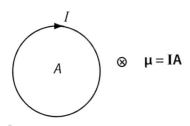

Here is the definition of the magnetic moment: take a circular loop of wire with area A carrying a current I. The quantity IA is called its magnetic moment. The symbol usually used for it is μ (Greek *mu*). This is a different use of the symbol μ from the one that we introduced earlier in this chapter. The magnetic moment is a vector quantity whose direction is defined to be perpendicular to the loop. We still have to decide between the two possible directions. Do it as before, using the first right-hand rule: curl the fingers of the right hand in the direction of the current. The thumb then points in the direction of the magnetic moment.

How is it that the electron has a magnetic moment? Where is the loop, where is the current?

The electron's magnetic moment is not associated with a current through a loop, but with another kind of motion, namely the electron's spin. Each electron has spin angular momentum. It is always there, regardless of any other motion of the electron. And the angular momentum and the magnetic moment that is associated with it always have the same size.

A tennis ball, even if it isn't going anywhere, has angular momentum if it spins. That is more or less the situation for an electron, except that you can't stop it from having this angular momentum and you can't change the amount. Both the spin angular momentum and the spin magnetic moment are fixed properties of each electron.

The electron, however, is not a *classical* entity that follows classical rules. It is not a little ball spinning about its axis. One indication of the nonclassical rules that it follows is that it has this intrinsic angular momentum, that it always has it, and always with the same magnitude.

Electron Spin and Its Orientation

The magnitude of the spin angular momentum of the electron is $\frac{\sqrt{3}}{2}\hbar$, where $\hbar$ ("h bar") is Planck's constant, h, divided by 2π. (In a notation that is used for other atomic angular momentum values this can be written $\sqrt{s(s+1)}\hbar$, where $s = \frac{1}{2}$.) The component of this angular momentum along the direction of a magnetic field is $\frac{1}{2}\hbar$.

A charge that has angular momentum also has a magnetic moment. The spin magnetic moment has a component along the direction of the magnetic field, either in the direction of the field ("up") or in the opposite direction ("down"). Its magnitude is $\frac{e\hbar}{2m}$ (where e is the electronic charge), and this amount of magnetic moment is called a *Bohr magneton*.

(a) Draw a diagram of the vector **S** representing the spin angular momentum, and its component S_z along the magnetic field. Do this for both possible orientations. What is the angle between **S** and S_z?

(b) Show that the units of the Bohr magneton are those of a magnetic moment.

Ans.:
(a) $S_z = \frac{1}{2}\hbar$, $S = \sqrt{\frac{1}{2}(\frac{1}{2}+1)}\hbar = \frac{\sqrt{3}}{2}\hbar$, and $\cos\theta = \frac{0.5}{0.866}$, $\theta = 54.7°$.

(b) The units of h and $\hbar$ are Js. The units of $\frac{e\hbar}{2m}$ are therefore $\frac{C}{kg}$ Js.

$J = Nm$ and $N = \frac{kg\,m}{s^2}$. We can substitute $N = \frac{kg\,m}{s^2}$ to get $Js = Nms = \frac{kg\,m}{s^2}ms$ and $\frac{C}{kg}Js = (\frac{C}{s})m^2$, which are the units of IA and therefore the units of a magnetic moment.

The electron is a tiny magnet. It has a magnetic moment, and this is what characterizes a magnet. Its *spin* magnetic moment is responsible for almost all of the strong magnetic properties of matter such as the *ferromagnetism* of iron.

An electron can have additional angular momentum if it is in an atom in orbit about a nucleus. Together with this *orbital* angular momentum there is an *orbital magnetic moment*. The effects of the orbital magnetic moments of atoms are generally smaller than those associated with the spin magnetic moments, and are usually overshadowed by them.

All matter contains electrons, and we can ask why the magnetic properties are only occasionally large. The answer is that it's just as for the charge properties: sometimes two magnetic moments cancel because they are in opposite directions. In other cases magnetic moments may be randomly oriented so that their vector sum is zero. The real surprise is that for some materials the cancellation is not complete. The best-known case is that of iron. An iron atom has a net magnetic moment because the magnetic moment vectors of the electrons in the atom do not cancel.

In a magnetic field atoms with spin magnetic moments tend to line up with their magnetic moments parallel to the field. In some materials the magnetic moments remain aligned even when the external magnetic field is taken away. A piece of material in which this happens is a *permanent magnet*.

(b) $\mu = iA$, where $A = \pi r^2$. If the time for one revolution is T, the current is $i = \frac{q}{T}$. T is given by $2\pi r = vT$, so that $i = \frac{q}{T} = \frac{q}{2\pi r/v} = \frac{qv}{2\pi r}$. μ is then $(\frac{qv}{2\pi r})(\pi r^2) = \frac{qvr}{2}$.

(c) $\frac{\mu}{L} = \frac{qvr}{2mvr} = \frac{q}{2m}$.

We see that the ratio between the magnetic moment, μ, and the angular momentum, L, of a charge moving in an orbit depends only on the mass and the charge of the orbiting object. In other words, it is independent of the size of the orbit and the speed of the moving charge. This makes it a fundamental and interesting quantity. $\frac{q}{2m}$ is the *classical* value of the gyromagnetic ratio.

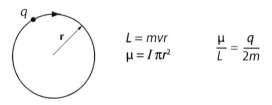

$L = mvr$
$\mu = I\pi r^2$
$\frac{\mu}{L} = \frac{q}{2m}$

One of the indications that the electron is not a classical object is that it has a spin angular momentum, and that its value is always the same. Another is that the spin's gyromagnetic ratio, the ratio between the spin magnetic moment and the spin angular momentum, does not have the classical value $\frac{e}{2m}$, but that it is instead twice as large, equal to $\frac{e}{m}$.

The Gyromagnetic Ratio

An object with mass m and charge q is in a circular orbit with radius r at a speed v.

(a) What is its angular momentum, L, in terms of these quantities?

(b) What is its magnetic moment?

(c) What is the magnetic moment divided by the angular momentum? This ratio is called the gyromagnetic ratio. Express this quantity in terms of q and m only. (Remember that the current is the charge passing by per second, i.e., the amount of charge passing in a given time divided by that time.)

Ans.:
(a) $L = mvr$.

Magnetic materials

When a material is put in a magnetic field, the atoms and molecules of which it is composed experience forces and torques as a result of their magnetic moments. The magnetic moments may be *intrinsic*, i.e., those that the atoms and molecules have even if there is no magnetic field. In addition, there are always also the magnetic moments that are there because there is a magnetic field and that are absent without it.

The two behave quite differently. An intrinsic magnetic moment (one that is always there, like the spin magnetic moment) experiences a torque that tends to line it up with the field and will strengthen it. This effect is called *paramagnetism*, and a material in which it predominates is *paramagnetic*.

A magnetic moment that is created by a field points in the direction opposite to the field and weakens it. (This is shown by Faraday's law, which is described in the next section.)

If the intrinsic magnetic moments are those of the separate individual atoms, the effects are usually quite small. They can, however, be very large if the atomic magnetic moments interact with each other so strongly that they line up parallel to one another even when there is no external magnetic field. This is what happens in iron and other *ferromagnetic* materials.

The total magnetic moment of a piece of iron is, however, usually small or zero. The material breaks up into *domains*, regions in which the atomic magnetic moments are parallel to each other, but in each domain with a different orientation, so that the net magnetic moment is zero or small. In an external field the magnetic moments of the different domains tend to line up and can enhance the magnetic field by a factor of several thousand.

10.3 Generating electricity: motional emf and Faraday's law

The motional emf

We already know that there is a magnetic force on a charge moving in a magnetic field. Can we use this fact to push the electrons in a wire or rod to one side, just as electrons in a battery accumulate at one end?

Look at a metallic rod or wire moving sideways in a magnetic field. Each of the charges in the rod experiences a force, the positive charges toward one end and the negative charges toward the other. The charges separate. The net effect is that one end of the rod becomes positively charged and the other negatively. While it is moving in the magnetic field the rod is *polarized*.

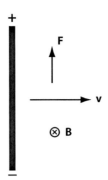

There is now a difference in electric potential, ΔV, between the two ends of the rod. The moving rod acts like a battery. A battery also has two ends, or terminals, with a difference of electric potential between them. The potential difference between the ends of the rod is there as long as the rod continues its motion in the magnetic field.

The difference in electric potential that is created is analogous to the difference in gravitational potential that is produced when an object is lifted. The gravitational potential energy of a lifted object can be changed to kinetic energy when the object returns to its lower position. Here too, the potential energy (this time the electric potential energy) can change to another form when the charges return to where they have lower potential energy.

With a battery this happens when we connect a wire from one terminal to the other. There is then a current from the positive terminal through the wire to the negative terminal. (Since it is the negatively charged electrons that move, their motion is in the opposite direction, from the negative terminal, through the wire outside the battery, to the positive terminal.) As a result of the collisions of the electrons with the ions of the wire the internal energy of the wire increases and the wire heats up. The electric potential energy is transformed to internal ("thermal") energy.

If we do the same with the rod moving in a magnetic field, we have to be careful that the wire through which the charges return is stationary, i.e., not also moving with respect to the magnetic field.

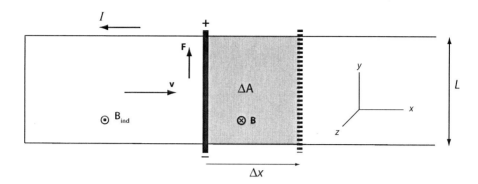

In both the battery and the moving rod there is a transformation of energy to electric potential energy from some other form of energy. In the battery it is the stored internal ("chemical") energy that is liberated by the chemical reactions within it. In the moving rod it is its kinetic energy that is transformed.

The amount of energy that is transformed, divided by the charge that moves, is the *emf*. It represents the energy, for each coulomb that is moved, which is changed to electric potential energy from some other kind of energy.

The figure shows a rod, free to slide from left to right along stationary rails while it continues to make electrical contact with them. There is a magnetic field, B, into the plane of the paper. The rod and the charges within it are being pushed to the right by an external force that is not shown, and they move to the right with velocity v. Because the charges in the rod are moving in a direction perpendicular to the field, there is a force on them, perpendicular both to their velocity and to the field.

The force on an amount of charge, Q, moving with a velocity, v, at right angles to the field B, is QvB. The positive charges experience a force in one direction and the negative charges a force in the opposite direction. The electrons move toward the end marked with a minus sign. The effect is the same as if an equivalent amount of positive charge were to move along the rod in the other direction toward the end marked with a plus sign. If an amount of charge, Q, is moved a distance L by a force, F, the work on it (force times distance) is $FL = QvBL$. The work per unit charge is $\frac{FL}{Q}$, which is equal to BLv. This is the potential difference, or emf, that is created. For this case it is called the *motional induced emf*, $\mathcal{E}$, equal to BLv.

The charges stay separated, and the induced emf continues to exist, as long as the rod or wire continues to move in the field. The figure shows the return circuit. The rod moves along a loop, the rest of which is stationary.

The emf in terms of the change in the flux: Faraday's law

We can now put the relation for the emf in another form which turns out to be much more generally valid. As the wire moves, the area of the loop changes by an amount ΔA, equal to $L\Delta x$ in a time Δt, the change in t.

The rate at which the area changes, $\frac{\Delta A}{\Delta t}$, is equal to $L\frac{\Delta x}{\Delta t}$, or Lv, where v is the rod's velocity. Instead of BLv for the motional emf, we can write $B\frac{\Delta A}{\Delta t}$.

We called the product of the electric field, E, and the area perpendicular to it, the electric flux. Similarly, we call the product of the magnetic field, B, and the area perpendicular to it, the *magnetic flux*, for which we use the symbol Φ (Greek capital *phi*). The magnitude of the motional induced emf can then be written as $\frac{\Delta \Phi}{\Delta t}$.

In our example the velocity, $v = \frac{\Delta x}{\Delta t}$, the rate of change of the area, $\frac{\Delta A}{\Delta t}$, and the rate of change of the magnetic flux, $\frac{\Delta \Phi}{\Delta t}$, are all constant. To encompass the case where this is not so, we use the notation $\frac{d\Phi}{dt}$, representing the rate of change with time of the magnetic flux through the loop, regardless of the particular nature of the variation.

We have not yet considered the direction of the induced emf. Look again at the figure that shows the motional emf. The rod moves to the right (in the x direction) in the external magnetic field, which points into the paper. The area of the current-carrying loop increases by ΔA as

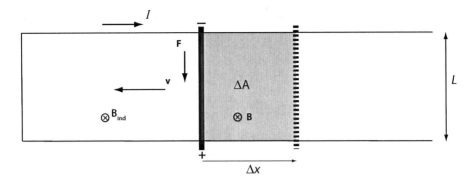

the rod moves. The force on positive charges is along the rod, in the y direction, and there is a current, the *induced current*, counterclockwise around the loop. It gives rise to an *induced magnetic field*, which here is up, out of the paper, in the z direction. (Use the first right-hand rule.) We see that it opposes the *increase* of the original downward magnetic flux.

What happens if the rod moves to the left? Now ΔA and the flux through the loop decrease. The force on positive charges in the rod is in the negative y direction and the other end of the rod becomes positive. The induced current is clockwise. Its magnetic field (the induced magnetic field) is downward, into the paper. It *opposes the decrease* of the original downward flux. It tries to keep the flux there.

In both cases there is a change in the original magnetic flux in the loop. In both cases there is an induced magnetic field that opposes the *change* in the original flux. In the first case it is in the direction opposite to that of the original field. In the second case it is in the same direction. In both cases it opposes the *change* of the original flux. The direction can be incorporated in the statement for the induced emf by using a minus sign: $\mathcal{E} = -\frac{d\Phi}{dt}$.

The statement that the induced field opposes the change in the original field is so important that it is given its own name. It is called *Lenz's law*.

The rod moves in the field so as to produce a magnetic force. Now look at the same event from the reference frame in which the rod is at rest. This time the charges are not moving, so that there is no magnetic force. Nevertheless, experiments show that the emf is still there!

In fact, we would be very surprised if this were not so. We do not expect the laws of physics to depend on the particular reference frame, or coordinate system, from which a phenomenon is observed. More formally, we call this the *principle of relativity*, a cornerstone of the special theory of relativity.

The relation $\mathcal{E} = -\frac{d\Phi}{dt}$ is called *Faraday's law*. It is valid regardless of how the flux changes, regardless, that is, of whether the rod or the loop of wire actually move in a particular coordinate system. It goes farther than the relation for the motional emf, and cannot be derived from the force law.

Faraday's law describes how the motion of wires in a magnetic field can be used to "generate electricity." It shows how the kinetic energy of the wires can be transformed to electric potential energy in an *electric generator*. This is how the overwhelming majority of the electric energy that we use is produced.

EXAMPLE 11

Go to the PhET website (http://phet.colorado.edu) and open the simulation *Faraday's Electromagnetic Lab*.

(a) Choose "bar magnet." Check "show field" and "show compass."

The field is shown by little compass needles. The red end points in the direction of the field. Move the magnet and the compass. See how the compass needle's orientation compares with that of the field vectors. "Flip polarity" and observe the effect.

(b) Choose "pickup coil." Check "show electrons." Move the magnet in and out of the coil and watch the response of the light bulb and of the electrons. Move the magnet slowly to see the direction of motion of the electrons. Increase the area of the loop and observe the effect. A voltmeter can replace the light bulb by clicking on the meter's picture on the right.

When the north pole is moved toward the coil the magnetic flux through the coil increases. What do you observe to be the direction of motion of the electrons? (Clockwise or counterclockwise as seen from the right?) Remember that the motion of the electrons is opposite to the direction of the induced current. What is the direction of the magnetic field (the induced field) produced by the induced current?

What does Lenz's law predict for the induced field? Do the prediction and the observation agree?

As the middle of the magnet comes to the coil, the flux through the coil decreases. What does Lenz's law predict for the direction of the induced field now? What is the direction of the current to produce this field? What is the motion of the electrons? What is your observation for the motion of the electrons? Does it agree with the prediction of Lenz's law?

Here is the answer to part (b). The answers to the other parts follow similarly from Faraday's law and Lenz's law.

As the magnet's north pole enters the loop, the electrons are seen to move CCW so that the induced current is CW and the induced magnetic field is to the left. It counteracts the change in the flux, which here is the increase in the flux to the right.

As the magnet's middle enters the loop, the field and flux decrease. This time the electrons move CW so that the induced current is CCW and the induced field is to the right. It again counteracts the change, which this time means that it counteracts the decrease in the flux. It is to the right, in the same direction as the original flux, so as to counteract its decrease.

(c) Choose "electromagnet." Check "show electrons" and "show compass." Check to see that the relation between the direction of the current and that of the magnetic field is in accord with what you expect.

Select "AC." There are two sliders on the "current supply" box that allow you to change the amplitude and the frequency. Try them out.

(d) Choose "transformer." Check "show field" and "show electrons." Select "DC." Is there an induced current? Explain.

Move the magnet coil (the *primary*) toward the *secondary* (the pickup coil). Predict the direction of the field, the current, and the electron motion induced in the secondary. Which of these quantities can you observe? Are they in accord with your prediction?

Click on "AC" and observe the effect. Look at the light bulb and then switch to the voltmeter. Use the five ways in which the simulation allows you to change the voltage in the secondary. What are they?

(e) Choose "generator." Check "show compass."

Move the compass near the magnet.

Turn on the "faucet" by turning the knob (drag the small brown cylinder on the knob) to a fairly low frequency (near 30 rpm). Watch the compass, the light bulb, and the electrons. Replace the light bulb by the voltmeter.

There are four ways to change the current in the pickup coil in this simulation. What are they? Try them. There is another that is not available here. What is it?

EXAMPLE 12

A circular loop of wire with an area of 10^{-3} m² is perpendicular to a uniform magnetic field of .1 T.

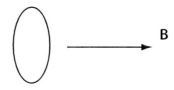

(a) The magnetic field is switched off and smoothly goes to zero in 0.2 s. What is the induced emf in the loop during this time?

(b) The loop is turned through 90° in 0.2 s. What is the induced emf during this time?

(c) The loop is flipped through 180° in 0.2 s, so that it is again perpendicular to the field, but now facing in the opposite direction. What is the induced emf this time?

Ans.:

(a) $\Delta\Phi = (0.1)(10^{-3}) = 10^{-4}$ Tm². $\Delta t = 0.2$. $\frac{\Delta\Phi}{\Delta t} = \frac{10^{-4}}{0.2} = 5 \times 10^{-4}$ V. The direction of the induced emf is such as to produce an induced current with an induced magnetic field that opposes the *change* in the original field. Here it opposes the decrease of the original field and is therefore in the same direction as the original field. The direction of the induced current is related to the direction of the induced magnetic field by the first right-hand rule. This is also the direction of the induced emf.

(b) After the loop is turned, the flux through it is zero. Hence the change of flux is the same as in part (a), and so is the induced emf.

(c) This time the flux, as seen from the loop, changes direction. Sitting on the loop you see the flux changing from $10^{-4}\,\mathrm{Tm}^2$ up to $10^{-4}\,\mathrm{Tm}^2$ down, i.e., a change that is twice as much as before, equal to $\Delta\Phi = 2 \times 10^{-4}\,\mathrm{Tm}^2$. The average induced emf is therefore twice as large as before, or 10^{-3} V.

EXAMPLE 13

emf induced in a turning loop; electric generator

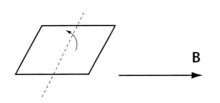

A square loop of wire with sides of 20 cm is suspended so that it can turn about an axis along its centerline. It rotates at a rate of 5 revolutions per second in a uniform magnetic field of 1.2 T.

(a) What is the maximum value of the flux through the loop?

(b) What is the flux as a function of time?

(c) What is the emf induced in the loop?

Ans.:
(a) The flux has its maximum value when the loop is perpendicular to the field. It is then $\Phi_{\max} = BA = (1.2)(0.2^2) = 0.048\,\mathrm{Tm}^2$.

(b) Each revolution of the loop turns it through 2π radians. Five revolutions are 10π radians, so that the angular velocity, ω, is 10π radians per second.

The flux varies between $\Phi_{\max}$ and zero. It is at its maximum when the loop is perpendicular to the magnetic field and is zero when it is parallel to the field. Two of the sides of the loop remain perpendicular to the field. The others rotate at an angle θ to the field. The flux through the loop is equal to $\Phi = \Phi_{\max} \sin\theta$. The angle θ is equal to ωt, so that $\Phi = \Phi_{\max} \sin\omega t$.

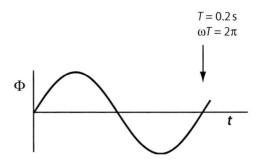

Here we have assumed that $\Phi = 0$ at $t = 0$, i.e., that at $t = 0$ there is no flux through it at that moment. We could also have used $\Phi = \Phi_{\max} \cos\omega t$, which would have the same time variation, but with $\Phi_{\max}$ at $t = 0$.

(c) The magnitude of the emf is given by $\mathcal{E} = \frac{d\Phi}{dt}$. To calculate the emf we have to know the derivative of $\sin\omega t$.

Look first at the derivative of $y = \sin x$. It is $\frac{dy}{dx} = \cos x$.

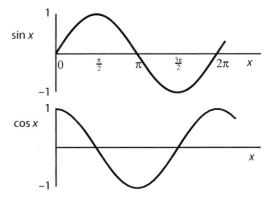

You can see from the diagram that $\sin x$ has a slope of 1 at $x = 0$, a slope of zero at $x = \frac{\pi}{2}$ and $\frac{3\pi}{2}$, and a slope of -1 at $x = \pi$. These values provide a check on the quoted relation that $\frac{dy}{dx} = \cos x$.

The derivative of $y = \sin\omega t$ is $\frac{dy}{dt} = \omega\cos\omega t$. The factor ω shows that as expected, when the rate of rotation is larger, the rate of change of y is higher.

Finally, we can write the induced emf as $\frac{d\Phi}{dt} = \omega\Phi_{\max}\cos\omega t$, equal to $(10\pi)(0.048)\cos 10\pi t$, or $1.51\cos 34.1t$ V.

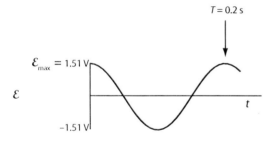

The induced emf varies between 1.51 V and −1.51 V at a rate given by the angular velocity of 10π or 34.1 radians/s, i.e., at a frequency (= $\frac{\omega}{2\pi}$) of 5 s^{-1} or 5 Hz.

We have gone into some detail in this example because of its great importance. An electric generator consists of a coil of wire with many loops, rotating in a magnetic field. This is how almost all of the electricity that we use is generated.

We also see that the induced emf of a rotating coil varies *sinusoidally*. This is one of the reasons why alternating voltage and alternating current (AC) are universally used for household and industrial distribution.

A direct current (DC) generator, i.e., one where the current is always in the same direction, can also be constructed from loops rotating in a magnetic field. However, the direction of the current has to be reversed each time the loop rotates through 180°. This can be accomplished by a *commutator*, like the one that we discussed in connection with the motor, when it was also necessary to reverse the direction of the current after each half turn.

10.4 Summary

The *magnetic force* is an interaction between electric charges over and above the *electric force* between them. It is there when the charges are moving with respect to each other. It can be described either as an interaction between moving charges or as an interaction between currents.

We describe the interaction in two steps: the current (or the moving charge) creates a magnetic field, and another current (or moving charge) in the field experiences a magnetic force.

The magnitude of the magnetic force between two long parallel currents is $\frac{F_m}{L} = k'\frac{I_1 I_2}{r}$. They attract if they are in the same direction and repel if they are in opposite directions.

The magnitude of the magnetic field of a long current, I_1, is $B = k'\frac{I_1}{r}$.

The direction of the magnetic field of a current-carrying wire is given by the first right-hand rule: grasp the wire with your right hand. The fingers curl in the direction of the magnetic field lines.

The magnitude of the force on a current I_2 in a magnetic field is $F = BI_2L$, if B is perpendicular to I_2 and L. If it is not, the relation is still true if for B we use only the component of the magnetic field that is perpendicular to I_2 and L.

The direction of the magnetic force on a current is given by the second right-hand rule: point the fingers of the right hand in the direction of the current, I. Bend the fingers so that they point in the direction of the magnetic field. The thumb then points in the direction of the force.

The shorthand notation that incorporates both the magnitude and the direction of the magnetic force on a current is $\mathbf{F} = I\mathbf{L} \times \mathbf{B}$. In this relation (the *cross product*) the magnitude of $\mathbf{F}$ is $ILB \sin \theta$.

The magnetic torque on a loop with current I and area A is $\tau = I\mathbf{A} \times \mathbf{B}$. (Since the magnetic moment of the loop is $\mu = I\mathbf{A}$, we can also write $\tau = \mu \times \mathbf{B}$.) This is the principle of the electric motor.

The magnetic force on a moving charge is $\mathbf{F} = Q\mathbf{v} \times \mathbf{B}$. If the velocity and the field are perpendicular to each other, $F = QvB$, and the charge moves in a circle in the plane perpendicular to B, with $\frac{mv^2}{r} = QvB$, where $\frac{mv^2}{r}$ is the centripetal force, which here is provided by the magnetic force, QvB.

To determine the direction of the magnetic field of a solenoid coil the fingers in the direction of the current; the thumb points in the direction of B. (First right-hand rule.)

Ampere's law: follow a path around a current. For each step multiply the component of B parallel to the path by the length of the path. The total is $\mu_0 I$. For a circular path $(B)(2\pi r) = \mu_0 I$.

The elementary magnetic object is a current loop, or *magnet*. The electron is a tiny magnet, although there is no loop in that case. The electron (as well as other particles) has angular momentum even when it is not going anywhere. It is called its *intrinsic angular momentum* or *spin*. The magnetism of the electrons leads to the various magnetic properties of materials.

The magnetic moment of a current loop is $\mu = IA$.

The gyromagnetic ratio is the ratio of the magnetic moment to the angular momentum.

A paramagnetic material contains permanent magnetic moments that tend to line up parallel to a magnetic field.

In a diamagnetic material magnetic moments are induced by a magnetic field in accord with Faraday's law. Their direction is opposite to that of the magnetic field.

A ferromagnetic material consists of domains in each of which the magnetic moments are lined up in the same direction. In a magnetic field the domains line up to produce strong magnetic moments.

The vast majority of the electric energy that we use is generated by the motion of wires in magnetic fields. A *motional emf* arises (an electric potential difference appears) between the ends of a wire moving in a magnetic field. A more inclusive and general description is provided by *Faraday's law*.

Motional emf: when a wire of length L moves in a magnetic field, B, with speed v, and $L, B,$ and v are perpendicular to each other, the induced motional emf is $\mathcal{E} = BLv$.

Faraday's law: there is an induced emf in a loop when the magnetic flux through the loop changes. Its magnitude is $\frac{d\Phi}{dt}$.

In a loop the induced emf produces an induced current, which produces an induced magnetic field. The direction of the induced magnetic field is such that it opposes the *change* in the flux, $\Delta\Phi$, that brought it about.

In a coil rotating about its diameter in a magnetic field, the emf varies sinusoidally. This is what happens in an alternating current (AC) generator.

10.5 Review activities and problems

Guided review

1. Two transmission wires are horizontal and parallel to each other, one above the other. The upper one carries a current of 100 A to the right and the lower one carries a current of 60 A in the same direction. How far apart do they need to be if the force on each is to be no more than 10^{-3} N on each meter?

2. A long wire carries a current of 10 A. Where and in what direction does a second, equal and parallel current have to be so that the magnetic field 10 cm from the first wire is zero?

3.

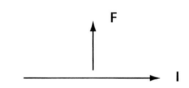

A wire whose length is 50 cm, carrying a current of 8 A in the x direction, experiences a magnetic force of 0.1 N in the y direction.
(a) What are the magnitude and direction of the smallest uniform magnetic field at the wire?
(b) What other magnetic fields at the wire can give rise to the same force?

4.

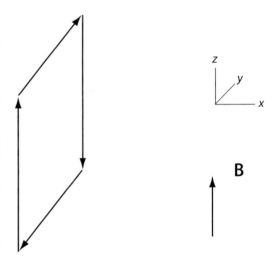

A square loop (sides 20 cm, $I = 12$ A) lies in the vertical (yz) plane. Looked at from the right, the current is clockwise. There is a uniform magnetic field of 0.5 T in the vertical (z) direction.

(a) What are the forces on each side?
(b) What is the torque on the loop?

5. In what orientation of the loop of the previous question would there be no torque?

6. (a) In what orientation would the loop of the previous questions experience a torque half as large as in question 4? Make a sketch that shows the magnetic field and the loop with its currents and the normal to the loop.

(b) Show the vectors representing the area **A** (in the direction of the normal) and the direction of the vector τ, the torque. Is the direction of the torque as found from the forces the same as that found from the relation $\tau = I\mathbf{A} \times \mathbf{B}$ in this and in the previous two questions?

7. (a) From the forces on the current loop in Example 7, determine the direction of the vector representing the torque exerted by the magnetic field.

(b) Determine the direction of the vector $I\mathbf{A} \times \mathbf{B}$ and compare it to the result of part (a). (The vector **A** is related to the current in the loop by a rule like the first right-hand rule: the fingers of the right hand point along the current and the thumb gives the direction of the area vector.)

8. Looking down, an electron is observed to move clockwise in a horizontal circle. What is the direction of the magnetic field in which it moves?

9. A charged particle moves in a circle in a magnetic field.

(a) Starting with the force relation show that r is proportional to v, and that
(b) the number of revolutions per second is the same for all values of the speed, v.
(c) A cyclotron has a magnetic field of 1 T and a radius of 20 cm. To what energy can it accelerate protons?

10. A coil whose diameter is 5 cm and whose length is 1.2 m is wound with 15,000 turns of wire. What is the magnetic field inside the coil when it carries a current of 20 A?

11. A "flip coil" is used to measure a magnetic field. It is a square loop of wire whose sides are 2 cm. It starts out with its axis (the normal) parallel to the field, and is then flipped through 180° in a time of 0.1 s. The induced emf is observed to be 10^{-3} V. What is the magnitude of the field?

12. In Example 12 the loop is initially perpendicular to the magnetic field, i.e., the normal to the loop is parallel to the field. Now consider the case where the loop is initially parallel to the field (with its normal perpendicular to the field).

(a) The loop turns through 90° in 0.2 s to the position in the figure of the example. What is the average emf during that time?
(b) The loop turns through an additional 90°, again in 0.2 s. What is the emf this time?
(c) What is the average emf when the two motions of parts (a) and (b) are combined, i.e., when the loop turns through 180° in 0.4 s?

13. A loop of wire whose area is 0.2 m² rotates in a uniform magnetic field of 0.9 T at a rate of 12 revolutions per second. It starts (at $t = 0$) with its normal perpendicular to the field.

(a) What is the flux through the loop at $t = 0$ and after the loop has rotated through 90°, 180°, and 270°?
(b) Sketch the graphs of Φ and $\mathcal{E}$ as a function of time. On your graphs mark $\Phi_{max}, \mathcal{E}_{max}$, and the time T for one revolution.

Problems and reasoning skill building

1. (a) A positively–charged particle travels horizontally with a velocity v. It enters a region with a uniform magnetic field such that it travels in a horizontal circle. Make a sketch of v, B, and the path with one of the possibilities for the direction of B.

(b) What is the other possibility for the direction of B?
(c) What paths would a negative particle follow in the same fields as in parts (a) and (b)?

2. Two particles with the same mass, traveling with the same velocity, enter a region with a uniform magnetic field, such that they move in circular paths. The first particle has charge q_1 and for the second $q_2 = 2q_1$. What is the ratio $\frac{r_2}{r_1}$ of the radii of the two paths?

3. A charged particle moves in a circle in a uniform magnetic field. An electric field is now turned on, in a direction opposite to that of the

magnetic field. What is the path of the particle now?

4. A loop of wire is in the plane of the paper. It carries a clockwise current.

(a) What is the direction of the magnetic field at its center?

(b) The area of the loop is $10\,\text{cm}^2$ and the current is 2 A. What are the magnitude and direction of the loop's magnetic moment?

(c) An additional magnetic field, horizontal and to the right, of $0.5\,\text{T}$ is turned on. What are the magnitude and direction of the torque on the loop? Through what angle will it turn until the torque is zero? What is the direction of its magnetic moment then?

5. Each of the three long parallel wires (i), (ii), and (iii) carries a current of 5 A.

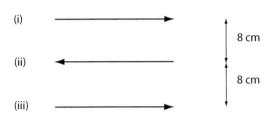

(a) What is the force on the middle one?

(b) What are the magnitude and direction of the forces per meter on each of the other two?

6. A magnetic field is perpendicular to a loop of wire. Looking down on the loop, what is the direction of the induced current (clockwise or counterclockwise) when the field is

(a) up and steady
(b) up and increasing
(c) up and decreasing
(d) down and increasing
(e) down and decreasing

7. A solenoid is surrounded by a loop of wire. A current is switched on in the solenoid in the clockwise direction when seen from the right.

(a) What is the direction of the induced current? What can you say about the length of time during which there is a current in the loop?

(b) What is the direction of the induced current when the current in the solenoid is switched off?

8. A long straight wire carries a current of $I_1 = 12\,\text{A}$ into the plane of the paper. A second wire is 10 m from the first and parallel to it.

(a) The magnetic field is zero at a point between them in the plane containing both wires, 2 m from the first. What are the magnitude and direction of the current I_2 in the second wire?

(b) Part (a) can be answered by first finding the field of the first current. What is the advantage of not doing that and first developing a relation between I_1 and I_2?

9. A house has a floor area of $120\,\text{m}^3$ and four walls, each of which has an area of $25\,\text{m}^3$. One wall faces north, one south, one east, and one west.

The earth's magnetic field there has a horizontal component of $2.4 \times 10^{-5}\,\text{T}$ and a vertical component, down, of $4.8 \times 10^{-5}\,\text{T}$.

What is the magnetic flux outward through each wall and through the floor? What is the total flux outward through all the walls, the floor, and the ceiling?

10. A circular coil of wire has a radius of $0.15\,\text{m}$ and a resistance per unit length of $3.8 \times 10^{-2}\,\Omega/\text{m}$. It is perpendicular to a magnetic field that increases from zero to $0.55\,\text{T}$ in $1.5\,\text{s}$. What is the electric energy dissipated in the wire during the 1.5 s?

11. Looking down on its path, a 1 MeV proton travels clockwise with a radius of 10 cm. What are the magnitude and direction of the field in which the proton moves?

12. A loop of wire whose area is $10^{-2}\,\text{m}^2$ rotates about its diameter in a uniform magnetic field of $0.8\,\text{T}$ at a rate of 15 revolutions per second. It starts out (at $\theta = 0$) with its normal parallel to the field.

(a) What is the average emf induced in the loop as it turns through $90°$ from its starting position?

(b) Repeat part (a) for $180°$ and $360°$.

Multiple choice questions

1.

————————▶ 5 A

10 A ◀————————

Two long parallel wires carry steady currents in opposite directions. The current in the upper one is 5 A to the right and the current in the lower one is 10 A to the left. The ratio of the magnetic force on I_1 to that on I_2 is
 (a) +1
 (b) −1
 (c) +2
 (d) −2

2. For the same wires and currents as in the previous question, the ratio of the magnetic field at I_1 created by I_2 to that at I_2 created by I_1 is
 (a) +1
 (b) −1
 (c) +2
 (d) −2

3. Both uniform magnetic and electric fields can do all but which one of the following on a charged particle?
 (a) accelerate it
 (b) exert a force on it
 (c) change its direction
 (d) increase its kinetic energy

4. A proton and an alpha particle ($q = 2e, m = 4m_p$) move with the same speed in circles in a uniform magnetic field. The ratio of their radii $\frac{r_\alpha}{r_p}$ is
 (a) 4
 (b) 2
 (c) 1
 (d) 0.5

5. A proton and an alpha particle move with the same kinetic energy in circles in a uniform magnetic field. The ratio of their radii $\frac{r_\alpha}{r_p}$ is
 (a) 4
 (b) 2
 (c) 1
 (d) 0.5

Synthesis problems and projects

1.

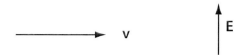

An alpha particle travels in the x direction with a velocity of 10^3 m/s. It enters a region with an electric field in the y direction, of 50 N/C.

(a) What are the magnitude and direction of a magnetic field such that the particle continues undeflected?
(b) Replace the alpha particle by an electron, and repeat.

2. An electric generator has 500 turns of wire, each with an area of 0.05 m². It rotates with a frequency of 60 revolutions per second in a magnetic field of 1.2 T. What is the maximum emf that it produces?

3. In a *mass spectrometer* a particle with mass m and charge q is first accelerated through a potential difference V. It then enters a region where a uniform magnetic field (B) causes it to move in a circle, with radius r.
 (a) Develop a relation for r in terms of q, m, V, and B.
 (b) What is the radius for a proton that is first accelerated through a potential difference of 300 V and then moves in a circle in a magnetic field of 0.05 T?
 (c) Use proportional reasoning to find the radius for an alpha particle.
 (d) Again using proportional reasoning, by what factor will the radius be different for an electron compared to that for a proton?

4.

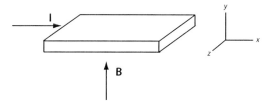

A flat horizontal strip carries a current in the x direction. There is a uniform magnetic field in the y direction (up, perpendicular to the strip).
 (a) The electrons, moving in the negative x direction, are deflected. What is the direction of the magnetic force on them?
 (b) The electrons move toward one side of the strip, which becomes negatively charged, while the opposite side becomes positively charged. This is called the *Hall effect*, and the potential difference at right angles to the current is called the *Hall voltage*.

In addition to the magnetic force at right angles to the current there is now also an electric force. What is its direction?

The build-up of charge continues until the electric force and the magnetic force are equal, so that there is then no further deflection of the electrons as they move through the strip.

(c) Consider the possibility that the current is carried not by electrons but by positive charges. What is the direction of the magnetic force on them?

(In semiconductors the current can be carried by "holes" in the ocean of electrons. The holes act as if they were positive charges. The Hall effect can distinguish between currents carried by holes and by electrons.)

(d) Show how the measurement of the Hall voltage can be used to determine the average velocity of the electrons. (The electrons have large velocities. When there is no current they move randomly in all directions, colliding with the ions of the crystalline lattice. Their average velocity is then zero. When there is a current there is a much smaller additional component of their velocity in the direction opposite to the current. Its average is called the *drift velocity*.)

5. A transformer consists of a primary coil surrounded by a secondary coil. An alternating current in the primary coil induces an alternating current in the secondary coil.

(a) For one quarter cycle the primary current is in the clockwise direction and increases. What is the direction (clockwise or counterclockwise) of the secondary current during that time?

(b) For the next quarter cycle the primary current is in the same clockwise direction and decreases. What is the direction of the secondary current in this part of the cycle?

(c) Describe the direction of the primary and secondary currents and how they change for the following two quarter cycles.

(d) Sketch the primary current as a function of time and underneath it the secondary current. (Use clockwise as positive and counterclockwise as negative.) Write down Faraday's law. Are your graphs consistent with the minus sign in it?

6. In a loudspeaker a coil is attached to a cardboard cone, which can move in the field of a permanent magnet. Describe its production of sound.

7. Describe the action of a magnetic (or "dynamic") microphone.

CHAPTER 11

Waves: Mechanical and Electromagnetic

What is a wave?
> *Different kinds of waves*

What can waves do? Describing waves and their properties
> *Interference*
> *Pictorial and mathematical description*

Sound and musical scales
> *From the source and through the medium*
> *Standing waves*
> *Resonance*
> *Scales*
> *The Doppler effect*

Maxwell's great contribution: electromagnetic waves
> *Maxwell's equations*
> *The electromagnetic spectrum*
> *The capacitor and the energy of the electric field*
> *The propagating fields*

Observing interference of light
> *Young's double slit experiment*
> *The diffraction grating*
> *Single slit diffraction*
> *Thin films*
> *The Michelson interferometer*
> *Coherence*

Reflection and refraction
> *The laws of reflection and refraction*
> *Mirrors*
> *Lenses*
> *The thin-lens relation*
> *Total internal reflection*
> *Resolution*
> *Camera and eye*

The magnifying glass
Microscope and telescope

Where Einstein started: electromagnetism and relativity

The ether and the speed of light
Kinematics of the special theory of relativity: time dilation and length contraction
Dynamics of the special theory: $E = mc^2$
Magnetism and electricity: inseparable, but interchangeable

When we think of transporting mechanical energy, we might first think of a ball or bullet. The kinetic energy that we give one of these projectiles at one end of its path is available at the other end. There is another way. Think of a group of people standing close to each other. Someone pushes the first one, let's call him Robert. That makes him lean over and push the person next to him, who, in turn, pushes the next one, and so on. The last one, call him Richard, falls over when his neighbor pushes him. Robert never touches Richard, but the initial push starts the sequence of forces from one to the next, through the line of people, from Robert to Richard.

We have described a single disturbance, or pulse. If it is a continuing disturbance, back and forth, or up and down that is transmitted, we call it a wave.

We first explore mechanical waves, characterized by forces and displacements. We then come to electromagnetic waves, where no material object moves and only the electric and magnetic fields change as they chase each other through empty space.

11.1 What is a wave?

Different kinds of waves

Think of a quiet lake. The water surface is smooth and horizontal. The gravitational force acts on each part of it. If some of the water is higher than the rest, its weight pulls it to a lower level. If some of it is lower, water from higher levels will tend to flow there. Only when all of the surface is at the same level is the water quiet.

Now look at a stone falling in the lake. Where it hits the water, it pushes it down. The stone sinks out of sight, but the water is no longer in equilibrium. The depression is still there. The gravitational force now acts to restore the equilibrium configuration. Water fills the depression. At the moment when the low part is filled, the water level is still changing. The water moves further and overshoots, and there is now a raised portion of the surface. It experiences a downward force, again back to the horizontal equilibrium. Once more it overshoots. The water surface at the point where the stone fell oscillates up and down, and would continue to do so if frictional forces did not cause the oscillations gradually to diminish and eventually to disappear.

As the water moves up and down in the same place, there is another motion that develops. Each time the water moves down, it pushes some of the rest of the water out of the way sideways, so that the neighboring region moves up. The water there, in turn, pushes on the water further away, so that the oscillation is propagated outward, away from the original point (the source) where the stone started it.

When we look at the water we see an up-and-down oscillation at the source and also at every other point that has been affected. At each point there is a motion of the water, up and down, as a function of time.

The second variation that we see is as a function of distance, away from the source in every direction along the surface. We can take a snapshot and see the oscillation at a particular moment in time.

If you think about what you actually look at when you see a wave, it is likely to be something different still. Your eye follows a point on the wave, a high point or a low point (a crest or a trough) as it moves away from the source along the direction of propagation of the wave.

Each of these variations is characteristic also for other waves. There are waves along a guitar string, sound waves in air, and elastic waves through a solid or liquid material. All of these are *mechanical* waves. In each case an equilibrium situation is disturbed. A restoring force acts to reestablish the original equilibrium, but the motion continues: there is an overshoot. The restoring force continues to act, changing direction so that it is always toward the equilibrium configuration.

EXAMPLE 1

Go to the PhET website (http://phet.colorado.edu) and open the simulation *Wave on a String*. Set "damping" to zero. (This eliminates friction and other dissipative forces.) Select "oscillate" and "no end." This will allow the string on the screen to move up and down as if the string continued on to infinity. Change the frequency and the amplitude. Push "pause" and then "step."

(a) Describe the motion of one of the green particles. Use the "step" feature.

(b) What is the relation between the direction of motion of the particles and the direction of propagation of the wave?

Ans.:
(a) The particles oscillate along the vertical direction in simple harmonic motion.

(b) In this case the motion of the particles is in the direction perpendicular to the direction of propagation of the wave. This kind of wave is called a *transverse wave*.

(In other kinds of waves, such as sound waves in air, the particle motion is along the same direction as the wave propagation. These waves are called *longitudinal* waves.)

Some features are common to all waves, but each kind has its special characteristics that depend on the forces that act to restore the equilibrium and on the medium (the water, the string, the material) in which the waves propagate.

A sound wave in air, for example, is quite different from a wave on the water surface, but both share properties that are common to all waves.

Let's see how a sound wave in air gets started. It can be by the back-and-forth motion of your vocal cords or of the cone of a loudspeaker, or by the vibration of a string, as in a guitar or violin. In each case the air is pushed and disturbed from its equilibrium. In turn, it pushes on the neighboring region, which pushes further, and so on along the propagating wave.

If the sound wave reaches an ear, it pushes on the eardrum, and makes it move back and forth. Knowledge of this vibration is transmitted to the brain, and is perceived by it as sound.

Along the wave there is the motion of air molecules, back and forth, at each point, superimposed on their random thermal motion. There is a second quantity that varies in space. As the air is pushed at the source, the air pressure increases in the region around it. When the oscillation continues at the source, and the vocal cord or string draws back, the opposite occurs, and the pressure decreases. Where the motion of the molecules causes them to be closer together, the air pressure increases; where they move farther apart, the pressure decreases. We can describe the wave either as a variation in space and time of the motion of the molecules or as a variation of the air pressure.

EXAMPLE 2

Go to the PhET website and open the simulation *Wave Interference*. Select "Sound" at the top, "one speaker," "no barrier," speaker "on," and "grayscale." Change the amplitude and the frequency with the sliders, and observe the speaker and the wave. You can use the "pause-step" function with the buttons at the bottom of the screen. Click on "show graph."

What varies as the wave propagates? What do the light and dark bands represent?

Click on "add detector." Place the detector in front of the speaker by clicking on it and moving it with the mouse. The graph can be moved out of the way of the waves. Now you can see what varies and how it varies.

Click on "particles." The markers allow you to follow the motion of one particle at a time. How do the particles move? What does the eye follow more naturally?

Ans.:

The back-and-forth motion of the speaker pushes the air particles. It causes regions of high pressure and low pressure to move outward (to radiate) from the speaker. The light regions are seen to represent high pressure and the dark regions low pressure.

The pressure varies sinusoidally with time.

Each particle oscillates along the direction of propagation of the wave. Where the particles bunch together the pressure is high. The eye more naturally follows the path taken by the points of maximum pressure.

11.2 What can waves do? Describing waves and their properties

Interference

If I throw a baseball at you, there's no doubt that it carries energy. When it hits you, some of its energy is transferred to you. If two baseballs come at you, you feel the energies transferred from each one. They add up.

The way waves add is very different. When two waves arrive at a point their effects sometimes add, but they can also cancel.

We can see that with the water waves that we looked at when we started. As the wave passes, it alternately raises and lowers the water level. Another similar wave, passing at the same place, coming from another direction, will, by itself, have the same effect. But what happens when both get to the same place at the same time? One possibility is that each one, separately, would cause the water level to rise. When both arrive together, the effects add, and the water level rises twice as high as it does with just one of the waves.

But it is also possible that when both waves get to the same place, one of them tends to make the water level rise, but the other one acts to lower it. The effects again combine, but if one of the waves, by itself, would cause a crest, and the other a trough, the two will cancel. The water level neither rises nor falls.

This possibility of cancellation is peculiar to waves. It is an essential feature that distinguishes waves from baseballs or other thrown objects. The way that waves superimpose, sometimes resulting in reinforcement and sometimes in cancellation, is called *interference*.

Pictorial and mathematical description

Let's describe a wave in detail with some mathematics. We can think of a water wave as it travels along a quiet lake, but it could also be some other kind of wave, such as a sound wave in air or through a solid material.

Here are two graphs of a wave.

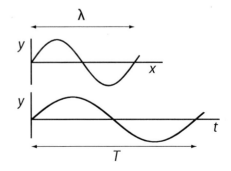

They look similar, but each tells something quite different. The first is a snapshot. It shows the height of the water as a function of the displacement along a line in the lake. It could go on, along the x-axis, repeating after a distance λ (Greek *lambda*), called the *wavelength*.

The second graph shows what happens at one point on this line as the water level rises and falls when the wave passes. It shows the wave at this point as a function of time. It repeats after a time T, called the *period*.

In the time the wave travels the distance λ, the point at $x = 0$ goes through one full cycle, from $y = 0$, up, down, and back to zero. This variation is shown by the second of the two graphs. We see that the wave travels a distance λ in the time T. The wave speed is therefore $\frac{\lambda}{T}$.

One more quantity characteristic of the wave is the *frequency*. If it takes $\frac{1}{4}$ s for the wave to repeat ($T = \frac{1}{4}$ s), then four full cycles of the wave pass in every second. The frequency, f, is then $4\,\text{s}^{-1}$ or 4 hertz (4 Hz). In general, $f = \frac{1}{T}$. We see also that the wave speed is given by $v = f\lambda$.

EXAMPLE 3

Go to the PhET website and open the simulation *Wave on a String*. Set "damping" to zero, "oscillate," and "no end." Select "rulers" and "timer."

Use the timer to measure how long it takes the wave to travel one wavelength. This is done most easily by going to "pause," start timer, then "step" until the wave has moved one wavelength.

How does this time (the *period* of the motion) change as the frequency is changed? Note that the number above "frequency" is proportional to the frequency, but is not the actual frequency.

Ans.:
The period, T, is the time it takes for the wave to travel one wavelength. It is the reciprocal of the frequency, f. The speed is $v = \frac{\lambda}{T} = f\lambda$.

EXAMPLE 4

Go to the PhET website and open the simulation *Wave Interference*. Select "Water," "One Drip," "No Barrier." Check "measuring tape" and "stopwatch." Set the f-slider near the middle. Pause (bottom button) and measure the wavelength with the measuring tape. Use the stopwatch to measure the period. (Pause, start, reset, and step.)

Repeat for two smaller and two larger settings of the frequency.

Make a graph of your data of λ vs. T. What quantity on your graph represents the speed of the water wave? What is your result for the speed?

What is the advantage of using the graph compared to using the individual results?

Ans.:
Since $\lambda = vT$, we expect a straight line through the origin, whose slope is the wave speed. The advantage of the graph is that drawing the straight line averages the results. Here the speed is about 2.3 cm/s.

We still haven't written down a mathematical description of the two graphs, y as a function of x and y as a function of t. We need a mathematical function that repeats. Here is one: the sine function $y = \sin\theta$. We can show how this function varies with the help of a circle whose radius is one unit.

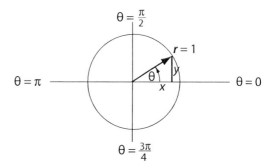

The radial vector (**r**) starts horizontally, to the right. Let the angle that it makes with the horizontal be θ, measured in radians. As this angle increases, we can look at the quantity $y = \sin\theta$. It is the vertical component of the vector **r**. ($\sin\theta = \frac{y}{r}$, and since the magnitude r is 1, $\sin\theta = y$.)

As θ goes from 0 to 90°, i.e., from 0 to $\frac{\pi}{2}$ radians, the distance y ($= \sin\theta$) goes from 0 to 1. In the next quadrant, as θ goes from 90° to 180°, or $\frac{\pi}{2}$ to π radians, y goes from 1 to zero. In the third quadrant, θ goes from π to $\frac{3\pi}{2}$ and y from 0 to -1, and finally, as θ goes from $\frac{3\pi}{2}$ to 2π, y returns from -1 to zero. The full cycle is 2π *radians*, and as θ continues beyond this value, the cycle repeats. We see that $\sin 0 = 0$, $\sin\frac{\pi}{2} = 1$, $\sin\pi = 0$, $\sin\frac{3\pi}{2} = -1$, and $\sin 2\pi = 0$, after which the sine function repeats.

That's not quite what we need. The function $\sin\theta$ repeats each time θ increases by 2π. We

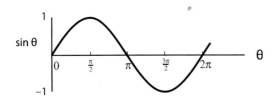

want a function that repeats when x is increased by λ. This will happen if we use the function $y = \sin \frac{2\pi}{\lambda}x$. When x in this function reaches λ, $y = \sin 2\pi$, which is equal to zero, and then the function repeats.

What about the variation with time? It is similar. The function $y = \sin \frac{2\pi}{T}t \ (= \sin 2\pi ft)$ repeats after a time T.

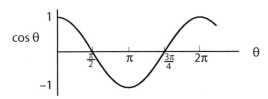

The cosine function $y = \cos \theta$ looks exactly like the function $y = \sin \theta$, except that it starts with the value $\cos 0$, which is equal to 1. It is shifted by $\frac{\pi}{2}$ radians from the sine function. We can also say that there is a *phase difference* of $\frac{\pi}{2}$ radians between the two curves, or that they are *out of phase* by $\frac{\pi}{2}$ radians.

We can add one more feature to our description. The sine and the cosine vary between the values $+1$ and -1. The functions $A \sin \theta$ and $A \cos \theta$ vary between $+A$ and $-A$, where the maximum value, A, is called the *amplitude*.

The amplitude is particularly important because it is related to the energy of the wave. Let's see what that relation is. Look again at the time variation $y = A \sin \frac{2\pi}{T}t$. As the value of y changes from zero to $+A$, to zero, to $-A$ and back, it moves with simple harmonic motion, as if it were on a spring with spring constant k. The energy of such a spring is $\frac{1}{2}kA^2$. This is the energy of the wave, proportional, as we see, to the square of the amplitude.

We can also define the *intensity* of the wave. It is the amount of energy that is transported by the wave per second, divided by the area through which it passes. Its SI unit is therefore the watt per square meter, $\frac{W}{m^2}$. Since it is proportional to the energy, it is also proportional to A^2.

Finally, we can combine the space and time variations in a single relation, $y = A \sin(\frac{2\pi}{\lambda}x - \frac{2\pi}{T}t)$. At the time $t = 0$ this relation becomes the space variation $y = A \sin(\frac{2\pi}{\lambda}x)$. At this time (and at any other time) y varies *sinusoidally* with x. Similarly, at the point $x = 0$ (or at any other point) y varies sinusoidally with T.

Think of the quantity in brackets as an angle in radians. As it goes through its cycle, from zero

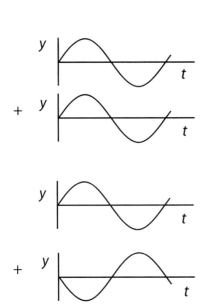
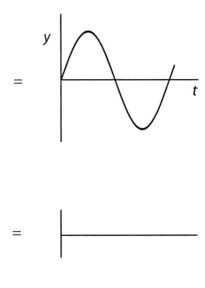

to 2π, y goes from zero to A, back to zero, to $-A$, and to zero again. Each crest ($y = A$), each trough ($y = -A$), and each of the points between represents a point with a certain *phase*. For a constant phase the quantity in brackets is constant. For a phase of zero, for instance, $\frac{2\pi}{\lambda}x - \frac{2\pi}{T}t = 0$, or $\frac{x}{\lambda} = \frac{t}{T}$, or $\frac{x}{t} = \frac{\lambda}{T}$. We see that a point with constant phase moves with the speed $\frac{\lambda}{T}$, as we saw before. This also shows again that as we watch a crest, a trough, or any point with a particular phase, it moves with the wave speed $\frac{\lambda}{T}$ or $f\lambda$.

When two similar waves (same A, λ, T) arrive at a point, they might be *in phase*, i.e., with their crests arriving at the same time. In that case they reinforce each other, resulting in a variation at that point between $+2A$ and $-2A$. This is called *constructive interference*. If they are water waves, they will combine to have a crest twice that of each wave alone. If they are sound waves, the pressure will be twice that of a single one.

But if they arrive so that the crest of one ($y = A$) occurs at the same time as the trough ($y = -A$) of the other, they will cancel. This is called *destructive interference*. At the point where two water waves interfere destructively, the water will not be displaced. If two sound waves interfere destructively, there will be no sound.

EXAMPLE 5

What is the relation that describes a wave with a wavelength of 3 m, a period of 4 s, and an amplitude of 20 cm?

Ans.:
$y = 0.2 \sin(\frac{2\pi}{3}x - \frac{2\pi}{4}t)$ or $y = 0.2 \sin(2.09x - 1.57t)$

(We are not explicitly putting the units into this relation. But note that the "3" in the first fraction is a distance, 3 m, and the "4" in the second fraction is a time, 4 s, so that the quantity in the brackets has no units at all.)

It could also be the cosine function $y = 0.2 \cos(2.09x - 1.57t)$.

This function has the value $A = 0.2$ m when $x = 0$ and $t = 0$, while the sine function is equal to zero when $x = 0$ and $t = 0$. The way in which y varies with x and t is the same in the two cases. One starts at the origin with the value zero, the other with its maximum value. Except for the phase difference between them the two graphs look the same.

EXAMPLE 6

What is the frequency of a wave described by the relation $y = 4 \sin(2x - 3t)$, where all quantities are in SI units?

Ans.:
Here the coefficient of t is equal to 3 s. In the general equation it is $\frac{2\pi}{T}$ or $2\pi f$. Comparing the two we see that $2\pi f = 3$ and $f = \frac{3}{2\pi} = 0.48\,\text{s}^{-1} = 0.48\,\text{Hz}$.

11.3 Sound and musical scales

From the source and through the medium

How is a sound wave generated? If we sing a note, we cause our vocal cords to vibrate. If the sound comes from a string instrument it is the string that vibrates. The frequency is determined by the length of the string, by its mass, and by the tension. With a wind instrument, such as a flute or clarinet, it is the length of the vibrating air column that determines the frequency, and with it the "pitch" of the note (from squeaky high to booming low) that is produced. Quite generally, when the source of a wave is a vibrating object, the frequency of the wave is equal to the frequency of the vibration of the source.

The wave then propagates through a *medium*. For a wave along a string the string is the medium, for a sound wave in air it is the air. As the wave travels through the medium, it is no longer connected to the source, and its properties no longer depend on the source (except for the frequency given to it by the source). In the medium there is a displacement from equilibrium and a restoring force back to the equilibrium configuration. The restoring force depends on the medium. For a wave along a string, for example, it is determined by how tight the string is, i.e., by the tension of the string.

The velocity of the wave depends on the restoring force and also on the mass of the material that is displaced. We see that the velocity is determined by the properties of the medium in which the wave travels. For a string under tension with a force F, and mass per unit length, $\frac{M}{L}$, the wave velocity is $\sqrt{\frac{F}{M/L}}$.

Once the frequency and the velocity are fixed, the value of the wavelength follows from $v = f\lambda$. In the cases that we have described it is a dependent variable that takes its value from the frequency of the source and from the velocity as determined by the medium. The two together determine the magnitude of the wavelength in the medium.

Standing waves

The waves that we have been talking about are waves that transport energy away from the source. They are *traveling waves*. If they encounter a barrier they are (completely or partially) reflected. This happens, for instance, when a water wave or a sound wave hits a wall, or when a wave along a rope arrives at the end of the rope where it is tied to a wall or tree.

Such a reflection can give rise to a quite different type of wave. The most direct demonstration is to take a string or rope, fixed at one end, and to shake the other end up and down. If the frequency at which you shake the string is just right, there will be a *standing wave* on the string: each point on the string goes up and down in simple harmonic motion, while the amplitude of the variation varies from point to point.

We can see in more detail what happens mathematically. We'll need the trigonometric relation that says that $\sin(A+B) = \sin A \cos B + \cos A \sin B$, and the one that says that $\sin(A-B) = \sin A \cos B - \cos A \sin B$, so that $\sin(A+B) + \sin(A-B) = 2 \sin A \cos B$.

Let's talk about a wave on a string, traveling to the right, described by $y = A \sin(\frac{2\pi}{\lambda}x - \frac{2\pi}{T}t)$. The reflected wave goes to the left, but is otherwise identical, with the same values of A, λ, and T, so that it is described by $y = A \sin(\frac{2\pi}{\lambda}x + \frac{2\pi}{T}t)$. Their sum represents the standing wave. We can write this sum, using our trigonometric relation, as $y = 2A \cos\frac{2\pi}{T}t \sin\frac{2\pi}{\lambda}x$.

What does this standing wave look like? The variation in space is described by $\sin\frac{2\pi}{\lambda}x$. The rest is the amplitude: $2A \cos\frac{2\pi}{T}t$. We have a sine wave in space, whose amplitude goes up and down sinusoidally with time.

You can see why the frequency with which you move the free end of the string has to be just right. The fixed end is a point with no vibration at all. (Such a point is called a *node*.) The end that you are holding is a point of maximum vibration. (It is called an *antinode*.) The distance between two adjacent nodes is $\frac{1}{2}\lambda$. The distance between a node and an adjacent antinode is $\frac{1}{4}\lambda$. The length of the string between the antinode at your hand and the node at the fixed end can therefore be $\frac{1}{4}\lambda$, or $\frac{1}{4}\lambda$ plus some number of half wavelengths. Standing waves are possible only at these wavelengths.

The velocity is, as usual, determined by the mass of the string and its tension. Since v is fixed, and only certain wavelengths (λ) are possible, it follows from $v = f\lambda$ that only certain discrete frequencies are possible.

The relation between the length of the string and the wavelength is even simpler when both ends of the string are fixed. The length of the string can then only be a whole number of half wavelengths.

Look at the unexpected and significant result that we have come to. We started with quantities, the frequency and the wavelength, that seemed to be able to take on any value. We described a situation where the ends are constrained, either by keeping one or both fixed or by holding one end and causing it to oscillate up and down. The conclusion that we came to is that then only certain values of the wavelength and frequency are possible. We have gone from continuous quantities to discrete or *quantized* values. What is it that brought about this change?

In an infinitely long string there would be no reflection and no limit to the possible values of the wavelength or frequency. The discreteness of

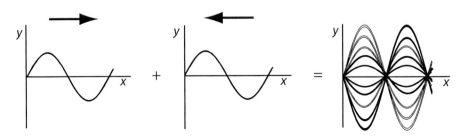

the values, i.e., the quantization of the spectrum of allowed values of the frequency, comes about as a result of what we do at the ends of the string. In other words, it comes about as a result of what are called the *boundary conditions*. We will see in the next chapter that there are strong similarities to the quantization of the energy spectrum of atoms, i.e., to the way in which the discrete allowed energies of atoms arise.

EXAMPLE 7

Go to the PhET website and open the simulation *Wave on a String*. Set the tension to its maximum and "damping" to zero. Select "oscillate" and "fixed end." Set the frequency to 50 and the amplitude to 1 or 2. Observe the standing wave. Why is it called a "standing wave"?

Try to find another combination of frequency and tension that also leads to a standing wave.

Ans.:
The points where there are maxima and minima (the antinodes) and the points where there is no motion (the nodes) remain in place. There is no wave propagation.

It is not easy to find other combinations of frequency and tension that produce a standing wave. This is mainly because the simulation does not allow these two quantities (or at least one of them) to be varied continuously. The boundary conditions determine what happens at the ends. They can be nodes, antinodes, or something between the two. The distance between two nodes is one-half wavelength. The pattern of nodes and antinodes has to "fit" between the two ends. (Here is another combination that you can try. It comes close: tension at 0.8 of the maximum value and frequency setting at 25.)

We have talked about waves on strings because they are so visible. Vibrating air columns have similar characteristics, and are particularly interesting because they form the basis of musical wind instruments. In a flute or recorder (the wooden, now often plastic flute that was common in the renaissance and baroque eras and that had a strong revival in the twentieth century) the air columns are set into vibration by the breath of the player. In the clarinet and the oboe the mouthpiece has reeds that vibrate and transmit their motion to the air column.

Standing waves can be set up in air columns, similar to those on oscillating strings. Here also the boundary conditions determine the relation between the length of the column and the wavelength of the standing wave. There are two quantities that vary sinusoidally. One is the pressure, the other is the displacement, i.e., the distance by which the air moves back and forth. We will use the pressure. There is a pressure node at the open end of the tube where the pressure is (approximately) equal to that of the air outside the tube. There is a pressure antinode at the closed end.

In a tube either open at both ends or closed at both ends, i.e., with the same boundary conditions at the two ends, the length of the vibrating air column is a whole number of half wavelengths. In a tube closed at one end and open at the other it is $\frac{1}{4}\lambda$ plus some whole number (including zero) of half wavelengths.

EXAMPLE 8

A string is 90 cm long and is fixed at both ends. The speed of waves along the string is 135 m/s. What are the three longest possible wavelengths and the corresponding frequencies of standing waves on this string?

Ans.:
There are nodes at each end of the string. The distance between nodes is $\frac{1}{2}\lambda$. Hence the three values are given by $L = \frac{1}{2}\lambda$, $L = (2)(\frac{1}{2}\lambda)$, and $L = (3)(\frac{1}{2}\lambda)$, or $\lambda = 2L, L$, and $\frac{2}{3}L$, i.e., 1.80 m, 0.90 m, and 0.60 m.

The corresponding frequencies are 75 Hz, 150 Hz, and 225 Hz.

EXAMPLE 9

An air column, 60 cm long, is open at one end and closed at the other. What are the three longest wavelengths and the corresponding frequencies of standing waves on this air column?

Ans.:
There is a node at one end and an antinode at the other. The distance between one antinode and the nearest node is $\frac{1}{4}\lambda$. The next distance is $\frac{1}{4}\lambda + \frac{1}{2}\lambda$, or $\frac{3}{4}\lambda$. The one after that is $\frac{1}{4}\lambda + \lambda$ or $\frac{5}{4}\lambda$, so that $\lambda = 4L, \frac{4}{3}L$, and $\frac{4}{5}L$, i.e., 2.40 m, 0.80 m, and 0.48 m.

To find the corresponding frequencies we have to know the wave velocity. The velocity of sound in air at 20°C (68°F) is 344 m/s. With this value the frequencies are $\frac{344}{2.40} = 143\,\text{Hz}$, $\frac{344}{0.80} = 430\,\text{Hz}$, and $\frac{344}{0.48} = 717\,\text{Hz}$.

Resonance

Frequencies at which a system such as a string or an air column can vibrate easily are called its *natural* or *resonant* frequencies. It requires a relatively small amount of energy to produce sustained oscillations at these frequencies.

Here is a demonstration of a natural frequency: sit on a table or high chair so that your leg dangles. Let it swing freely from the knee. Now make it swing, first at half the frequency and then at twice the frequency. This illustrates that there is a natural frequency that is easy to bring about. Other frequencies are quite unnatural, and require more effort.

Another example is that of two similar musical strings near each other, with the same length and under the same tension, so that they have the same natural frequency. When one is made to sound, the other will "resonate," i.e., it will vibrate also. The first string is the source of the waves. The air between them transmits the waves. It serves as the coupling between the two. The second string receives the wave and is pushed back and forth so as to vibrate also. The amplitude of its vibration will be largest at its own natural or resonant frequency. Some instruments, such as the sitar, have a set of strings that vibrate only through this mechanism.

Scales

What makes a pleasing sound? We are now asking a question about how sounds are perceived, i.e., how our brain interprets the vibrations received by the ears. We are therefore leaving the objective description provided by physics for the subjective question of how we *feel* when we hear certain sounds. This is different for different people and depends on what we are famliar with.There are nevertheless some guidelines. It turns out that two notes together sound "special" to us when the ratio of their frequencies is small. If the ratio is two, for example, we say that the two notes are an *octave* apart. We can subdivide the octave interval, and this is done differently in different cultures. The western eight-note scale has two versions, each of seven intervals. In the *major scale* the intervals are in the ratios to the first note of $1, \frac{9}{8}, \frac{5}{4}, \frac{4}{3}, \frac{3}{2}, \frac{5}{3}, \frac{15}{8}, 2$. In the *minor scale* the ratios are $1, \frac{9}{8}, \frac{6}{5}, \frac{4}{3}, \frac{3}{2}, \frac{8}{5}, \frac{9}{5}, 2$.

These notes are sometimes called the members of the *natural* scales. A problem arises if we want to start the scale with notes of different frequencies. In the *C-major* scale we start with C. The next note, D, has a frequency $\frac{9}{8}$ as large. The other notes follow with their frequencies, D, E, F, G, A, B, and again C, now an octave higher than at the starting point. If we want to start with G, to get the *G-major* scale, we get a whole new set of frequencies. This is, in fact, what happens when people sing together, or play instruments that produce notes whose frequencies can be varied continuously, such as those in the violin family. But what do we do if we want to use instruments, such as the piano, where each key produces a sound with a fixed frequency?

The scale that is used for pianos is a compromise. We divide the octave into 12 equal intervals, so that the frequency of each note is larger than that of the previous one by $\sqrt[12]{2}$ or 1.0595. The resulting notes of what is called the *equal tempered scale* do not sound as "pure" as those of the natural scales, but with these fixed intervals we can start our scales with any note. *A capella* singers (without instruments) and string quartets are not constrained by the compromise, and can play the notes of the natural scales. In the equal tempered scale the intervals can be *half notes* with ratios of 1.0595 or *whole notes* with ratios of 1.0595^2 or 1.1225. The natural scale is approximated by using five whole-note intervals and two half-note intervals. In the C-major scale, for instance, the half-note intervals are between *E* and *F* and between *B* and *C*.

The Doppler effect

When a police car comes toward you with its sirens on and then passes you, you hear a change in the pitch (the frequency) of the siren's sound. That's the Doppler effect, first suggested for lightwaves from stars by Christian Doppler in 1842, and studied for sound waves soon after.

Let's see how it comes about. Think of a bus stopped near a store. Passengers leave it, and immediately start walking toward the store. The bus is our source, "emitting" passengers. The store is the receiver.

Suppose a passenger leaves the bus every 5 s and walks at a velocity of 1.2 m/s toward the store. One passenger arrives at the store every 5 s. The frequency of passengers leaving as well as arriving is $\frac{1}{5}$ or 0.2 passengers per second. In 5 s a passenger goes $(1.2)(5) = 6$ m, so that the passengers are 6 m apart as they walk.

Now let's get to the interesting part. Let the bus move toward the store at a speed of 0.4 m/s. The passengers leave the bus, one every 5 s, as before, and walk toward the store with the same speed as earlier. The frequency with which they leave the bus is still the same, but the frequency with which they arrive at the store is different. Each successive passenger has a smaller distance to go from the bus to the store. In the 5 s between passengers leaving the bus, the bus travels $(5)(0.4) = 2$ m. Instead of being 6 m apart as they walk they are only 4 m apart. Still moving at 1.2 m/s, they now arrive at intervals of $\frac{x}{v} = \frac{4}{1.2}$ or 3.33 s, i.e., with a frequency of $\frac{1}{3.33} = 0.3$ passengers/second.

EXAMPLE 10

Go to the Java Applet on the Doppler effect, e.g., at http://lectureonline.cl.msu.edu/~mmp/applist/doppler/d.htm.

(a) Click on the gray rectangle to make a blue dot appear. This is the source of the waves. You can stop the wave pattern by clicking on "s." It will resume when you click on "s" again. (Think of the dot as representing the siren of a police car, and the pattern representing the spread of the sound waves.) Stop the pattern and estimate the wavelength on the screen.

(b) Use the mouse to drag an arrow toward the right-hand bottom corner. Use a velocity less than $\frac{v}{v_s} = 1$. You are the observer of the waves at that corner. What has happened to the wavelength, frequency, and velocity that you observe?

Ans.:
While the source moves toward you the waves arrive with greater frequency. The speed of sound depends on the medium and does not change. The wavelength (the distance between successive lines on the wave pattern) decreases.

You can see that if the bus moves in the opposite direction, the frequency at the store (the "receiver") is smaller. The same considerations can be used to describe the frequency changes that occur when the receiver moves toward or away from the source.

The situation is analogous when a wave is emitted by a moving source. When it arrives at a stationary receiver, the frequency is larger when the source moves toward the receiver and smaller when it moves away. Motion by the receiver has a similar effect.

The applications of the Doppler effect are of great importance. The original paper of 1842 described the possibility of measuring the speed of stars with respect to the earth. However, the Doppler effect for light was not observed until 1901. Later (in 1912) it was shown that light from other galaxies was shifted to lower frequencies ("redshifted"). The observations led to *Hubble's law*, formulated by Edwin Hubble in 1929, which says that the universe is expanding at a rate that is now called *Hubble's constant*.

The reflection of microwaves ("Radar") is widely used for the location of weather patterns. Rain, snow, sleet, and hail, even insects and dust reflect the electromagnetic waves, and the Doppler effect allows the speed of the reflecting objects to be determined. Other applications are the measurement of the speed of moving cars with microwaves, using a *Radar detector*, and of the rate of blood flow.

11.4 Maxwell's great contribution: electromagnetic waves

Maxwell's equations

They are known as *Maxwell's equations* and are the fundamental relations of electromagnetism, but Maxwell didn't invent them. All four of them were there before him. Maxwell saw that one of them was incomplete, added one more term, and so opened up a whole new world. Let's see what they are.

The first is Gauss's law, equivalent, as we know, to Coulomb's law: the flux of the electric field out of any closed surface is proportional to the net charge inside it.

The second is Gauss's law for magnetic fields, easy to write down, since there seem to be no isolated poles, no *monopoles*: the magnetic flux out of any closed surface is zero.

The third is Faraday's law, $\mathcal{E} = -\frac{d\Phi}{dt}$. The induced emf around any loop is equal to the rate of change of the magnetic flux through the loop. We can also describe it differently: the emf drives the current in a circuit; it is another way to talk about the electric field that makes the charges go around. The magnetic flux depends on the magnetic field. In other words, the electric field that is created is proportional to the rate of change of the magnetic field.

The fourth is the one we have to look at in more detail. In its incomplete form it is Ampere's law. For any path around a current, I, the length of the path multiplied by the component of the magnetic field along the path is equal to $\mu_0 I$.

Now what did Maxwell see? He said suppose the wire is not continuous. There can still be a current, either for a short time, or one going back and forth (an alternating current, or "AC").

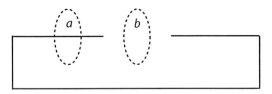

What happens as we take loops at a or b? Experimentally, there is a magnetic field at b as well as at a. If Ampere's law is to hold, the result must be the same regardless of whether the path encircles the wire or the empty space. There could also be a more strangely shaped path that is neither clearly around the wire nor around the part where there is no current. There must be something else, something like a current, also in the broken part!

He showed what it was and gave it the name *displacement current*. (The name is not very helpful, since it doesn't tell you anything about the concept.)

Let's connect the ends of the wires to large plates. (Assume that they are infinitely large.) The electric field between them, E, is equal to $\frac{\sigma}{\epsilon_0}$. If there is (temporarily) a current I, it is equal to $\frac{dQ}{dt}$ or to $\epsilon_0 A \frac{dE}{dt}$.

While there is a current in the wire, charge is moving to and from the plates. In the space

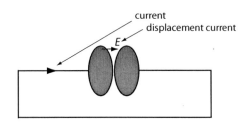

between the plates there is then a changing electric field. The quantity $\epsilon_0 A \frac{dE}{dt}$ is equal to the current in the wire leading to and from the plates. By adding this term to I there are now two terms that together remove the earlier discontinuity. In each part of the circuit there is either a current or a displacement current. The field times the path, $2\pi r B$, is now equal to $\mu_0 (I + \epsilon_0 A \frac{dE}{dt})$.

The adding of this one term to the fourth of the equations was so significant that the whole set is now known as *Maxwell's equations*.

What is it that is so important about the term with $\frac{dE}{dt}$? It is like Faraday's law, but with the role of the electric field and the magnetic field reversed. The two together lead directly to the existence of electromagnetic waves.

Let's see how this comes about. Faraday's law shows that an emf, and therefore an electric field, can be produced by a changing magnetic field. The magnetic field is produced in the whole region, not just at a point. The term in $\frac{dE}{dt}$ (the displacement current) shows that, in turn, a changing electric field produces a magnetic field, again in the surrounding region.

If the field (B or E) changes with time like a wave (i.e., sinusoidally), the slope, or derivative ($\frac{dB}{dt}$ or $\frac{dE}{dt}$), will also change with time in a similar way. Faraday's law shows that a changing magnetic field gives rise to a changing electric field. The fourth equation now shows that a changing electric field produces a changing magnetic field. The two equations together show that a change in either one of the fields gives rise to a changing field of the other kind. The changing magnetic field produces a changing electric field, leading, in turn, to a changing magnetic field, which leads to a changing electric field, and so on. The two kinds of fields chase each other through space, each giving rise to the other, propagating on their own, cast loose from their source, continuing their journey as an electromagnetic wave.

Of course this can be shown rigorously, mathematically. The velocity of the waves can be calculated from the equations, and is $\frac{1}{\sqrt{\mu_0 \epsilon_0}}$. Since $\frac{\mu_0}{2\pi} = k'$ and $\frac{1}{4\pi\epsilon_0} = k$, $\mu_0 \epsilon_0 = \frac{k'}{2k} = \frac{10^{-7}}{9 \times 10^9}$ or $\frac{1}{9 \times 10^{16}}$ SI units, and $c = \frac{1}{\sqrt{\mu_0 \epsilon_0}}$ is 3×10^8 m/s.

These are not the units or quantities that were used in Maxwell's days, but the result is the same. Maxwell knew (from experiments by others) that the speed of light was about 3×10^8 m/s. He not only showed that the equations predict the existence of electromagnetic waves, but it also became clear that lightwaves must be just this kind of wave.

Look at what a remarkable result this is: μ_0 and ϵ_0 are the constants that are needed to describe the forces between charges at rest and between currents. Now they are seen to describe also the velocity of electromagnetic waves, which evidently include lightwaves.

Does this mean that electromagnetic waves can be generated with only charges and currents? Yes, Hertz showed that, but not until much later, in 1888, more than two decades after Maxwell's work. Today every radio and TV transmittor, every microwave oven and cell phone system do just that.

The electromagnetic spectrum

Electromagnetic waves can have wavelengths larger than hundreds of meters and smaller than the diameter of a proton. They are united in their speed through empty space, $c \ (= f\lambda)$, equal to 3×10^8 m/s. They can be classified by their frequency, or equivalently by their wavelength, and all form part of the *electromagnetic spectrum*. We give names to the different regions of the spectrum, but the way we do it is not very consistent. Some, like visible light, are characterized by their receiver, and some, like gamma rays, by their emitter. The regions overlap, and their boundaries are not precise.

Light is what we see. The wavelengths to which the eye is sensitive range roughly from 400 nm at the small-wavelength, high-frequency (violet) limit to 700 nm at the large-wavelength, low-frequency (red) end of the visible spectrum.

All matter consists of charges, and all matter is in motion. In gases the atoms move and collide with other atoms. In solids they oscillate about their equilibrium positions. This is their *thermal* motion. The accelerating charges lead to changing electric fields, hence to changing magnetic fields, and so to electromagnetic waves. Every object radiates electromagnetic waves with a range of frequencies that depends on its temperature. The sun and the other stars are so hot that they emit all parts of the visible spectrum and appear white. The part of the spectrum emitted by hot bodies with frequencies larger than those that are visible is called ultraviolet, and the part with lower frequencies is called infrared. A hot stove may emit radiation with frequencies into the visible part of the spectrum and be "red hot," but all objects, regardless of their temperature, emit infrared radiation. This is so for everything around us, including our own bodies.

Gamma rays are electromagnetic waves emitted by nuclei. The name "x-rays" is given to the radiation emitted by atoms. The frequencies of x-rays are generally smaller than those of gamma rays, but there is some overlap. (We will see later that the frequency of the radiation is related to its energy. The greater frequency of gamma rays is related to the much greater spacing of the energy levels of nuclei compared to those of atoms. This is because nuclei are held together by the strong nuclear force, while the energy that holds the electrons to the nucleus in atoms is the much weaker electric force.)

The term *x-rays* is also used for the electromagnetic waves that are emitted by electrons that are accelerated when they are not bound in atoms. The reason is that both kinds are emitted by an *x-ray tube*, where a beam of electrons is stopped when it hits a target made of tungsten or another heavy metal. (There is more about this in the next chapter.) A particularly strong x-ray source is the *synchrotron*, a device in which electrons are accelerated by a magnetic field so that they move in a circle.

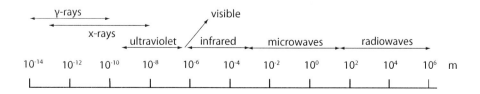

Electrons moving back and forth in a circuit or in an antenna give rise to *radio waves*. They usually have wavelengths from hundreds of meters down to a few meters. The alternating current that we normally use has a frequency of 60 Hz, and generates electromagnetic waves with a correspondingly long wavelength. Microwaves are generated by electrons oscillating in vacuum tubes and have smaller wavelengths, of the order of centimeters.

The capacitor and the energy of the electric field

We introduced the capacitor in the example at the end of Chapter 8. It consists of two plates, one carrying a charge Q and the other a charge $-Q$, with a potential difference V between them. If we use the approximation that the area of each plate, A, is large, and the distance, d, between them is small, then the electric field, E, in the volume between the plates is uniform, and $V = Ed$. The surface charge density, σ, is equal to $\frac{Q}{A}$. The *capacitance* is defined as $\frac{Q}{V}$. It is the amount of charge that can be stored on one of the plates for each volt of potential difference between them. It is equal to $\frac{\sigma A}{Ed}$, and since the field between two large (assumed infinite) plates is $\frac{\sigma}{\epsilon_0}$, it is equal to $\frac{\epsilon_0 A}{d}$. We see that the capacitance depends only on the geometric configuration, i.e., on the size of the plates and the distance between them, and not on the amount of charge or the potential difference.

Consider what happens as we charge a capacitor by connecting it to a battery. Charge transfers from one plate to the other. At first there is no potential difference, and it takes little work to transfer the first bit of charge, ΔQ. As more charge accumulates it takes more and more work to transfer an amount ΔQ, as the work, $V \Delta Q$, becomes larger. To transfer the amount Q takes an amount equal to the area under the curve of V against Q. This is equal to $\frac{1}{2}QV$.

This is the energy stored in the capacitor. It can also be written as $\frac{1}{2}CV^2$. If we substitute the expression for the capacitance, $C = \frac{\epsilon_0 A}{d}$, and $V = Ed$, we see that the energy, U, is $\frac{1}{2}\epsilon_0 A d E^2$, and the energy per unit volume, $\frac{U}{Ad}$, is $\frac{1}{2}\epsilon_0 E^2$. This expression for the energy per unit volume, or the *energy density*, holds also for other electric fields.

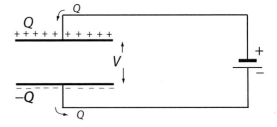

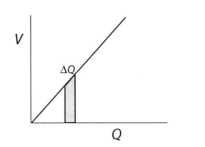

It is possible to show also that the energy density of the magnetic field in a long solenoid is $\frac{B^2}{2\mu_0}$. This is also the energy density in any other magnetic field.

The propagating fields

Look again at a loop of wire in which a current is induced by a changing magnetic field. The induced current is there as a result of the induced emf. There is an induced electric field along the wire of the loop in accord with Faraday's law. The electric field is in the plane of the loop, while the magnetic field is at right angles to the plane of the loop: **E** and **B** are perpendicular to each other.

In an electromagnetic wave **E** and **B** are at right angles to each other and perpendicular to the direction in which the wave travels. (The vector **E** × **B** is at right angles to both fields and is in the direction of propagation.)

As the wave travels, both **E** and **B** oscillate, in phase, both in space and in time. In other words, as the wave passes a particular point, the magnitude E at the point varies from its maximum value E_{max} to 0 to $-E_{max}$ and back with the frequency of the wave, as $E = E_{max} \sin \omega t$. At the same time the magnitude B of the magnetic field varies from B_{max} to 0 to $-B_{max}$ and back, in phase with E. Similarly, a snapshot of the fields at a particular moment shows the sinusoidal variation as a function of z, the direction of propagation, i.e., both $E_{max} \sin kz$ and $B_{max} \sin kz$, along the direction of propagation of the wave.

In an electromagnetic wave both kinds of fields travel together and both carry energy. This is one of the most important features of the waves: the fields have energy, and the electromagnetic waves transport that energy.

Every electromagnetic wave, whether it is a radio wave, microwave, lightwave, x-ray, or gamma ray, consists of electric and magnetic fields moving through space together. Each wave carries energy. This is how we get the energy from the sun that makes our life on earth possible. This is how we get the signals to our radios, TV sets, and cell phones, the radiation that warms our skin, and that which we see with our eyes.

How do we know the waves are there? We don't see the fields or feel them directly. We know about them only when they exert forces on electric charges. We have to return to the beginning: in an electric field there is a force on a charge. In an electromagnetic wave there is an oscillating electric field. A charge in the path of such a wave is set in motion, oscillating under the action of the force provided by the wave's electric field. This is what happens when an electromagnetic wave is detected, whether it is in an antenna, on our skin, or in the eye.

When we discussed what we mean by the word "real" (in Chapter 5) we shied away from a definition of reality. Regardless of the definition, there are few who would doubt that electric fields, magnetic fields, and their joint dance through space are real.

11.5 Observing interference of light

Young's double slit experiment

That light waves are electromagnetic waves became clear from Maxwell's work in about 1862. But the wave properties of light had been established by 1801, principally when Thomas Young showed that light could exhibit the properties of interference.

Wave properties become apparent primarily when the waves encounter an opening or an obstacle that is not too much larger than their wavelength. For visible light this is between about 400 nm for violet light and 700 nm at the red end of the spectrum. These distances are so small that we are normally not aware of the wave properties of light, and it takes special efforts to observe them. Young's double slit experiment provides the most famous and direct demonstration of the interference of light, and hence of the wave nature of light. Here is the experiment.

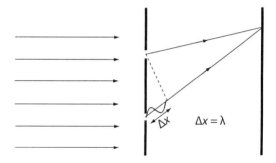

Two narrow openings, or slits, are illuminated by a source of light of a single color. Each then acts as a new source from which light spreads out. When the waves from the two slits arrive at a point on a surface or screen, they have traveled different distances. One of them has gone further than the other by an amount Δx. If Δx is just one wavelength long, the crests of the two waves will arrive at the same time.

There is then constructive interference, and there will be a bright line on the screen with the addition of the light from the two slits. The same is true if Δx is equal to 2, 3, or some other whole number of wavelengths. But if, for some place on the screen, Δx is $\frac{1}{2}\lambda$, the crest of one wave arrives there at the same time as the trough of the other, and there is destructive interference. The two waves cancel, and there is no light on the screen.

The result is that there are alternating bright and dark lines on the screen. They are the result of the interference of the lightwaves from the two slits. This pattern could not be there if light did not have the properties of waves. If you imagine shooting "bullets" of light rather than waves, there would be no interference, and no such pattern of bright and dark lines.

The spacing between the lines gives us a measure of the wavelength of light. The experiment shows that it ranges from about 400 nm to about 700 nm, and that each value of the wavelength in that range corresponds to a particular color. For the interference pattern to be clearly seen, the source of light needs to be of a single wavelength and color, i.e., to be *monochromatic*.

EXAMPLE 11

Go to the PhET website and open the simulation *Wave Interference*. Select "Light." Click on "show screen" and observe the waves.

Select "two slits" and move the barrier close to the source. (You can drag it with the mouse or move the pointer in the set of tools on the right. You may also want to change the slit width.) Again observe what you see. Click on "intensity graph" and look at the result. How does the pattern change when you change the wavelength?

Ans.:

The pattern on the screen shows the alternating bands of light and dark that result from the interference of the light from the two slits. The intensity graph shows the variation of the intensity along the screen. The distance between successive minima becomes smaller when the wavelength decreases.

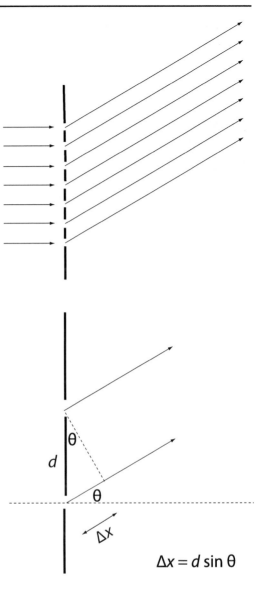

$\Delta x = d \sin \theta$

The diffraction grating

There are other experiments that show the interference of light, and we will explore some of them. The first is the diffraction grating. The idea is the same as in Young's double slit, but there are many slits.

Again each slit acts as a source of waves. As in the double slit experiment, the interference pattern is determined by the path difference between the waves from the different slits. Look at the direction at the angle θ to the original direction of the light where the path difference between the light from the first slit and that from the next one is one wavelength. Then for the next one it is 2λ, then $3\lambda, 4\lambda$, and so on. The light from all of the slits will interfere constructively. If the distance between adjacent slits is d, we see that the angle θ is given by $\sin \theta = \frac{\lambda}{d}$. There is also constructive interference from all the slits when $\sin \theta = \frac{2\lambda}{d}$, and, in general, for $\sin \theta = \frac{n\lambda}{d}$, where n is an integer. (For all other angles there is destructive interference. Since there are many slits, there is always another one from which the light is out of phase by half a wavelength.)

For light with a single wavelength there is constructive interference at the values of θ for which $\sin \theta = \frac{n\lambda}{d}$. If the beam consists of light with different wavelengths the angle for constructive interference is different for each. The grating then separates the light, and allows us to see the range of wavelengths that are present. The distribution of colors is called a *spectrum*. More precisely, a spectrum is a graph of the intensity of radiation as a function of wavelength (or another quantity related to the wavelength).

Single slit diffraction

A diffraction grating could also be called an "interference grating." The word "diffraction" is commonly used instead for some interference phenomena, such as the pattern that occurs when light passes through a single slit. In that case alternating dark and bright bands occur because of the interference of light from different parts of the slit.

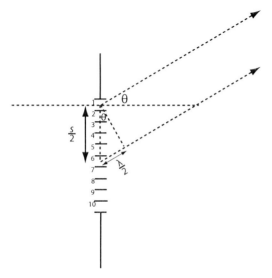

To see this we divide the slit into parts. The figure shows a slit of width s divided into 10 strips. We're going to compare the light from the five slits of the upper half to the light from the five strips of the lower half. Let's look at the first strip in each of the halves, i.e., the first and the sixth of these strips, at an angle such that there is destructive interference of the light from the two. The two strips are separated by a distance $\frac{s}{2}$. For destructive interference the path difference Δx is $\frac{1}{2}\lambda$, so that $\sin\theta = \frac{\frac{1}{2}\lambda}{\frac{1}{2}s}$, or $\sin\theta = \frac{\lambda}{s}$.

At this angle there is also destructive interference from the second and seventh strips, as well as for the third and eighth, the fourth and ninth, and the fifth and tenth. We see that at this angle there is destructive interference for every part of one-half of the slit and a corresponding part of the other half. At this angle, therefore, there is a dark band on the screen.

A similar analysis can be made for other angles and other path differences. The result is that there are again bands of bright and dark fringes, as in the double-slit experiment, but their origin and the angles at which they occur are different.

We can extend the same considerations to another situation. We have looked at light stopped by a barrier, except for the single slit in it. Suppose we reverse the arrangement, so that light continues everywhere *except* for the space that was occupied by the slit. We might guess that the same kind of pattern will occur, and it does! Other obstacles in the path of a light beam can also give rise to *diffraction patterns*.

The lines or bands that we see in Young's double slit experiment, with a diffraction grating, and in single-slit diffraction, have the shapes of the strips. A similar effect occurs when sunlight passes through the spaces between the leaves of a tree. The pattern of light and shadows is to a large extent the result of diffraction, and shows the image of the sun. We are not usually aware of that, and it is startling to see the effect at the time of a solar eclipse, when the source of the light, the sun, is no longer round. When the sun is half obscured, for example, the light between the shadows, still consisting of images of the sun, is seen to be made up of half circles.

Thin films

A common interference effect occurs when light is reflected from both the bottom and the top of a thin film.

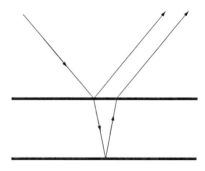

For a beam incident at right angles to the film, the light reflected from the bottom goes farther by a distance $2d$, where d is the film thickness. The nature of the interference depends on the relation between this path difference and the wavelength of the light. Here we have to know that the wavelength in the film is different from that in air: it is smaller by the factor n, the index of refraction, i.e., $\lambda_{film} = \frac{\lambda_{air}}{n}$. (On the figure the light rays are shown at angles different from 90°, so that the rays reflected from the top and from the bottom are separated.)

Since light of different colors has different wavelengths, the conditions for constructive and destructive interference will be different for different colors. The colors seen in soap films, and in oil films on water, are interference patterns. They come about because of the differences in wavelength, because the path difference is different at different angles, and as a result of thickness variations in the films.

The Michelson interferometer

A fascinating application of interference that had a profound influence on the history of physics is the Michelson interferometer. It allows measurements of distance with great precision, because they are made in terms of units equal to the wavelength of light. It can also be used to compare the velocities of light in two directions at right angles to each other. This was done in 1887 in the *Michelson–Morley experiment*. The experiment showed that there was a fundamental difference between the behavior of lightwaves and mechanical waves, indicating that Newtonian mechanics could not adequately describe electromagnetic phenomena. It paved the way for the development, 18 years later, of the special theory of relativity.

A half-silvered mirror lets some of the light from a monochromatic source through to mirror #1. Some of it is reflected to mirror #2. The light returns from mirror #1, and some of it is reflected toward the eye of the observer. Some of the light that is reflected from mirror #2 goes through the half-silvered mirror to the eye. The two rays combine and give rise to interference, the nature of which depends, as before, on the path difference. If the path difference is equal to a whole number of wavelengths, there is constructive interference. If it is a whole number of wavelengths plus a half wavelength, there is destructive interference.

The interferometer can be used to measure distances in terms of the wavelength of light. If one of the mirrors, #1 or #2, is moved through $\frac{1}{4}\lambda$, the path difference changes by $\frac{1}{2}\lambda$. Constructive interference changes to destructive interference, and what the eye sees as it looks along the line of the diagram changes from bright to dark.

At an angle to the lines of the light path in the diagram the distances are larger. What is observed by the eye is an interference pattern consisting of a series of concentric rings, alternately bright and dark. As one of the mirrors is moved, the rings move inward or outward. A bright ring is replaced by a dark ring when one of the mirrors is moved through a distance of $\frac{1}{4}\lambda$.

Coherence

In all our examples of ways to exhibit the interference of light there was a single source, with a split into different paths from the source to a place where they recombined. You might ask why we cannot observe interference with beams from two separate light bulbs, instead of going through the more or less elaborate splitting procedure. The answer lies in the nature of the radiation emitted from a hot object, such as the filament in a light bulb. The atoms of the filament emit flashes of light, first one, then another, and another, randomly and independently. The flashes come from different places and at different times.

This is very different from the beams that we described earlier, which come from the same source, split into two or more parts along different paths, and then recombine. The two different parts in Young's double-slit experiment, for example, start out together (in phase) and

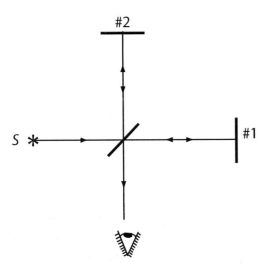

then are out of phase because one goes further than the other by the path difference Δx.

The flashes from the hot filaments arrive at different times, and so the light from two separate light bulbs cannot lead to interference. Is it ever possible to have interference effects from two separate light sources? They have to have the same wavelength and there has to be a definite phase relationship between the two sources. This is possible with *lasers*. A laser emits a beam of light with a single wavelength, rather than the random flashes with a range of wavelengths emitted by a hot object. The light from two similar lasers can therefore lead to interference patterns.

The light from lasers is referred to as *coherent*, while the light from a hot body, such as a light bulb filament, is *incoherent*.

11.6 Reflection and refraction

The laws of reflection and refraction

Normally, we are used to thinking of light as traveling in straight lines. In the absence of the openings and obstacles that lead to interference this is, in fact, what happens. The straight-line rays can still change direction. One way is by *reflection*, as from a mirror. Another is by *refraction*, the change of direction that occurs when a ray goes from one medium to another.

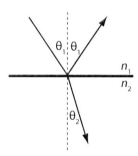

The figure shows an incident ray, a reflected ray, and a refracted ray. The incident ray makes an angle θ_1 with the normal to the surface (a line perpendicular to the surface), which is called the angle of incidence. The angle of the reflected ray with the normal, also θ_1, is called the angle of reflection. The fact that the two angles are equal is called *the law of reflection*.

The angle of a ray changes when it goes from one medium to another in which the speed of light is different. In empty space, i.e., in a vacuum, the speed of light is $c = 3 \times 10^8$ m/s. In any other medium it is $\frac{c}{n}$, where n is called the *index of refraction*.

If the angle with the normal in one medium, with index of refraction n_1, is θ_1, and in the second medium the values are n_2 and θ_2, then $n_1 \sin \theta_1 = n_2 \sin \theta_2$. This is the *law of refraction*, or *Snell's law*.

These are all of the fundamental principles of the subject called *geometric optics*. There are, however, many important and interesting applications. We can use the law of reflection to study different kinds of mirrors. The law of refraction leads to an understanding of lenses and optical instruments, such as the microscope and the telescope.

Mirrors

How do we see ourselves in a plane mirror?

The figure shows a mirror, and a person, Lucy, in front of it. It also shows two rays, coming from her nose, and reflected by the mirror.

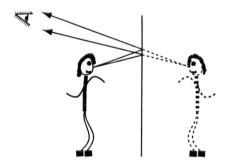

The two rays are looked at by another person. They come from Lucy's nose, but they *seem* to come from a point behind the mirror. That point is called the *image* of the point on her nose where the rays originate. The rays only seem to come from there. There are no actual rays from her at the image. This kind of image is called a *virtual* image.

With a parabolic mirror rays coming to the mirror parallel to the axis are reflected so that they cross at a point on the axis called the *focus*. The fact that they do that is a geometric property of a parabola. It is much easier to make spherical mirrors than parabolic mirrors. Rays parallel to the axis of a spherical mirror do not come to a focus *exactly*. But for rays near the axis they do so approximately. Let's see how this comes about.

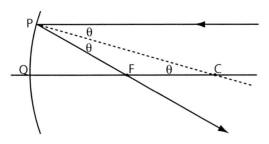

Look at a ray that comes to the mirror parallel to the axis. It comes to the mirror at a point P, and is reflected toward the axis, which it crosses at the point F, the *focus*. The line from P to the center of curvature of the mirror (C) is the radius of the spherical surface. It is perpendicular to the mirror, i.e., it is the normal to the surface, so that both the incoming and the reflected rays make the same angle, θ, with it. The angle of the line PC with the axis is also θ. The triangle PCF has two angles equal to θ so that it is isosceles and its sides PF and FC have the same length.

The axis crosses the mirror at the point Q. The lengths PC and QC are the same, since they both go from the mirror to its center of curvature. The lengths PF and QF are not the same, but we will make the approximation that they are. The smaller the diameter of the disc of the mirror is compared to its radius of curvature, the more closely this will be true. In this approximation it doesn't matter how far P is from the axis—all the rays coming to the mirror parallel to the axis cross the axis at the point F. Moreover the *focal length* $QF = f$ is seen to be equal to half the radius of curvature. (In fact, QF changes as θ changes, so that the point where the reflected ray crosses the axis also changes.)

A spherical mirror can produce an image that is different from the virtual image of a plane mirror. In the diagram the rays originate at the *object*, and after reflection they converge at the *image*. This time the rays actually go to the image. It is a *real* image. It can be caught on a piece of paper or on a screen. This is not so for a virtual image, a point on the other side of the mirror from which the rays seem to come.

We can see where the image is by following some representative rays. A ray coming to the mirror parallel to the axis (1) is reflected so as to go through the focus. Conversely, a ray through the focus (2) is reflected so that it is then parallel to the axis. A ray along the line through the center, C, of the spherical surface (3) is reflected so that it returns along the same line.

The figure shows an object at a distance further from the mirror than C, and a real image, smaller and inverted, between the focus and the center.

Since the rays follow the same path in either direction, we can interchange the image (I) and the object (O), and see that for an object between the focus and the center there is a real, inverted, and magnified image.

There is a simple relation between the object distance d_O, the image distance d_I, and the focal length f. The figure shows that the height of the object (h_O) and the height of the image (h_I) are related in two ways. First, we see that because the two angles marked φ (Greek *phi*) are the same, $\frac{h_O}{d_O - f} = \frac{h_I}{f}$, or $\frac{h_O}{h_I} = \frac{d_O - f}{f}$.

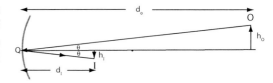

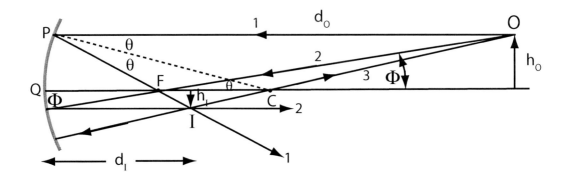

11.6 Reflection and refraction

A ray from the object to Q goes to the image, forming two similar triangles that show that $\frac{h_O}{d_O} = \frac{h_I}{d_I}$ or $\frac{h_O}{h_I} = \frac{d_O}{d_I}$.

Equating the two expressions that are each equal to $\frac{h_O}{h_I}$, we get $\frac{d_O - f}{f} = \frac{d_O}{d_I}$, which can be written as $\frac{d_O}{f} - 1 = \frac{d_O}{d_I}$. Dividing each term by d_O we get $\frac{1}{f} - \frac{1}{d_O} = \frac{1}{d_I}$ or $\frac{1}{d_I} + \frac{1}{d_O} = \frac{1}{f}$.

This relation also holds for mirrors that curve outward, or *convex* mirrors, if f is taken to be negative, and for virtual images if d_I is then taken to be negative. (Mirrors that curve inward are called *concave*.)

We also see that the *magnification* $\frac{h_I}{h_O}$ is equal to $\frac{d_I}{d_O}$. We will take h_O to be positive above the axis. To avoid an extra negative sign we will take h_I to be positive when it is below the axis, as it is when the image is real and inverted.

What happens when the object is between the mirror and the focus? The figure shows the representative rays. A virtual and magnified image is produced, right-side up. This is what happens in a magnifying mirror like those used in bathrooms for shaving and makeup.

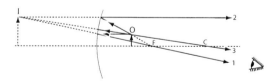

EXAMPLE 12

A concave mirror has a radius of curvature of 30 cm.
(a) An object is 90 cm from the mirror. What is the position of the image? What kind of image is it? What is the magnification?
(b) Repeat part (a) for $d_O = 20$ cm.
(c) Repeat part (a) for $d_O = 10$ cm.

Ans.:
(a) $f = \frac{1}{2}r = 15$ cm.
$\frac{1}{d_I} = \frac{1}{f} - \frac{1}{d_O} = \frac{1}{15} - \frac{1}{90}$ so that $d_I = 18$ cm.
[We can use $\frac{1}{f} - \frac{1}{d_O} = \frac{d_O - f}{fd_O}$ so that $d_I = \frac{fd_O}{d_O - f} = \frac{(15)(90)}{90 - 15} = 18$ cm.]
$\frac{h_I}{h_O} = \frac{d_I}{d_O} = \frac{18}{90} = 0.20$.

The image is real and inverted, 18 cm in front of the mirror. The magnification is 0.20, i.e., the size of the image is 0.20 times that of the object.

(b) $\frac{1}{f} - \frac{1}{d_O} = \frac{1}{15} - \frac{1}{20} = \frac{4-3}{60} = \frac{1}{60}$. $d_i = 60$ cm.
Magnification = $\frac{d_I}{d_O} = 3$.

The image is real and inverted, 60 cm in front of the mirror. The magnification is 3, so that the image is three times as large as the object.

(c) $\frac{1}{d_I} = \frac{1}{15} - \frac{1}{10} = \frac{2-3}{30} = \frac{-1}{30}$. $d_i = -30$ cm.

The image is virtual, right-side up, 30 cm behind the mirror. The magnification is -3. The image is again three times as large as the object, but not inverted.

Lenses

Lenses are among the most familiar optical devices. They are used as pocket magnifiers, camera lenses, and eyeglasses. In combination they are the constituents of telescopes and microscopes. Lenses usually have spherical surfaces. Each ray's path can be followed and analyzed by using Snell's law. Computers now allow this to be done much more efficiently than earlier, and this has led to the development of complex lenses, consisting of many parts, even in inexpensive cameras.

It is also possible to use an approximate analysis similar to that which we have used for mirrors. Just as in that case, it is most accurate for rays close to the lens axis. Lenses, like mirrors, have focus points. There are two of them, one on each side of the lens. We will start with lenses that are thicker in the middle, called *convex* or converging lenses. (Lenses that are thinner in the middle are called *concave* or diverging lenses.)

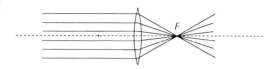

The figure illustrates that rays arriving parallel to the axis are refracted so as to cross at the focus. This leads us to two representative rays: one that comes to the lens parallel to the axis (1), and is refracted so as to go through the focus, and a second that comes through the focus (2), and is refracted so as to leave the lens parallel to the axis. A third representative ray goes through the center of the lens (3). For a sufficiently thin lens, and for rays close to the axis, this ray goes through the lens, to a good approximation, without changing direction.

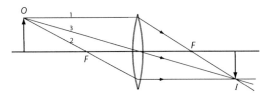

The figure shows the three representative rays coming from an object and converging to an image. In this case the image is real, i.e., the rays actually go to the image. It is inverted and reduced in size. Since the rays can follow the same path in both directions, the image and object can be interchanged. If this is done the image is still real and inverted, but it is magnified.

The situation is quite different for an object between the lens and one of the focal points. In that case the rays, after being refracted by the lens, diverge. There is no real image. After they go through the lens, however, the rays seem to come from a point on the same side of the lens as the object. This is the virtual image. It is right-side up and magnified.

Concave lenses, by themselves, produce only virtual images, right-side up and reduced in size, regardless of the position of the object.

EXAMPLE 13

Go to the PhET website and open the simulation *Geometric Optics*. Select "principal rays," "curvature 0.8," "refractive index 1.87," and "virtual image."

Move the object back and forth and observe what happens to the image and the magnification. What happens as the object moves into the region between the focus and the lens?

Ans.:

When the object is outside the focus the image becomes larger as the object moves closer to the focus. When the object moves past the focus the image becomes virtual and magnified. It is then on the same side of the lens as the object.

The thin-lens relation

Let's look once more at the representative rays that show the formation of a real image by a thin convex lens. Call the heights of the object and the image h_O and h_I. Call the distances from the object and the image to the lens d_O and d_I. Look at the two similar triangles whose sides are h_O and d_O for the one on the left and h_I and d_I for the one on the right. They both contain the angle marked θ, so that $\frac{h_O}{d_O} = \frac{h_I}{d_I}$ or $\frac{h_I}{h_O} = \frac{d_I}{d_O}$.

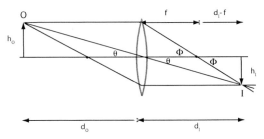

Let f be the *focal length*, i.e., the distance from either focus to the lens. A second pair of similar triangles, both with the angles marked φ, has sides h_O and f between the lens and the focus on the right and h_I and $d_I - f$ between the same focus and the image. They show that $\frac{h_O}{f} = \frac{h_I}{d_I - f}$ or $\frac{h_I}{h_O} = \frac{d_I - f}{f}$.

Combining the two relations we get $\frac{d_I - f}{f} = \frac{d_I}{d_O}$ or $\frac{d_I}{f} - 1 = \frac{d_I}{d_O}$. Divide by d_I to get $\frac{1}{f} - \frac{1}{d_I} = \frac{1}{d_O}$, or $\frac{1}{d_O} + \frac{1}{d_I} = \frac{1}{f}$.

This *thin-lens relation* describes not only the situation that we started with, for a convex lens with a real image. It can also be used for a concave lens if the focal length is taken to be negative. For a virtual image d_I needs to be taken to be negative.

Just as for the spherical mirror, the magnification $\frac{h_I}{h_O}$ is equal to $\frac{d_I}{d_O}$.

EXAMPLE 14

Thin lens calculation

A thin convex lens has a focal length of 15 cm.

(a) An object is 90 cm to the left of the lens. What are the position and nature of the image? What is the magnification?

(b) Repeat part (a) for $d_O = 20$ cm.

(c) Repeat part (a) for $d_O = 10$ cm.

Ans.:

The calculations are the same as for the previous example with the spherical mirror. Since the focal length is the same in the two cases, and the object distances are the same, the answers are the same. The only differences are that the real images (parts a and b) are on the right-hand side of the lens and the virtual image (part c) is on the left-hand side of the lens.

Total internal reflection

Look at what happens when a ray in one medium comes to a boundary with a medium whose index of refraction is smaller. The angle of reflection is then larger than the angle of incidence. If the angle of incidence, θ_1, is now gradually increased, the angle of refraction, θ_2, becomes larger, and eventually reaches 90°. The angle of incidence at that point is called the critical angle θ_{cr}. What happens if the angle of incidence is increased still further? There is then no refracted ray at all! The ray is reflected back into the first medium. This is called *total internal reflection*.

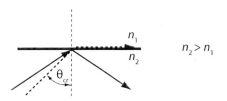

EXAMPLE 15

What is the critical angle at the interface of glass with an index of refraction of 1.5 and air?

Ans.:
Snell's law is $n_1 \sin\theta_1 = n_2 \sin\theta_2$. Use the subscripts 2 for air and 1 for glass. In air n_2 is close to 1. The critical angle is $\theta_{cr} = \theta_1$ when $\theta_2 = 90°$ so that $1.5 \sin\theta_{cr} = 1$. Hence $\sin\theta_{cr} = \frac{1}{1.5} = 0.67$ and $\theta_{cr} = 42°$.

Light entering a glass or plastic cylinder at one end is totally reflected each time it hits the wall of the cylinder at an angle greater than the critical angle, until it emerges at the other end. This is the principle of *fiber optics*. The attenuation of the light as it travels along a thin fiber can be very small, and light signals can be carried by optical fibers with much less loss than for electric signals along copper wires.

Bundles of thin fibers, a few μm in diameter, can transmit images from one end to the other, even when they are bent. This has led to a number of important medical applications. One is the exploration of the inside of the human body, as for the stomach with an *endoscope* and the colon with a *colonoscope*. The images provided by fiber optics can be combined with miniaturized instrumentation for *arthroscopic surgery*, which requires only very small incisions.

EXAMPLE 16

The figure shows a beam of light entering a 45° prism made of glass ($n = 1.5$). What is the subsequent path of the beam?

Ans.:
Since the angle of incidence is 45°, and therefore larger than the critical angle of 42°, the beam is totally reflected. It is reflected a second time at the lower surface and emerges in the direction opposite to its original direction.

Resolution

When light passes through a hole, or *aperture*, or through a lens, what we see is a diffraction pattern. When the aperture is large compared to the wavelength, we are not usually aware of that. The diffraction effects become larger and more apparent when the aperture is small or when we are trying to resolve very small or distant objects. They provide a limit to the validity of geometric optics. When they become observable it is no longer appropriate to consider light to be propagated as rays that travel through the aperture in straight lines.

Camera and eye

In a camera a lens produces a real image, which is then recorded by a film or other "sensor."

Even in simple cameras the lens is made of several components, designed to reduce the deficiencies of single lenses. "Chromatic aberration," for instance, refers to the fact that for each color (wavelength) there is a different index of refraction and focal length, so that the image of a blue object is formed at a different place than the image of a red object. This effect can be partially compensated for with a lens made of two or more components.

Lenses generally have spherical surfaces because other shapes are much more difficult to make. Rays that come from the axis to such a lens at a point far from the axis do not come to an image at the same distance as rays close to the axis. This is called "spherical aberration." In addition distortions can occur from objects that extend far from the axis.

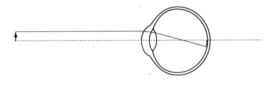

The same considerations apply to the eye. The lens of the eye forms an image on the *retina*. The retina contains the sensors that send signals to the brain via the *optic nerve*. In contrast to the lens of a camera the eye's lens can change its curvature and hence its focal length under the action of the muscles that surround it. In other words, it can *adapt* so as to create images of objects that are at different distances from it.

In cases where the eye's lens cannot adapt correctly, it is usually possible to to use external lenses ("eyeglasses") to correct for the deficiencies of the lens of the eye. In a *nearsighted* eye the rays converge more than is required to form an image on the retina so that the image is in front of the retina. A diverging lens in front of the eye can correct for that. In a *farsighted* eye the rays from an object do not converge sufficiently, and are refracted so that they head toward an image behind the retina. A converging lens is used to correct this problem. The strength of eyeglasses is usually measured in *diopters*, equal to $\frac{100}{f}$, where f is the focal length in cm.

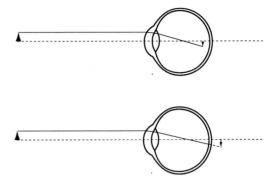

The magnifying glass

When a lens is used as a magnifying glass, the object is between the lens and one of the focal points. The virtual image is on the same side of the lens as the object, right-side up and magnified. When the object is moved closer to the focus, the image becomes larger.

You can make the image huge by letting the object get really close to the focus. The magnification ($\frac{h_I}{h_o}$) can be as large as you like. So why does that not give you all the magnification that you would ever want? The downside is that as the image becomes larger it moves further away. Since you have to stay on the other side of the lens from it, the fact that it gets so much larger doesn't do you any good. The image doesn't *appear* to become larger.

We need a new criterion to describe what happens. The magnification $\frac{h_I}{h_o}$ isn't very useful here. What matters is the *angle* that the image subtends at the eye. That's $\frac{h_I}{\ell_I}$, where ℓ_I is the distance from the image to the eye. You want this angle to be as large as you can make it. What the lens does for you is to make this angle larger than the angle that the object subtends at your eye without the lens.

How large is that angle without the lens? Can't you make it larger just by moving your eye close to the object? Yes you can, but there's a limit. At some distance the eye can't focus any more. Your eye has a *near point*, and when you move an object closer than that it looks blurred.

That gives us our criterion: we need to use the *angular magnification*. The best we can do without the lens is to put the object at the near point. If this point is a distance s from the eye, the angle subtended by the object at the eye is $\frac{h_o}{s}$. The lens makes it possible for both the object

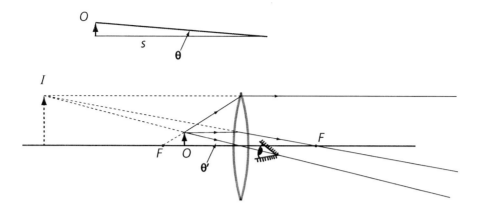

and the image to subtend the angle $\theta' = \frac{h_o}{d_o}$. The angular magnification is then $\frac{\theta'}{\theta} = \frac{h_o}{d_o}\frac{s}{h_o} = \frac{s}{d_o}$.

We can express this relation in terms of d_I by using $\frac{1}{d_o} + \frac{1}{d_I} = \frac{1}{f}$, or $\frac{1}{d_o} = \frac{1}{f} - \frac{1}{d_I}$ so that $\frac{\theta'}{\theta} = \frac{s}{d_o} = s(\frac{1}{f} - \frac{1}{d_I})$. When the object is at the focus ($d_o = f$), $\frac{1}{d_I} = 0$, and the angular magnification is $\frac{s}{f}$.

It can be made a little larger by moving the image to the near point, so that $d_I = -s$. (The distance to a virtual image is negative!). In this case the angular magnification is $\frac{\theta'}{\theta} = s(\frac{1}{f} - \frac{1}{-s}) = \frac{s}{f} + 1$.

Microscope and telescope

Some of the most interesting applications of lenses are those that use them in combinations where the image formed by one lens is the object for a second lens. Two examples are the microscope and the telescope. In a microscope one lens is near the object and is called the *objective*. It forms a real enlarged image. This image is the object for a second lens, the *eyepiece*, which forms a virtual enlarged image of it.

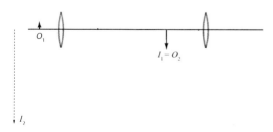

A telescope can also be constructed of an objective that produces a real image, which is then observed with a magnifying glass. There are, however, some different requirements. The object is far away and you want the first image to be as large as possible. This is best done with a lens that has a long focal length. The image is inverted, which may not be a problem for astronomical observations but needs to be corrected for use on earth. A third lens can be put between the objective and the eyepiece to change the orientation of the image. The resulting *spyglass* has the disadvantage that it is very long. The modern equivalent is the *prism binocular*. Two right-angled prisms are inserted between the objective and the eyepiece. This shortens the distance between the two lenses and also inverts the image.

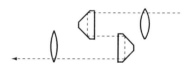

11.7 Where Einstein started: electromagnetism and relativity

The ether and the speed of light

Sound waves are propagated by the vibration of atoms or molecules in air or other materials, water waves by the motion of the water. Both are examples of mechanical waves in a medium that is disturbed as the wave passes. Sound cannot travel through empty space. Electromagnetic waves are different. The generation of electric fields from changing magnetic fields, described by Faraday's law, and the generation of magnetic fields by changing electric fields do not require a material, a *medium*. This is confirmed by the fact that electromagnetic radiation reaches us through empty space from the sun and other stars, while the sound of the violent explosions that take place there cannot reach us. This seems perfectly natural to us.

It didn't seem so natural in the nineteenth century, when every other known kind of wave required a medium whose constituents could move back and forth. So a new one was imagined, the *ether*, as a medium for electromagnetic waves. It was not a very comfortable idea right from the beginning. It would have to be there even where nothing else seemed to be, and it would have to be able to vibrate with the almost unimaginably high frequencies required by the high speed of the waves. For example, at one end of the visible spectrum the frequency is about 10^{15} Hz, and the ether would have to be able to vibrate that many times per second.

There was a more subtle and more interesting difficulty. Was the ether attached to the earth or at rest with respect to the sun or some other part of the universe? Just as the speed of sound waves is measured with respect to the medium in which they propagate, and is different in systems moving with respect to it, so it was expected for the electromagnetic waves as they propagated in the ether. Experiments of great sophistication and precision were made to detect motion with respect to the ether, but no evidence for such a motion was ever found.

The most famous and decisive of these experiments was done by A. A. Michelson, who had already made the most precise measurement of the speed of light, and who now recruited his colleague Edward Morley to collaborate with him. It compared the speed of light along two directions at right angles to each other. We would expect a difference depending on the angle between the earth's motion and the motion of the ether. However, no evidence for any such "ether wind" was detected. The experiment led to the startling conclusion that irrespective of the direction and the time of day or year, the speed of light was the same.

This result is entirely in accord with Maxwell's equations. They lead to the same speed, c, with no indication of any dependence on a particular coordinate system. This is fundamentally different from the way mechanical waves, like sound and water waves, behave. A sound wave on a moving train has different speeds with respect to the train and with respect to the ground.

That the different kinds of waves should behave so differently was not at all understood at the time, and all attempts at reconciliation over the next 17 years failed.

Only Einstein, in 1905, followed Maxwell's equations seriously when he said that the speed of electromagnetic waves is c ($= \frac{1}{\sqrt{\epsilon_0 \mu_0}} = 3 \times 10^8$ m/s) regardless of the frame of reference. He was the only one not to try to make lightwaves conform to the behavior of waves in materials, i.e., to mechanical waves such as sound waves. He took as his point of departure the simple fact that Maxwell was right, that the speed of electromagnetic waves was c, and that it remained so, regardless of any velocity of the frame of reference with respect to which the waves might be observed.

The results were startling. If c is constant, the dimensions, i.e., the observed lengths of objects, are no longer constant and independent of the motion of the object with respect to the observer. They are subject to the *Lorentz contraction* by the factor $\sqrt{1 - \frac{v^2}{c^2}}$, when determined by an observer moving with a velocity v with respect to the object. Time intervals also depend on the relative motion, and are larger by the reciprocal of the same factor, $\frac{1}{\sqrt{1 - \frac{v^2}{c^2}}}$.

Kinematics of the special theory of relativity: time dilation and length contraction

Let's see how these results come about. We will do two "thought experiments." For the first we start with two mirrors a distance L apart. A beam of light with its velocity c goes back and forth between them. This is a primitive clock, marking time intervals Δt between reflections, so that $L = c\Delta t$ and $\Delta t = \frac{L}{c}$.

Look at this clock while it flies by at a speed v. The light beam has to go further between reflections. If its speed continues to be c, the time interval between reflections must be different. Call it $\Delta t'$.

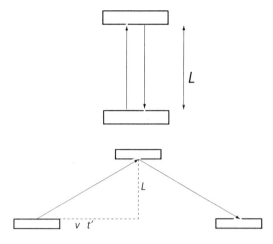

The beam has to go along the hypotenuse of the triangle whose sides are L and $v\Delta t'$, i.e., a distance $\sqrt{L^2 + (v\Delta t')^2}$. It takes a time $\Delta t'$ to traverse this distance, so that $\sqrt{L^2 + (v\Delta t')^2} = c\Delta t'$, or $L^2 + v^2(\Delta t')^2 = (c\Delta t')^2$.

Dividing by c^2 we get $\frac{L^2}{c^2} + \frac{v^2}{c^2}(\Delta t')^2 = (\Delta t')^2$, or $\frac{L^2}{c^2} = (\Delta t')^2(1 - \frac{v^2}{c^2})$.

But $\frac{L}{c} = \Delta t$, so that $(\Delta t)^2 = (\Delta t')^2(1 - \frac{v^2}{c^2})$, or $\Delta t' = \frac{\Delta t}{\sqrt{1-(v^2/c^2)}}$.

The denominator is less than one, so that $\Delta t'$ is greater than Δt. The time interval in the moving system is greater, and we refer to the phenomenon as *time dilation*.

Δt is the time interval in the system in which the observer is at rest with respect to the clock. It is called the "proper time interval." In all other systems the observer is moving with respect to the clock, and the time interval ($\Delta t'$) is longer.

For the second thought experiment we look at a spaceship, with its astronaut, A, traveling between points in two cities on earth, where you are the observer. As the spaceship travels between the two cities, A, at rest with respect to his clock, records the time interval Δt. You, on earth, see the longer time interval, $\Delta t' = \frac{\Delta t}{\sqrt{1-(v^2/c^2)}}$. Both you and A measure the same relative velocity v.

What about the distance between the two cities? We'll call that distance, as you observe it, L. You can take your time measuring it because the cities are at rest in your coordinate system. That's why we call this the "proper distance." As you watch the spaceship fly by at the speed v, you see that it takes a time $\Delta t'$ to traverse the distance L, so that $v = \frac{L}{\Delta t'}$.

A, in the spaceship, sees the two cities pass her in the shorter ("proper") time interval Δt. The relative velocity is still v. The distance between the cities must therefore, as she sees it (L'), be smaller than the proper distance, L, which is the one observed by you on earth: $v = \frac{L'}{\Delta t}$.

We see that the proper distance (L) goes with the longer time interval ($\Delta t'$), while the proper time interval (Δt) goes with the shorter distance (L'), and $L' = v\Delta t = L\frac{\Delta t}{\Delta t'}$, or $L' = L\sqrt{1 - \frac{v^2}{c^2}}$. The flying astronaut sees the cities closer to each other, i.e., the distance between them is *contracted*. We speak of *length contraction* and *time dilation*. This goes both ways: the length of the spaceship is seen by you as smaller than by A.

Dynamics of the special theory: $E = mc^2$

These are the kinematic results. In addition there are the even more interesting and important dynamic aspects, the ones that deal with mass, energy, and momentum, and that lead to $E = mc^2$.

We will quote some of the results for the dynamic quantities, momentum, and energy, that are also part of Einstein's special theory of relativity. The "relativistic momentum," p, is no longer mv, but is $\frac{mv}{\sqrt{1-\frac{v^2}{c^2}}}$.

The most startling result is that the *total energy* is $E = \frac{mc^2}{\sqrt{1-\frac{v^2}{c^2}}}$, of which $\frac{mc^2}{\sqrt{1-\frac{v^2}{c^2}}} - mc^2$ is the relativistic kinetic energy. The rest, mc^2, is called the *rest energy*.

We can derive a relation between energy and momentum from these relations: from the energy relation we have $\frac{E}{c^2} = \frac{m}{\sqrt{1-\frac{v^2}{c^2}}}$, which, from the momentum relation, is equal to $\frac{p}{v}$, so that $\frac{v}{c} = \frac{pc}{E}$.

We can then write $E^2 = \frac{m^2c^4}{1-\frac{v^2}{c^2}}$ as $E^2 = \frac{m^2c^4}{1-\frac{p^2c^2}{E^2}}$, and finally $E^2 - p^2c^2 = m^2c^4$, or $E^2 = p^2c^2 + m^2c^4$.

Can we reconcile the expression for the relativistic kinetic energy with the Newtonian form that we are used to? Let's see how it changes when the velocity changes. We can write it as

$$K = mc^2 \left(\frac{1}{\sqrt{1-\frac{v^2}{c^2}}} - 1 \right),$$

or $mc^2[(1-\frac{v^2}{c^2})^{-1/2} - 1]$.

We can expand the square-root part by using the *binomial theorem*, in the form $(1+x)^n = 1 + nx + \cdots$ [This is the most useful part and is sufficient here. The next terms are $[\frac{n(n-1)}{2}]x^2$: and $[\frac{n(n-1)(n-2)}{3!}]x^3$. Here $x = -\frac{v^2}{c^2}$ and $n = -\frac{1}{2}$, so that $(1-\frac{v^2}{c^2})^{-1/2}$ becomes $1 + (-\frac{1}{2})(-\frac{v^2}{c^2}) = 1 + \frac{v^2}{2c^2}$, and the kinetic energy becomes $mc^2(1 + \frac{v^2}{2c^2} - 1)$, or $\frac{1}{2}mv^2$! The next terms are successively smaller and smaller, and we see that as $\frac{v^2}{c^2}$ becomes smaller, the expression for the kinetic energy gets closer and closer to the familiar classical expression, $\frac{1}{2}mv^2$.

Let's look at $E = mc^2$. It is a real equivalence of mass and energy. If I heat an object and so increase its internal energy, and then measure its mass, I have every reason to expect that the measurement will lead to a greater mass than when it was cold. Of course the difference, equivalent to the increase in the internal energy divided by c^2, is so small that there is no hope of actually making the measurement.

It takes 13.6 eV to pull the electron away from the proton in a hydrogen atom. Adding this energy (the *binding energy*) is equivalent to increasing the mass, so that once you have separated the two particles you have that much more mass. Since the mass of the hydrogen atom (almost entirely that of the proton) is equivalent to about 937 MeV, the change corresponds to one part in 10^8, still very small.

In the nuclear realm the difference becomes significant. Pulling the proton away from the neutron in the deuteron, the simplest nucleus consisting of more than one particle, takes about 2 MeV, or about 1 part in 1000.

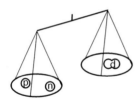

Is there a process in which all of the mass disappears? Yes, that can happen also. Let a positron hit a material. It slows down as it makes collisions, and eventually it comes to rest (or nearly so) near an electron. The two annihilate. Both disappear. That means they're gone, completely and forever. What happens to the energy that is equivalent to the mass that disappears? It is now that of two photons, each with an energy of about $\frac{1}{2}$ MeV. The electron and the positron are gone, but the energy equivalent to their mass is still there as pure photon energy.

Small wonder that Einstein's 1905 paper "*On the Electrodynamics of Moving Bodies*" on what came to be known as the *special theory of relativity* was met with skepticism and even with hostility. In spite of the strange and counterintuitive conclusions, however, it became clear as time went on that the theory was right, that it described experiments and observations correctly, and that it necessitated a revolutionary reappraisal of the concepts of time and space, of mass and energy. Now, more than a hundred years later, it has been so widely and completely verified that it has become a cornerstone of science without which many modern developments are unthinkable.

In nuclear and particle physics the theory plays a fundamental role. The relativistic relations are essential for the understanding, design, and operation of particle accelerators such as cyclotrons. Without a knowledge of the equivalence of mass and energy there could be no understanding of nuclear processes and transformations, including those in nuclear reactors and the fusion reactions that light the stars. Among applications that impinge directly on the experiences of nonscientists are the *Global Positioning Systems* that are able to provide remarkably precise information on location by signals from earth satellites.

The theory is so much part of the fabric of today's science that it is often regarded as part of "classical" physics, i.e., that part of physics that is no longer questioned as to its correctness within its realm of applicability.

Magnetism and electricity: inseparable, but interchangeable

The fact that electricity and magnetism are closely related was already known from the discoveries of Oersted, Faraday, and others, but our knowledge of that synthesis was enormously expanded by Maxwell when he combined them in what came to be known as Maxwell's equations, and showed how they cooperate in electromagnetic waves. Later, in 1905, a higher level of understanding was achieved by Einstein. Even as a student, Einstein was already on the path that led to the theory of relativity when he tried to imagine what would be observed by someone riding at the crest of an electromagnetic wave.

We will look here at the much simpler ride, moving along with the charges of a current. This will allow us to see how the electric and magnetic forces change, one into the other, depending on the frame of reference in which they are observed.

Imagine two parallel lines with equally spaced charges, at rest. The two lines repel, as usual, as described by Coulomb's law. Now look at the same two lines from a frame of reference that moves to the left, so that all the charges are seen to move to the right, parallel to the two lines of charge. The two lines are now currents, and there is a magnetic force of attraction between them.

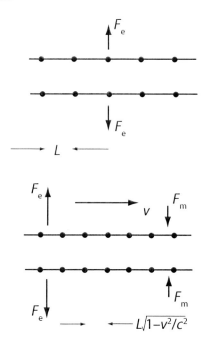

How can we reconcile this conclusion with the expectation that the actual, observed, measured force cannot depend on the frame of reference? In Newtonian, classical physics there seems to be no way out. It takes the theory of relativity to resolve the dilemma. It tells us that the spacing between the charges is now changed. All distances parallel to the direction of motion are reduced by the Lorentz contraction. There are now more charges in any given length along the lines, and the electric force per meter is therefore greater. But the greater electric repulsion between the charges, together with the magnetic attraction between them, now that they are moving, add up to the same net force as before, when there was only an electric force and the charges were at rest.

We could also take the point of view that we know the electric and magnetic forces. We could then show that to get the same result, independent of the frame of reference, there must be a Lorentz contraction, and we can calculate how large it has to be.

Alternatively, we can start from the Lorentz contraction by the factor $\sqrt{1-\frac{v^2}{c^2}}$ and calculate how much larger the electric force becomes. This gives us the magnitude of the attractive force that is required so that the total force remains unchanged. In other words if, somehow, we already know that there is such a thing as the Lorentz contraction, we can show that there must be an additional force, an attractive force between two lines of moving charges. This force is, of course, the magnetic force, and it can be seen to be a consequence of the existence of the electric force together with the theory of relativity.

There is an interesting feature of this story. It is sometimes thought that the kinematic consequences of the special theory of relativity are observable only when the relative velocities are close to the speed of light. This is not so in the present example. Think of the electrons moving in a copper wire carrying a current. They move very fast, but their average velocity, called the *drift velocity*, is of the order of a few cm/s or even smaller for ordinary currents. This is the relative velocity of the two frames of reference that we have used. The phenomena of magnetism arising from currents, including all the forces between currents in wires and coils, are seen to be the consequence of these very small relative velocities.

As we reflect on the place of the theory of relativity we can look back to the view of science before it was developed. The great achievement of Newton was to create *classical mechanics*, the theory that seemed to structure all of physics, all of science, and perhaps all of knowledge.

Understanding a phenomenon meant to know its *mechanism*, i.e., the way it comes about in accord with Newton's laws. When Newton's laws came in conflict with the more recently discovered laws of electromagnetism, as demonstrated, for example, by the Michelson–Morley experiment, it was natural to look for flaws in the newer theory. How astonishing then that it was Maxwell's synthesis of electricity and magnetism that was given new meaning by the theory of relativity, and that it was the electromagnetic equations that survived unchanged. Newton's mechanics, on the other hand, and his views of time and space, no longer hold their place as the bedrock of science after Einstein's fresh look and bold reexamination.

11.8 Summary

A wave along a string or a water wave can be seen to move fast and far, but the pieces of the string and the parts of the water are only oscillating in place. Although the string and the water do not move very far, energy and momentum are transported. One particle pushes on its neighbor, and that's how the wave propagates. Mechanical waves also include sound waves, produced by the oscillation of strings or vocal cords or columns of air.

The propagation of electromagnetic waves is more subtle. A changing electric field produces a changing magnetic field, which in turn produces a changing electric field. The disembodied fields propagate even when there are no material bodies. That's how energy reaches us from the sun through the empty space between.

Waves reach us when we hear or see, or when we feel the warmth of the sun or the stove.

In a wave there is a quantity that oscillates in space and in time. The motion repeats in space after one wavelength (λ), and in time after one period (T).

The frequency is the reciprocal of the period. $f = \frac{1}{T}$. The wave's speed is $v = \frac{\lambda}{T} = \lambda f$.

When two waves come together, their amplitudes add. (They "interfere.") They can produce constructive interference or destructive interference. Two waves are "in phase" when they move up and down together. Their space dependence has a difference in phase by an angle θ when one varies as $\sin x$ and the other as $\sin(x - \theta)$. The sine function and the cosine function are out of phase by $\frac{\pi}{2}$ radians.

The intensity of a wave (in watts/m^2) is the amount of energy transported per second divided by the cross-sectional area through which it passes.

The expression $y = A \sin(\frac{2\pi}{\lambda}x - \frac{2\pi}{T}t)$ represents a wave.

The frequency of a wave is determined by the vibration of its source. The speed of a (mechanical) wave depends on the properties of the medium through which it travels. The two properties on which the speed depends are the displaced mass and the restoring force.

A standing wave is produced by two traveling waves in opposite directions. In a standing wave there is an oscillation at each point but no propagation. The amplitude varies (sinusoidally) with distance.

On a string fixed at both ends, standing waves are possible only with certain frequencies and wavelengths, namely those for which $L = n\frac{\lambda}{2}$. At a fixed end there is a node. At a free end there is (approximately) an antinode. The frequencies corresponding to $L = n\frac{\lambda}{2}$ are the natural or resonant frequencies. In an air column also, if the boundary conditions are the same at both ends (closed or open), $L = n\frac{\lambda}{2}$. If one end is open and the other closed, $L = \frac{\lambda}{4} + n\frac{\lambda}{2}$.

Two notes whose frequencies differ by a factor of two are said to be an octave apart. The octave interval is divided into smaller intervals to produce a *scale* of notes.

When the source of a wave and the receiver move with respect to each other, the frequency observed by the receiver is changed. This is the *Doppler effect*.

Maxwell showed that Ampere's law is incomplete when there is a break in a circuit. An additional term (the *displacement current*) then has to be added to the current. The displacement current is proportional to $\frac{dE}{dt}$. There is now a term similar to Faraday's law, but with the rate of change of the electric field instead of the magnetic field. The two laws together lead to the existence of electromagnetic waves.

All electromagnetic waves have the same speed, c, in a vacuum. They can have widely different frequencies and wavelengths.

The energy density of an electric field is $\frac{1}{2}\epsilon_0 E^2$ J/m^3. In a magnetic field it is $\frac{B^2}{2\mu_0}$.

In an electromagnetic wave electric and magnetic fields vary together (in phase with each other). The fields are at right angles to each other and to the direction of propagation.

In Young's experiment light travels to two slits, each of which acts as a source of light. The light from the two slits combines at a screen to form regions of light (constructive interference) and darkness (destructive interference). In a diffraction grating light from many slits recombines. With a single slit light from different parts of the opening leads to regions of constructive and destructive interference.

Light reflected from the top and the bottom of a thin film combines and gives rise to interference effects, leading to the colors of soap and oil films.

In the Michelson interferometer a light beam is split into two parts that travel at right angles to each other. The interference effects when they recombine can be used for precision measurements in terms of the wavelength of light. It was also used in the Michelson–Morley experiment to look for a difference in the speed of light depending on its direction.

Light from the hot filament of an incandescent light bulb is "incoherent." It consists of tiny randomly occurring flashes with different wavelengths. Light from a laser is "coherent." It consists of waves of a single frequency and wavelength, all in phase.

Law of reflection: the angle of incidence is equal to the angle of reflection.

Law of refraction, or Snell's law: $n_1 \sin \theta_1 = n_2 \sin \theta_2$.

The image formed by a plane mirror is "virtual." No rays actually go to such an image or come from it. When we look at the mirror, rays *seem* to come from the virtual image.

Rays coming to a spherical mirror parallel to the axis are reflected toward the focus. (This is approximately so for a spherical mirror and exactly so for a parabolic mirror.) Rays coming through the focus are reflected parallel to the axis. The distance from the mirror to the focus is half the distance to the center of curvature of the mirror.

A spherical mirror produces a real inverted image if the object is further from the mirror than the focus. For a closer object the image is virtual and enlarged.

Thin lens: rays coming to a lens parallel to the axis are refracted toward the focus. Rays coming through the focus are refracted parallel to the axis. Rays through the center continue in a straight line. With these three rays we can describe where an image is formed.

Thin-lens relation: $\frac{1}{d_O} + \frac{1}{d_I} = \frac{1}{f}$. For a virtual image d_I is negative, and for a diverging (concave) lens f is negative.

Total internal reflection results when a light ray is at the boundary to a medium with a smaller index of refraction, and the angle of incidence is so large that Snell's law leads to an angle of refraction greater than 90°. (This occurs in binoculars and in the filament of a fiber optic cable.)

The lens of a camera produces an image on a film or sensor. The lens of the eye produces an image on the retina. Nearsightedness and farsightedness occur when the lens of the eye cannot adapt sufficiently and the image is formed in front of the retina or behind it.

To use a converging lens as a magnifying glass, the object is between the lens and the focus, so as to produce a virtual enlarged image. This is also so for the eyepiece of a microscope or a telescope. In both there is a second lens that forms a real image of the object, which then is the object looked at with the eyepiece.

The speed of light, c (in a vacuum), is always the same. This assumption leads to length contraction, $L' = L\sqrt{1 - \frac{v^2}{c^2}}$, and time dilation, $\Delta t' = \frac{\Delta t}{\sqrt{1 - \frac{v^2}{c^2}}}$.

Einstein's special theory of relativity also leads to $E = mc^2$: the mass of an object corresponds to an energy (the rest energy) equal to

mc^2. The rest energy of an object whose mass is 1 kg is 9×10^{16} J.

The relativistic momentum is $\dfrac{mv}{\sqrt{1-\frac{v^2}{c^2}}}$. The total energy is $\dfrac{mc^2}{\sqrt{1-\frac{v^2}{c^2}}}$ and the kinetic energy is the total energy minus the rest energy. When v is much less than c, the kinetic energy $\dfrac{mc^2}{\sqrt{1-\frac{v^2}{c^2}}} - mc^2$ approaches $\tfrac{1}{2}mv^2$.

The binding energy of a system is the energy required to separate it into its components.

11.9 Review activities and problems

Guided review

1. Go to the PhET website and open the simulation *Wave Interference*. Select "Sound" and "on." Set a low frequency and a large amplitude. Compare "Grayscale" and "Particles."
   (a) Select "Particles" and "show markers." Describe the motion of one of the particles with a red x. You may want to use the "pause-step" feature at the bottom of the screen.
   (b) What is the relation between the direction of motion of the particles and the direction of propagation of the wave? What is that relation for a wave on a string?
   (c) Click on "Show graph." How is the pressure related to the color of the "grayscale"? Click on "Add Detector" twice. Set the two detectors a half wavelength apart and watch the pressure readings.

2. Go to the PhET website and open the simulation *Wave Interference*. Select "Sound," speaker "on," and "grayscale."
   (a) Increase and decrease the frequency and observe what happens to the spacing of the rings.
   (b) How does the wave speed vary?

3. Go to phet.colorado.edu and to *Wave on a String*. Set "damping" to zero, "oscillate," "no end," and check "rulers" and "timer."
   (a) Set the frequency at 35. Use the timer to measure the period by measuring the time for the wave to travel one wavelength. Use the pause/play-step button. What is the frequency? What is the ratio of the frequency setting to the measured frequency? Measure the wavelength and calculate the wave speed.
   Repeat this for three other frequencies between 25 and 65. What are your conclusions about the wave speed and the frequency setting?

4. Go to the PhET website and open the simulation *Wave Interference*. Select "Sound" and check "measuring tape" and "stopwatch."
   (a) Select five different frequencies. For each of them pause and measure the wavelength. Measure the period with the timer. (Use the step feature at the bottom of the screen.)
   (b) Plot your data so as to determine the wave speed. What do you need to plot so that v is the slope? Look up the speed of sound in air and in metals and compare them to your experimental value. Is the sound traveling in a gas or in a metal?

5. (a) Write an equation for a wave with a frequency of 2 Hz, a wavelength of 30 m, and an amplitude of 75 cm.
   (b) Find the wave speed.

6. The equation for a certain wave is $y = 15 \cos(\tfrac{\pi}{5}x - 40\pi t)$. What are the wavelength, frequency, period, and wave speed?

7. Go to the PhET website and open the simulation *Wave on a String*. Set the tension to its maximum and "damping" to zero. Select "oscillate" and "fixed end." Check "rulers" and "timer." Set the frequency to 50. Since this is a resonant frequency, and so requires little energy to keep it going, the amplitude of the oscillator can be quite small. Set it at 1 or 2.
   (a) Measure the wavelength by measuring the distance between points with the same phase. (Two successive points with the same phase are a wavelength apart.) (Use "pause.")
   (b) Measure the frequency with the timer by measuring the time for some number of cycles. How does it compare to the frequency setting?
   (c) What is the number of half-wavelength segments equal to the length of the string? What is the effect of the boundary condition at the right-hand end?
   (d) Calculate the wave velocity from your measurements.
   (e) Select "pulse." (Increase the amplitude.) Measure the wave speed with the timer and the ruler. Compare it to the value of part (d).

8. A string, 1.2 m long, is fixed at both ends. The wave speed along the string is 220 m/s. What are the three lowest frequencies of standing waves that can be set up?

9. An organ pipe contains an air column 80 cm long, open at one end and closed at the other. What are the three lowest frequencies of standing waves that can be set up in the air column?

10. Go to the Java Applet on the Doppler effect as in Example 10, e.g. at http://lectureonline.cl.msu.edu/~mmp/applist/doppler/d.htm.

(a) Click on the gray rectangle to make the blue dot (the source) appear. Pause the pattern, and estimate the wavelength (the distance between successive circles).

(b) Drag an arrow with the mouse, representing a velocity less than the sound velocity, v_s, toward the bottom right corner. What would you observe happening to the wavelength, frequency, and wave speed, standing at the upper left corner? What would you hear there, compared to the sound from the stationary source?

11. Go to the website phet.colorado.edu and open *Wave Interference*. Select "Light," "show screen," "no barrier," and "measuring tape."

Pause and measure the wavelength with the tape. Leave the tape in place and change the color using the wavelength slider. Measure the wavelength again with the tape. Measure the wavelength at the two ends of the spectrum. What are the approximate limits of the wavelength of visible light?

12. A concave mirror has a focal length of 20 cm.

(a) What is its radius of curvature?

(b) At what distance from the mirror must an object and an image be so that the magnification is 1?

(c) Where is the image for an object distance of 15 cm? What kind of image is it?

(d) For what range of object distances is the image larger than the object? For what part of this range is the image real and for what part is it virtual?

13. Go to the PhET website and open *Geometric Optics*. Select "principal rays," "curvature 0.8," "refractive index 1.87," "diameter 0.3," and "virtual image."

(a) Check "ruler" and use it to measure the focal distance.

(b) Select an object distance larger than the focal length and measure it. Measure the image distance.

(c) Calculate the image distance from the object distance with the thin-lens equation and compare the result with your measurement.

(d) Measure the magnification. Calculate the magnification and compare the result with your measurement.

(e) Move the object until the magnification is one. What are the object and image distances? What is their relation to the focal length?

(f) Use the thin-lens formula to find the answers to part (e) and compare them to the results of your measurement.

14. Go to phet.colorado.edu and open *Geometric Optics*, "curvature 0.8," "refractive index 1.87," "diameter 0.3," and "virtual image."

(a) Use the ruler to measure the focal distance.

(b) Select an object distance smaller than the focal length and measure it. Measure the image distance.

(c) Calculate the image distance from the object distance with the thin-lens equation and compare the result with your measurement.

(d) Measure the magnification. Calculate the magnification and compare the result with your measurement.

15. You are swimming in a quiet lake. You put your head under water and look up. What do you see?

16. What is the constraint on the index of refraction of the prism to produce the reversal of the path described in Example 16.

Problems and reasoning skill building

1. A wave travels along a long (assume infinite) rope under constant tension. The rope is marked off in 1 m intervals. At the 0 m mark the rope is observed to reach its maximum transverse displacement every 8 s. The distance between maxima at any instant is 25 m. Write a function that describes this wave, assuming that it has its maximum displacement at $x = 0, t = 0$.

2. A sinusoidal wave can be described by its frequency, speed, and wavelength. Which of these

quantities depends on the medium which on the source, and which on both?

3. You have a rope that is 4 m long. You move one end up and down to create a standing wave. What are some possible wavelengths of standing waves
   (a) when the other end is fixed
   (b) when the other end is free to move.

4. A plastic coating ($n = 1.4$) is used to reduce the amount of light reflected from a lens whose index of refraction is 1.5. What is the thickness required to cause destructive interference of light in the middle of the spectrum ($\lambda = 550$ nm)?

5. (a) A traveling wave is described by $y(x, t) = 0.2 \sin(\pi t - \frac{\pi}{2}x)$. What are the five physical quantities that you can determine from this relation and what are their values?
   (b) Make a graph of $y(x, 0)$.
   (c) Make a graph of $y(t)$ at $x = 2$ m.
   (d) Which of your graphs is a snapshot of the wave? What is the other graph?

6. John is shaving, using a concave mirror 20 cm in front of him. The virtual image is 60 cm from the mirror.
   (a) What is the mirror's focal length?
   (b) What is the magnification?

7. A lightbulb emitting light in all directions is 3 m below the surface of a lake. What is the shape and size of the illuminated area as seen from above?

8. Light of wavelength 550 nm passes through a single slit. Find the angle to the first intensity minimum for the following two slit widths:
   (a) 2×10^{-4} m
   (b) 2×10^{-6} m
   (c) Explain how you knew right away that one of the angles was going to be so much larger for one of the answers.

9. Two sound sources are 3 m apart and emit waves in phase. A detector is on a line perpendicular to the line joining the sources, and directly in front of one of them, 4 m away. The speed of sound is 340 m/s.
   What are the two lowest frequencies that lead to destructive interference at the detector?

10. Look at the diagram of the Michelson interferometer. The source emits yellow light with wavelength 620 nm. You look at the pattern (as shown by the eye in the figure) and see a bright yellow spot in the middle. Mirror 2 is now moved until the bright spot is replaced by a dark spot. Through what distance has the mirror moved?

11. A concave mirror projects the image of a slide on a wall. Its focal length is 30 cm. The slide is 35 cm from the mirror.
    (a) How far from the wall should the mirror be?
    (b) What is the magnification?
    (c) For the image to be right-side up, what should the orientation of the slide be?

12. A standing wave on a 0.8 m string is described by the relation $y = 0.02 \sin 5\pi x \cos 100\pi t$ SI units.
    (a) What are the amplitude, frequency, and wavelength?
    (b) Draw the standing wave pattern. Which harmonic is this? (Counting the lowest frequency as 1, what is the number of this one?)
    (c) The mass of the string is 0.008 kg. What is the tension in the string?

13. Three successive resonance frequencies of an air column are 75, 125, and 175 Hz.
    (a) Does this column have one end open and one end closed, or both ends open?
    (b) What is the fundamental (lowest) frequency?
    (c) Draw the wave pattern for the 75 Hz wave.

14. You are holding a 1.6 m solid brass bar by a clamp at its middle. You strike the bar so that it resonates with its longest wavelength.
    (a) Draw the standing wave pattern.
    (b) The frequency is 1000 Hz. Describe what would have to be different for the frequency to be 2000 Hz.
    (c) Determine the speed of sound in the brass bar.

15. For a demonstration of standing waves a string with length 2.5 m is attached to an oscillator operating at 60 Hz. The other end of the string passes over a pulley to a hanger where various masses can be attached to vary the tension in the string. Transverse standing waves are set up in the string with nodes at the pulley and at the oscillator. The mass per unit length of the string is 8 g/m.

(a) What must the tension be for the string to vibrate with its lowest frequency?
(b) What tensions are needed for the three next higher frequencies?

16. Yellow light with $\lambda = 620$ nm enters an oil film at right angles. The index of refraction of the film is 1.3.
(a) What is the wavelength in the film?
(b) What is the thickness for destructive interference?

17. A spaceship travels away from the earth for 10 days at a speed of 8000 m/s with respect to the earth as measured from earth. What is the difference between that time and the time measured by the astronaut in the spaceship?
Since the difference is small, it is not useful to calculate each time interval separately and then take the difference. Instead use the first two terms of the binomial series in the form $(1+x)^n = 1 + nx + \cdots$ (valid when x is much smaller than 1.) For example, with $x = -\frac{v^2}{c^2}$ and $n = \frac{1}{2}$, $(1 - \frac{v^2}{c^2})^{\frac{1}{2}} = 1 - \frac{1}{2}\frac{v^2}{c^2}$.

Multiple choice questions

1. Which of the following is not a kind of electromagnetic radiation?
   (a) gamma rays
   (b) x-rays
   (c) ultraviolet rays
   (d) electrons
   (e) microwaves

2. A tube is closed at one end and open at the other. It is placed in front of a loudspeaker that is playing the sound generated by a variable frequency audio oscillator. The frequency is slowly increased from 0. The first frequency at which standing waves are generated is 325 Hz and the next is 975 Hz. The speed of sound is 340 m/s. What is the length of the tube?
   (a) 20 cm
   (b) 30 cm
   (c) 40 cm
   (d) 50 cm
   (e) 10 cm

3. A 0.40 m string is clamped at both ends. The lowest frequency for a standing wave is 325 Hz. What is the wave speed?
   (a) 340 m/s
   (b) 260 m/s
   (c) 813 m/s
   (d) 130 m/s
   (e) 406 m/s

4. Ocean waves with a wavelength of 120 m are arriving at the rate of 8 per minute. What is their speed?
   (a) 8.0 m/s
   (b) 16 m/s
   (c) 24 m/s
   (d) 30 m/s
   (e) 4.0 m/s

5. A stretched string, fixed at both ends, is observed to vibrate in three equal segments when it is driven by a 480 Hz oscillator. What is the fundamental (lowest) frequency of standing waves for this string?
   (a) 480 Hz
   (b) 320 Hz
   (c) 160 Hz
   (d) 640 Hz
   (e) 240 Hz

6. A 2 m string, fixed at both ends, is observed to vibrate as a standing wave in six segments. The wave speed is 45 m/s. What is the frequency?
   (a) 270 Hz
   (b) 140 Hz
   (c) 200 Hz
   (d) 68 Hz
   (e) 34 Hz

7. An organ pipe, open at both ends, has successive resonances at 150 Hz and 200 Hz when the velocity of sound in air is 345 m/s. What is its length?
   (a) 5.25 m
   (b) 5.75 m
   (c) 2.76 m
   (d) 4.90 m
   (e) 3.45 m

8. A wave is described by the relation $y = 0.15 \sin(\frac{\pi}{8}x - 4\pi t)$ SI units. What is the first positive value of x where y is a maximum at $t = 0$.
   (a) 16 m
   (b) 8 m
   (c) 13 m
   (d) 2 m
   (e) 4 m

9. A standing wave on a string is described by $y = 0.080 \sin 6x \cos 600t$ SI units. What is the distance between successive nodes?
 (a) 0.24 m
 (b) 0.08 m
 (c) 0.02 m
 (d) 0.52 m
 (e) There is insufficient information.

10. Assume that the human vocal tract can be thought of as a tube open at one end. Assume that the length of the tube is 17 cm and that the speed of sound is 340 m/s. What are the lowest two resonant frequencies?
 (a) 500 Hz, 1500 Hz
 (b) 500 Hz, 1000 Hz
 (c) 1000 Hz, 2000 Hz
 (d) 1000 Hz, 3000 Hz
 (e) 1500 Hz, 2500 Hz

11. Sound travels at 340 m/s in air and at 1500 m/s in water. A sound of 256 Hz is made in water. If the 256 Hz sound is made in air, which of the following is true:
 (a) The frequency remains the same and the wavelength is shorter.
 (b) The frequency remains the same and the wavelength is longer.
 (c) The frequency is lower and the wavelength is longer.
 (d) The frequency is higher and the wavelength stays the same.
 (e) Both the frequency and the wavelength remain the same.

12. Two strings, fixed at both ends, are 1 m and 2 m long, respectively. Which of the following sets of wavelengths, in m, can represent waves on *both* strings?
 (a) 0.8, 0.67, 0.5
 (b) 1.33, 1.0, 0.5
 (c) 2.0, 1.0, 0.5
 (d) 2.0, 1.33, 1.0
 (e) 4.0, 2.0, 1.0

13. A string, 4 m long, vibrates according to the relation $y = 0.04 \sin(\pi x) \cos(2\pi t)$ SI units. The number of nodes is
 (a) 1
 (b) 2
 (c) 3
 (d) 4
 (e) 5

Synthesis problems and projects

1. Go to the PhET website and open the simulation *Fourier: Making Waves*.

 Go to "discrete." The first row shows the amplitudes of the first 11 harmonics (f, 2f, 3f, ... 11f). Each can be dragged up or down. The decomposition of a complex wave into its components is called *Fourier analysis*, and their addition is *Fourier synthesis*. The components are called *Fourier components*.

 The second row shows the separate waves coresponding to the selection you make in row 1. Start by experimenting with just one component and its amplitude at a time.

 The third row shows the sum of the waves that are shown in the second row.

 (a) Start with $A_1 = 1$ and $A_{10} = 0.3$ to see how two waves add when the frequencies are quite far apart. Try other combinations.

 (b) See how close you can come to a "square wave" (up—straight across down—straight across and up again) with just three harmonics (A_1, A_3, A_5). What are the amplitudes? (Set A_1 to 1, and then drag A_3 up to its best point for what you want to do. Then A_5. Go back to A_3 and adjust it.) Select "Math form" and "Expand sum" to see the equations of the waves.

 Select "function: square wave" to see what can be done with 11 components. Compare this with your selection: what kind of improvement can you get when there are more components?

 (c) Select "function: wave packet." This shows a combination of waves that add up to a pattern confined to a narrow region in space. You can change the x and y amplitudes with the buttons to the right of the graphs. You can see a better wave packet when you choose "discrete to continuous." Start with $k_1 = \frac{\pi}{4}$, $k_0 = 12\pi$, and $\sigma_k = 2\pi$. Experiment with others values.

 (d) Go to "wave game." It asks you to match a given wave by selecting the components. Start with level 1 (very easy) and go on to level 4. Continue as far as you can.

2. You have some string and you want to know how fast a pulse or wave travels along the string. You have a pulley, a meter stick, a scale, some masses, and a hanger for the masses.

What can you do with this equipment to predict the speed of a pulse or wave on the string? List any assumptions that you make.

What additional equipment do you need to test your prediction?

3. Parallel rays to a lens with spherical surfaces do not come together exactly at a focus. This is called "spherical aberration." The *aperture* of a camera is an opening with variable diameter in a disk to limit the part of the lens that is used. In general, a sharper image is produced when the aperture is small because spherical aberration is then smaller. However, the smaller the aperture the greater are the diffraction effects.

An aperture has a diameter of 1 mm. The film is 5 cm from the lens. What is the distance from the central maximum of the diffraction pattern to the first minimum? Use $\lambda = 500$ nm.

The distance to the first minimum produced by a point object characterizes the resolution of the lens. A second image at this distance from the center can just barely be seen separately on the film or screen. The limit imposed by diffraction effects can be overcome by using radiation with a smaller wavelength. This is done, for example, in an *electron microscope*.

4. Calculate the electric and magnetic forces for the two lines of charge near the end of Section 11.7 to show that the forces in the stationary and the moving systems are the same.

5.

Light enters a 45° prism made of glass with an index of refraction of 1.50 at an angle 5° below the horizontal. Describe the subsequent path: draw a ray diagram.

CHAPTER 12

Quantum Physics

The "old" quantum physics
>   *The photoelectric effect*
>   *Bohr's vision of the atom*
>   *Bohr's calculation*
>   *Suddenly, photons everywhere*
>   *The Compton effect*
>   *Pair production and annihilation*
>   *X-rays*

The new synthesis
>   *Photons and electromagnetic waves*
>   *Complementarity*
>   *The photons and their guide*
>   *Beyond light: electron waves and the new synthesis*
>   *From de Broglie to Schrödinger*
>   *How does the Schrödinger equation work? What can it do?*
>   *Representing waves*
>   *Finally the best we have: the Schrödinger equation*
>   *What does the Schrödinger equation tell us?*
>   *Heisenberg and the uncertainty principle*

Order in the universe: the elements
>   *The beginning: hydrogen and its quantum numbers*
>   *Hydrogen shows the way: the other elements*
>   *The origin of order: the Pauli exclusion principle*
>   *Atomic structure: beyond the Coulomb field*
>   *The ℓ dependence and the order of the elements*
>   *The order changes*

While it is never safe to affirm that the future of Physical Science has no marvels in store even more astonishing than those of the past, it seems probable that most of the grand underlying principles have been firmly established and that future advances are to be sought chiefly in the rigorous application of these principles to all the phenomena which come under our notice.

It is here that the science of measurement shows its importance—where quantitative results are more to be desired than qualitative work. An eminent physicist has remarked that the future truths of Physical Science are to be looked for in the sixth place of decimals.

—Albert Abraham Michelson

How ironic that it was Michelson who said this. He thought of himself as a pillar of classical physics, devoting his life to measuring the speed of light more and more accurately. Yet the Michelson–Morley experiment showed that light behaves in a completely unexpected way. It was a signpost on the way, an indication, as seen from our present vantage point, that our understanding of the natural world would be fundamentally different in the future.

There were others. X-rays were discovered by Röntgen in 1895. Today we know that they are electromagnetic waves emitted by atoms or by accelerated charges. Radioactivity was discovered in 1896 by Becquerel. Radioactive emissions come from the nuclei of atoms, but the existence of nuclei was not even suspected at that time. In 1897 J. J. Thomson found that the constituents of a beam of what we now call electrons always have the same ratio of charge to mass, and hence he is usually credited with the discovery of the electron.

In the first part of the twentieth century there were further discoveries in atomic and nuclear physics, one piled on the other. Some of the experiments seemed to contradict well-established knowledge. This was especially true for the observations that showed light to have the seemingly irreconcilable properties of both waves and particles. Attempts to eliminate the difficulties had only limited success. New insights and understanding came, sometimes slowly, sometimes quickly, but the first quarter of the century is now seen as a time of transition.

The resolution came in 1925, with the development of quantum mechanics and its synthesis of the wave and particle aspects, not only of light and other electromagnetic radiation, but also of matter and all of its constituents.

12.1 The "old" quantum physics

The year 1900 saw the discovery that would later be seen as leading to the most profound and influential changes in our view of nature. Like the Michelson–Morley experiment it was not just new, but showed that what was thought before was fundamentally flawed. It took years for the implications of the new idea to be understood and to become part of the mainstream of science.

As was true for Michelson, Max Planck saw himself as firmly rooted in the past, and believed that he was just applying Maxwell's work to a different situation. He tried to look at the details of the electromagnetic radiation emitted by hot bodies, such as a red-hot stove, the white-hot wire in a light bulb, or the sun. Experiments had shown that the radiation and its distribution of wavelengths are closely the same at any given temperature, regardless of the nature of the hot material.

He succeeded in developing a formula that describes the way the intensity of the radiation is distributed among the different wavelengths, in other words, the *spectrum* of the radiation. There was, however, a totally unexpected feature. The formula seemed to imply that the particles that make up a hot body, and whose vibrations lead to the emission of the radiation, cannot vibrate with just any frequency, but only with a discrete set of definite, equally spaced frequencies, each of which corresponds to a definite, discrete energy of the vibrating system. There was a lowest frequency, f. All other possible frequencies were multiples of it, Nf. Each was proportional to the corresponding energy of the vibrating body: $E \propto Nf$, where N is a whole number. The proportionality constant is given the symbol h, so that $E = Nhf$. h is a constant of nature and it came to be known as Planck's constant.

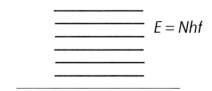

The idea that only certain frequencies, and therefore only certain energies are possible is called the *quantization* of energy. It was so at variance with anything that had been previously thought of that no one quite knew whether and how to take it seriously. Einstein, five years later, was the first to take the result to the next logical step. If the emitting material can have only certain definite energies, then in keeping with the law of conservation of energy, the emitted radiation can also not have any arbitrary energy, but must consist of definite amounts (*quanta*) of energy. Einstein introduced the idea of this definite amount, or quantum, of electromagnetic energy, later called a *photon*. Its energy is hf, where this time f is the frequency of the radiation and h is Planck's constant.

The new concept of the photon made it possible to understand the until then mysterious *photoelectric effect*. This is the emission of electrons from a material that is illuminated by electromagnetic radiation, and it had until then been quite incomprehensible. Einstein showed that the energy of each photon of the radiation (hf) is given to the electron as kinetic energy, except for the amount of energy that is necessary to liberate the electron from the material of which it initially forms a part.

Succeeding years brought further seemingly compelling evidence that a beam of light or other electromagnetic radiation consists of photons, particles with energy hf and momentum $\frac{hf}{c}$.

It was clear that the experimental evidence for the existence of photons was revolutionary. It was quite unclear how it could be reconciled with the electromagnetic theory, which was firmly supported by other experiments that demonstrated clearly that light was propagated as waves.

The photoelectric effect

When doors open automatically as we come near, or when an alarm is set off when we cross a certain line, it probably happens because a light beam shining on a *photocell* is interrupted. The light shines on a piece of metal in the photocell and causes electrons to be emitted from the metal. This is called the photoelectric effect, and it has been known since the late nineteenth century.

At that time it was confidently expected that like other electric phenomena it would be explained by the electromagnetic theory, which had shown light to be electromagnetic waves. The general features of an explanation seemed clear. An electromagnetic wave has an oscillating electric field. As the wave hits a metal, the electrons in the metal find themselves in this field, and experience an oscillating force. We would expect that if this force is strong enough it can shake some electrons loose, and so explain how the photoelectric effect comes about.

Experiments, however, turned up puzzles and discrepancies. The first is that the effect depends on the color of the light. When the frequency is too small (the wavelength too large), no electrons are liberated, regardless of how much light there is.

The second puzzle concerns the energy of the electrons. It is measured by examining how large a voltage is required to stop the electrons. This is like measuring the kinetic energy of a marble by seeing how high a hill it can climb before it stops. The expectation was that more light would lead to a higher energy of the electrons. The results showed, however, that for a given frequency (i.e., for a particular color) of light, the energy of the electrons does not change, no matter what the intensity of the light is. Only the number of emitted electrons is affected. On the other hand, the energy changes when the frequency of the light is different.

These features can be seen in the two kinds of graphs shown here. Each graph in the figure below shows the current of photoelectrons for light of a given frequency. The upper curve is for a greater light intensity. Here V is the voltage between the metal that emits the photoelectrons and an electrode close to it. When V is negative on this graph, it tends to stop the emitted electrons. Beyond some negative value of V

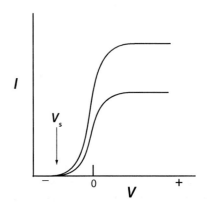

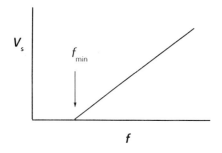

there is no current. This is the "stopping potential," V_s, and it measures the electron energy. When an electron is stopped in an electric field, its kinetic energy is converted to electric potential energy. The charge of the electron times the stopping potential is equal to this potential energy, and therefore it is also equal to the kinetic energy that the electron has before it is stopped.

We see that the magnitude of the current is greater for the greater light intensity of the upper curve, but the energy of the electrons, as measured by the stopping potential, remains the same.

The second graph shows the stopping potential as a function of the frequency of the light. It is even more unexpected. Below a certain frequency no electrons are emitted, and for higher frequencies the electron energy, as measured by the stopping potential, increases linearly with frequency. With different metals the results are similar, but with different minimum frequencies. The consistent pattern of a "threshold" necessary for electron emission was strange and unexpected.

Eventually the puzzling features were explained by Einstein in 1905, but at a startling expense. The explanation did not use the electromagnetic theory. Instead Einstein used the concept of quantized energy, so reluctantly invented by Planck five years earlier. He said that the emitted light consists of bundles of energy (the photons) that behave like particles and not at all like waves.

Planck had considered only the material that emits the waves. To fit the experiments he had been forced to the conclusion that only certain energies were possible for the material as it emitted the waves. He did not really seem to trust this revolutionary discovery, and it was not carried further in the next few years. Einstein realized that if the emitting material could exist only with certain energies, then the emitted light would also have to exist in bundles of definite energy. When the emitting material goes from one definite energy to another, the light carries away the difference in energy, and must also consist of definite amounts or *quanta* of electromagnetic energy, which were subsequently called "photons."

The explanation of the photoelectric effect now becomes amazingly simple. Each photon can liberate an electron. It takes a certain amount of energy to do that, different for different materials, depending on how strongly the electrons are attached, or *bound*. The amount of energy to free the electrons—their "binding energy" E_B—is also the minimum energy that a photon must have if an electron is to be emitted. Any extra energy is given to the electron as kinetic energy.

If the photon energy is hf, the kinetic energy of the electron is K, and the smallest energy that it takes to liberate the electron is E_B, then $hf = K + E_B$. This is the result for which Einstein was awarded the Nobel Prize, but not until 17 years later. (For metals the smallest amount of energy that will liberate electrons is called the *work function*).

The equation was immediately successful. It explained the photoelectric effect and established the existence of photons. The question of what happened to the wave theory, which was supported by many experiments, and which had seemed to be firmly established until that time, was left unanswered.

Go to the PhET website and open the simulation *Photoelectric Effect*.

You can vary the material of the target. You can also vary three quantities with sliders: the light intensity, the battery voltage, and the wavelength.

Choose sodium.

(a) Set the intensity at 35% and the wavelength at about 150 nm. Check "current vs battery voltage."

Vary the battery voltage with the slider on the battery on both sides of zero and observe the first graph, current against voltage. Compare the graph to the corresponding graph in the text. Why are the shapes somewhat different there? Observe the electrons in each of the voltage regions. At what battery voltage does the current go to zero? This is the "stopping potential." What is the energy of the photoelectrons?

Increase the intensity to 60%. What are the stopping potential and the photoelectron energy now?

(b) Check the boxes to see the other two graphs. What do you see on the second graph as you change the battery voltage? What do you see on the third graph? Does it correspond to your prediction?

(c) Uncheck the boxes for the second and third graphs so that you see only the first. Set the wavelength at about 300 nm. Again vary the battery voltage. What are the stopping potential and the photoelectron energy? What is the frequency of the light?

What do you expect for the graph of photoelectron energy vs. light frequency? Turn on the third graph vary the wavelength, and see whether your prediction was correct.

(d) Why is it that below a certain frequency there are no photoelectrons? What is that frequency for sodium? What quantity in the Einstein photoelectric equation can you calculate from that frequency? What is it, from your graphs, for sodium?

(e) Put the intensity again at 35% and set V at 0.
Vary the wavelength and observe the third graph, electron energy against frequency.

(f) Choose zinc. Find its work function from the graphs of the simulation.

EXAMPLE 1

In zinc the minimum amount of energy that it takes to remove an electron is 4.3 eV.

(a) What minimum energy and frequency must a photon have to produce a photoelectron?

(b) What is the wavelength of that photon?

(c) A photoelectron is produced by a photon that has an energy twice the minimum. What is the energy of the photoelectron?

Ans.:

(a) The minimum energy of the photon is 4.3 eV. This energy is equal to hf. Planck's constant, h, is equal to 6.63×10^{-34} Js. Before we can use the relation $E = hf$, we have to convert the value of the energy to SI units: $E = (4.3)(1.6 \times 10^{-19}) = 6.9 \times 10^{-19}$ J. Now we can use $f = \frac{E}{h} = \frac{6.9 \times 10^{-19}}{6.63 \times 10^{-34}} = 1.04 \times 10^{15}$ s^{-1} or 1.04×10^{15} Hz.

(b) $\lambda = \frac{c}{f} = \frac{3 \times 10^8}{1.04 \times 10^{15}} = 2.9 \times 10^{-7}$ m $= 290$ nm.

(c) If the photon energy is twice 4.3 eV, the energy above 4.3 eV goes to the photoelectron. Here this energy is 4.3 eV.

Bohr's vision of the atom

The greatest triumph of the early quantum theory came with Bohr's model of the hydrogen atom in 1913. There had been an earlier atomic model by J. J. Thomson that envisioned electrons stuck in a ball of positive charge. This model could not predict any of the observed atomic properties and had to be abandoned.

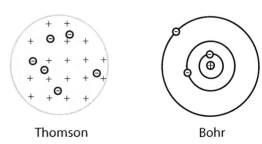

Thomson   Bohr

Bohr's "planetary" model of electrons in orbit around a nucleus gave a more realistic view of atomic structure. It led to the correct allowed, quantized, energies of the hydrogen atom and to values of the atom's size that are closely related to what we know today. It also has the advantage that it uses only some elementary mechanics and Coulomb's law.

The Bohr model is easy to visualize and is relatively simple. As a result it has continued to have a life far beyond the time in 1925 when it was replaced by a model based on *quantum mechanics*, the modern theory of matter, which requires more complicated mathematical methods. The picture of electrons moving in circles or ellipses about the nucleus is still widely used to illustrate what atoms are like, in spite of the fact that it is obsolete and even misleading.

It is therefore a legitimate question whether there is any value in going into detail about the

Bohr model. We will do so, briefly, because of its historical importance and because it allows us to introduce some important concepts.

We need not stay with Bohr's original sequence and terminology, and will use modern terms wherever possible. We will emphasize which aspects are still regarded as correct and which have to be discarded.

As we show where it falls short, we will be able to introduce some of the features of today's quantum mechanics that are startlingly different, but have proved themselves to be enormously more powerful in their description and analysis of atoms and their combinations.

Here are the basic ingredients of the Bohr model of the atom.

1. *The atom consists of a nucleus and electrons.*

   The existence of the nucleus was known from the work of Rutherford. He bombarded atoms (in a gold foil) with radioactive particles (alpha particles, which he later showed to be helium nuclei). By analyzing the way they bounced off, he showed that almost all of the atom's mass is concentrated in a minute piece in its middle, the nucleus.

2. *The only force acting between the nucleus and the electrons is the electric force, as described by Coulomb's law.*

   The gravitational force is too weak, and the nuclear forces, which were then unknown, act only over much smaller distances. The realization that the electric force is the only force that acts between the nucleus and the electrons (as well as between atoms), and that it has precisely the same form there as in the macroscopic domain, is still fundamental to all our understanding of atomic phenomena.

3. *The atom can exist only in states with certain total energies or "energy levels."*

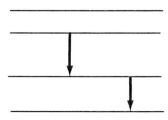

   This is as in Planck's original hypothesis. The allowed energies can be described in an *energy level diagram*. Each allowed energy is represented by a horizontal line. The lines are spaced along a vertical scale that is proportional to the energy.

4. *The atom can make transitions between energy levels with the emission or absorption of energy.*

   One way it can do this is by the emission or absorption of a photon, a quantum of electromagnetic energy, hf, equal to the energy difference between the levels before and after the emission or absorption, so that energy is conserved. The emitted photon has exactly the energy equal to the energy difference between the initial and final energy levels. Similarly, a photon is absorbed only if its energy is exactly equal to the energy difference between the initial and some final level. If there is no such level the photon is not absorbed, i.e., a collection of such atoms is *transparent* to photons of that frequency and energy.

Each of these statements remains as valid today as it was when Bohr used them in 1913. Bohr, however, imagined definite mechanical orbits for the electrons. Today we know that there are no circular or other geometric orbits, and that we can only determine the probability of finding the electron in a particular region or at some particular distance from the nucleus.

Now we come to the actual calculation of the energy levels for the simplest atom, that of hydrogen. According to Bohr's initial model its single electron moves in a circle around the proton at its center.

Bohr knew that there was a problem with the orbits. An electron moving in a circle is accelerated (with the centripetal acceleration $\frac{v^2}{r}$), and Maxwell's theory shows that an accelerated charge radiates, and so loses energy. This happens, for instance, in modern synchrotrons, where electrons race around a circular track, and which are used as sources of intense electromagnetic radiation. Bohr simply said ("postulated") that this doesn't happen in atoms, and that the energy remains constant as long as an electron moves in one of the orbits allowed to it.

This was his major contribution, together with the prescription for finding the allowed orbits, or, as we would prefer to say today, the allowed energy levels. He did this by making a further postulate, known as the *quantization condition*, that said that the angular momentum of the electron (mvr) could only be a multiple of Planck's constant divided by 2π, i.e., $\frac{h}{2\pi}$, today written as $\hbar$ (pronounced "h-bar"). In other words, $mvr = n\hbar$, where the *quantum number*, n, can take on the values 1, 2, 3, and so on, up to infinity.

He tried to justify this choice, but the major justification came from the fact that the quantization condition could be used, together with Coulomb's law, to calculate the allowed energy levels, and that they came out correctly, i.e., in accord with the experimentally observed radiation from hydrogen.

Here is the result: the allowed energies of the hydrogen atom are described by the formula $E_n = \frac{E_1}{n^2}$, where $E_1 = -13.6\,\text{eV}$. The negative sign follows from the choice of the reference level as $E = 0$ when the electron and the nucleus are far apart, i.e., when $\frac{1}{r} = 0$. When the hydrogen atom is in one of the allowed, stable states, energy needs to be given to it to bring it to this reference level. In other words, its energy needs to be raised to bring it to zero. All the allowed energies are therefore negative.

Each value of the *quantum number, n*, corresponds to one of the allowed values of the energy. For $n = 2$, the energy is $+\frac{13.6}{4}$ eV, or $-3.4\,\text{eV}$, and so on. The radius of the atom is $n^2 r_1$, where r_1, the radius of the atom when it is in its lowest (ground-state) energy level, is 0.53×10^{-10} m or 0.053 nm. (A unit that is often used for atomic distances is the *Angstrom*, equal to 10^{-10} m, so that the ground–state radius of the hydrogen atom according to Bohr is 0.53 Å.)

As n increases, and the energy gets closer to zero, the radius becomes larger. When n goes to infinity, r also goes to infinity, and the electron and the proton are completely separated.

Bohr's assumption of circular electron orbits was wrong. His method of calculating the energies was wrong. The values of angular momenta of his quantization condition turned out to be wrong. But the notion of energy levels is correct, and he got the right values for them for the hydrogen atom. His model also led to values for the radii of the orbits, and so to the atomic size. The picture of atoms with exact radii later had to be abandoned and replaced by the statements of quantum mechanics, which allow us to say only what the probability is of finding the electron at a given distance from the nucleus. The probabilities turned out to have maxima at or close to the Bohr orbits, so that the model was basically successful in describing the atomic size.

The successes were so impressive that it was difficult to doubt that Bohr had shown the correct path. Over the next decade attempts were made by Bohr and others to extend the model to atoms with more than one electron and to describe other atomic properties. Success seemed tantalizingly close, but more and more assumptions had to be made, and it proved more and more difficult to find agreement between theory and observation. It was eventually realized that radically new ideas were required, and they came, starting in 1924, in a torrent of activity and creativity, and with astonishing success.

Bohr's calculation

Here is the Bohr model calculation for the energy levels and the radii of the hydrogen atom.

Bohr's electron moves in a circle, so that there must be a force on it toward the center (a centripetal force) equal to $\frac{mv^2}{r}$. In this case the force is the electric force of attraction to the proton, $\frac{ke^2}{r^2}$, where e is the magnitude of the charge on the electron or the proton.

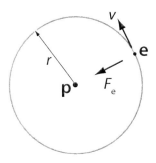

The first equation is therefore

$$F = \frac{ke^2}{r^2} = \frac{mv^2}{r}$$

which we can rewrite as $v^2 = \frac{ke^2}{mr}$.

The second equation is the quantization condition, which says that the angular momentum of the atom is a multiple of $\hbar$:

$$mvr = n\hbar$$

so that $v = \frac{n\hbar}{mr}$.

We can square it and set the two relations for v^2 equal to each other, to get

$$\frac{ke^2}{mr} = \frac{n^2 \hbar^2}{m^2 r^2}$$

which we can solve for r, to get

$$r = n^2 \frac{\hbar^2}{kme^2}$$

For $n = 1$, the radius is $r_1 = \frac{(1.055 \times 10^{-34})^2}{(9 \times 10^9)(0.91 \times 10^{-30})(1.6 \times 10^{-19})^2}$, which is 0.53×10^{-10} m or 0.53 Å.

The other radii are $r_n = n^2 r_1$, four times as large for $n = 2$, nine times as large for $n = 3$, and so on for other values of n.

Now we can find the energies. Instead of calculating the kinetic energy and the potential energy separately, we see that the Coulomb's law relation can also be written $\frac{ke^2}{r} = mv^2$, where the left side is the negative of the potential energy and the right side is twice the kinetic energy, $2K$. We will follow the established practice and call the potential energy U in the rest of this chapter

so that $-U = 2K$. The total energy is $E = U + K$ or $U - \frac{U}{2}$, so that $E = \frac{U}{2}$ or $\frac{-ke^2}{2r}$.

With our earlier value for r this is equal to

$$E = -\frac{ke^2}{2} \frac{kme^2}{n^2 \hbar^2}$$

or

$$-\frac{1}{n^2} \frac{k^2 me^4}{2\hbar^2}$$

For $n = 1$ this is $-\frac{(9 \times 10^9)^2 (0.91 \times 10^{-30})(1.6 \times 10^{-19})^4}{(2)(1.055 \times 10^{-34})}$ J or -13.6 eV. The other energies can be written as $E_n = (\frac{1}{n^2})E_1$, so that they are $-\frac{13.6 \text{ eV}}{4}$ for $n = 2$, $-\frac{13.6 \text{ eV}}{9}$ for $n = 3$, and so on, up to $E = 0$ as n approaches infinity.

The lowest state ($n = 1$) is the *ground state*. The others are the *excited states*. The atom will, unless something prevents it from doing so, go to the lowest state that is available to it. As it goes to a lower state it gives up the difference in the energy. One way to do this is to emit a photon. Let's look at a transition from the state $n = 3$ to the state $n = 2$. The energies of the excited states are $\frac{E_1}{4} = -3.4$ eV for $n = 2$ and $\frac{E_1}{9} = -1.51$ eV for $n = 3$. The photon energy is $\frac{E_1}{9} - \frac{E_1}{4}$ or -1.51 eV $- (-3.4$ eV$)$, which is about 1.9 eV. We can also calculate the corresponding frequency and wavelength from the fact that the photon energy is hf and the wavelength is $\frac{c}{f}$. (Just remember that you have to pay attention to the system of units: in the SI system you first have to change from eV to joules, and the wavelength then comes out in meters. The frequency is in hertz [Hz], where 1 Hz is one vibration per second.)

As n increases, and the energy gets closer to zero, the radius ($n^2 r_1$) gets larger. When n and r go to infinity, the electron and the proton are completely separated, and both the kinetic energy and the potential energy go to zero.

Our choice for the reference level leads to the fact that when the electron is closer to the nucleus, so that its potential energy is smaller than at the reference level, both the potential energy and the total energy are then negative.

To take the atom from its ground state to the state where the electron and proton are infinitely far from each other, we have to give it 13.6 eV. This, the energy to destroy the atom, is also called its *binding energy*.

To remove the electron is called to *ionize* the atom. The hydrogen atom's *ionization energy* is therefore also 13.6 eV.

EXAMPLE 2

(a) What is the total energy and what are the potential and kinetic energies of a hydrogen atom in the state with $n = 4$?

(b) What are the energy, frequency, and wavelength of the light emitted by a hydrogen atom in the transition from the state with $n = 4$ to the state with $n = 2$?

Ans.:

(a) For $n = 4$ the energy is $-\frac{E_1}{4^2} = -0.85$ eV. The potential energy is twice as large and the kinetic energy is 0.85 eV.

(b) The energy is $\Delta E = E_4 - E_2 = -0.85 - (-3.40) = 2.55$ eV.

The frequency is $\frac{\Delta E}{h}$. In SI units the energy is $(2.55)(1.6 \times 10^{-19} \text{ J}) = 4.08 \times 10^{-19}$ J, and $f = \frac{\Delta E}{h} = \frac{4.08 \times 10^{-19}}{6.63 \times 10^{-34}} = 6.15 \times 10^{14}$ Hz. The wavelength is $\frac{c}{f} = \frac{3 \times 10^8}{6.15 \times 10^{14}} = 4.88 \times 10^{-7}$ m = 488 nm (= 4880 Å).

Suddenly, photons everywhere

The Bohr model's success confirmed the existence of the photons that Einstein had introduced when he explained the photoelectric effect. We have already discussed the photoelectric effect: the photon hits a material and gives all of its energy to an electron. Part of the energy is used to liberate the electron, i.e., to overcome its binding energy. Whatever is left over is given to the electron as kinetic energy.

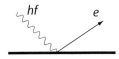

$$hf = K + E_B$$

where the left side is the photon energy and the right side is the kinetic energy, K, of the electron plus its binding energy. (The binding energy, E_B, is the energy needed to separate the electron from the material of which it is originally a part.) This is Einstein's photoelectric equation. In the initial experiments the electrons were ejected from metals in which the binding energy of the electrons that are released is typically of the order of a few eV.

The Compton effect

A second interaction between a photon and an electron is the Compton effect. This time the photon gives only part of its energy to an electron and gives the rest to a new photon. The energies are related by

$$hf = hf' + K_e$$

where hf is the energy of the original photon, f' is the frequency of the new photon, which is also called the recoil photon, and K_e is the electron's kinetic energy.

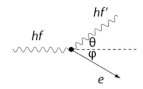

This experiment was done with x-rays with frequencies two or three orders of magnitude greater than those of visible light. The photon energies are therefore also greater (of the order of keV), and the electron's binding energy is so small by comparison that it is usually neglected. The Compton effect can therefore be analyzed as a collision between a photon and an electron in which both energy and momentum are conserved.

The result is a relation between the wavelength, λ, of the original photon and the wavelength, λ', of the new photon. Their difference is

$$\lambda' - \lambda = \Delta \lambda = \frac{h}{m_e c}(1 - \cos \theta)$$

This result, the *Compton effect equation*, is startling and incomprehensible to anyone who expects the radiation to follow Maxwell's electromagnetic theory. The photon collides with an electron and its energy changes. Its frequency and wavelength change. It is as if a beam of blue light were suddenly to turn red. Once again we are confronted with the particle properties of photons.

EXAMPLE 3

In the Compton effect the most probable angle for the recoil photon is 180°. What is the energy of the Compton electron for this angle when the energy of the original photon is equal to $E_0 = m_e c^2 = 0.51$ MeV, the rest energy of an electron?

Ans.:

For $\theta = 180°$, $\cos \theta = -1$, and $(1 - \cos \theta) = 2$.
$\lambda = \frac{c}{f} = \frac{hc}{hf} = \frac{hc}{E}$.

The Compton equation for this angle can now be written as

$\frac{hc}{E'} - \frac{hc}{E} = \frac{2h}{m_e c}$ or $\frac{1}{E'} - \frac{1}{E} = \frac{2}{E_0}$.

We can solve for E' by writing $\frac{1}{E'} = \frac{1}{E} + \frac{2}{E_0} = \frac{2E + E_0}{EE_0}$ so that

$E' = \frac{EE_0}{2E + E_0}$.

For $E = E_0$, E' is equal to $\frac{E_0^2}{3E_0} = \frac{1}{3}E_0$.

The energy of the electron is $E - E'$, which is $\frac{2E_0}{3}$.

For all other angles the electron energy is less. It ranges from zero to $\frac{2}{3}E_0$.

For higher photon energies the ratio of the electron energy to the photon energy (which here is $\frac{2}{3}$) is closer to one.

Pair production and annihilation

If a photon has still larger energy, of the order of MeV, it can create an electron–positron pair. (The positron is like an electron, but positively charged.) This time the particles are not just knocked out of a material or given some momentum in a collision. They are not there at all to start with, and are created in the process that is called *pair production*. The creation uses up an energy equal to the rest energy of the two particles, which is 1.02 MeV, which is therefore the minimum energy that a photon must have if this process is to occur. If the photon has additional energy it is given to the two particles as kinetic energy.

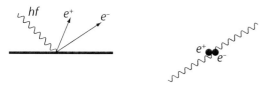

The photoelectric effect, the Compton effect, and pair production are three processes that are initiated by photons. There are also processes that go in the other direction, initiated by electrons and resulting in photons. One is *pair annihilation*. If an electron and a positron find themselves near each other, they will *annihilate*, i.e., they will both disappear, and give their energy to two photons.

This usually happens when a positron (for example, after being emitted by a radioactive substance) slows down and comes approximately to rest near an electron. The momentum of the two particles at rest is zero, and that is why a single photon cannot be created. It takes two photons, moving off in opposite directions, to satisfy the law of conservation of momentum.

EXAMPLE 4

Most detectors and counters of high-energy photons do not detect or count the photons directly. Instead they count the electrons emitted through one of the *photon-electron interactions*.

A photon has an energy of 2 MeV. What are the three processes by which this photon can lose energy with the emission of one or more electrons? What is the range of energies of the resulting electrons in each case?

Ans.:

The three processes are the photoelectric effect, the Compton effect, and pair production. In the photoelectric effect the photon loses all of its energy. A few eV are used to liberate the electron, but most of the 2 MeV go to the photon.

In the Compton effect part of the energy goes to the electron and part goes to the recoil photon. The Compton-effect equation shows that the largest difference between the initial photon and the recoil photon occurs when the angle between them is 180°. This, then, is the angle between them for which the electron energy is largest. The previous example showed that the energy of the recoil photon for this angle is $E' = \frac{EE_0}{2E + E_0}$. The corresponding electron energy is $E_e = E - E' = E - \frac{EE_0}{2E + E_0} = \frac{E(2E + E_0) - EE_0}{2E + E_0} = E\frac{2E}{2E + E_0}$.

For a 2-MeV photon this is $\frac{4}{4.51}E = 1.77$ MeV. The electron can have any energy from zero up to this maximum.

In pair production 1.02 MeV $(=2E_0)$ is used to create the pair. The rest, 0.98 MeV, is shared as kinetic energy between the electron and the positron.

EXAMPLE 5

A positron and an electron are at rest next to each other. What happens?

Ans.:
The positron and the electron disappear. They *annihilate*. In their place two photons appear, going in opposite directions, each with the energy $E_0 = 0.51$ MeV.

X-rays

The final process that we will consider is x-ray production. Actually there are two processes. They occur together when a beam of electrons hits a target material. (In commercial x-ray tubes it is usually a metal high up in the periodic table of elements, because the emission of x-rays is then more likely.)

When the electrons hit the target, they are stopped. This means that they undergo an acceleration, and we already mentioned in connection with the Bohr model that accelerated electrons are expected to radiate. In x-ray production this actually happens. The result is easily understood in terms of photons. The electrons can give the photons various amounts of energy, but the maximum energy that a photon can get is equal to the energy of an electron in the beam. The emitted photons have a maximum energy and frequency and a corresponding minimum wavelength. All lower energies are possible, and the process is therefore called *continuous x-ray production*.

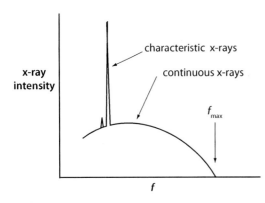

The second process that occurs simultaneously is the result of the fact that the electrons can raise the atoms of the target material to higher energy levels or knock electrons out completely. The target atoms will come back to the ground state with the emission of photons. The photons have the energies characteristic of the energy-level structure of the target atoms and are therefore called *characteristic x-rays*. Because the target is a heavy metal, the x-ray energies are in the range of 1000 eV (keV) rather than the eV of the hydrogen atom.

EXAMPLE 6

The electrons in an x-ray tube are accelerated by a potential difference of 12 kV in an electron gun. What are the minimum and maximum energies, frequencies, and wavelengths of the x-rays that are produced?

Ans.:
The maximum x-ray energy is 12 keV and the minimum is zero. The minimum frequency is zero and the maximum wavelength is infinite. The maximum frequency is $\frac{E}{h} = \frac{(12000)(1.6 \times 10^{-19})}{6.63 \times 10^{-34}} = 2.9 \times 10^{18}$ Hz. The corresponding wavelength is $\frac{c}{f} = \frac{3 \times 10^8}{2.9 \times 10^{18}} = 1.03 \times 10^{-10}$ m $= 0.103$ nm.

12.2 The new synthesis

Photons and electromagnetic waves

There was an overwhelming amount of evidence that showed that light and other kinds of electromagnetic radiation consist of photons, each with a definite amount of energy and a definite amount of momentum, both proportional to the frequency.

But what does it mean to have a photon with a certain frequency? Frequency and wavelength are the attributes of waves, not of particles. What happened to the electromagnetic waves that had so successfully explained the phenomena of light earlier? There seemed to be no way that a particle theory could explain interference and polarization.

What a dilemma! More and more phenomena were discovered that could be explained only by assuming that light consists of photons. At the same time the phenomena of interference and polarization seemed to make it clear that light is a wave phenomenon. Two theories seemed to be necessary to describe light, and the two were quite incompatible.

So which is it? Does light consist of particles or waves? Is it just one of the two, or both, or neither one? The question is at the heart of the difference between classical physics, i.e., Newtonian mechanics and the electromagnetism of Maxwell, and *modern* physics.

Faced with incontrovertible evidence for light waves as well as for photons there seemed to be no way out. The established view of the time became that it had to be both, with light propagating as waves and being emitted and absorbed as photons. It didn't seem to make sense, but no one knew how to resolve the dilemma. There was good reason to believe that important elements to understanding were missing, but that is easier said looking back from our present vantage point.

Complementarity

Bohr tried to transform ignorance into virtue by enshrining it in a new philosophical principle called *complementarity*. He said that both aspects of light exist. They represent *complementary* views that are different and even opposite and incompatible. Only by looking from different viewpoints and accepting both can the whole range of phenomena surrounding the concept of light be seen and understood.

Having established a principle meant that it could, perhaps, be applied to other phenomena. Bohr tried to do this. He pointed to complementary approaches in realms far from physics. There is reason and emotion, thought and sentiment, justice and love. The roles of actor and spectator are complementary, in that it seems that you can only be one or the other, but with a fuller view resulting from being both. And think about being both a student and a teacher! Bohr suggested that different cultures are complementary, in that differences in background lead to profound differences in points of view, with a higher synthesis coming with knowledge and experience of the complementary modes of life and thought.

The photons and their guide

Today we can make a beam of light with such small intensity that what we observe is only one, or just a few photons. Each can be separately detected by a *counter*.

Is light really just photons, or is it photons sometimes and waves at other times? Bohr said that both were necessary, and represented complementary points of view. Only with the inclusion of both apparently incompatible sets of phenomena could the full range and meaning of the concept of light be understood.

Today there is a better answer, consistent with all observations and free of the earlier contradictions. This, then, is the synthesis of the wave and particle descriptions: light is an electromagnetic wave. It moves through space as a vibration of electric and magnetic fields. The electric and magnetic fields travel together just as Maxwell described them. The experimental evidence consists of Young's double slit experiment and all the other phenomena that are the result of interference effects.

What Maxwell didn't know is that the field is *quantized*. When it interacts with a screen or anything else, it can do so only in definite amounts, each of which is a quantum of the electromagnetic field. *The photon is a quantum of the electromagnetic field.*

We don't know where a photon will appear on the screen. If there is only one it can be anywhere. Only when there are many of them can we predict what the pattern on the screen will be.

The intensity of light can be described and calculated from the wave theory just as was true before photons were ever thought of. Maxwell showed it to be proportional to the square of the electric field. Now, however, we have to interpret the intensity as telling us how many photons are likely to be observed: *the square of the electric field at any point on the screen is proportional to the probability that a photon can be observed there.* We need both concepts. *The wave tells us where the photons are likely to be.*

The electromagnetic wave, with its propagating electric and magnetic fields, is a ghostly presence, guiding and deciding what can be "seen," but it is not itself directly observed. What we observe are the photons, the quanta of the field. Whenever the wave hits something, as for example in the photoelectric effect, it can lose energy. But it can do so only in pieces, in the *quanta* of the field. We then observe photons, with their definite energy and momentum.

We can't tell precisely where any one of them will be. But together they form the patterns that can be calculated from the wave theory, and that Maxwell had described and calculated earlier.

The fields are continuous. The photons are not. They are discrete. They are the *quanta* of the electromagnetic field.

Are the photons there before they are observed, before the interaction takes place? No, only the wave propagates. It fills space and can go through two slits at the same time. Light is not a beam of photons. They are there only when the wave loses energy by hitting a screen or some other material.

So is light still sometimes a wave and sometimes a particle? Is it still "both"? In some ways it may still seem to be so, but what we have now is a description that encompasses the properties of waves *and* particles. It is neither of the two, but more than both. The new picture that we have of what happens is more complex, and richer than either one or than the two taken separately.

Beyond light: electron waves and the new synthesis

The reconciliation of photons with electromagnetic waves received its greatest impetus from the realization that the same considerations apply to electrons, and even further, to parts of atoms, and eventually to all particles and their combinations.

Before 1905 light had been thought to consist of electromagnetic waves. From then on it became clear that it also had the attributes normally associated with *particles*. Now (from 1925 on) electrons, the quintessential particles, were seen to have the wave attributes that had never before been considered appropriate for them.

We called an electromagnetic wave a ghostly presence that can exchange energy only in quanta. Similarly there are waves that describe the behavior of the electrons. They are no more and no less ghostly than the electromagnetic waves that we are so familiar with, and that we have come to know and learned to use in their various forms.

We can make double-slit experiments, just like Young's double-slit experiment, with electrons. This time again there is a wave that travels through space and through the two slits. Again there is a field—this time it is the "electron field." We can observe it when it interacts with a counter or with a television screen. And again the field can transfer energy only in discrete amounts, as the quanta that we call electrons.

We are so used to electromagnetic waves that in spite of what we have called their ghostly nature, we don't think of them as particularly mysterious or obscure. Perhaps it's time we gave equal rights and recognition to the waves that guide the observation of electrons and other particles of matter. Our mistrust is shown by the fact that we don't even have a good name for them. They are sometimes called *matter waves*. Today's materials science, the physics and chemistry that shapes our lives, is unthinkable without a knowledge of these waves. After more than three quarters of a century they are part of our common heritage, but still not part of common awareness and knowledge.

The quantum-mechanical description requires a new attitude toward the physical universe, incorporating uncertainty and probability in essential, inescapable ways. Some of the physicists most closely identified with bringing about the new view battled against this feature, including Einstein, Schrödinger, and de Broglie. In this they were largely unsuccessful, but many physicists today believe that the last word may not yet have been written about this question.

Let's see how the great change in attitudes began. We're back in 1924. We know about photons with their energy, E, equal to hf, and momentum, p, equal to $\frac{E}{c}$, so that for them $E = pc$. We also know that we can talk about the wavelength, λ, of the radiation, equal to $\frac{c}{f}$. This leads to $E = hf = \frac{hc}{\lambda}$. We can combine the two expressions for E to give $pc = \frac{hc}{\lambda}$ or $\lambda = \frac{h}{p}$.

Along comes graduate student Louis de Broglie who says suppose, just suppose, that this relationship works not only for photons but also for electrons. No one had ever considered the possibility that wave phenomena and a wavelength could conceivably be associated with electrons. And we wouldn't remember this stab in the dark either if it hadn't turned out to be one of the great seminal ideas of the twentieth century. De Broglie got his Ph.D. and a Nobel Prize, and h/p is called the de Broglie wavelength.

12.2 The new synthesis / 273

Did de Broglie have any evidence? Yes he did, of a sort. He said that if there are waves associated with electrons, then they can form standing waves around a proton at a distance r from the proton if the circumference, $2\pi r$, is equal to a whole number of wavelengths. Imagine a wave traveling around the proton. It goes around once. It goes around a second time. If the second time around the maxima and minima are exactly where they were the first time (are *in phase*), the waves will add, i.e., *interfere constructively*. If, on the other hand, the second time around, the wave's maxima fall behind the previous ones, they will fall further behind the next time, and the next time, until they cancel, i.e., *interfere destructively*. Only when the circumference, $2\pi r$, is exactly equal to a whole number of wavelengths, $n\lambda$, where n is a whole number, can the wave continue as a *standing wave*, as in the figure. Since $\lambda = \frac{h}{p}$ or $\frac{h}{mv}$, the criterion can also be written as $2\pi r = \frac{nh}{mv}$ or $mvr = n\hbar$, exactly the quantization condition that Bohr had written down in 1913 when he invented his model.

Where did Bohr get this relation? There was an attempt at justification, but by itself it would hardly have seemed convincing, except that *it worked*, which means that combined with some elementary mechanics and Coulomb's law it led to a set of values for the possible energies of a hydrogen atom which were *correct*, which were, in other words, experimentally observed. De Broglie's hypothesis led to the same relation. This suggested, at the very least, that there might be something to de Broglie's radically new approach.

Further evidence followed: Clinton Davisson and Lester Germer, working at the Bell Telephone Laboratories, then in New York City, saw some strange peaks in the intensity of an electron beam reflected from a nickel crystal when the angle was varied. These were soon interpreted as being the result of interference between the reflections from successive atomic planes of waves whose wavelength was $\frac{h}{p}$. In other words, they observed the interference pattern of electron waves as they were reflected from different planes of nickel atoms.

Shortly afterward George P. Thomson demonstrated what we now know as *electron diffraction*, the pattern produced by an electron beam when it passes through a thin crystal. (G. P.'s father, J. J. Thomson, is credited with the discovery of the electron by showing that a beam of electrons has a definite ratio, $\frac{e}{m_e}$, of charge to mass. It is sometimes said that he described the particle properties of electrons while his son demonstrated the electron's wave properties.)

The beams used by Davisson and Germer and by G. P. Thomson were *monochromatic*, i.e., they consisted of electrons of a single energy and momentum, and the experiments showed them to correspond to a single wavelength.

Even in the simplest atoms the electrons do not all have the same energy. In fact, what seemed to be de Broglie's strongest point, the explanation of the quantization condition, turned out to be, at best, suggestive of a better approach. The attempts to take the Bohr model and its quantization condition further had already limped along for more than a decade. But now everything was about to change, when following de Broglie, quantum mechanics was developed.

From de Broglie to Schrödinger

De Broglie's hypothesis worked perfectly for the electron beams of Davisson and Germer, and G. P. Thomson, with their single wavelengths, single momentum, and single energy. But Bohr had already shown that in atoms the electrons can have different energies and exist in different quantized energy levels. Was there now a way to describe the atomic electrons using de Broglie waves?

This is the step that Erwin Schrödinger took in 1925, opening up the field of *quantum mechanics*, the vastly successful framework for all modern theories of matter.

De Broglie's waves, with their single wavelength, momentum, and energy, represent motion without interactions, motion in accord with Newton's first law. Schrödinger's leap into the unknown created an analog to Newton's second law. He showed how the electrons behave when they interact with other particles. In the hydrogen atom, for example, the electron interacts with the proton as a result of their electric interaction. Instead of the concept of forces, however, Schrödinger used that of potential energy. For forces like the gravitational and electric forces that depend only on position, we can use force or potential energy. Either of the two concepts, whichever is more convenient in a particular situation, can be used to describe what happens. In the hydrogen atom, for example, the force between the electron and the proton is $\frac{ke^2}{r^2}$ and the potential energy is $-\frac{ke^2}{r}$, where k is the constant in Coulomb's law, e is the magnitude of the charge on the electron and on the proton, and r is the distance between the two. (For the potential energy we have, as usual, taken the reference level to be where r becomes infinitely large, i.e., where $\frac{1}{r} = 0$.)

Schrödinger started with a relation, or "wave equation," that describes the de Broglie waves. It is usually expressed in terms of the wavelength, λ, but because the wavelength is related to the momentum, p, by $\lambda = \frac{h}{p}$, it can also be expressed in terms of the momentum, or in terms of the kinetic energy, $K = \frac{p^2}{2m}$.

With de Broglie waves we are still talking about electrons moving with constant wavelength, momentum, and kinetic energy, unaffected by forces. But at this point we can ask whether a similar kind of relation, a new wave equation, might work when there are forces, and a potential energy, U. This is the crucial step that Schrödinger took. He said that if there are both a kinetic energy, K, and a potential energy, U, then the total energy is $E = K + U$, and we can substitute the quantity $E - U$ for K in the former wave equation.

The resulting equation is the *Schrödinger equation*, and it created a revolution.

How does the Schrödinger equation work? What can it do?

Our objective is to show you the Schrödinger equation and how it leads to results. Most of what we know about atoms (not just hydrogen atoms), as well as about molecules and solids and a lot more is obtained this way. We won't do any of the calculations for atoms and more complicated systems, but we will show examples that use some of the methods and illustrate the nature of the results that can be obtained.

Suppose we want to know about the hydrogen atom. We'll assume that the nucleus is fixed and concentrate on the electron. The "input" is the relation between potential energy and position, which for the electron in the hydrogen atom is $U = -\frac{ke^2}{r}$. We substitute this relation for U in the Schrödinger equation. The equation now leads to the "output" or "solution." That's a *wave function* with an amplitude that's usually called Ψ (capital Greek *Psi*). It tells you where the electron is likely to be. We can't find out where it is *exactly*. We can only determine the probability of finding it in a given region.

It turns out that the equation has such solutions only for certain values of the total energy, and these turn out to be the allowed, quantized energy levels of the hydrogen atom that we are looking for! In other words, we can find out as much as it is possible to know about where the electron is for each value of the energy, and we can find the *spectrum* of the possible energy levels.

Not only that, but the values of the orbital angular momentum come out right this time, and once we learn how to deal with the equation and add a few extra features, so do all the other atomic and molecular properties. Every place where the Bohr theory was stuck now becomes accessible. This includes the description of atoms beyond hydrogen in the periodic table of elements, for which the potential energy is no longer simple because there is then more than one electron.

To follow our program we will first review how all kinds of waves can be represented by sine functions. We will then apply this knowledge to electron waves by substituting the de Broglie wavelength $\lambda = \frac{h}{p}$, and show how this leads to the Schrödinger equation. Finally we will show for some examples how the Schrödinger equation leads to the wave functions and the energy levels, i.e., to the knowledge about where a particle is and what energy values it can have.

The main difficulty is that there are symbols that you are probably not used to (like Ψ). Don't be put off by them! There is some algebra, but you will see that it isn't any more difficult than what you have already worked with. We would like to encourage all of you to read the following sections before you go on to the Heisenberg

Uncertainty Principle. Some of you may not go through all the details, but we think that you will be happy to see that some of the most modern ideas in physics are not at all mysterious.

We have not yet written down the famous equation, so that our talk about it is so far somewhat vague. We now want to get more serious. This wonderful part of physics, which plays such a profound role in all we know about the world of atoms, and in much of modern technology, has largely remained hidden from a majority of the population. We would like you to take a closer look at it, and although it may mean some courage on your part, you won't want to miss it.

Representing waves

The Schrödinger equation is a *wave equation*, and before we develop it in detail we will review the way waves of all kinds are described mathematically and with graphs. We already know how to represent waves that repeat in space after one wavelength with sines and cosines. (We won't deal with the repetition in time.) The figure shows a graph of sin x as a function of x. Below it is the graph of cos x as a function of x, which looks the same, except that the starting point is different. And below it is a graph of $-\sin x$, i.e., the first graph upside down.

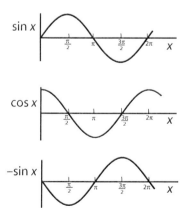

We can see that a special feature of the first two graphs is that at every value of x the middle graph gives the slope of the graph above it. We see, for example, that when sin x is horizontal, so that its slope is zero, at $\pi/2, 3\pi/2$, etc., cos x is zero. When sin x is zero, its slope is one, and so is the value of cos x. The slope of sin x vs. x is cos x vs. x.

Another way of saying exactly the same thing is "the derivative with respect to x of sin x is cos x."

Underneath both graphs is the graph of $-\sin x$ vs. x, and you can see that it gives the slope of the graph just above it, of cos x. This can also be expressed by saying that the derivative with respect to x of cos x is $-\sin x$. It is also the "second derivative" of sin x. We see that the function sin x has the property that its second derivative (the slope of the slope) is equal to the negative of itself.

There is a special notation for these statements. For the first derivatives, $\frac{d}{dx}(\sin x) = \cos x$ and $\frac{d}{dx}(\cos x) = -\sin x$. For the second derivative of sin x, $\frac{d^2}{dx^2}(\sin x) = -\sin x$ (read $d2dx2$ of $\sin x = -\sin x$.) And if $y = \sin x$, $\frac{d^2y}{dx^2} = -y$. You can stay with "slope" instead of "derivative," and "slope of the slope" for the second derivative. They mean the same thing.

For our electron wave we'll need the slightly more general function $y = A \sin kx$. The amplitude, A, and the *wave number*, k, can have values to suit the particular wave that we want to consider. For this function the slope is $kA \cos kx$. We can write the relation between the function and its slope as $\frac{dy}{dx} = kA \cos kx$. We can now look at the slope of $kA \cos kx$. It is $-k^2A \sin kx$, which we see to be $-k^2y$. This is the first derivative of $kA \cos kx$ and the second derivative of $y = A \sin kx$. We can write it as $\frac{d^2y}{dx^2} = -k^2y$.

These functions repeat when kx is increased by 2π. In other words, $A \sin kx = A \sin (kx + 2\pi)$ or $A \sin kx = A \sin k(x + \frac{2\pi}{k})$. This shows that the function repeats when x is increased by $\frac{2\pi}{k}$.

But this is just what we mean by a wavelength. We see that the wavelength, λ, is equal to $\frac{2\pi}{k}$. The quantity k is called the wave number. (The symbol is the same as the one we use for the constant in Coulomb's law, but it is a quite different quantity. It should be easy to see from the context which one we're talking about at any one time.)

That's it for the mathematics. The rest is just some talk and playing around with what you already know.

First the talk. $\frac{d^2y}{dx^2} = -k^2y$ is a relation between a function (y) and its second derivative ($\frac{d^2y}{dx^2}$). It's an equation. We can call it the wave equation. It contains a derivative, so it's called a *differential equation*. $y = A \sin kx$ is one function that is a solution of this equation. What does that mean? Only that if you substitute $A \sin kx$ for y in the equation, the left-hand side is equal to the right-hand side.

All this was well known long before the twentieth century, and had been used successfully to describe all kinds of waves, such as waves on a string as on a violin or guitar, or sound waves, or light waves.

Finally the best we have: the Schrödinger equation

What is it that oscillates in an electron wave? There is nothing that moves, as on the string or in the sound wave. There isn't a simply observable physical property that changes, as the pressure does in a sound wave and the electric and magnetic fields in an electromagnetic wave. The electron wave tells us where the electron is. Not exactly, but, after we square it, it tells us what the probability is of observing the electron in a given spot.

Part of the mystery that still seems to surround the concept of electron waves comes from the fact that we have never found a good word for whatever it is that changes. It continues to be called *the wave function*. It would be nice to have a symbol that would make the waves seem less remote, but that hasn't happened either. The symbol that is universally used for them is Ψ, the Greek *psi*.

Now for the playing around. For all waves $k = \frac{2\pi}{\lambda}$. For the de Broglie waves we also have $\lambda = \frac{h}{p}$, so that $k = \frac{2\pi p}{h}$ or $\frac{p}{\hbar}$ and $p = \hbar k$.

We need the k that appears in the wave equation, but we want to express it in terms of the kinetic energy, K, which is $\frac{1}{2}mv^2$ or $\frac{p^2}{2m}$. For the de Broglie waves we can write it as $K = \frac{\hbar^2 k^2}{2m}$, so that $k^2 = \frac{2mK}{\hbar^2}$.

The same old differential equation can now be written to say that the slope of the slope, or the second derivative, $\frac{d^2\psi}{dx^2}$, is equal to $-k^2\Psi$, i.e., $\frac{d^2\psi}{dx^2} = -\frac{2mK}{\hbar^2}\Psi$ or $-\frac{\hbar^2}{2m}\frac{d^2\psi}{dx^2} = K\Psi$.

This is all just playing around with symbols, and has so far accomplished nothing except to recast the wave equation for a noninteracting electron. Now, however, comes the leap of faith. Let the electron interact with other charged particles, e.g., a positively charged nucleus, so that there is a potential energy. What happens if we assume that the equation also holds when there is such a potential energy, U, so that the total energy, E, is $K + U$, and K is equal to $E - U$? We rewrite the relation once more, but with $E - U$ instead of K. That gives us $-\frac{\hbar^2}{2m}\frac{d^2\psi}{dx^2} = (E - U)\Psi$, which we can also write as $-\frac{\hbar^2}{2m}\frac{d^2\psi}{dx^2} + U\Psi = E\Psi$.

This is what we've been leading up to. *This is the Schrödinger equation.*

What does the Schrödinger equation tell us?

The Schrödinger equation doesn't look anything like Newton's second law, but that is what it corresponds to. It looks, in fact, more like the law of conservation of energy. If we think of the term with the derivatives as representing the kinetic energy, then it says "K"$\Psi + U\Psi = E\Psi$, which looks reassuringly familiar. Newton's second law comes to life when we say what the forces are. The same thing happens here when we specify the potential energy. Here is our prime example: let $U = \frac{-ke^2}{r}$, so that it describes the potential energy of an electron in the field of a proton. (Before we can do that we have to generalize our one-dimensional Schrödinger equation to three dimensions by adding derivatives with respect to y and to z.) The equation now becomes the key that unlocks the secrets of the hydrogen atom.

That key is so different from the classical, Newtonian one that it took a while to figure out how to use it, and there are aspects of the procedure and the results that remain controversial to this day.

Let's go back for a moment to the waves on a guitar string. For them $y = A \sin kx$. We saw that $k = \frac{2\pi}{\lambda}$. Now we know, quite without any new ideas, that if the string has a definite length, waves on the string can exist only with certain definite wavelengths, corresponding to definite musical notes.

The situation is similar for the Schrödinger equation. First put in the potential energy. The solutions are a set of wave functions, Ψ, now not just sines and cosines. They are still functions of the space coordinates. The wave function tells us whatever there is to know about where the particle that the equation describes is located. That information is different and less exact than in the classical case. It tells us the probability of finding the particle at any given place or in any given region.

There is a second piece of information that we get. There are solutions only for certain definite values of the quantity E, the total energy. Each value of E corresponds to a possible, allowed value of the energy, i.e., to an energy level of the system.

For the hydrogen atom the allowed energies come out to be exactly those of the Bohr model, but the angular momentum which Bohr had wrong is now correct, and a whole lot of other stuff, which Bohr couldn't even get close to, is also there. Most importantly, we can describe the properties of the other, more complicated, atoms as well as combinations of atoms in molecules or solids.

How does the wave function tell us where the particle that it describes is to be found? For a wave on a string the energy is proportional to the square of the amplitude. For an electromagnetic wave, in which the electric field varies in time and space, the energy is proportional to the square of the magnitude of the field. When an electromagnetic wave hits something, the square of the electric field in any small region of space is proportional to the number of photons to be found there. It is proportional to the probability of finding a photon there. That is also the way to use the wave functions that are the solutions of the Schrödinger equation. When the matter wave interacts with a screen or with a counter or with anything else, the square of the magnitude of the wave function, Ψ^2, in a given region is proportional to the probability of finding our particle in that region.

Here, once more, is what we need to do to determine how a particle moves when there are forces on it. We represent the forces by the way the potential energy changes as the particle moves. For the electron in the hydrogen atom, for example, it is $-\frac{ke^2}{r}$. For electrons in other atoms it cannot be written down so simply because there are other electrons, but there is still some function that describes the electric potential energy. We put the appropriate potential energy in the Schrödinger equation and try to find a function that is a solution to the equation. This also gets us the values of the energy, E, for which solutions exist. They are the values of the energy levels. For each such value there is a *solution* (or perhaps several). It is a function, the *wave function*, that when substituted in the equation makes the right-hand side equal to the left-hand side. When we square this function we get another function of position, and this is the function that gives us the probability that we can find the electron in any given region.

The information that we get about the position of our particle is not exact. We get only a probability. That's the best that we can do. But it's a lot. It allows us to describe the structure of atoms, their combinations, and much more. Quantum mechanics is the framework on which our knowledge of the structure of matter is built.

EXAMPLE 7

The "square well"

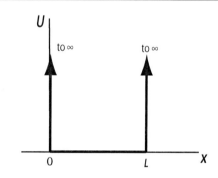

An electron can be represented as being in an *energy well* that is described by $U = 0$ between $x = 0$ and $x = L$ and by $U = \infty$ outside the well, everywhere else. This may seem like an unrealistic problem. It is, however, the simplest model for a number of physical situations. It can, for example, be used for the nucleus, with nucleons replacing the electrons. To keep the problem as simple as possible we limit it to one dimension. The extension to three dimensions is straightforward. The problem also illustrates that the quantization comes from the confinement of particles to a limited region of space.

(a) What are the wave functions?

(b) Sketch the wave functions for the first three energy levels.

(c) What are the energy levels of the system?

(d) Draw an energy level diagram.

Ans.:

(a) The particle cannot be outside the well, where the potential energy is infinitely large. The wave function must therefore go to zero at the well's boundaries. Inside the well the Schrödinger equation is

$$-\frac{\hbar^2}{2m}\frac{d^2\Psi}{dx^2} = E\Psi$$

where we have used the x component only, and the potential energy $U = 0$.

Since $U = 0$, there is only kinetic energy, and $E = \frac{1}{2}mv^2 = \frac{p^2}{2m}$. We can therefore write the equation as $\frac{d^2\Psi}{dx^2} = -\frac{2mE}{\hbar^2}\Psi$, where the right-hand side is equal to $-\frac{p^2}{\hbar^2}\Psi$, which is equal to $-k^2$, where k is the wave number $\frac{2\pi}{\lambda}$, i.e., the equation is $\frac{d^2\Psi}{dx^2} = -k^2\Psi$.

We know that this equation has solutions that are sine and cosine functions. Both have the property that the slope of the slope (the second derivative) is proportional to the negative of the function itself. We need a function that is equal to zero at $x = 0$. We cannot use the cosine function since $\cos 0 = 1$. The sine function, $\sin x$, on the other hand, is zero at $x = 0$. The wave functions therefore have the form $\Psi = A \sin kx$, where we still have to determine the possible values of k.

The wave function also has to go to zero at $x = L$. We can therefore immediately see what the wave functions are. They are sine functions that have the value zero at both ends of the well. There is no restriction on how often they can go to zero inside the well, so there is an infinite number of such wave functions. The largest wavelength is such that half the wavelength is equal to L, i.e., that $L = \frac{1}{2}\lambda$. For the next one $L = \lambda$. For all of them $L = n\frac{\lambda}{2}$. Each wave function is described by a value of the *quantum number n*. Since $k = \frac{2\pi}{\lambda}$ and $\lambda = \frac{2\pi}{k}$, $L = \frac{n}{2}\frac{2\pi}{k}$ or $k = \frac{n\pi}{L}$.

We can see that also from the relation for the value zero of the wave function at $x = L$, $\sin kL = 0$. This is so for $kL = 0$ and also for $kL = \pi$, or 2π, 3π, or, in general, $n\pi$. Since $k = \frac{2\pi}{\lambda}$, this gives $\frac{2\pi}{\lambda}L = n\pi$, or $L = n\frac{\lambda}{2}$, and $k = \frac{n\pi}{L}$, as before.

We can now write the wave functions as $\Psi = A \sin \frac{n\pi}{L}x$. There is a different one for each value of n.

(c) To see the values of the energies, we have to go back to E. $E = \frac{1}{2}mv^2 = \frac{p^2}{2m} = \frac{\hbar^2 k^2}{2m}$. We can now substitute the quantized values of k, $k = \frac{n\pi}{L}$, to get $E = \frac{\hbar^2}{2m}\frac{n^2\pi^2}{L^2}$ or $E = n^2\frac{\hbar^2\pi^2}{2mL^2}$.

(b)

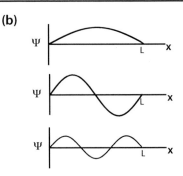

(d)

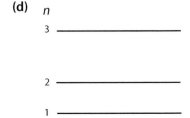

For a particle with a mass m and a well of width L, the energy levels are $n^2 E_1$, where E_1 is the energy of the lowest state (the ground state) $E_1 = \frac{\hbar^2\pi^2}{2mL^2}$. Since the problem specifies that $U = 0$ inside the well, this energy is entirely kinetic.

EXAMPLE 8

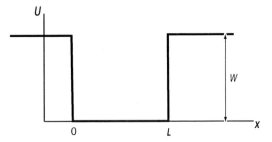

An electron is in an energy well that is described by $U = 0$ between $x = 0$ and $x = L$, and by $U = W$, where W is constant, outside the well, everywhere else.

(a) Describe the form of the wave functions inside and outside the well for values of the energy less than W.

(b) Sketch the wave functions for the first three energy levels.

Ans.:

(a) Inside the well the wave functions are similar to the ones in the previous example, but this time the values at $x = 0$ and $x = L$ are not zero. They can be a combination of sines and cosines. We can see this from the fact that there is, this time, a Schrödinger equation with a different kind of solution also outside the well.

(b)

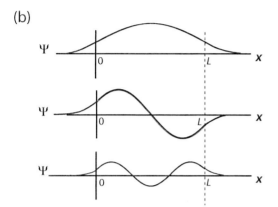

Outside the well the Schrödinger equation is $-\frac{\hbar^2}{2m}\frac{d^2\Psi}{dx^2} + W = E\Psi$. Since E is smaller than W, we can write this as $\frac{\hbar^2}{2m}\frac{d^2\Psi}{dx^2} = (W - E)\Psi$, where both sides are positive.

This time, outside the well, we need a function for which the slope of the slope (the second derivative) is proportional to the function itself and not its negative. Such a function is e^{ax}. The function e^x has the property that its slope is equal to the function itself. For e^{ax} the slope is ae^{ax} and the slope of the slope is $a^2 e^{ax}$.

Since a can be either positive or negative, the wave function outside the well can either increase or decrease exponentially. We have to discard the exponential increase, which goes to infinity and would lead to unphysical results. (The particle would be infinitely far away all the time.) What remains is the exponential decrease. It has to be fitted to the sinusoidal wave function inside the well in such a way that the function and its slope are continuous, in other words that there are no breaks and no kinks. (A break in the wave function would mean that it would have two values and therefore two probabilities there. A kink would mean two values for the slope, and hence two possible values for other physical quantities.)

This result is quite different from the classical result. Newtonian mechanics cannot envision a situation where the potential energy is larger than the total energy so that there is a negative kinetic energy. Here there is a rapidly decreasing but finite probability that this occurs. If we think of the "well" as a box, we see that the particle has a finite probability of being outside the box for all finite heights of the box.

EXAMPLE 9

This time there is a "box" with a finite "wall," as in the previous example, but further outside the energy E is larger than the potential energy U. Draw a sketch of a possible wave function.

Ans.:

In the well or box the wave functions for energies less than the height of the walls are sines as before. In the wall section they are exponentials as in the previous example. There is now a third region, farther outside. Since E is greater then U in this region, the wave functions are again sinusoidal, but with a smaller amplitude.

The fact that there is a solution outside, in this third region, means that there is some chance that

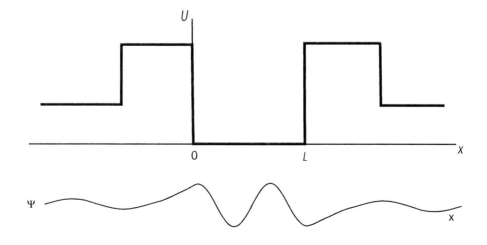

the particle can be found there. In other words, the particle can "leak out" or "tunnel" out of the box.

This could not occur in Newtonian mechanics. Since the particle could never be in the wall region, where U is greater than E, corresponding to a negative kinetic energy, there is no way for the particle to overcome the barrier represented by the wall of the box, and it would have to stay inside. The tunneling out is possible according to quantum mechanics. The observation of tunneling is a strong confirmation of the correctness of quantum mechanics. It describes the tunneling of alpha particles from the nucleus in radioactive decay and the overcoming of barriers by electrons in solids.

EXAMPLE 10

What is the Schrödinger equation for a particle in simple harmonic motion?

Ans.:

The potential energy of a particle moving along the x-axis in simple harmonic motion is $\frac{1}{2}kx^2$.

This potential energy is substituted for U in the Schrödinger equation (using only the x component) to get

$$-\frac{\hbar^2}{2m}\frac{d^2\Psi}{dx^2} + \frac{1}{2}kx^2\Psi = E\Psi$$

The solution leads to the energy levels $(n + \frac{1}{2})hf$. The spacing of the levels is the same as that of Planck, but the lowest level is not at $E = 0$, but at $E = \frac{1}{2}hf$. That there is a lowest energy that is not zero (a "zero-point energy") is characteristic for quantum mechanics. It shows that there is some motion even at the lowest possible energy.

Heisenberg and the uncertainty principle

Quantum mechanics was developed almost at the same time, independently, by two people. Werner Heisenberg used mathematical methods quite different from those of Schrödinger, and at first it seemed that the two approaches had little in common. Shortly afterward, however, Schrödinger showed that in spite of the fact that they look so different, the two theories are, in fact, equivalent.

Heisenberg emphasized a feature of quantum mechanics that has come to be known as the *Heisenberg uncertainty principle*. Consider first a de Broglie wave with a single wavelength, λ. It is not possible to say where the wave is. It extends through space. The momentum, on the other hand, equal to $\frac{h}{\lambda}$, is precisely known.

It is possible to add waves with different wavelengths, and so to construct a "wave packet" that is in a limited region of space. To do this it is necessary to use waves with different wavelengths and the wave's momentum is then no longer known exactly. With an infinite series of waves, each with a different wavelength, it is possible to describe a localized particle exactly, but the wavelength, and hence the momentum, is then completely unknown.

Now let's look at what happens between these two extremes. Suppose you try to measure the position of an electron. You can shine light on it, i.e., you can send a photon toward it. By seeing the photon bounce off the electron you can determine the electron's position. But with light of wavelength λ, the position can be determined only with an uncertainty Δx of about λ. (With a microscope you can't see anything smaller than the wavelength of the light that you are using. You don't know where a wave is, closer than with an uncertainty of about its wavelength.) The photon that strikes the electron has a momentum $\frac{h}{\lambda}$, and therefore transfers an amount of momentum of about this magnitude to the electron. The momentum it gives up can be this much or less, depending on the angle. The momentum of the electron, Δp, is therefore uncertain to about this extent. We see that the product of the two uncertainties is equal to about h. This is a general result. Light with a smaller wavelength will reduce Δx, but the photon will have more momentum, and so increase Δp. The product $\Delta x \Delta p$ remains the same.

Note that the uncertainty here is intrinsic, i.e., it is not related to just how we measure the quantities. There is no way to get around the limitations of the Heisenberg uncertainty principle.

EXAMPLE 11

Show that the ground-state energy of the hydrogen atom is determined approximately by the uncertainty principle.

Ans.:

To show this we will look at the extent of the electron's wave function as the uncertainty in the distance. In other words, we will take Δx to be about equal to the radius r. We will use the Bohr model to estimate r. (Of course the Bohr model was developed more than a decade before the uncertainty principle, and was considered to be definite.) The lowest possible magnitude of the electron's momentum is its value in the ground state. We don't know its direction, and can consider the uncertainty Δp to be equal to the magnitude of the momentum in the ground state. The kinetic energy is equal to the negative of the total energy, $K = -E = \frac{k^2 m e^4}{2\hbar^2}$. K is equal to $\frac{p^2}{2m}$, so that $p^2 = \frac{k^2 m^2 e^4}{\hbar^2}$ and $p = \Delta p = \frac{kme^2}{\hbar}$. For r we will use the smallest Bohr radius, $\frac{\hbar^2}{kme^2}$, and we see that $\Delta x \Delta p = \hbar$, in keeping with the expectation from the uncertainty principle.

We have to remember that the uncertainty principle is itself not exact. The product $\Delta x \Delta p$ is *of the order* of $\hbar$, but may be larger or smaller. It has been shown that the smallest value that the product can have is $\frac{\hbar}{2}$.

12.3 Order in the universe: the elements

The beginning: hydrogen and its quantum numbers

We now know how to write down the Schrödinger equation for the hydrogen atom. We use $\frac{-ke^2}{r}$ for the potential energy, U, in it. We are not going to go through the solution process here, but we will write down some of the results.

The energy of the hydrogen atom depends only on the quantum number n, and is equal to $\frac{E_1}{n^2}$, where $E_1 = -13.6$ eV.

The orbital angular momentum was already quantized in the Bohr model by his quantization condition $mvr = n\hbar$. The solution of the Schrödinger equation for the hydrogen atom also leads to the conclusion that it is quantized, but with a different relation. It is related to a second quantum number, ℓ, and the orbital angular momentum is equal to $\hbar\sqrt{\ell(\ell+1)}$.

A third quantum number, m_l, describes how the atom can orient itself in a magnetic field.

These relations are not just pulled out of a magic hat, but are the result of perfectly well-defined mathematical procedures that need to be used to *solve* the Schrödinger equation and that existed and were known long before Schrödinger.

The numerological rules for the three quantum numbers, n, ℓ, and m_l, are the unambiguous mathematical consequences of the Schrödinger equation.

There is a fourth quantum number (m_s) for the electron in a hydrogen atom, which is not derivable from the Schrödinger equation. Like the other quantum numbers it is related to a dynamic property, in this case the electron's spin angular momentum. Its existence was discovered, even before the advent of quantum mechanics, when it was realized that there were twice as many possible states (for the two possible spin states) as could be accounted for without it.

Does that mean that it has no theoretical justification or explanation? No, the situation is much more interesting than that. It is also the natural outcome and consequence of an equation, but one that goes beyond the Schrödinger equation. Unlike it, it is consistent with the special theory of relativity. (In the Schrödinger equation we use $K = \frac{1}{2}mv^2$, which is equal to the kinetic energy only when v is much less than c, the speed of light.)

This equation, the Dirac equation, developed by P.A.M. Dirac in 1930, is not a generalization of the Schrödinger equation, but rather a quite different equation whose primary achievement is that it describes properties of the electron, including those that are direct consequences of the theory of relativity. These include the spin angular momentum, the spin magnetic moment, and their ratio, the gyromagnetic ratio. These properties cannot be derived from the Schrödinger equation. If we use the Schrödinger equation to find the energy levels of hydrogen, these properties have to be included separately.

The spin angular momentum of the electron always has the same magnitude. It can be written as $\hbar\sqrt{s(s+1)}$ so as to look like the orbital angular momentum, but s can have only the single value $\frac{1}{2}$. The component of this spin angular momentum along the direction of the magnetic field can take on two values, $m_s \hbar$, where m_s is equal to $+\frac{1}{2}$ or $-\frac{1}{2}$.

EXAMPLE 12

What is the angle between the spin angular momentum and the magnetic field in the ground state of hydrogen?

Ans.:

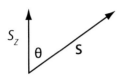

The spin angular momentum vector is S. Its magnitude, S, is $\hbar\sqrt{s(s+1)}$, where s is the quantum number that is always equal to $\frac{1}{2}$, so that $S = \hbar\sqrt{\frac{1}{2}(\frac{1}{2}+1)} = 0.866\hbar$. Its component along the magnetic field is $S_z = \frac{1}{2}\hbar$. The relation between the two is $S\cos\theta = S_z$ or $0.866\hbar \cos\theta = 0.5\hbar$, so that $\cos\theta = \frac{0.5}{0.866}$ or $\theta = \cos^{-1}\frac{0.5}{0.866} = 54.7°$.

Let's review the hydrogen atom quantum numbers. Here they are, together with the related mechanical properties:

(1) n can vary from one to infinity. The energy is $\frac{E_1}{n^2}$, where $E_1 = -13.6$ eV. The other quantum numbers do not influence the energy. (This is true only in the hydrogen atom!)

(2) ℓ can go from zero to $n-1$. The magnitude of the orbital angular momentum is $\hbar\sqrt{\ell(\ell+1)}$.

(3) m_l can vary between $-l$ and l. It describes the component of the orbital angular momentum along the direction of the magnetic field. This component is equal to $m_l\hbar$.

(4) m_s can be either $\frac{1}{2}$ or $-\frac{1}{2}$. The component of the spin angular momentum in the direction of the magnetic field is $m_s\hbar$.

Each possible set or combination of these quantum numbers distinguishes a particular *state* of the atom with a different wave function. The energies of some of the states may be the same, and this is true in hydrogen (only!) for all of the states with the same value of the quantum number n. Similarly, the orbital angular momentum is the same for all of the states with the same value of the quantum number ℓ.

EXAMPLE 13

What is the sequence of energy levels in the hydrogen atom?

Ans.:
In the hydrogen atom the energy is determined only by the quantum number n. But for each value of n there is more than one state, each with its different wave function. For the ground state, with $n=1$, there are two states, one with $m_s = \frac{1}{2}$ and one with $m_s = -\frac{1}{2}$. Both have $\ell = 0$ and $m_l = 0$. For $n=2$, ℓ can be zero or one. For $\ell = 0$, $m_l = 0$, and there are again the two states with the two different values of m_s. Still for $n=2$, there are also the states with $\ell = 1$. For them m_l can take on the values 1, 0, and -1. For each there are the two possible values of m_s, so that there are six states, for a total of eight for this value of n.

The next higher levels are for $n=3$. In hydrogen all have the same energy, with two states for $\ell = 0$, six states for $\ell = 1$, and 10 states with $\ell = 2$ ($m_l = 2, 1, 0, -1$, and -2, each with the two values of m_s), for a total of 18 states with this value of n. For $n=4$ there are 14 more states, and so on.

n

3                     $\ell = 0$ $\ell = 1$ $\ell = 2$

2    $\ell = 0$ $\ell = 1$

1    $\ell = 0$

$\ell = 2$: $m_l = -2, -1, 0, 1, 2$, $m_s = \pm\frac{1}{2}$ (10 states)
$\ell = 1$: $m_l = -1, 0, 1$, $m_s = \pm\frac{1}{2}$ (6 states)
$\ell = 0$: $m_l = 0$, $m_s = \pm\frac{1}{2}$ (2 states)

Hydrogen shows the way: the other elements

We can now look at what happens in atoms other than hydrogen. The first obstacle arises immediately. In the hydrogen atom we have just two particles, the proton and the electron. In all other atoms we have a nucleus and more electrons. Even in the old, classical, Newtonian scheme we don't know how to deal with more than two particles at a time. At least we don't know how to do it *exactly*.

What works sufficiently, most of the time, is to use a model that considers just one electron at a time, and to think of it as moving in a steady field, which is the average field whose sources are the nucleus and all of the other electrons. We know that these other electrons are moving around in ways that are impossible to know or to follow, but we will use the blur that they represent as if it were a steady background for the one electron on which we choose to focus. We can then talk about the wave function of that one electron in the field of the nucleus and all the other electrons.

The origin of order: the Pauli exclusion principle

We now have to consider the principle that gives rise to the hierarchy of the elements and leads to their diversity and to their order in the periodic table of the elements. Without it there would not be the vast variety of properties among the elements and their combinations. The world would be a much more homogeneous and bleaker place, without a chance for us to exist. Here it is:

Each electron in an atom must be characterized by a different set of quantum numbers. This is the *Pauli exclusion principle*, discovered by Wolfgang Pauli in 1925.

Without the Pauli principle all of the electrons in an atom could be in the lowest energy level, with the quantum numbers that the electron has in the ground state of the hydrogen atom. The atoms would be much more similar to one another than they actually are. The differences between the atoms would be minor, and combinations between them little different from element to element.

With the Pauli exclusion principle the electrons must have different sets of quantum numbers. They must move at different distances from the nucleus, with different energies and angular momenta, and be differently affected by magnetic fields. This is what underlies the richness among the elements and of their combinations, and the variety and complexity of chemistry and biology.

Atomic structure: beyond the Coulomb field

We can now go to the atoms of the periodic table of elements beyond hydrogen. They have nuclei that have more protons, so that there are also more electrons in the neutral atoms.

Each time a proton is added to the nucleus it changes the electric field and the electric potential for all the electrons. If this were all, we could deal with it very simply, because all that does is to change the amount of charge of the nucleus. The field would still be proportional to $\frac{1}{r^2}$, one that is called a *Coulomb field*. But we also add electrons, in an unknown, complicated way, at some distance from the nucleus.

The electrons exert forces on each other, and as they move, the electric field changes with time in unpredictable ways. If we want to calculate what happens, we have to resort to approximations. The usual approach is to look at one electron at a time, in the field created by the nucleus and all the other electrons. We don't know that field from moment to moment, but we can get a pretty good idea of its average over time and use that. This average field is no longer the Coulomb field of a single charge at the center. It is no longer proportional to $\frac{1}{r^2}$.

EXAMPLE 14

What are the quantum numbers that describe the two electrons in the ground state of the neutral helium atom?

Ans.:

Helium is the second element in the periodic table. Its atomic number is 2. n is still one. ℓ and m_l both have to be zero. Only m_s distinguishes the two electrons. The quantum numbers are $1, 0, 0, \frac{1}{2}$ and $1, 0, 0, -\frac{1}{2}$.

The number of protons in the nucleus (the atomic number, Z) increases by one for each step up the periodic table of elements. For each added proton another electron is added to the neutral atom. Where do these electrons go? The new atom will normally be in its *ground state*, the state with the lowest possible energy. Each added electron will therefore go into the configuration (the *state*) where it has the lowest possible energy, and this is determined by the set of its quantum numbers.

We know what happens when the electric field is exactly that of a single charge. The allowed states are then those of the hydrogen atom. For this case we know the sequence of energy levels. We know that the energy depends only on the quantum number n. The states with the lowest energy are all the possible states with $n = 1$ (2 of them). The next lowest are the eight with $n = 2$, all with the same energy ($\ell = 0$ with two different values of m_s, and $\ell = 1$ with 3 different values of m_l, each with two different values of m_s), and so on.

We have already seen that even with $Z = 2$ the electric field is more complicated than a Coulomb field. We now need to know what the sequence of energy levels is in atoms other than hydrogen. In addition to its dependence on n, the energy will then also depend on the value of the quantum number ℓ. Indeed, this ℓ dependence is fundamental to the structure of the periodic table of the elements.

The ℓ dependence and the order of the elements

If all states with a given value of n, or with a given value ℓ, are filled, the resulting distribution of electrons is spherically symmetrical. We talk about a filled *shell* (with a particular value of n) or a filled *subshell* (with a particular value of ℓ). In helium, for example, the $n = 1$ shell is filled with its two electrons. Element number 3, with three protons in its nucleus, is lithium. Look at the third electron. It will have $n = 2$, but which value of ℓ will correspond to the lowest energy, and therefore to the ground state?

Higher angular momentum represents motion farther away from the nucleus. We can see how this comes about by going back to how the Bohr model was extended before quantum mechanics was invented by proposing that the orbits could be elliptical. The "biggest" ellipse is a circle, and circular orbits represent the maximum angular momentum and the highest value of ℓ.

Bohr orbits for different angular momenta:

angular momentum and ℓ

lowest ($\ell = 0$) ————————

intermediate ⬭

highest (circular orbit) ◯

In this description lower angular momentum corresponds to elliptical orbits. But what is an orbit with zero angular momentum and $\ell = 0$? There is then no rotation at all, just back-and-forth motion of the electron. It is straight-line simple harmonic motion right through the nucleus. The fact that the straight-line orbit can be at any angle leads to he conclusion that the state with $\ell = 0$ is spherically symmetrical.

Bohr's way of looking at the atom, with its exact paths for the electrons, is not correct, but it gives a partial view of some of the properties of atoms and provides a picture that can help to clarify some of the concepts.

The following is true in Bohr's model and remains so in the correct quantum-mechanical description. The lower the angular momentum, the more time the electron spends near the nucleus. An electron in the $\ell = 0$ state spends the most time near the nucleus where the attraction is strongest and the energy the lowest. With higher angular momentum the electron spends more time further away. The attraction is then smaller and the energy higher.

This continues to be so as more electrons are added, with each, as always, going to the lowest available energy level. Each time we go from one element to the next, we add a proton to the nucleus. The corresponding added electron goes into the state with the lowest value of angular momentum, i.e., with the lowest value of ℓ that is available.

The sequence is simple, but only up to $Z = 18$ (Ar). By this time we have the two elements with $n = 1$, the two elements with $n = 2$ and $\ell = 0$, the six elements with $n = 2$ and $\ell = 1$, the two elements with $n = 3$ and $\ell = 0$, and the six elements with $n = 3$ and $\ell = 1$. The energies are different from those of the hydrogen atom, not only because the nuclear charge is different, but because of the ℓ-dependence of the energy. Nevertheless the sequence is still that of the energy levels of the hydrogen atom.

The order changes

As we move further along in the periodic table, the sequence changes. We might have expected to go on with $n = 3$, $\ell = 2$, but this does not happen. The low (zero) angular momentum of $n = 4$, $\ell = 0$ causes the next electron to be so close (on the average) to the nucleus, hence so strongly attracted, that this state has the lowest energy, lower than the state with $n = 3$ and $\ell = 2$. It is therefore the one that is filled first. Only after that do we get to the electrons with $n = 3$ and $\ell = 2$.

EXAMPLE 15

(a) What is the sequence of energy levels in the argon atom?

(b) What is the sequence of energy levels in the potassium atom? The calcium atom?

(c) Sketch the sequence of energy level schematically for the elements up to $Z = 36$, showing the dependence of the energy on n and ℓ.

Ans.:

(a) Argon is element number 18. In its ground state its 18 electrons are in the 18 lowest-energy states. First are the two in the shell with $n = 1$. Then there are the eight in the shell with $n = 2$, of which two are in the subshell with $\ell = 0$ and six in the subshell with $\ell = 1$. Then there are eight in the shell with $n = 3$, first two in the subshell with $\ell = 0$ and then six in the subshell with $\ell = 1$.

(b) In potassium ($Z = 19$) the first 18 electrons have the same values of n and ℓ as in argon (but with different energies!). The nineteenth is in the lowest state with $n = 4$, which is the one with $\ell = 0$ (and $m_l = 0$).

The sequence continues with calcium, with the twentieth electron in the second state with the same n and ℓ as potassium (and the other value of m_s). Only then come the 10 elements in which the subshell with $n = 3$ and $\ell = 2$ is gradually filled. After that we go back to the "regular" sequence, with $n = 4$ and $\ell = 1$. That gets us to $Z = 36$.

(c)

				$\ell = 1$
			$\ell = 2$	$\ell = 0$
			$\ell = 1$	
			$\ell = 0$	
		$\ell = 1$		
		$\ell = 0$		
	$\ell = 0$			
n:	1	2	3	4

$\ell = 2$: $m_\ell = -2, -1, 0, 1, 2$, $m_s = \pm \frac{1}{2}$	(10 states)
$\ell = 1$: $m_\ell = -1, 0, 1$, $m_s = \pm \frac{1}{2}$	(6 states)
$\ell = 0$: $m_\ell = 0$, $m_s = \pm \frac{1}{2}$	(2 states)

The interruption of the sequence of values of n has remarkable consequences. We have gone to states with $n = 4$ before completing the possibilities with $n = 3$. But higher values of n, as in the Bohr model, still correspond to larger (average) radii. Going back to $n = 3$ after already having put electrons in the shell with $n = 4$, with its larger radius, means that an *inner* shell, one with a smaller average radius, is then being filled.

When atoms combine to form molecules or solids, it is the outer electrons (the *valence electrons*) that are primarily involved. Filling an inner shell doesn't affect the outside of the atom very much, so that the way the atom engages in chemical combinations is not much changed. If electrons are added to an inner shell as we go from one element to the next, therefore, the neighbors in the periodic table are much more similar to each other than in the more normal sequence where electrons are added to the outermost shell. They are called *transition elements*.

Another feature of transition elements with an inner shell that is only partially full is that they often have extraordinary magnetic properties. This is because for them the angular momentum and the magnetic moment do not add up to zero as they do for filled shells or subshells.

The first group of transition elements is the one beyond calcium, from scandium ($Z = 21$) to nickel ($Z = 28$). It includes iron ($Z = 26$), the most important magnetic element. A similar situation occurs several times in the periodic table. The rare-earth elements ($Z = 58$ to 71) are chemically so similar to one another that the early investigators couldn't even separate them.

The filling of levels continues. Each successive element has a ground state that represents the lowest available energy consistent with the Pauli principle. Each electron can be described by a solution of the Schrödinger equation (with the spin component added) in the electric field that is created by the nucleus and all the other electrons.

Why does the periodic table end? What causes elements to cease to exist beyond about $Z = 92$? Here we have to leave atomic physics and go to the question of what holds the nucleus together. Already for Z greater than 83 the nuclei are unstable, *radioactive*. Some have very long lifetimes, of the order of billions of years, as is true for two of the uranium isotopes. Others are continually produced as radioactive *daughter products* of these long-lived nuclei. Some elements with values of Z beyond 92 have been produced artificially, most notably neptunium and plutonium (Z equal to 93 and 94), which have also been detected in small quantities in nature. There are about 10 more with higher values of Z, most with extremely short lifetimes, falling apart mostly by the process of spontaneous fission.

We have already touched on this question in the first chapter. The stability of nuclei depends on the competition between the actions of the nuclear force and the electric force. The nuclear force attracts all nucleons to each other, but acts only between neighboring nucleons. The electric force acts only between the protons, and repels them from each other. But its range is longer. Each proton repels all of the others. As we go to larger and larger nuclei, eventually the disruptive electric force wins, and the periodic table of the elements comes to a natural end.

12.4 Summary

At the end of the nineteenth century it seemed reasonable to assume that the basic laws of physics were known. The mechanics of Newton and the electromagnetism of Maxwell were the central parts of what we now call *classical physics*. Together with the laws of force for gravitation, electricity, and magnetism they were expected to describe and "explain" all of the phenomena of the physical world.

The discovery of x-rays (1895), radioactivity (1896), and electrons (1897) pointed in new directions. The Michelson–Morley experiment (1887) and the discovery of quantization by Planck (1900) revealed contradictions with the seemingly established classical laws.

The problem highlighted by the Michelson–Morley experiment was resolved by the Special Theory of Relativity in 1905, which resulted from a fresh examination of the fundamental nature of space and time. It was another 20 years before the quantum theory was put on a firm footing by what is now called *quantum mechanics*. The result was the modern understanding of matter in all its forms.

When a body is heated its atoms gain energy of vibration. Planck showed that only certain frequencies of vibration are possible, in accord with the relation $E = Nhf$, where N is a whole number and h is a constant, later named Planck's constant.

A photon is a *quantum* of electromagnetic energy. It behaves in many ways like a particle. It has speed c, energy hf, and momentum $\frac{hf}{c}$. In the *photoelectric effect* a photon of energy $E_{ph} = hf$ hits a material and is absorbed by it. An electron is released if the photon energy is at least as large as the electron's binding energy, E_B. Any leftover energy is given to the electron as kinetic energy, K: $E_{ph} = K + E_B$ or $hf = K + E_B$.

The Bohr model of the atom is part of the "old" quantum theory and leads to some incorrect conclusions. It does, however, lead to the correct spectrum of quantized energies for the hydrogen atom. It was superseded by quantum mechanics in 1925.

It was the first model of an atom consisting of a nucleus surrounded by electrons. It was also the first model to include the quantization of atomic energies and transitions between its *energy levels*. In the transition between levels with energies E_i and E_f the energy difference $E_i - E_f = \Delta E$ can be given to a photon with energy $E_{ph} = hf$ so that $\Delta E = hf$.

The simplest form of the Bohr model is that of a hydrogen atom with one electron moving in a circle about a nucleus consisting of a single proton. There are two relations that characterize the model. One is the *force equation* $F_e = k\frac{e^2}{r^2} = m\frac{v^2}{r}$. The other is the *quantization condition* $mvr = n\hbar$. The combination of these two relations leads to expressions for the possible values of the speed, the radius, and the energy of the atom.

The radius is $n^2 r_1$, where $r_1 = 0.53 \times 10^{-10}$ m. The energy is $\frac{1}{n^2}E_1$, where $E_1 = -13.6$ eV. One electron volt (eV) = 1.6×10^{-19} J. n is a *quantum number* that can take on all integral numbers from one to infinity.

In the Compton effect a photon collides with an electron in an elastic collision, i.e., with the conservation of kinetic energy and momentum. A new photon is created, with a smaller energy and hence smaller frequency and larger wavelength. The wavelength is longer by $\Delta \lambda = \frac{h}{m_e c}(1 - \cos \theta)$.

The rest energy of an electron is 0.51 MeV. (1 MeV = 10^6 eV). When a photon with energy greater than twice this amount (1.02 MeV) hits a material, it can create an *electron pair*, i.e., an electron and a positron. (A positron is similar to an electron, but with a positive charge.)

Conversely, when a positron comes to rest near an electron, the two annihilate, i.e., they disappear, and their energy is given to two photons. The two photons move in opposite directions in order to conserve momentum.

When electrons with a single kinetic energy, K, hit a target and are decelerated, *continuous x-rays* are emitted, with energies from zero to K. In addition, the electrons can raise the energy of the target atoms to higher levels from which they then fall back to the ground state with the emission of *characteristic x-rays*.

Electromagnetic waves propagate just as Maxwell showed. They can go through two slits and recombine to show *interference effects*. But the electromagnetic field can interchange energy only in discrete amounts, as quanta of the field, called photons.

This synthesis shows that the quantized electromagnetic field accounts both for the propagation as electromagnetic waves and for the emission and absorption of electromagnetic energy as photons.

In Maxwell's theory the *intensity* of an electromagnetic wave is proportional to the square of the amplitude of the wave. We can now interpret the intensity in a region as proportional to the probability of finding a photon in that region.

The same reconciliation and synthesis of the points of view of particles and waves as for electromagnetism applies also to electrons and to other particles. The "particles" propagate as waves and show interference effects, as shown by the groundbreaking experiments of Davisson and Germer and of G. P. Thomson. The field associated with them can interchange energy only in discrete steps, i.e., as *quanta* whose properties are those of particles, i.e., of electrons and the other particles that we know.

Just as the square of the amplitude of an electromagnetic wave in a region is proportional to the probability of finding a photon there, so the square of the amplitude of an "electron wave" is proportional to the probability of finding an electron there. The symbol that is used for the matter wave amplitude is Ψ (Greek capital *psi*).

The general wave equation is $\frac{d_2 \Psi}{dx^2} = -k^2 \Psi$, where $k = \frac{2\pi}{\lambda}$.

For de Broglie waves: $k = \frac{2\pi}{h}p = \frac{p}{\hbar}$, and since $K = \frac{1}{2}mv^2 = \frac{p^2}{2m}$, $k^2 = \frac{p^2}{\hbar^2} = \frac{2mK}{\hbar^2}$, and the wave equation is $-\frac{\hbar^2}{2m}\frac{d^2\Psi}{dx^2} = K\Psi$.

If, with Schrödinger, we let $K = E - U$, we get the Schrödinger equation $-\frac{\hbar^2}{2m}\frac{d^2\Psi}{dx^2} + U\Psi = E\Psi$.

The value of Ψ^2 in a region is proportional to the probability of finding the electron (or other particle) there. For any expression for the potential energy, U, there are solutions of the Schrödinger equation only for certain values of the energy, E, and these are the allowed values of the energy.

The *Heisenberg uncertainty principle* says that the position and momentum of a particle cannot both be known exactly at the same time. The products of their uncertainties $\Delta x \Delta p$ cannot be less than (approximately) h.

The Schrödinger equation for the hydrogen atom leads to the same energy levels as the Bohr model, $E_n = \frac{1}{n^2} E_1$, with $E_1 = -13.6$ eV. Other properties, as for example the angular momentum, are incorrectly predicted by the Bohr model, but the results derived from quantum mechanics are in accord with all observations.

In atoms beyond hydrogen there are more protons in the nucleus and more electrons in the neutral atom. Each energy level of the atom is characterized by a different set of four quantum numbers. The quantum number n determines the energy. The quantum number ℓ determines the orbital angular momentum, which is equal to $\sqrt{\ell(\ell+1)}\hbar$. ℓ can be between zero and $n-1$. There are two more quantum numbers, m_ℓ, which can be from $-\ell$ to ℓ, and m_s, which can be $+\frac{1}{2}$ or $-\frac{1}{2}$.

In the ground state of an atom the energy levels are occupied by electrons, one for each level, starting from the level with the lowest energy. The fact that there can be only one electron occupying a level is called the *Pauli exclusion principle*.

With lower values of the orbital angular momentum (and ℓ) an electron spends more time near the nucleus. It is then more strongly attracted to the nucleus and its energy is lower. Therefore energy levels with lower ℓ are filled first.

In hydrogen only n determines the energy. In other atoms it is both n and ℓ. The quantum number n also determines the size of the atom. In the sequence of atoms in the periodic table of elements the levels are not always filled in the order of values of n. Sometimes inner levels (with lower n) are left empty and filled only in atoms further up in the table of elements (with greater atomic number). This happens for the *transition elements*. Transition elements with partially filled inner levels often have extraordinary magnetic properties.

12.5 Review activities and problems

Guided review

1. Photoelectrons are observed to be emitted by a cesium surface with an energy of 1.5 eV. The work function (the minimum energy to release an electron) of cesium is 2.1 eV. What are the energy, frequency, and wavelength of the photons that give rise to this emission?

2. (a) What are the total energy, the kinetic energy, and the potential energy of a hydrogen atom in the state with $n = 3$, according to the Bohr model?

(b) What are the energy, frequency, and wavelength of the light emitted by a hydrogen atom in the transition from the state with $n = 5$ to the state with $n = 3$?

3. A photon with energy 1 MeV hits a metal surface. It gives rise to a Compton recoil photon, which travels in the direction opposite to that of the original photon. What are the energies of the recoil photon and of the Compton electron?

4. A beam of 1-MeV photons hits a cesium surface.

(a) What is the energy of the photoelectrons?

(b) What are the maximum and minimum energies of the Compton electrons? What is the energy of the Compton electron with the most probable energy?

5. In positron emission tomography the patient is injected with a chemical substance that contains a positron-emitting isotope. The injected substance is designed to travel preferentially to specific parts of the body, e.g., to cancerous regions.

(a) After a positron is emitted, it slows down as a result of collisions with the surrounding atoms. What happens to it after that?

(b) What feature makes it possible to determine by measurements outside the body where the positron emission takes place?

(c) Why does the injected material have to contain positive rather than negative beta rays?

6. In a cathode–ray tube, such as those used in oscilloscopes and in old computer and television monitors, electrons are accelerated by an "electron gun" and then hit a fluorescent screen.

In one tube the electrons are accelerated by a potential difference of 20,000 V. What are the ranges of the energies, frequencies, and wave-lengths of the x-rays produced at the screen?

7. One of the simplest models of a nucleus is that of an energy well whose size is the experimentally known nuclear size. What is the lowest energy of a neutron in a one-dimensional well whose size is 3×10^{-15} m?

8. (a) The solution of the Schrödinger equation outside a finite well is proportional to $\Psi = e^{-ax}$, so that $\frac{d^2 \Psi}{dx^2} = a^2 \Psi$. Substitute this solution in the Schrödinger equation, $\frac{\hbar^2}{2m} \frac{d^2 \Psi}{dx^2} = (W - E)\Psi$, and so find a in terms of $W, E,$ and m.

(b) Find the value of a for a neutron in a well for which $W = 100$ MeV in an energy level $E = 80$ MeV. (Remember to use SI units for the energies and all other terms.)

(c) What is the ratio of the value of Ψ at $x = 10^{-15}$ m to that at $x = 0$?

(d) What is the ratio of the probabilities of finding the neutron at these two places?

9.

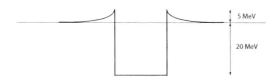

One of the first applications of quantum mechanics to the atomic nucleus was the successful description of alpha–particle radioactivity.

The diagram shows the cross section of a three-dimensional energy well as a model for the nucleus. There is a rectangular part in the center and a part that decreases as $\frac{1}{r}$ outside. The rectangular part represents an approximation to the attraction of a particle inside the nucleus as a result of the nuclear force. The $\frac{1}{r}$ part represents the electrostatic (Coulomb) repulsion outside.

Assume that the nucleus consists of alpha particles inside the well, with the various available energies.

(a) Qualitatively describe the fate of alpha particles in the nucleus as described by this model for the following ranges of the alpha particle energy inside the well:
(i) less than 20 MeV,
(ii) between 20 and 25 MeV,
(iii) more than 25 MeV.

10. (a) Write down the Schrödinger equation for the electron inside the energy well of Examples 7 and 8.

(b) In what way are the wave functions similar in the two cases?

(c) In what way are they different?

11. Show that the ground state energy of the infinite square well (Example 7) is given approximately by the Uncertainty Principle.

12. Define the four quantities **L**, **S**, ℓ, and s.

13. In hydrogen the states with the same value of n and different values of ℓ have the same value of the energy. What is different about them?

14. What are the quantum numbers that describe the electrons in the ground state of the neutral lithium ($Z = 3$) atom?

15. (a) What feature of the figure for Example 15 leads to the fact that the sequence of energy levels is different for hydrogen and for elements with $Z > 20$?

(b) Which level on this diagram represents the transition elements? Explain.

Problems and reasoning skill building

1. What is the smallest energy a photon can have that will allow it to be absorbed by a hydrogen atom in its ground state?

2. Ultraviolet light causes sunburn and damage to the skin. What characteristic is responsible for the fact that this is so for ultraviolet light but not for visible light?

3. An electron and a neutron travel at the same speed. What is the ratio of their de Broglie wavelengths, $\frac{\lambda_e}{\lambda_n}$?

4. The uranium atom has 92 times as many electrons as the hydrogen atom, but its radius is only about $3\frac{1}{2}$ times larger. Explain.

5. Atoms of a certain element have three energy levels. They are observed to emit photons with energies 1 eV, 2 eV, and 3 eV. Draw two possible energy level diagrams.

6. A hypothetical atom has three atomic energy levels. They are at $-1.5, -2,$ and -3 eV.
   (a) What is its ground state energy?
   (b) What is its ionization energy?
   (c) What is the lowest energy of photons that can be emitted by this atom?
   (d) What are the energies of photons that can be emitted by this atom when it is in its ground state?

7. Suppose an electron could be in a nucleus whose diameter is 3×10^{-15} m. What would the depth of an energy well have to be to make this possible? To get an approximate answer, calculate the ground-state energy of an electron in an infinite well of this diameter.

8. The diffraction pattern on a screen for a certain setup is the same for an electron beam as for light whose wavelength is 495 nm. What is the speed of the electrons?

9. It takes a photon of 1.4×10^{-18} J to break a certain molecule apart into its atoms. What are the wavelength and frequency of the photon? What kind of photon is this?

10. An electron is accelerated through a potential difference of 300 V.
    (a) What is the energy of the electron in eV and in J?
    (b) What is its wavelength?

11. An electron's kinetic energy is larger by 1% than $\frac{1}{2}mv^2$.
    (a) What is its energy in eV?
    (b) What is its wavelength?

12. A photon hits a hydrogen atom and is absorbed. The photon momentum is 6.45×10^{-27} kgm/s. To which energy level (i.e., to which value of n) is the atom raised from its ground state?

13. The lowest binding energy of the electrons in barium (its work function) is 2.48 eV.
    (a) What are the threshold energy, frequency, and wavelength of photons that can lead to the emission of photoelectrons from barium?
    (b) What is the range of wavelengths of visible light that can give rise to photoelectrons from barium?

14. Sketch the wave functions for each of the three regions of Guided review Question 9.

Multiple choice questions

1. In the annihilation of a positron and an electron at least two photons are produced because of
   (a) energy conservation,
   (b) momentum conservation,
   (c) $E = mc^2$,
   (d) the difference in the masses of the positron and the electron.

2. Compared to the initial photon, the photon scattered by the Compton effect has a larger
   (a) energy
   (b) momentum
   (c) frequency
   (d) wavelength

3. The number of electrons emitted in the photoelectric effect is proportional to the following characteristic of the incident light:
   (a) intensity
   (b) energy
   (c) momentum
   (d) wavelength

4. A hypothetical atom has three energy levels at $-1.5, -2,$ and -3 eV. The atom, in its ground state, can absorb photons with the following energies (in eV):
   (a) 1, 1.5, 3
   (b) 0.5, 1, 2
   (c) 0.5, 1, 1.5, 2, 3
   (d) 1, 1.5, 2, 2.5, 3

Synthesis problems and projects

1. Go to the PhET website and open the simulation *Models of the Hydrogen Atom*.
   Click on *Experiment*.
   Click on the button under "Turn on the gun." You will see "photons" moving toward a "box" that contains hydrogen. Inside the box

is a question mark to indicate that we are trying to determine the nature of the hydrogen atoms.

Click on "Show Spectrometer" and "white". Some of the photons that you see moving upward on the screen are absorbed by the hydrogen in the box. The hydrogen then emits new photons. The spectrometer detects the photons that are emitted by the hydrogen and displays their wavelengths. Move the slider at the bottom to "fast" so that the photons accumulate more quickly.

After the spectrum has developed, write down, as best you can, the wavelengths at which photons are detected by the spectrometer. Note that the scale is linear only in the central region, from 380 nm to 780 nm, i.e., in the visible region.

Click on *Predictions*.

Here the question mark is replaced by various models of the atom. For each case you can observe the behavior predicted by the model.

(a) Click on the "Billiard Ball model": the atom is a hard ball without any internal structure. The photons (or any other particles) make only elastic collisions. That means that the total kinetic energy is the same before and after the collision.

Describe why there is or is not any radiation emitted according to this model.

(b) In the "Plum Pudding Model" electrons are embedded (like plums or raisins) in a pudding (or cake) of positive charge. The electrons can be given some energy to move back and forth in the atom. They would radiate this energy away, but not with the spectrum of discrete wavelengths that are registered by the spectrometer.

(c) Rutherford showed that the positive charge is concentrated in a tiny nucleus rather than spread out throughout the atom. (See the simulation "Rutherford Scattering" in the next problem. There is also a description of nuclei in Chapter 1 and at the beginning of Chapter 13.)

Move the slider to its slowest position. Click on the "Classical Solar System" model. Click on "Show electron energy level diagram." Here the electrons move around the nucleus as do planets around the sun. Since the electrons are accelerated ($a = \frac{v^2}{r}$) the electromagnetic theory predicts that they radiate energy rapidly. (To reset, go to an earlier model.)

What happens to the atom? Why?

(d) Now click on the "Bohr model" and look at the prediction. How did Bohr deal with the failure of the Classical Solar System model? What does his model say about the stability of the atom? What does it indicate quantitatively about the angular momentum of the atom? What are the two relations that together let you calculate the energy levels of the atom? How can you calculate the energies that may be emitted by the atoms from the energy levels? How can you calculate the frequencies and wavelengths of the emitted radiation?

(e) Does the de Broglie model change anything that we said in the previous paragraph (d) about the angular momentum and the energy levels? What was de Broglie's contribution?

(f) Which quantity do the Bohr and de Broglie models describe correctly? Which do they describe incorrectly?

(g) What are the predictions of the Schrödinger model about the energy levels? About the angular momentum?

(h) List some other successes that the Schrödinger model has with quantities that the other models either describe incorrectly or cannot deal with.

(i) Look at the line on the spectrum with the largest wavelength in the central region (where the scale is linear). It corresponds to a transition between two energy levels of the atom. Try to identify the values of n of the two energy levels. (You will have to guess what the values are and then calculate to see whether you guessed correctly.)

Do the same for the level pairs that give rise to the radiation at the next lower wavelengths. (Your experience with the first pair should make this easier.)

2. Go to the PhET website and open the simulation *Rutherford Scattering*.

Select "Rutherford Atom" from the tabs at the top. You see a gold atom with its 79 protons and 118 neutrons.

"Turn on the gun" by clicking on the "0" on the left. Alpha particles come from the bottom and are deflected by the nucleus. Click on "Show traces" to see their paths.

Change the alpha-particle energy by moving the slider from "min" to "max." What changes do you observe? Explain.

Change the target nucleus to $^{44}_{20}$Ca (20 protons and 24 neutrons). What change to you observe? Explain.

Select "Plum Pudding Atom" from the tabs at the top. Turn on the gun and "Show traces." What do you observe? Explain your observation and contrast it with the observations for the Rutherford Atom.

The atom is about 14 cm across on the screen. What would be the order of magnitude of the diameter of a gold nucleus on the screen to the same scale?

3. Describe positron emission tomography. (Try the article in newworldencyclopedia.org.)

4. What are the energies, according to the Bohr model, of a singly ionized helium atom, i.e., an alpha particle with a single atomic electron?

5. A muonic hydrogen atom is one in which the atomic electron is replaced by a negative muon. The negative muon has the same charge as the electron, but 207 times the mass. The other properties of the two particles are the same.

What, according to the Bohr model, are the allowed energies?

6. Bohr suggested that truth and clarity are complementary. Explain what he may have meant, with examples.

CHAPTER 13

The Nucleus: Heart of the Atom

Henri Becquerel, Marie Curie, and the beginning of nuclear physics

What is the universe made of? The stable nuclei and their binding energy
- *Nuclear structure*
- *Nuclear energies*
- *Binding energy*

Radioactivity
- *Alpha decay*
- *Beta decay*
- *Half-life*
- *Stability and decay series*

Biological effects
- *Ionization and excitation*
- *Units*
- *Background radiation*

The nuclear force
- *The liquid drop model and beyond*

Observing radioactive radiations
- *The ionization chamber and the Geiger counter*
- *Scintillations*

Nuclear reactions
- *Neutron reactions and fission*
- *Fusion reactions*
- *The energy of stars*
- *The beginning of the universe and the cosmic background radiation*

Particles

The existence of the atomic nucleus was discovered in 1911. The next quarter of a century saw the flowering of nuclear physics, with a gradual growth in the understanding of nuclear structure, nuclear radiations, and nuclear reactions. The composition of the nucleus was seen to determine the element to which an atom belongs. With the possibility of nuclear transformation, elements no longer needed to be considered permanent and immutable.

The first nuclear transmutations that were observed were the ones that occur spontaneously in radioactive materials. The large amount of energy released in radioactivity was a puzzle from the beginning. Eventually the large-scale release of energy in stars was also understood to be the result of nuclear reactions. But it was in 1939, with the discovery of fission, that nuclear physics left the sheltered domain of the physics laboratory. The first nuclear reactor began operation in 1942, and the first atomic (really nuclear) bombs were detonated in 1945. The genie was out of the bottle, and we are still trying to learn how to live with it.

13.1 Henri Becquerel, Marie Curie, and the beginning of nuclear physics

In 1896 Henri Becquerel wrapped a photographic plate so that no light could get to it together with a uranium compound outside the package. It was to be an experiment on x-rays, which had been discovered by Röntgen a few weeks earlier. He meant to investigate the possibility that sunlight would somehow stimulate the uranium to emit x-rays, which would then pass through the packaging and expose the plate, so that it would be darkened after being developed. The darkening did indeed take place, but did so even in the absence of any light on the uranium. Becquerel's hypothesis was that in the absence of any external energy source there would be no radiation coming from the uranium, but somehow that's what was happening. It was clear that a new and previously unsuspected process was taking place.

This discovery of the *radioactivity* of uranium can be said to mark the birth of nuclear physics. At that time, however, this was not apparent, because it was not until 15 years later that it was realized that the atom had a nucleus.

The subject of radioactivity was taken up by Marie Curie when she started the work toward her doctorate, at a time when no woman in Europe had achieved this degree. She had come to Paris from her native Poland in 1891 as Maria Sklodowska to study physics. Four years later she married Pierre Curie, who was eight years older, a physicist who already had important and lasting accomplishments to his credit. Their first child was born in 1897, and near the end of that same year she began the work toward her doctor's thesis. Within a few weeks she showed that the uranium-containing mineral pitchblende emitted more radiation than the uranium itself. She concluded that this material must contain another substance more radioactive than the uranium. This was the fundamental discovery that led directly to her later successes.

The first report to the French Academy of Sciences carried only her name as author, and later she was quite firm in specifying that the paper was about her idea and her work. Later that year both Curies showed that there were, in fact, two such substances, namely the two elements that they named polonium and radium. Marie Curie took on the work of isolating them, a task that proved to be formidable and took four more years.

For many people it was quite difficult to accept the fact that a fundamental discovery was made by a woman, a beginning graduate student, alone, at a time when she was closely associated in her personal and professional life with a senior experienced physicist. Subsequently the names of Pierre and Marie Curie were closely linked, but the major credit was often given to him. Both of them received the Nobel Prize in physics for 1903 (jointly with Becquerel), but it eventually became known that a number of physicists had nominated only Pierre. In 1911, the same year that she received her second Nobel Prize (this time in chemistry), Marie Curie was rejected for membership in the Academy of Sciences.

Eventually her reputation became strong and widespread. But in the popular mind she was identified much more with the four-year work of purification, boiling, precipitating, and separating ever more concentrated samples of the new elements, perhaps because that seemed closer to the accepted view of "women's work." It was, however, the more abstract and intellectual achievement of predicting the existence of the new substances that was her most significant contribution.

There are two further features of the initial work that were fundamental for the development of the field. One is that she understood that the radiation is independent of chemical combinations or other molecular properties, that is, in other words, associated with the atoms themselves. (Later it would be realized that it comes from the atomic nucleus.) The other is that the radiation can be used to identify a material even when the sample is so small that chemical methods cannot be used.

During the next few years it was shown that there are three kinds of radiation that were called alpha, beta, and gamma rays. In time they were identified as helium nuclei, electrons, and high-energy photons, respectively.

The crucial discovery that most of the mass of an atom is concentrated in a nucleus at its center came in 1911. It was based on an experiment in the laboratory of Ernest Rutherford in England, in which a beam of alpha particles was directed at a gold foil. The unexpected result was that some of the alpha particles were deflected through large angles, with some even reflected back along their original direction. It was known that the atom consists of electrically-charged particles, but they were thought to be uniformly distributed throughout its volume. The electrical force on the positively-charged alpha particle would then be relatively small, too small to cause the large deflections. Later Rutherford would describe the observation by saying "It was almost as incredible as if you fired a 15-inch shell at a piece of tissue paper and it came back and hit you." To see why he said this, it helps to remember how vast the distance between the nucleus and the electrons is compared to the nuclear size.

Rutherford was able to show that the deflection is consistent with a model of the atom in which the positive charge and almost all the mass is concentrated in a nucleus at a point inside the atom. Collisions between the alpha particle and the gold nucleus can occur when the distance between them is very small. The electric force on the alpha particle can then, in accord with Coulomb's law, be very large, large enough to deflect the alpha particle through the observed large angles.

The discovery of the nucleus led soon afterward to the development of Bohr's model of the atom, but a more complete knowledge of the structure of the nucleus became possible only after the discovery of the neutron in 1932.

13.2 What is the universe made of? The stable nuclei and their binding energy

Nuclear structure

The discovery of the neutron was essential for the understanding of the structure of nuclei. Here is a review of the basic results that emerged.

Nuclei are composed of protons and neutrons, collectively called *nucleons*. The number of protons, Z, is called the *atomic number*, and it determines the position in the periodic table of elements. In a neutral atom this is also the number of electrons in the atom. (An atom in which the number of electrons is not equal to Z is electrically charged and is called an *ion*.)

The number of neutrons in a nucleus is its *neutron number*, N. A particular species of nucleus, characterized by its values of Z and N, is called a *nuclide*. For a particular value of Z, i.e., for a particular element, there may be nuclides with different values of N, each called an *isotope* of the element.

The number of nucleons in a nucleus, $Z + N$, is called its *mass number, A*, because the nucleons represent almost all of the mass of an atom. The notation for the corresponding nuclide is $^{A}_{Z}X$, where X is the symbol for the element whose atomic number is Z.

The nuclei and nuclides of hydrogen have special names. "Ordinary" hydrogen (^{1}H), whose nucleus is the proton, comprises 99.985% of natural hydrogen. The rest is the isotope ^{2}H, with $Z = 1, N = 1$, and $A = 2$. It is called *deuterium* or *heavy hydrogen*, and its nucleus is called the *deuteron*. Water whose hydrogen is deuterium is called *heavy water*.

There is a third isotope, ^{3}H, *tritium*, whose nucleus consists of a proton and two neutrons. It is radioactive and is not part of natural hydrogen.

EXAMPLE 1

The atomic number of oxygen is 8. The most abundant isotope of oxygen is oxygen 16, with the symbol $^{16}_{8}$O, or just ^{16}O, since the fact that we are talking about oxygen automatically means that the atomic number is 8.

What is the composition of a nucleus of ^{16}O and of the other naturally occurring isotopes of oxygen, ^{17}O and ^{18}O?

Ans.:

The atomic number of oxygen is 8, so that all isotopes of oxygen have 8 protons in the nucleus. Oxygen 16 has $A = 16$ and consists of 16 nucleons. Eight of them are protons and the rest (also eight) are neutrons. The other isotopes, ^{17}O and ^{18}O, have 9 and 10 neutrons, respectively.

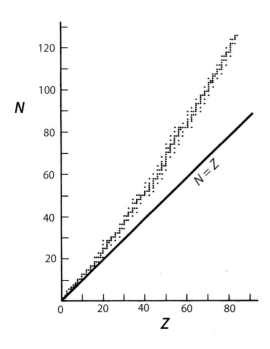

Some nuclides are stable and continue to exist indefinitely. Some break apart (*decay*) spontaneously and are *radioactive*. During the decay process they emit radiation, most commonly alpha and beta particles, and are transformed into nuclides of other elements. All of the elements with Z greater than 83, such as uranium ($Z = 92$) and radium ($Z = 88$), are radioactive and have no stable isotopes.

The figure shows all the stable nuclides on a graph of N against Z. It reveals some important characteristics of nuclei and of the forces within them.

We see that on this graph the stable nuclides are in a narrow band called the *region of stability*.

That nuclei exist at all shows that there must be a *nuclear force* that acts to attract nucleons to each other. Since nuclei contain protons, the nuclear force inside the nucleus must be stronger than the electric force. The electric force acts only between the positively charged protons, but it acts in the opposite direction, repelling them from each other. The fact that the region starts out, for small Z, along the line $Z = N$ leads us to believe that nature favors equal numbers of protons and neutrons, and that the nuclear force acts equally between neutrons and protons.

The region of stability then deviates from the line $Z = N$, with N gradually getting larger and larger than Z. We can understand this as a result of the electric force. As the number of protons in the nucleus increases, the electric repulsion between them plays a greater and greater role. Neutrons are not affected by the electric force and are therefore favored. The result is that the ratio of neutrons to protons becomes larger as the number of protons increases.

In the region of stability the nuclear force of attraction between the nucleons predominates. The electric force of repulsion between the protons, which acts to disrupt the nucleus, eventually results in the end of the region of stability at $Z = 83$.

A closer examination of the naturally occurring nuclides shows that even numbers are favored for both Z and N. There are only four nuclides with odd Z and odd N, and they are near the origin of the graph: ^{2}H, ^{6}Li, ^{10}B, and ^{14}N. On the other hand there are 49 nuclides with odd Z and even N, 51 nuclides with odd N and even Z, and 157 nuclides with even Z and even N.

Nuclear energies

The relation $E = Mc^2$ was developed by Einstein in 1905 as a result of the special theory of relativity. Even at that time, when there was no knowledge of the existence of nuclei, he suggested that the release of energy accompanying the decay of radium was so large that it might provide a test of the mass-energy equivalence. He realized that the necessary precision for the measurement of mass was then not available. A test and verification involving nuclear energies was made, but not until 1932, shortly after the discovery of the neutron.

Since $c = 3 \times 10^8$ m/s, Einstein's relation shows that 1 kg of mass is equivalent to the enormous amount of energy 9×10^{16} J. The mass unit that is commonly used for atoms and nuclei is the atomic mass unit, u, defined as $\frac{1}{12}$ the mass of a neutral atom of ^{12}C, or 1.660×10^{-27} kg. Its energy equivalent is 931.5 MeV. The mass of the proton is 1.007 u, equivalent to 937.5 MeV.

Since mass and energy are equivalent, mass and energy units can be used interchangeably, and we can say simply that the mass of the proton is 937.5 MeV.

Binding energy

The *total binding energy* of a nucleus, E_b, is the amount of energy that would have to be supplied to it to separate it into its separate A nucleons. We can write the process as

$$^{A}_{Z}X + E_b \rightarrow Zp + Nn$$

On the left side is the original nucleus, and we add the binding energy to it. On the right side are the separated Z protons and N neutrons. We can translate the relation into an energy-balance equation:

$$m_X + E_b = Zm_p + Nm_n$$

which says that the mass of the original nucleus plus the binding energy is equal to the sum of the masses of the separated Z protons and N neutrons. In the application of this and similar equations it is, of course, necessary to use the same units for each term. They can be mass units, such as kg or u, or they can be energy units, such as J or MeV.

EXAMPLE 2

$m_p = 1.007276$ u and $m_n = 1.008665$ u. The total binding energy of ^{16}O is 127.6 MeV. What is the mass of a ^{16}O nucleus?

Ans.:
$Z = 8$ and $N = 8$. Use atomic mass units: $8m_p = 8.058206$ u. $8m_n = 8.06932$ u. The sum is 16.127526 u.

We need to subtract the binding energy. First convert it to u: 1 u = 931.49 MeV. 127.6 MeV = .136985 u. The mass of the oxygen nucleus is $m_{(^{16}O)} = 8m_p + 8m_n - E_b = 16.127526 - 0.136985 = 15.990535$ u.

(Tables of masses usually give the masses of the neutral atoms, which are greater by the masses of the electrons. For oxygen this means the addition of eight electron masses or 0.00439 u.)

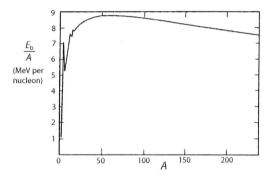

The figure shows one of the most informative and interesting graphs in nuclear physics. On the vertical axis is the binding energy divided by the mass number, $\frac{E_b}{A}$, and on the horizontal axis is the mass number, A. It is determined by plotting a point on the graph for each stable nuclide. The line closely follows the experimental values. On this graph the higher the point for a given nuclide, the greater is its *binding energy per nucleon* and the more stable the corresponding nucleus.

We see that from about $A = 10$ on, the curve varies by only about $\frac{1}{2}$ MeV on either side of 8 MeV. In other words, the binding energy per nucleon is, to a first approximation, about constant in this region. This is very different from the binding energies of atoms, where the arrangement of electrons in shells and subshells leads to a hierarchical structure with large variations of the binding energy. In contrast, the nucleus is seen to be quite egalitarian.

The situation is similar to that of molecules in a drop of water, where each is equivalent and each is bound equally to the others. The model of the nucleus that behaves this way is therefore called the *liquid drop model*.

It is easy to understand the downward trend on the right-hand side of the graph. As we go to the right, more and more protons are in the nucleus, and the disruptive effect of the electric force of repulsion between them makes itself felt more and more as it decreases the binding energy, i.e., the energy that needs to be added to disrupt the nucleus.

The most stable nuclides are seen to be in the vicinity of $A = 50$, so that they are near $Z = 26$, which is the atomic number of iron. At the left we see peaks for $A = 4, 12$, and 16, i.e., for the alpha particle and two of its multiples, indicating that the alpha particle (the helium nucleus, ^4_2He) is a particularly stable configuration. The nucleus with the smallest binding energy is the deuteron. It takes only about 2 MeV to separate its proton and neutron, so that $\frac{E_b}{A}$ for it is about 1 MeV.

EXAMPLE 3

Estimate how much energy it would take to decompose a nucleus with $A = 100$ into its separated nucleons.

Ans.:
The figure shows $\frac{E_b}{A}$ to be about 8.6 MeV. The total binding energy is therefore about $(8.6)(100) = 860$ MeV.

EXAMPLE 4

Think of the different ways that 200 nucleons can arrange themselves. Which combination is more stable—a single nucleus with $A = 200$ or two nuclei, each with $A = 100$?

Ans.:
For $A = 200$ the figure shows $\frac{E_b}{A}$ to be about 7.8 MeV, for a total binding energy of about 1560 MeV. From the answer to Example 3 we see that the total binding energy for two nuclei of 100 nucleons each would be twice 860 MeV, or 1720 MeV. The larger number means a larger binding energy and a more stable nucleus.

13.3 Radioactivity

Unstable, radioactive nuclides exist in the vicinity of the region of stability. The radioactive nuclides undergo spontaneous transformations toward greater stability. They change into nuclides that have larger binding energy and are either stable, or at least closer to the region of stability. This can happen by several processes. The main ones are the emission of alpha particles, or alpha decay, and the emission of beta particles, or beta decay. The two are fundamentally different, and we will discuss them in turn.

Alpha decay

The alpha particle was shown by Rutherford and his co-workers to be the nucleus of ^{4}He. A radioactive nucleus that emits an alpha particle transforms into a daughter nucleus whose atomic number is smaller by 2 as a result of the loss of two protons, and whose mass number is smaller by 4 as a result of the loss of four nucleons. On the figure of N against Z, where each box represents a nuclide, this corresponds to a move two steps down and two to the left, i.e., a move along the line $N = Z$ at 45° to the two axes.

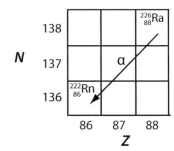

Examples of alpha-emitting nuclides are ^{238}U, ^{235}U, and ^{226}Ra. The process can be written in the form of a relation such as $^{226}\text{Ra} \rightarrow {}^{222}\text{Rn} + \alpha$. The amount of energy that is transformed into kinetic energy may be included:

$$^{226}\text{Ra} \rightarrow {}^{222}\text{Rn} + \alpha + 4.76 \text{ MeV}$$

In this form it represents an equation that describes the mass and energy transformations quantitatively. On the left is the mass of the parent nucleus. On the right are the masses of the daughter nucleus and the alpha particle as well as the released kinetic energy:

$$M_{Ra} = M_{Rn} + M_\alpha + 4.76 \text{ MeV}$$

where the masses are those of the nuclei of the isotopes ^{226}Ra and ^{222}Rn.

In any calculation the units of each term must, of course, be the same. (Usually atomic mass units are used.) If all the masses are known, the released energy can be calculated. In other cases the released energy may be measured, and an unknown mass calculated.

EXAMPLE 5

The nucleus of ^{226}Ra is heavier than the nucleus of ^{222}Rn. What is the mass difference, in kg, in atomic mass units and in MeV?

Ans.:
We can turn the relation between the masses around to say that $M_{Ra} - M_{Rn} = M_\alpha + 4.76$ MeV. The difference is the mass of the alpha particle plus the binding energy of 4.76 MeV. The mass of the alpha particle is 4.00150 u. The extra binding energy, in atomic mass units, is $\frac{4.76}{931.5} = 0.00511$ u. The difference is therefore 4.00661 u or 3732 MeV, or 6.653×10^{-27} kg.

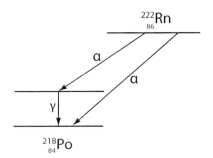

The emission of the alpha particle is a kind of explosion. Energy is released as a result of the transformation of internal energy to kinetic energy. Momentum is conserved, and the daughter nucleus recoils with a momentum equal in magnitude and opposite in direction to that of the emitted alpha particle. In our example the mass of the nucleus is about 56 times that of the alpha particle, so that its velocity and kinetic energy are correspondingly smaller, and are often neglected.

If the daughter nucleus is in its ground state, all of the released energy becomes kinetic energy. The nucleus may also be left in an excited state. In that case the excited state subsequently goes to the ground state with the emission of a photon. A photon emitted by a nucleus is called a *gamma ray*. The figure shows an example of an energy-level diagram or *decay scheme* for the alpha decay of ^{222}Rn. If instead of going directly to the ground state, the daughter nucleus is left in the excited state, a gamma ray with an energy of 0.51 MeV accompanies the alpha decay.

Beta decay

Beta particles were identified as electrons in 1902, even before the identity of the alphas was established in 1908. The alpha particles were soon seen to have definite energies, equal to the difference in the energies of the parent and daughter nuclei. It was confidently expected that this would also be true for the betas. It was therefore a major surprise when it was gradually realized that this was not so.

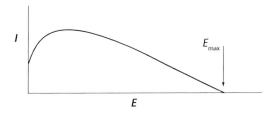

It was found that the emitted electrons have a continuous range of energies, from very low to a definite maximum. And it is that maximum energy that is equal to the difference in the energies of the parent and daughter nuclei.

The problem was that in spite of diligent efforts, no other particles were observed. It seemed therefore that the beta decay process violated those cornerstones of science, the law of conservation of energy and the law of conservation of momentum. Wolfgang Pauli, in 1930, made the hypothesis that there is another particle that carries away the missing energy, even though none had been observed. That particle (now called the *neutrino*) would have to be able to traverse large amounts of material without interacting with it as it passes through.

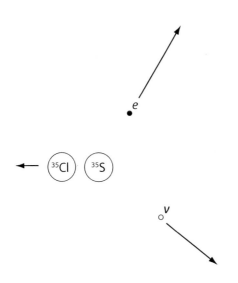

The neutrino hypothesis allowed the preservation of the laws of conservation of energy and momentum. For the next quarter of a century all efforts to gain direct evidence for neutrinos failed. Nevertheless they seemed so necessary and accounted for so many details of the beta–decay process that their existence came to be accepted. Eventually, in 1956, neutrinos were observed directly.

In the meantime, in 1932, a new particle, the *positron*, was discovered. It has the same mass and most other properties as the electron, but it is positively charged. Two years later it was shown (by the Curies' daughter Irène and her husband Frederic Joliot) that it can also be the result of beta decay.

Positive beta decay is also accompanied by a neutrino. Electrons and positrons are *antiparticles* of one another. (We saw in Chapter 12 that their meeting leads to annihilation.) Similarly the neutrinos emitted in positive and negative beta decay are antiparticles of one another. The one accompanying β^- decay is called the *antineutrino*, $\bar{\upsilon}$ ("nu-bar," where the bar indicates an antiparticle), and the one accompanying β^+ decay is called the neutrino, υ. Generically, both are usually just called neutrinos. Much later other kinds of neutrinos were discovered, and when it is necessary to distinguish those that accompany beta decay they are called *electron neutrinos*.

An example of a relation representing β^- decay is

$$^{35}_{16}S \rightarrow\ ^{35}_{17}Cl + \beta^- + \bar{\upsilon} + Q$$

Here the symbol Q represents the released energy that is shared as kinetic energy by the three decay products, namely the electron, the antineutrino, and the residual nucleus. Because of its greater mass, the kinetic energy of the recoiling nucleus is even smaller than for alpha decay and it is shared almost entirely by the electron and the neutrino. Sometimes the electron gets all the energy, sometimes the neutrino gets it all, and most of the time each gets part of it. The energy of the electron can therefore be any amount between zero and the maximum energy, Q.

EXAMPLE 6

The mass of the nucleus of ^{35}S is heavier than that of ^{35}Cl by 0.00073 u. What can you say about the energies of the emitted beta rays?

Ans.:
When a beta particle is emitted by ^{35}S, some of the mass that the sulfur nucleus loses is that of the emitted electron (0.000549 u). Most of the rest is that of the ^{35}Cl "daughter" nucleus. The remaining mass is shared by the electron and the neutrino as kinetic energy. Here it is .00073 − .00055 = .00018 u, which is equivalent to 0.17 MeV. The kinetic energy of the beta particle can be anywhere between zero and 0.17 MeV.

An example of β^+ decay (the one found by the Joliot–Curies) is

$$^{30}P \rightarrow\ ^{30}Si + \beta^+ + \upsilon + Q$$

In alpha decay the emitted particle is a constituent of the parent nucleus. But there are no electrons in the nucleus—they are created at the time of the disintegration. In β^- decay one of the neutrons in the nucleus changes into a proton, and an electron and an antineutrino are emitted.

The neutron decays into a proton even when it is free, outside a nucleus, in accord with the relation

$$n \rightarrow p + e + \bar{\upsilon} + Q$$

with the energy balance

$$m_n = m_p + m_e + m_\upsilon + Q$$

The neutrino mass, m_υ, is much smaller than that of the electron. For many years it was thought to be zero, but more recent experiments indicate that it has a very small mass. We can neglect it most of the time. The known masses of the neutron, the proton, and the electron show Q for this case to be 0.78 MeV.

EXAMPLE 7

Can the proton decay into a neutron? Explain.

Ans.:
The transformation of a proton into a neutron, together with a positron and a neutrino, cannot occur for a free proton, since the proton's mass is smaller than that of the neutron. There is no internal energy that can be given away and still leave a positive value for Q.

If, on the other hand, the proton is part of a nucleus, the internal energy is shared by all the nucleons, and a radioactive decay can take place in which the proton changes to a neutron. This is what happens in β^+ decay. Of course it can occur only if the mass of the initial nucleus is larger than the sum of the masses of the daughter nucleus and one electron.

There is an alternative to β^+ decay that also leads to a daughter nuclide with Z less by one and N greater by one: instead of emitting a positron a nucleus may absorb one of the atom's orbiting electrons. In this process, called *electron capture*, the positron does not need to be created, so that less energy is required. On the other hand, since the atomic electrons spend most of their time far from the nucleus, the probability of electron capture is usually much less than that of positron decay when both are possible. Electron capture is primarily important when there is enough energy for it to occur, but not enough for positron decay.

What is the difference in the energy that is required for the two processes to take place? Write down the relation for positron decay as $P \rightarrow D + \beta^+ + \upsilon + Q_1$, where P is the parent nucleus, D is the daughter nucleus, and Q_1 is the energy released in the process. The corresponding mass relation is $M_P = M_D + m_e + Q_1$, where M_P and M_D are the nuclear masses of the parent and daughter and m_e is the electronic mass, so that $Q_1 = M_P - M_D - m_e$.

For electron capture the relations are $P + e \rightarrow D + \upsilon + Q_2$ and $M_P + m_e = M_D + Q_2$ or $Q_2 = M_P - M_D + m_e$. We see that Q_2 is larger than Q_1 by $2m_e$! In other words, electron capture releases more energy than positron decay by 1.02 MeV. If the difference in energy between the parent's state and the daughter's state is positive, but less than 1.02 MeV, electron capture can take place, but positron decay cannot.

^{40}K is a nuclide that can decay by β^- and β^+ emission and by electron capture. It is 0.12% of naturally occurring potassium.

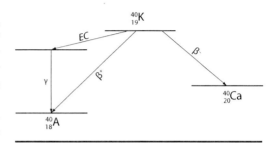

Half-life

Soon after the discovery of radioactivity it was found that the amount of radiation decreases with time, and that the change can be described by a *half-life*, $T_{1/2}$. After one half-life the rate at which particles are emitted decreases to one-half of the original value, after a second half-life it goes to 25%, and so on. The rate of particle emission is proportional to the amount of radioactive material. Hence the original amount of material decreases to $\frac{1}{2}$ after one half-life, to $\frac{1}{4}$ after two half-lives, to $\frac{1}{8}$ of the original amount after three half-lives, and so on. In general, after n half-lives

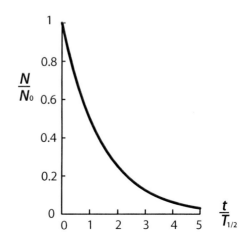

the amount of material as well as the rate of emitted particles decrease to 2^{-n} times their original value. The graph shows $\frac{N}{N_0}$ plotted as a function of time, as $\frac{t}{T_{1/2}}$. $\frac{N}{N_0}$ is the fraction of material remaining and $\frac{t}{T_{1/2}}$ is the time measured as a multiple of the half-life.

Intermediate values can be calculated from the *decay equation*

$$N = N_0 e^{-\lambda t}$$

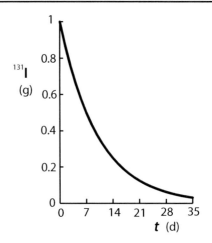

where N is the number of nuclei at time t and N_0 is the number at time $t = 0$. The exponent λ (called the decay constant) describes how quickly the number of nuclei decreases.

[We can also write this expression differently by using natural logarithms: if $e^x = A$, then the exponent, x, to which e has to be raised to give A, is the natural logarithm (ℓn) of A. Here $\frac{N}{N_0} = e^{-\lambda t}$, so that $\ell n \frac{N}{N_0} = -\lambda t$ or $\ell n \frac{N_0}{N} = \lambda t$].

After one half-life $t = T_{1/2}$ and $N = \frac{N_0}{2}$. We can substitute these values in the decay equation to get $e^{\lambda T_{1/2}} = 2$, which can also be written as $\lambda T_{\frac{1}{2}} = \ell n\, 2 = 0.693$, so that $\lambda = \frac{0.693}{T_{\frac{1}{2}}}$.

The number of disintegrations per second is called the *activity*. The SI unit for activity is the *becquerel* (Bq), equal to one disintegration per second. An older unit that is still widely used is the *Curie*, Ci, equal to 3.7×10^{10} Bq. It was originally defined as the activity of one gram of radium.

The activity is the rate at which the number of nuclei (N) changes. It is the slope on a graph of N against t, which we can write as $\frac{dN}{dt}$. If $N = N_0 e^{-\lambda t}$, then $\frac{dN}{dt} = -\lambda N_0 e^{-\lambda t}$, which is equal to $-\lambda N$. The relation $\frac{dN}{dt} = -\lambda N$ shows that the activity ($\frac{dN}{dt}$) is proportional to the amount of material (N).

EXAMPLE 8

(a) Draw a graph of the decay of ^{131}I, plotting the mass of the radioactive material as a function of time. At $t = 0$ there is 1 g of the iodine. Its half-life is 7 days.

(b) How much iodine is left after 2.5 half-lives?

(c) How long does it take for the activity to decrease to 10% of the original amount?

(d) Check your answers to parts (b) and (c) by looking at the graph in part (a).

Ans.:

(b) $N = N_0 e^{-\lambda t} = N_0 e^{\frac{-0.693}{T_{\frac{1}{2}}}t} = N_0 e^{-(0.693)(2.5)} = N_0 e^{0.173} = 0.177 N_0$. After $2\frac{1}{2}$ half-lives (or 17.5 days) 0.177 of the original amount or 0.177 g is left.

(c) For $\frac{N}{N_0} = 0.1$, $e^{\frac{-0.693}{T_{\frac{1}{2}}}t} = 0.1$, or $\ln 0.1 = -\frac{0.693}{T_{\frac{1}{2}}}t$, i.e., $-\frac{0.693}{T_{\frac{1}{2}}} = -2.30$, or $\frac{t}{T_{\frac{1}{2}}} = 3.32$: it takes 3.32 half-lives or 23.2 days for the original amount to decay to 10% of its original value.

Stability and decay series

In both kinds of beta decay the parent and daughter nuclides have the same mass number, A. β^- decay is represented by an increase in Z of one and a decrease in N of one, i.e., one step down and one step to the right. It allows a neutron-rich nuclide to get closer to the region of stability from

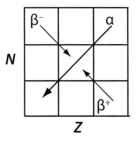

the left. Similarly β^+ decay occurs for nuclides to the right on the diagram so that they can get closer to the region of stability.

Alpha-particle decay occurs primarily for the nuclides at the upper end of the periodic table of elements, beyond the last stable one ($_{83}$Bi). In this region the nuclides form *decay series*, such as the one that starts with ^{238}U and proceeds through a number of alpha and beta decays until it reaches the stable nuclide ^{206}Pb, as shown on the diagram. One nuclide in the chain is ^{226}Ra with its half-life of 1600 years, followed by ^{222}Rn, whose half-life is 3.8 days. It leads to other alpha and beta emitters, including ^{210}Bi, which has a half-life of 22 years.

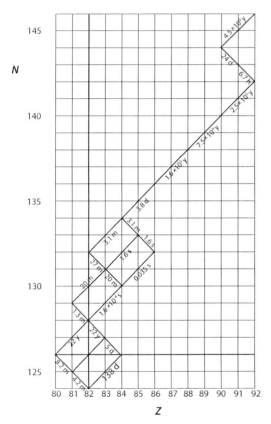

13.4 Biological effects

It was realized soon after the discovery of x-rays by Röntgen in 1895 and of α, β, and γ rays by Becquerel in 1896 that these radiations can cause "burns." Becquerel and Marie Curie developed skin lesions, and Pierre Curie exposed himself to radioactive sources purposely to study the effect. The damage is no surprise since even ultraviolet radiation, with its much less energetic photons, can cause sunburn and, as we know today, cancer.

The possibility of delayed effects was not, however, realized, and insufficient precautions were taken. Marie Curie lived until she was 67, but it seems clear that her health suffered from the effects of the radiation with which she worked. In the biography of her mother, her daughter Eve went so far as to quote what she calls the verdict of science: "Marie Curie can be counted among the eventual victims of the radioactive bodies which she and her husband discovered."

By the 1920s it had become evident that radioactive radiations could cause severe damage and death. The best-known case, which served to publicize the dangers, was that of the women who were employed at the U.S. Radium Corporation in New Jersey to paint watch dials with luminous paint containing radium. They pointed their brushes with their lips, and so absorbed the radioactive material, leading to bone destruction, bone cancer, blood disorders, and eventually 41 deaths among them. (Radium is in the same column of the periodic table of elements as calcium, and like it, once in the body it travels preferentially to the bones.)

At the same time various curative powers were ascribed to radium, and health tonics and medicines containing it were used until the 1950s.

Ionization and excitation

Since the particles emitted by radioactive materials have energies up to several MeV, they can cause the disruption of atoms and molecules that they encounter. The main effect is that charged particles can remove electrons from atoms just by passing close to them, as a result of the Coulomb force. In other words they can cause ionization of the atoms. In other cases they can give the atoms that they pass the energy to go to higher energy levels, from which they then return to the ground state. (This second effect was used in the luminous paints mentioned in the previous section. The radium was mixed with zinc sulfide, which is excited to higher energy levels by the alpha particles from the radium, and then emits visible radiation on its return to the ground state.)

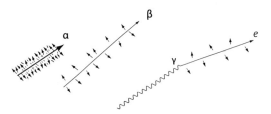

Charged particles, such as alpha and beta particles, can give rise to ionization and excitation directly. Gamma rays and neutrons don't, but they can do so indirectly by first giving some of their energy to charged particles. Gamma rays can give energy to electrons by the photoelectric effect, the Compton effect, and by pair production. Neutrons can give up energy by collisions with nuclei and through nuclear reactions.

Units

A unit that is used to measure the amount of ionization is the *roentgen (R)* (or röntgen). One roentgen is the amount that produces 2.58×10^{-4} coulombs of electrons per kg of air at standard temperature and pressure (STP), i.e., at 0°C and one atmosphere of pressure. (This amount of charge is an obsolete unit of charge called the *statcoulomb*.)

The amount of ionization that is produced is different for different materials. In air at STP it takes, on average, 34 eV of absorbed energy to produce one ion pair.

The roentgen measures the amount of radiation by measuring the ionization that it produces in a standard sample of air. Other units are needed to measure the effect of the radiation. The most important quantity that characterizes the effect of the radiation is the energy that is absorbed per unit mass. This is called the *absorbed dose*, and the SI unit for it is the *gray* (Gy), equal to 1 J/kg. An older unit is the *rad*, equal to 10^{-2} Gy.

Ionizing radiation causes biological damage principally by disrupting molecules that it passes. The absorbed energy per unit mass is a good guide for beta and gamma radiation, but alpha particles and neutrons cause more damage for the same absorbed dose. This has led to the introduction of an empirically determined quantity called the *quality factor* (QF) (also called the *relative biological effect,* RBE) to take into account the fact that some types of radiations cause more damage than others. It is equal to one for betas and gammas, to 20 for alphas, to 3 for slow neutrons, and to 10 for fast neutrons. When the absorbed dose (in Gy) is multiplied by the QF it becomes the *effective dose*, for which the SI unit is the *sievert* (Sv). (If the absorbed dose is measured in *rads* and is then multiplied by the QF, it becomes the effective dose in *rems*.)

The QF is larger for alphas because they lose their energy in a shorter distance, so that the ionization is more densely distributed. A 1 MeV alpha particle travels about 0.56 cm in air before it stops, and about 0.1% of that distance in tissue. Electrons with the same energy travel about 3.1 m in air.

Gamma rays do not travel a definite distance (or *range*) and then stop. Their intensity decreases exponentially. A 1 MeV gamma ray beam in air will be attenuated by half in about 75 m of air and in 0.85 cm of lead.

We see that an alpha source does not pose a threat as long as it is outside the body and at least a few cm from it. Once ingested, however, an alpha emitter can cause major damage. One of the most important examples is radon (^{222}Rn), which emits 5.5 MeV alpha particles as it decays to ^{218}Po with a half-life of 3.8 days. It is a decay product of uranium 238 that (in small quantities) is widely distributed in rocks and hence in building materials such as bricks. Radon is a gas, and if it is breathed in and decays in the lungs it leaves its daughter products, which emit further alpha, beta, and gamma rays. The alphas, in particular, because of their short range, cause dense ionization and hence the greatest damage.

Radiation to the whole body of 0.25 to 1 Sv causes changes, primarily in the blood, that the affected individual may not be aware of. "Radiation sickness" with malaise and vomiting will result from 1 to 3 Sv, but the person is likely to recover. Higher doses lead to widespread damage throughout the body, particularly to the gastrointestinal tract and the bone marrow. With a dose of 5 Sv the chance of recovery is about 50%. Above 6 Sv death is almost certain.

The study of cancer induced by radiation is made difficult by the fact that cancer is a common disease that is usually caused by other factors. Only massive doses lead to numbers that can be distinguished from those that would occur in their absence. Some of the most important evidence comes from survivors of the atomic bombs in Hiroshima and Nagasaki. This population has been studied over the years and the doses to which people were subjected have been estimated from knowledge of where they were at the time of the exposure.

Estimates based on these and other studies lead to cancer deaths attributed to radiation of the order of 5% per sievert. This means that in a population of 100 people exposed to 1 Sv there would be five excess cancer deaths. It must be stressed that this is a rough estimate. Needless to say, it depends on many factors, some dependent on the nature of the radiation and some on the irradiated individual. Furthermore the estimate applies to survivors of high doses, since it has not been possible to demonstrate directly that radiation below 0.5 Sv can produce cancer. Its use at lower doses is highly controversial. It has been called the *LNT hypothesis*, standing for "linear—no threshold." It implies that there is no minimum dose for radiation-induced cancer, and that the variation of the cancer incidence with dose (the *dose–response relationship*) remains linear to the lowest doses. The result would be that even the amount of background radiation to which everyone is exposed would lead to some development of cancer. Some groups recommend the use of the LNT hypothesis because it represents a conservative approach in the face of unknowns. Others believe that its conclusion that all radiation carries risks, regardless of dose, is unrealistic, unsupported by the data, and leads to large expenditures of public funds that are used to reduce amounts of radiation that have not been shown to be harmful.

Background radiation

The average natural background radiation in the United States is estimated to be about 3 mSv/year. Of this amount about 60% comes form radon, and the rest, in very roughly equal amounts, from cosmic radiation, radioactive materials in the earth, and radioactive material inside the body, principally ^{14}C and ^{40}K. Another 0.5 mSv is from man-made sources, principally medical x-rays and nuclear radiations. These numbers are obviously quite variable, depending on location and other factors.

13.5 The nuclear force

Before the discovery of the neutron, nuclei were thought to consist of protons and electrons. By 1921, however, it had become clear that electric forces were not strong enough to hold such a nucleus together. The eventual realization that nuclei consist of protons and neutrons made it definite that another kind of force is required. The only forces that were then known were the gravitational force, which is far too weak, and the electrical force, which causes the protons to repel each other and does not affect the neutrons.

At the time of his discovery of the nucleus, Rutherford showed that the scattering of alpha particles could be described by assuming only electrical forces between the alphas and a point nucleus. With this hypothesis he developed the *Rutherford scattering formula*, which was in very good agreement with the experiments. Further experiments during the following decade showed, however, that when the alphas have sufficient energy to come very close to the target nucleus, the formula no longer holds. A new force, the strong nuclear force, comes into play.

There is no simple formula for the nuclear force, similar to Newton's law of gravitation or Coulomb's law. Instead its primary characteristics are that it acts between nucleons regardless of whether they are protons or neutrons, that its range is so short that in a nucleus it acts only between neighboring nucleons, and that it is stronger than any other force.

The liquid drop model and beyond

We have already seen that the leading feature of the liquid drop model of the nucleus is that the binding energy depends only on the number of nucleons, so that the binding energy per nucleon is roughly constant.

As in a sphere composed of closely packed marbles, the number of nucleons is proportional to the volume of the nucleus, and we can call this part of the binding energy the *volume energy*. It is proportional to the number of nucleons, A, and hence to the nuclear volume, i.e., to the cube of the nuclear radius. Experiments show that the nuclear radius is about $1.2 \times 10^{-15} A^{\frac{1}{3}}$ m.

In the interior of the nucleus the nucleons experience forces from all sides around them. At the surface, however, there are nucleons only on one side. At the nuclear surface the nucleons are therefore less strongly bound than in the interior. We can describe this feature by adding a negative term to the binding energy. It is proportional to the surface of the nucleus, i.e., to the square of the nuclear radius, so that it is proportional to $A^{2/3}$.

These two terms, the volume energy and the surface energy, act in opposite directions. The volume energy is the main term that describes the attraction between the nucleons. The surface energy reduces this overall attraction.

To describe the binding energy curve further we can now add a term that takes into account the electric repulsion between the protons. This term is proportional to the square of the number of protons, Z, and in accord with Coulomb's law it is proportional to $\frac{1}{r}$, or to $A^{-\frac{1}{3}}$.

These three terms, the volume energy, the surface energy, and the Coulomb energy, can describe the binding energy curve quite closely. The relation that includes them is the core of what is usually called the *semiempirical binding energy relation*, first developed by C. F. von Weizsäcker.

The Semiempirical Binding Energy Relation

We can incorporate two further features in the formula. One is the tendency toward equal numbers of protons and neutrons. This "asymmetry term" is proportional to $(N-Z)^2$. It is zero when $N=Z$ and becomes larger with the difference between N and Z. In addition, this term turns out to be proportional to $\frac{1}{A}$.

Finally we can include the "odd–even" effect by including a term that is positive when both Z and N are even, negative when they are odd, and zero when one is even and the other odd. This term turns out to vary as $\frac{1}{\sqrt{A}}$.

The complete expression is

$$E_b = a_1 A - a_2 A^{\frac{2}{3}} - a_3 \frac{Z^2}{A^{\frac{1}{3}}} - a_4 \frac{(N-Z)^2}{A}$$

$$(\pm \; a_5 A^{-\frac{1}{2}} \text{ or zero})$$

With empirically determined coefficients this formula follows the observed binding energies to about $1\frac{1}{2}\%$ for nuclei with A greater than 20.

Both the tendency toward $N=Z$ and the odd–even effect are consequences of the Pauli exclusion principle. The proton and the neutron can be considered to be different states of the nucleon. Each has a spin angular momentum equal to that of the electron, with two possible spin states. In an alpha particle each of these four possible states is represented.

This is the origin of the extraordinary stability of the alpha particle and its high binding energy compared to that of its neighbors, as seen by the peak at $A=4$ on the binding energy curve. In accordance with the Pauli principle a fifth nucleon would have to be in a quite different state, with a higher energy, and hence a lower binding energy. It turns out, in fact, that there is no stable configuration with $A=5$. As we shall see, this fact has an enormous influence on the evolution and the composition of the universe.

In atoms each electron moves in the electric field of the nucleus and the other electrons. The electrons are in widely spaced states determined by the Pauli exclusion principle. There is a hierarchy of states with their different quantum numbers. Nuclei are very different. As in a liquid drop their components (the nucleons) form a much more egalitarian structure. Nucleons are close to each other, but not so closely packed that they cannot move. One consequence is that there is a shell structure in nuclei somewhat like the one in atoms. In atoms the shell structure plays a dominant role. Its effect on nuclear structure is much smaller. It does lead to the fact that there are other numbers of nucleons (in addition to the number four for the alpha particle) that are relatively stable compared to the average behavior. For some time their origin was not understood and they were called "magic numbers." Eventually they were shown to be the consequence of a shell structure within the nucleus. Although the shell structure does not dominate, as it does in atoms, it still plays a significant role, sufficient to cause the *magic numbers* 20, 28, 50, 82, and 126 for Z and N to be favored.

13.6 Observing radioactive radiations

The ionization chamber and the Geiger counter

How can we detect and measure radioactivity? The first and most direct method is to measure the charge liberated by the ionization that is produced. The particles enter an *ionization chamber* containing two plates with an electric field between them. The ions that are produced move toward the plates, constituting a current that is then measured.

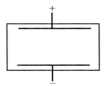

A related device was developed by one of Rutherford's associates, Hans Geiger, and is known as a *Geiger counter*. It is similar to the ionization chamber in that it has two electrodes with an electric field between them. The positive electrode is a wire along the center of a glass or metal tube and the negative electrode is a metallic cylinder surrounding the wire. The important difference is that the electric field at the positive wire is much higher than that at the negative cylinder. You can see this on the figure that shows

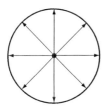

the electric field lines. They are much closer to each other at the wire in the center. An electron in the space between the electrodes is accelerated toward the wire. As it approaches the center the electron reaches an electric field that is so strong that it gives the electron sufficient energy to cause further ionization. The resulting electrons are also accelerated toward the wire and give rise to even more ionization. An avalanche of electrons arrives at the wire, triggered by just one initial ionization. The advantage is that just one initiating particle produces a pulse that is easily detected. The disadvantage is that each avalanche is the same, so that the counter gives no information about the energy of the original particle.

Scintillations

A very different method depends on the excitation to higher energy levels that is produced by the radiation. The material of the counter is chosen so as to have energy levels from which it returns to the ground state with the emission of photons in the visible part of the spectrum. This method was already used by the Curies to detect alpha particles. The source was placed close to a zinc sulfide detector, and the resulting flashes of light, or *scintillations*, were observed under a microscope in a darkened room.

Today's scintillation counters use the same principle, but they are much more sophisticated. The detector is a transparent crystal, and instead of the tedious observations of the light flashes, they are "seen" by a sensitive phototube called a photomultiplier.

The figure shows the crystal and the photomultiplier. A gamma ray that enters the crystal gives some or all of its energy to an electron. The electron slows down in the crystal and gives up some of its energy by excitation, i.e., by causing the crystal to go to higher energy levels. The crystal goes back to the lower energy levels and emits a visible light flash as it does so. (On the figure you see one of the photons of this light flash moving through the crystal to the photomultiplier.)

At the top of the tube, inside the glass, is a thin metal coating that serves as the photosensitive surface. The light that falls on it gives rise to electrons through the photoelectric effect. These electrons are accelerated toward another electrode that is at a higher potential by about 100 V. That gives them the energy to knock out several electrons when they hit. Each of them is then accelerated to a further electrode, again knocking out more electrons. After about 10 similar multiplications a single electron from the photosensitive surface produces about a million electrons at the last stage of multiplication.

A significant feature of the process is that the magnitude of the final pulse that is produced is proportional to the energy of the light flash that falls on the photosensitive surface. The energy of the light flash, in turn, is proportional to the amount of energy that is lost in the crystal by the radiation that is absorbed there.

When the combination of the crystal and the photomultiplier is used to measure the energy of the incoming radiation, it is called a *scintillation spectrometer*. It can be used to detect and measure the energy of gamma rays.

Because the crystal and photomultiplier are in a lightproof enclosure they "see" only the light produced in the crystal. The crystal is commonly sodium iodide, which is similar to sodium chloride (table salt). The higher atomic number of the iodine leads to a higher probability that inside the crystal the gamma ray is stopped by the photoelectric effect rather than by the Compton effect. (If the energy of the gamma ray is greater than 1.02 MeV there can also be pair production.)

Let's look at what happens when gamma rays with a single energy (say 1 MeV) enter the crystal and are absorbed there. Some will be absorbed by the photoelectric effect. The resulting photoelectron has all of the energy of the gamma ray except for the electron's binding energy of a few eV.

Do not confuse the photoelectric effect caused in the crystal by the gamma ray with the photoelectric effect caused by the visible light at the photosensitive surface of the photomultiplier. The first produces photoelectrons with the energy of the gamma ray (here 1 MeV) in the crystal. The second gives rise to photoelectrons with the energy of visible light, of the order of 1 eV, inside the photomultiplier.

Each gamma ray that is stopped in the crystal by the photoelectric effect gives rise to a light flash in the crystal of a single size. This leads, at the end of the multiplying process in the photomultiplier, to voltage pulses, all with the same size. That would make the measurement of the energy spectrum of the gamma rays straightforward. Some of the time, however, the gamma rays give up their energy through the Compton effect. In that case only part of the energy of the gamma ray is given to an electron. As a result a gamma ray with a single energy gives rise to electrons with a range of energies in the crystal. The light flashes produced in the crystal are therefore not all the same size, and neither are the voltage pulses at the output of the photomultiplier.

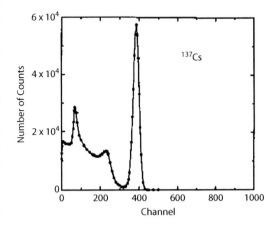

The figure shows the distribution of pulses initiated by a ^{137}Cs source that emits gamma rays with the single energy of 0.66 MeV. The number of the largest pulses is plotted on the right, with successively smaller pulses further to the left. The peak on the right is the "photopeak," resulting from the electrons that get their energy through the photoelectric effect. This is the full gamma-ray energy (except for the negligibly small binding energy of the electrons). The electrons that get their energy through the Compton effect have less energy and therefore are plotted further to the left. (There is more on scintillation spectra in the problems at the end of the chapter.)

13.7 Nuclear reactions

We have already discussed the spontaneous nuclear reactions of radioactivity. In 1919 Rutherford was the first to observe a nuclear reaction initiated by bombardment with another particle. He used alpha particles emitted in radioactivity to bombard nitrogen. The reaction was

$$^{14}N + \alpha \rightarrow {}^{17}O + p$$

Note that the number of nucleons (18) and the number of electronic charges (9) are the same before and after the reaction. Both quantities are *conserved*.

Alpha particles from polonium were also used in the discovery of the neutron. There had been speculation about the existence of such a particle for more than a decade, as well as unsuccessful experiments. In some of them neutrons were in fact observed, but wrongly thought to be high-energy photons. Finally, in 1932, following a flurry of activity in several countries, the reaction

$$^9Be + \alpha \rightarrow {}^{12}C + n$$

was correctly identified by Rutherford's former student James Chadwick in England.

A neutron source could then be constructed by mixing an α emitter, such as radium or polonium, with beryllium. The half-life of the neutron was later determined to be about 12 minutes.

By 1932 several devices had been built to accelerate electrically charged particles. The best known is the cyclotron, invented by Ernest Lawrence in Berkeley, California. The first reaction that was reported was

$$^7Li + p \rightarrow 2\alpha$$

first in England, by Cockroft and Walton, using a device that they called a "voltage doubler," which produced 0.7-MeV protons, and later in the same year with 1.22-MeV protons in Lawrence's cyclotron.

The charged particles, both the ones emitted in radioactivity and those that were given high kinetic energies in cyclotrons and other *particle accelerators*, were able to initiate nuclear reactions. To do so, however, they had to have enough energy to overcome the electric repulsion of the target nucleus, the so-called *Coulomb barrier*. Neutrons are not affected by the electric force and can come close to the nucleus even when they are moving slowly.

Neutron reactions and fission

The first person to exploit this feature in a series of experiments in which targets made of many different elements were bombarded by neutrons was Enrico Fermi in Rome.

The typical behavior that he and his collaborators found was that the neutron was absorbed by the target nucleus. In many cases the new nucleus then had too many neutrons to be stable, in other words, it was to the left and above the stability region. It would then be radioactive and emit a β⁻ particle to return to the stability region. The emission of the electron would, as usual, cause the daughter nucleus to have one more proton, and it would therefore be one step higher in the periodic table of elements than the initial target nucleus.

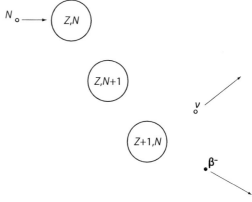

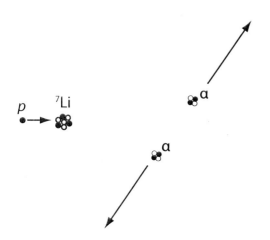

Fermi followed this behavior in a number of elements. It didn't take very long for him to ask what would happen if he went to the top of the periodic table. In 1934, soon after the start of his work with neutrons, he bombarded uranium, the naturally occurring element with the highest atomic number.

As in the other cases, he tried to identify the resulting material by studying its radioactive radiations and believed that he had discovered new *transuranic* elements with atomic numbers higher than uranium. He gave the supposed new elements names and tried to unravel the observed decay schemes, but this proved to be a difficult task because of the jumble of the resulting radiations.

Looked at from the vantage point of our present knowledge, it is strange and surprising that the mystery of what happens when uranium is bombarded with neutrons was not resolved until five more years had passed. It was 1939 when Otto Hahn and Fritz Strassmann in Germany showed by chemical means that one of the products of the bombardment seemed to be barium, element No. 56. They were hesitant to describe their result because in all nuclear reactions up to that time the changes had been no greater than that of an alpha particle, i.e., two steps in the periodic table of elements. Here was a change in the atomic number by 36!

It is clear on the binding energy curve that the nuclides near the middle of the horizontal axis have a higher binding energy per nucleon than those at the far right, and therefore represent more stable assemblies of the nucleons. However, all previous experience had been with nuclear reactions that resulted in the emission of a single particle, a proton, a neutron, an alpha particle, or no particle at all, as in the absorption of neutrons discussed earlier in this section.

The possibility of a split of the nucleus down the middle was so different that no one considered it until Hahn and Strassmann produced the direct chemical evidence. Even then they offered no explanation of how the barium might have been produced. It was Lise Meitner and Otto Frisch who within a short time described the process in detail in a paper that for the first time used the word "fission."

Lise Meitner had been working closely with Hahn in Berlin since they met in 1907, she as a physicist and he as a chemist. In 1938 she left Germany abruptly, forced out because of her Jewish background. Their collaboration continued by letter as well as it could. The experiments that constitute the discovery of fission came to fruition a few months after she left, but she is widely considered to have had a major share in their success. Nevertheless, the 1944 Nobel Prize in chemistry was awarded to Hahn alone.

The binding-energy curve lets us estimate the energy that is released in the fission process. The uranium is at the far right, with values of A of 235 and 238, and has a binding energy per nucleon of about 7.5 MeV. When it splits, the fission products are near the middle of the figure, where the binding energy per nucleon is in the neighborhood of 8.5 MeV. For each nucleon the binding energy goes up by about 1 MeV, for a total of more than 200 MeV each time a uranium nucleus undergoes fission. The nucleus doesn't just fall apart, it explodes, with the fragments sharing the released energy as kinetic energy.

A radioactive disintegration is also an explosion in which the products share the released energy as they separate, but the amount of energy that is released is usually of the order of at most a few MeV, much less than the energy that is released in fission.

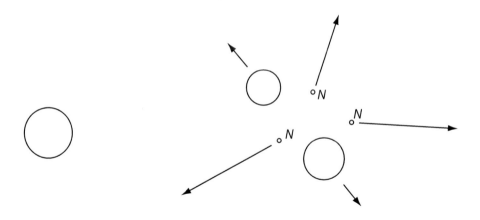

Now look again at the graph of N against Z. The farther up a nuclide is on the graph, the greater is the excess of neutrons over protons. The fission process leads to a mixture of nuclides widely distributed near the middle of the graph, where the neutron excess is quite a bit smaller. If, for example, $^{238}_{92}$U falls apart, with one of the fission products being $_{56}$Ba, the second must have an atomic number of 92 − 56, or 36, the atomic number of krypton. The neutron excess of stable barium is at most 26, and for krypton 14, while for the parent uranium it is 54. There are lots of neutrons to spare! What happens to them?

Some remain as part of the fission product nuclides. The number of extra neutrons is, however, so large that several fly off instantly as part of the fission process.

These neutrons may escape, or they may cause further nuclear reactions. If one of them hits another uranium nucleus, it may cause it to undergo fission. If enough neutrons are emitted each time, more than those that escape or cause reactions other than fission, there can be a *chain reaction*, with each fission process leading to further fissions. The chain reaction proceeds in a slow and controlled way in a *nuclear reactor*. It happens very quickly in an atomic bomb, which should, more correctly, be called a nuclear bomb.

EXAMPLE 9

Where are you most likely to find the fission products on the graph of N against Z? What happens to them?

Ans.:

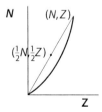

The heavier line on the figure represents the region of stability. Its curvature comes from the fact that the "neutron excess," $N - Z$, increases as Z goes up. The straight thinner line shows that if a heavy nucleus with N neutrons and Z protons splits in half, to products with $\frac{1}{2}N$ neutrons and $\frac{1}{2}Z$ protons, they are substantially above the stability region, with too many neutrons.

During fission a few neutrons are liberated and fly off separately. Negative beta decay reduces N by one and increases Z by one. It reduces the neutron excess and brings the nuclides closer to the region of stability. It is the most common form of radioactivity among fission products.

We know that neutrons can initiate nuclear reactions much more easily than protons or alpha particles because they are not deterred by the electric field, *the Coulomb barrier*, which repels positively charged particles. There is an additional, perhaps unexpected feature that comes into play: the slower a neutron moves, the more likely it is to cause nuclear reactions.

There are two reasons. One is that a slower neutron spends more time in the vicinity of the target nucleus and so has a greater chance to interact with it. The second one is that a slower neutron has a larger de Broglie wavelength. Where the neutron is cannot be localized more closely than by saying that it is within about a wavelength. In other words, the neutron will be able to initiate a reaction if it comes within a wavelength of the target nucleus.

The neutrons that are emitted in the fission process have kinetic energies of the order of 1 or 2 MeV. To enhance the probability of further fissions, and that there will be a chain reaction, it is advantageous to slow down the emitted neutrons. This may be done by using a material that has little chance of undergoing a nuclear reaction itself, but that can slow down the neutrons by letting them bounce around among its nuclei.

A material used for this purpose is called a *moderator*. It will make the neutrons slow down most efficiently if its nuclei have masses close to that of the neutron. Hydrogen is the obvious first one to consider. To have a sufficient density it is used in the form of water. A drawback is that the proton can absorb a neutron to form a deuteron. Heavy water (in which the hydrogen is the isotope ^{2}H) does not suffer from this limitation. It is present in all water, but deuterium is only 0.015% of natural hydrogen and is difficult and expensive to separate. The first nuclear reactor used carbon as a moderator in the form of graphite blocks alternating with the uranium fuel.

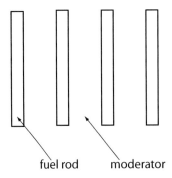

fuel rod   moderator

Natural uranium consists primarily of ^{238}U, with 0.72% of ^{235}U. Both can undergo fission when they absorb neutrons, but only the ^{235}U will do so with slow neutrons, i.e., neutrons that do not have an appreciable amount of kinetic energy. The slow neutrons have a much greater probability of causing fission. This makes ^{235}U the uranium isotope with the most favorable properties for a chain reaction. Its percentage in natural uranium is sufficient for a nuclear reactor that uses heavy water or carbon as the moderator. The majority of nuclear reactors today use water as a moderator, and in that case the uranium fuel needs to be *enriched* so as to contain a greater percentage (3–5%) of ^{235}U.

Later experiments showed that Fermi was right to look for elements beyond uranium when he bombarded uranium with neutrons. In addition to the fission products, which were responsible for most of the radioactivity, transuranic elements were also formed. Element 93 (neptunium, Np) and element 94 (plutonium, Pu) arise from the neutron absorption of ^{238}U:

$$^{238}\text{U} + n \rightarrow {}^{239}\text{U}$$

The ^{239}U is formed in an excited state and emits a γ-ray as it goes to its ground state. It is radioactive with a half-life of 23 min and emits a β particle:

$$^{239}\text{U} \rightarrow {}^{239}\text{Np} + \beta^- + \bar{\nu}$$

^{239}Np is also a β emitter. Its half-life is 2.4 days.

$$^{239}\text{Np} \rightarrow {}^{239}\text{Pu} + \beta^- + \bar{\nu}$$

Plutonium 239 is an alpha emitter with a half-life of 24,000 years. Just like ^{235}U it undergoes fission when bombarded with slow neutrons.

Fusion reactions

The binding-energy curve illustrates that the fission process leads to the release of energy: the binding energy per nucleon increases when the parent nucleus on the right side of the graph splits into two fission products near the middle. We see that the binding energy can also rise if we start from the left. This happens, for example, when helium 3 is bombarded by deuterons:

$$^{3}\text{He} + d \rightarrow \alpha + p + 18.3 \text{ MeV}$$

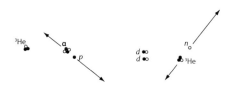

Such reactions are called *fusion* reactions. They would seem to be a great way to produce energy, but there are some major obstacles. The reaction of this example requires ^{3}He, which occurs in nature, but only to a minute extent (0.00014% of He). It can be made by nuclear reactions, but that is a slow and expensive process. There are other fusion reactions that use more common materials, such as

$$d + d \rightarrow {}^{3}\text{He} + n + 3.2 \text{ MeV}$$

and

$$d + d \rightarrow {}^{3}\text{H} + p + 4.03 \text{ MeV}$$

All fusion reactions share the difficulty that the nuclei repel, in other words that the Coulomb barrier has to be overcome. This can be done by using cyclotrons or other nuclear accelerators to give one of the initial materials sufficient energy. However, the machines require much more energy for their operation than is released in the reactions, so that their use does not lead to the production of net usable energy.

Another way to give the reacting nuclei more kinetic energy is to heat the material that contains them. With the growth of understanding of nuclear reactions in the 1930s, it became clear that this is how energy is produced in the sun and in the other stars.

To produce the temperatures required for fusion reactions on earth is formidably difficult.

The kinetic energies of the atoms become so high that electrons leave them. In other words, the atoms become ionized. (An assembly of ionized atoms is called a *plasma*.) The major problem is that the ions are positively charged and repel each other. It is very difficult to keep them together long enough and at a high enough concentration for the fusion reactions to occur with a net release of energy.

The energy of stars

The main process by which energy is released in the stars is the *proton–proton cycle* (or p–p cycle) of nuclear reactions, whose net effect is to produce helium from hydrogen with the release of energy. The binding-energy curve shows that the energy that is liberated is about 6.2 MeV per nucleon, or about 25 MeV for each alpha particle that is produced.

The basic reactions of the cycle are

$$p + p \rightarrow d + \beta^+ + \upsilon$$
$$p + d \rightarrow {}^3\text{He}$$
$$2\,{}^3\text{He} \rightarrow \alpha + 2p$$

If we multiply the first and second reactions by two, and then add all the nuclei on the left and on the right, we see that the net effect is to change four protons into an alpha particle, two positrons, and two neutrinos.

EXAMPLE 10

Use the masses to determine the energy that is liberated in the first of the reactions of the p–p cycle. The mass of the deuteron is 2.013558 u.

Ans.:
The mass–energy balance for this reaction is $2m_p = m_d + m_e + Q$, where Q is the energy that is released. (The neutrino mass is either zero or negligibly small compared to the other masses.) $Q = 2(1.007276\,\text{u}) - 2.013553 - 0.000549\,\text{u} = 0.000450\,\text{u} = 0.42$ MeV.

A feature of this cycle with vast consequences is that the first reaction proceeds at a much slower rate than the others. (This is related to the fact that this reaction involves electrons and neutrinos, which are not affected by the strong nuclear force. The reacton rate is therefore determined by the much weaker *weak force*.) Hence it, more than any of the others, determines the rate at which the sun and the other stars radiate energy. The timing of the life cycles of stars and of the evolution of the universe is determined to a large extent by this reaction.

The p–p cycle takes the synthesis of nuclei from individual nucleons to $A = 4$, with helium 4. It seemed at one time as if all elements might have been created one step at a time from a primordial "soup" of nucleons through fusion reactions. To start with, the ratio of helium to hydrogen, as it is observed in the sun and the other stars, can be accounted for quite closely by that assumption. There is, however, a crushing obstacle to the creation of heavier nuclei: there is no combination of five nucleons that is stable. The formation of nuclei, one nucleon at a time, stops with helium 4!

Not only that, but there is also no stable nuclide with $A = 8$. ${}^8\text{Be}$ can be formed, but decays in about 10^{-16} s. When the gradual buildup of nucleons to form elements beyond helium was first considered, the absence of nuclides with $A = 5$ and $A = 8$ seemed like a minor problem that later knowledge would surely be able to deal with. That turned out not to be the case. The obstacle presented by these holes in the chart of the nuclides has remained.

How then did we get the rest of the periodic table of elements? The main reaction that passes the $A = 5$ hole turns out to be the two-step process by which ${}^8\text{Be}$ is first made from two alpha particles. It exists long enough to lead to ${}^{12}\text{C}$ when bombarded by another alpha particle. We can write the sequence of reactions as $3\alpha \rightarrow {}^{12}\text{C}$. It occurs only at extremely high temperatures, larger than about 10^8 K, and they are found only in the interior of stars.

The beginning of the universe and the cosmic background radiation

To the best of our knowledge, the universe, as we know it, started at a particular moment, about 14 billion years ago, with a "big bang." The initial explosion created a mix of particles that soon (after a few minutes) became a mix of nucleons. Among the pieces of evidence that the universe originated in this way is the observed ratio of helium to hydrogen.

The most direct evidence is, however, the cosmic microwave background radiation, which was observed and identified in 1965. It had been predicted (by R. A. Alpher and R. Hermann in 1948) that the big bang should have resulted in radiation characteristic of the original temperature, which would then cool with the universe's expansion, so that the present spectrum would be like that of a radiating body at a few kelvins. The prediction was not, however, followed up for 16 years, partly because it was thought that the radiation would be too weak to be detected.

In 1964 R. H. Dicke suggested to his colleagues Peter Roll and David Wilkinson at Princeton University that they set up a microwave antenna to look for the radiation. They were getting their equipment ready when there was a phone call during a meeting of the group. When Dicke put the phone down he said: "Boys, we've been scooped!"

The call was from Arno Penzias at the Bell Laboratories in Holmdel, New Jersey, who, with Robert Wilson, had been testing a large and sensitive antenna. They observed radiation that they could not eliminate. It did not change with the direction in which the antenna was pointing, nor with the time of day.

Now anyone who has ever dealt with an electronic detection system, or at least turned on a radio, knows that there is always some unwanted background, referred to as "noise." The small amount of radiation seen by Penzias and Wilson would probably have been dismissed by most others. These two, however, thought that their antenna was so good that there should not be any spurious signals, no matter how small.

They puzzled over its possible origin, but had no clue until someone who knew of the Princeton group's activities suggested the fateful call. Two papers were published, one by Penzias and Wilson describing their observations and an adjoining one by the Princeton group on the likely source. Penzias and Wilson shared the Nobel Prize in physics for 1978.

A short time later the Princeton apparatus began to give results. It provided a point on the spectrum of the radiation, i.e., on the graph of intensity as a function of wavelength. Eventually the complete spectrum was measured and shown to have the form expected for the radiation from a hot body at 2.7 K. Since then, largely through the efforts led by Wilkinson, the cosmic microwave background radiation has turned out to be a rich and unique source of information on the early universe. With increased precision it was seen not to be completely isotropic, but rather full of hot and cold spots that have been called "the scars of creation." They are the remnants of the original inhomogeneities at the beginning of the evolution that led to our present nonuniform universe.

13.8 Particles

By the time the neutron was discovered in 1932, all the basic constituents of matter seemed to be assembled. Nuclei consist of protons and neutrons and atoms consist of nuclei and electrons. This, however, turned out not to be quite the whole story. First the positron was discovered (also in 1932) as the antiparticle to the electron. It was realized that antiparticles might also exist for the proton and neutron, and when higher-energy accelerators were available they were indeed observed in 1955 and 1956.

A completely different particle was predicted in 1935 by Hideki Yukawa. It was based on the quantum-mechanical picture of the electromagnetic force as being "mediated" by photons. This means that the force can be described as an *exchange* of photons, a sort of throwing back and forth of photons between the two bodies that exert the force on each other. Yukawa tried to think of a particle that would play the same role for the strong nuclear force. He showed that the mass of such a particle should be inversely proportional to the range of the force. For the electromagnetic force (with its infinite range) the mass of the mediating particle (the photon) is zero. For the nuclear force he predicted a mass about 250 times as large as that of the electron.

There was great excitement in 1937 when a particle (now called the *muon*) with 207 times the electron mass was found, and it was disappointing when experiments showed that it has none of the other properties of Yukawa's particle. In particular, it is totally unaffected by the strong nuclear force. We know now that it is a particle subject instead to the weak force. Its properties, except for its mass, are almost entirely the same as those of the electron. Its mean lifetime ($\frac{1}{\lambda}$) is 2.2×10^{-6} s and it decays into an electron, a neutrino, and an antineutrino.

The particle predicted by Yukawa was eventually observed in 1947. It is called the *pion*. It can be positive (π^+) or negative (π^-), with masses of 140 MeV and lifetimes of 2.6×10^{-8} s, decaying into a muon and a neutrino. There is also a neutral pion (π^0), with a mass of 135 MeV and a lifetime of 8×10^{-17} s, as it decays into two gamma rays.

Later (in 1983) particles mediating the weak force were identified, leaving only the particle mediating the gravitational force, the *graviton*, undetected.

Shortly after the pion was observed, other particles came on stage in bewildering profusion. They lived for small fractions of microseconds so that the thought that they were constituents of matter was hardly any longer appropriate. They did, however, provide a rich medium for experimental and theoretical exploration of regularities among them, to sort out what features are fundamental and which particles are truly "elementary."

Eventually it became possible to classify them in families. Among the "old" particles we still think of the electron as elementary. That means that it has, as far as we know, no smaller constituents. In fact, it seems to have no measurable size at all. It is part of the family called *leptons* ("the light ones"), a name given to small Judaean coins of biblical times. This name was more appropriate before the *tau* (τ) was discovered with a mass greater than that of the proton and neutron. The family consists of the electron, the muon, and the tau, each with its own kind of neutrino, six particles in all, plus their antiparticles. Their primary distinguishing feature is that they are subject to the weak but not the strong nuclear force. (There is also the gravitational, and, for the charged leptons, the electromagnetic force.) Insofar as the neutrinos have negligible mass, the weak force is the only one that they respond to. This explains why they hardly interact with anything and can go right through the earth without much chance of being stopped.

That the proton and neutron are not elementary particles is foreshadowed by their magnetic properties. Both have the same spin angular momentum as the electron. They both also have magnetic moments. It is not surprising that the proton has a magnetic moment, but its magnitude is unexpected: it does not seem to have any recognizable relation to the angular momentum. The neutron also has a magnetic moment, and this is very strange, since it has no charge and a magnetic moment is always associated with a moving charge. The measured magnetic moments of both make sense only if we accept that both the neutron and the proton are composite particles. Their constituents must be charged, although in the case of the neutron the charges must add up to zero.

The two nucleons are part of a family called *baryons* ("the heavy ones"). A third family is composed of the *mesons*, of which the pion is one. Both the baryons and the mesons are primarily affected by the strong nuclear force; together they are called *hadrons*. None of them is an elementary particle. Today we believe that they consist of particles that are elementary, called *quarks*. A meson is made of a quark–antiquark pair. A baryon consists of three quarks. We know of six kinds of quarks, each with its antiquark. Each quark has a fractional charge that has a magnitude either $\frac{1}{3}$ or $\frac{2}{3}$ that of the electron. A difficulty is that it seems to be impossible to observe quarks individually, i.e., outside the particles of which they are parts. They have been "detected" only by high-energy scattering experiments.

There will, no doubt, be further developments and perhaps surprises. New insights have been gained each time higher energies have been achieved. More powerful devices continue to be planned, with the expectation that we may continue to learn more about the fundamental interactions that underlie the structure of the universe.

13.9 Summary

The interaction of an atom with its neighbors is through the electrons, but there can be no atom without a nucleus. Think of a baseball team: all action is by the players, but the manager controls where they are and what they do.

Actually the nucleus has much tighter control than any baseball manager. With one more proton in it the electric field distribution is quite different, and so is the behavior of the electrons: we have a different element. But to change a nucleus is not so easy. It takes energies of the order of MeV, a hundred thousand to a million times as much as the energy to take away an electron.

To make it happen we need devices such as cyclotrons that can give charged particles energies that are sufficiently high to initiate *nuclear reactions*. Some nuclei disintegrate spontaneously. They are *radioactive*. By emitting alpha particles or beta particles they change into nuclei of other elements.

The nucleus of the atom determines which element an atom belongs to. Specifically it is the number of protons (Z), which is therefore called the atomic number. There can be different numbers of neutrons (N) in the nucleus (different isotopes), but different isotopes of the same element have the same chemical properties.

A given combination of Z and N represents a *nuclide*. Nuclides with low Z have values of Z and N that are close to each other. For higher values of Z there is a neutron excess because the electric repulsion between the protons favors neutrons over protons.

Some nuclides are unstable or *radioactive*. They fall apart (*decay*) spontaneously, with a characteristic half-life. They can emit alpha (α) particles (helium-4 nuclei consisting of 2 protons and 2 neutrons), beta rays or beta (β) particles (electrons), or gamma (γ) rays (photons emitted by a nucleus).

An energy-level diagram can represent the transition of a nucleus from a higher state to a lower state with the emission of a gamma ray. It can also represent a transition from one nuclide to another with the emission of an alpha particle. In that case Z decreases by 2 and N decreases by 2, while the alpha particle carries away 2 protons and 2 neutrons.

The emission of a beta ray carries away an electronic charge ($-e$) so that the remaining nucleus has one more positive charge ($+e$). It has one more proton and is therefore a nucleus whose Z has increased by one. The emission of the beta ray is accompanied by the emission of a neutrino. The energy that is released is shared as kinetic energy by the beta ray and the neutrino. (The kind of neutrino that accompanies electron emission is called an *antineutrino*.)

Each radioactive nuclide has its characteristic half-life. A radioactive sample decays exponentially.

The total binding energy of a nucleus is the energy required to separate it into its protons and neutrons. It can be measured in mass units or in energy units, since the two are related by the expression $E = mc^2$. c^2 ($= 9 \times 10^{16}$ J/kg) can be considered a conversion constant: 1 kg $= 9 \times 10^{16}$ J. Another mass unit is the atomic mass unit (u), equal to 1.66×10^{-27} kg; another energy unit is the MeV, equal to 1.6×10^{-13} J. Both these and other mass or energy units can be used in mass and energy relations, but each term must be in the same units. For example, in the relation $m_x + E_B = Zm_p + Nm_n$, we can use MeV, kg, or u as long as the same unit is used for each of the four terms.

We can compare the stability of different nuclides on a graph of binding energy divided by the mass number (or nucleon number, $A = N + Z$) as a function of A. The higher a nuclide is on this graph, the more stable it is. The semiempirical binding energy relation fits this graph with terms whose form is in keeping with theoretical considerations and whose size is determined empirically. Its main terms are the volume energy, the surface energy, and the Coulomb energy. There is also the asymmetry term and the odd–even term. In addition there is a shell structure, which is not dominant as it is for electrons in atoms, but leads to particularly stable numbers ("magic numbers") for Z and N.

A charged particle emitted by a radioactive substance usually moves very fast. When it hits another material it gives its energy in small steps to the atoms of the material and causes the electrons in the atoms to go to higher energy levels (*excitation*) or to leave the atom (*ionization*). On the other hand, when a gamma ray hits a material it loses its energy in one big step by the photoelectric effect, the Compton effect, or if the energy is larger than $2m_ec^2$ (1.02 MeV), by pair production.

The *roentgen* (R) is a unit that measures the amount of ionization produced in air. The *gray* (Gy) and the *rad* are units that measure the energy absorbed per unit mass. The *sievert* (Sv) and the *rem* are similar, but are adjusted by an empirical *quality factor* (QF) to take into account the greater effect of some kinds of radiation, primarily alpha particles and neutrons.

An *absorbed dose* of 5 to 6 Sv is likely to be fatal. At the much lower levels of the natural background (3 mSv/year) it is very difficult to establish radiation effects. The question of how to extrapolate from higher doses is controversial.

Charged particles can be detected and counted by various kinds of counters that make use of the excitation and ionization that the particles produce. Neutral particles and gamma rays first give energy to charged particles that can then be detected and counted when they give rise to excitation and ionization.

Protons, alpha particles, and other nuclei can be used to initiate *nuclear reactions*. They have to have enough energy to overcome the electric repulsion of the target nucleus. Cyclotrons and other devices can give the particles the energy to come sufficiently close to get within the range of the nuclear force.

Neutrons do not face this *Coulomb barrier*. They can initiate nuclear reactions much more easily. But first they must themselves be the result of nuclear reactions to be liberated from nuclei.

Fission reactions occur in the heaviest nuclides. The result is that the nucleus splits into two pieces. Some neutrons are emitted at the same time. This is because the neutron excess ($N - Z$) is greater for heavier nuclides. These neutrons can initiate further fission reactions and so cause a *chain reaction*.

In a nuclear reactor the neutrons are slowed down by a *moderator* so that they have only the kinetic energy of their thermal motion. This makes them very efficient at causing fission in those isotopes that can undergo fission with slow neutrons, i.e., with neutrons that have little or no kinetic energy. Two isotopes that can undergo fission when bombarded with slow neutrons are ^{235}U and ^{239}Pu.

Fusion reactions occur for the lightest nuclides. On the graph of E_B against A we see that the most stable nuclides are in the middle, near iron and nickel. Fission reactions proceed from the high-A end toward the middle, and fusion reactions from the low-A end. Energy is released in both cases. Neutrons can initiate fission reactions, but fusion reactions have to overcome the Coulomb barrier. One way to do this is to heat the reacting materials to very high temperatures. This is the source of the energy of stars.

The *microwave background radiation* is the best evidence that the universe as we know it started with an explosion (the "big bang") about 14 billion years ago.

The basic building blocks of matter, the protons, neutrons, and electrons, were known by 1932 when the neutron was identified. Other particles have been discovered since that time, and the pursuit of their properties and functions is an important branch of today's physics.

13.10 Review activities and problems

Guided review

1. What is the composition of a nucleus of ^{35}S? Of ^{238}U?

2. The mass of a $^{138}_{56}$Ba nucleus is 137.874492 u. What is its total binding energy in MeV? What is the binding energy per nucleon?

3. The binding energy per nucleon for an alpha particle is 7.1 MeV. How much energy is made available in a process that synthesizes an alpha particle from its separate nucleons?

4. Describe how the graph of $\frac{E_b}{A}$ vs. A shows the stability of the stable nuclei. On this graph, at which value of A do you find the least stable nucleus and where is the most stable nucleus?

5. The mass difference between the nuclear masses of ^{235}U and ^{231}Th is 4.006524 u.

(a) What is the kinetic energy (in MeV) of the alpha particle emitted in the decay of ^{235}U?

(b) What is the kinetic energy (in MeV) of the recoiling thorium nucleus? Can it be neglected?

6. The energy released in the β decay of ^{14}C is 0.16 MeV.

(a) What is the mass difference between the ^{14}C nucleus and its daughter nucleus (in MeV and in u)?

(b) What is the mass difference (in MeV) between the neutral atoms of ^{14}C and the neutral atom of the nuclide to which it decays?

7. What properties of the proton and the neutron cause one of these particles to be stable and the other to be radioactive?

8. A counter registers 10,000 counts per minute as it counts the radiation emitted by a sample of ^{131}I.

(a) What is the decay constant, λ, for ^{131}I?

(b) What is the closest whole number of half-lives after which the counter (in the same position) will register less than one count per minute?

9. (a) Why is β$^+$ radioactivity rare among fission products?

(b) Where on the graph of N vs. Z are you most likely to find alpha radioactivity? What is the direction of a line on this graph from the parent to the daughter nucleus?

10. Use the nuclear masses to determine the energy that is liberated in the second and third steps of the p-p cycle. What is the total energy that is liberated in one cycle? (m_{He3} = 3.014931 u)

Problems and reasoning skill building

1. ^{61}Cu is radioactive and emits β$^+$ particles.

(a) Write the decay relation that shows the parent nucleus and the products of the disintegration.

(b) Write the relation between the masses of the parent and daughter nuclei and the other decay products.

2. A nuclide (in its ground state) is observed to emit alpha particles with energies of 3.7 MeV, 2.5 MeV, and 1.8 MeV.

(a) Draw a possible energy-level diagram for this decay.

(b) What other radiations do you expect to be emitted? Give their energies and show them on the diagram.

3. (a) After how many half-lives does a radioactive source decay to 10% of its original amount? Do this approximately by raising 2 to successively higher powers.

(b) Do it more precisely by using the relation $N = N_0 e^{-\lambda t}$ or its equivalent, $\lambda t = \ln \frac{N_0}{N}$.

4. ^{35}S is radioactive and emits β$^-$ particles whose maximum energy is 0.17 MeV. What is the range of energies, from the smallest to the largest, of the electrons and of the neutrinos that are emitted in this process?

5. ^{90}Sr has a half-life of 30 years. 10 curies of this substance fall on a one-acre field. Draw a graph of the amount remaining against time for five half-lives.

6. Write the relations that describe the following radioactive transformations:

(a) the beta decay of ^{90}Sr,

(b) the beta decay of tritium,

(c) the alpha decay of radon,

(d) the positron decay of ^{22}Na.

7. A radioactive substance has a half-life of one day.

How many days will it take for a counter that originally registers 100,000 counts per minute to have an average count rate of one count per minute?

8. The mass of an atom of $^{212}_{84}$Po (i.e., that of its nucleus and all of the electrons) is 211.988842 u.

(a) What is the total binding energy of the nucleus?

(b) What is the binding energy per nucleon?

9. A deuteron and a tritium nucleus combine to form an alpha particle.

(a) What other particle is produced in this reaction?

(b) Use the masses to calculate the released energy.

10. ^{60}Co disintegrates by emitting negative beta rays with a maximum energy of 0.31 MeV.

(a) Write down the reaction relation.

(b) The beta ray emission leaves the daughter nucleus in an excited state from which it goes to the ground state with the successive emission of gamma rays, first one of 1.17 MeV and then one of 1.33 MeV. Draw the energy-level diagram.

(c) What is the difference in the masses of the parent and daughter nuclei in units of u?

11. The natural abundance of deuterium is 0.015%.

(a) What is the mass of heavy water in 1 m^3 of water?

(b) What is the number of deuterium atoms in this amount of water?

Multiple choice questions

1. The plot of N vs. Z for the stable nuclides deviates from the line $N = Z$ as a result of

(a) the electric force,
(b) the strong nuclear force,
(c) the weak nuclear force,
(d) the fission process.

2. The graph of $\frac{E_B}{A}$ vs. A for the stable nuclei has a peak at $A = 4$. This shows the following:

(a) The alpha particle is more stable than any other nucleus.
(b) The alpha particle is less stable than a nucleus with $A = 5$.
(c) The alpha particle is more stable than a nucleus with $A = 3$.
(d) The total binding energy of an alpha particle is greater than that of an iron nucleus.

3. Among radioactive radiations:

(a) Alpha particles have discrete energies, but beta and gamma radiations have continuous ranges of energies.
(b) Alpha and beta particles have continuous ranges of energies, but gamma rays have discrete energies.
(c) Neutrinos and negative beta particles have continuous ranges of energies, but positive beta particles have discrete energies.
(d) Alpha particles and gamma rays have discrete energies and beta particles have continuous ranges of energies.

4. There are 1000 protons and 1000 neutrons, each separate and with 0.1 MeV of kinetic energy, in a container. After one hour:

(a) There will be more neutrons than protons.
(b) There will be more protons than neutrons.
(c) There is no radiation that can escape the container.
(d) The neutrons will have changed into protons and the protons will have changed into neutrons.

Synthesis problems and projects

1. (a) What is the approximate whole number of grams in a mole of ^{226}Ra? (You can also look up the more precise atomic mass.)

(b) What is the number of nuclei in 1 g of ^{226}Ra?

(c) The half-life of ^{226}Ra is 1600 years. What is its decay constant, λ, in y^{-1}? What is it in s^{-1}?

(d) What is the activity (in Bq) of 1 g of ^{226}Ra? (This is the rate of disintegration, $-\frac{dN}{dt}$, that is also equal to λN.)

(e) Compare your result to the number of Bq in one curie.

2. A hypothetical fusion plant uses the d–t reaction $d + t \rightarrow \alpha + n + 17.6$ MeV. It produces 3000 MW (3×10^9 W) of thermal energy.

(a) What is the mass of heavy water that is required in one year to supply the deuterons?

(b) What is the mass of tritium that is required?

3. Tritium for the reactor of Problem 2 is made by the bombardment of ^{6}Li with neutrons.

(a) Write down the reaction relation.
(b) How many atoms and what mass of ^{6}Li are required for one year of operation?

4. What is the mass of ^{235}U that must undergo fission to produce the same amount of thermal energy as a a ton (2000 lb) of coal and a barrel of oil? (The heat of combustion of coal is 1.3×10^4 btu/lb and of oil is 6.5×10^6 btu/barrel).

5. What mass of natural uranium is required in one year for a pressurized water reactor that produces 1000 MW of thermal energy from the fission of ^{235}U?

6. The count rate from ^{14}C in a prehistoric wooden tool is found to be 6.5 counts per second. The mass of the tool is 220 g.

(a) Compare the ratio of ^{14}C to ^{12}C in this tool to the ratio found in living trees today, which is 1.3×10^{-12}.

(b) What is the age of the tool?

7. Radium 226 has a half-life of 1600 years. It emits an alpha particle with an energy of about 5 MeV.

(a) Write the relation that describes this reaction.

(b) What is the number of atoms in 1 mg (10^{-6} kg) of Ra?

(c) What is the total amount of energy (in J) that is emitted during one half-life by a source of radium whose initial mass is 1 mg?

8. The potassium content of the human body is about .12% and .012% of that is the radioactive isotope ^{40}K. The half-life of ^{40}K is 1.25×10^9 years. Each disintegration releases 1.3 MeV of energy, about 40% of which is absorbed by the body.

(a) What kinds of radiation are likely not to be absorbed by the body?

(b) What is the number of atoms of ^{40}K in a person whose mass is 70 kg?

(c) What is the decay constant of ^{40}K?

(d) What is the number of nuclei of ^{40}K that disintegrate in the body of this person every second? What is the activity in microcuries?

(e) What is the energy absorbed in the body per second and the energy per unit mass (in J/kg)? What is the absorbed dose in Gy and in rads? What is the number of Sv and rems?

9. (a) Use the information given in Problem 6 in this section to estimate the mass and the activity (in Bq and Ci) of ^{14}C in a 70 kg person.

(b) The beta particles emitted by ^{14}C carry away 40% of the energy release in the disintegration. Calculate the absorbed energy and the energy absorbed per kg in the 70 kg person.

10.

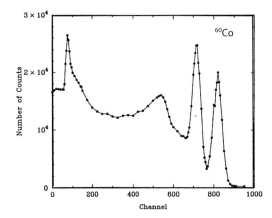

(a) The figure shows the scintillation spectrum for ^{60}Co. The two peaks at the right are the "photopeaks," resulting from the photoelectric effect produced by the two gamma rays at 1.17 MeV and 1.33 MeV in a sodium iodide crystal. Use the information on the graph together with that for the photopeak of ^{137}Cs in the text to draw a "calibration graph," namely a graph in which you plot the channel number on the vertical axis and the energy on the horizontal axis.

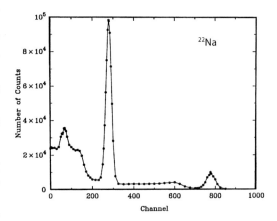

(b) The spectrum in the text, for ^{137}Cs, and the ones here, for ^{60}Co and ^{22}Na were recorded with the same settings of amplification and voltages, so that they have the same relation between channel number and energy. You can now find the energy corresponding to the two photopeaks on the ^{22}Na spectrum. The one on the right is for a gamma ray. Find the energy for it from the channel number on your graph.

(c) Do the same for the large peak at the lower energy. The fact that ^{22}Na is a positron emitter tells you the origin of this peak. Its energy represents the rest energy of the electron.

(d) If you know the speed of light, you can consider your measurement to be that of the mass of the electron. Calculate the mass of the electron from your result.

(e) If you decide that you already know the mass of the electron from other experiments, you can consider your measurement to be one that determines the speed of light. What is the primary result of the experiment?

11. In Chapter 12 we said that the most probable angle between the incoming and outgoing photons in the Compton effect is 180° and that

the energy of the "backscattered" photon is then $\frac{EE_0}{2E+E_0}$, where E_0 is the rest energy of the electron. This angle also corresponds to the maximum energy that the Compton electron can have.

(a) Calculate the maximum Compton electron energy for ^{137}Cs and compare it to the scintillation spectrometer plot for this nuclide.

(b) Calculate the energy of the backscattered photon corresponding to the ^{137}Cs gamma ray and find it on the scintillation spectrometer plot. The peak corresponding to this energy is called the "backscattering peak."

(c) What is the relation between the answers to parts (a) and (b)?

CHAPTER 14

Energy in Civilization

The flow of energy

Electric energy: what is it and what does it do for us?

DC and AC: transformer and generator

Energy storage
- *Batteries*
- *Supercapacitors*
- *Hydrogen*
- *Fuel cells*

Entropy and the second law of thermodynamics: the limits of energy transformation
- *Thermal energy*
- *The second law of thermodynamics*
- *Entropy*

Our addiction to fossil fuels
- *Availability*
- *The greenhouse effect*
- *The rate of energy increase*

Other sources of energy
- *Nuclear energy*
- *Fusion*
- *Solar energy*
- *Wind*
- *Biomass*
- *Energy from rivers, oceans, and the earth*
- *The sustainability transition*

Energy fuels our civilization. We use it to heat our homes and our food. As mechanical energy it turns the motors in our industries and households. It is necessary for transportation and communication.

Estimated U.S. Energy Use in 2008: ~99.2 Quads

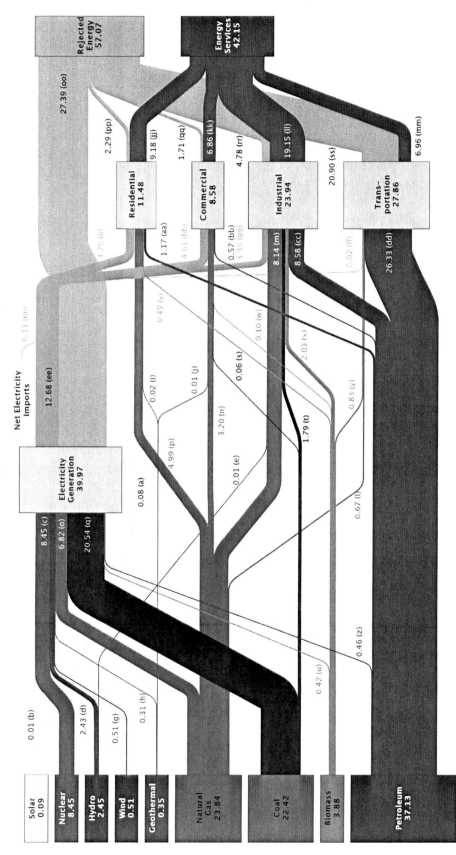

324 / Energy in Civilization

We say that we *generate* energy, but we know that we can only transform it from one form to another. The *source* of our energy is overwhelmingly the internal energy of the *fossil fuels*, coal, oil, and natural gas, whose carbon content consists of the remains of living organisms built up over millions of years. There is also the internal nuclear energy of uranium in our reactors, and the energy radiated to us by the sun.

In this chapter we explore the energy transformations that underlie our civilization. We also consider the limitations described by the second law of thermodynamics, and some of the obstacles that accompany our use of energy.

14.1 The flow of energy

The flow diagram (courtesy Lawrence Livermore National Laboratory) shows the path of energy in the United States from the sources on the left, to the way we use it on the right. The numbers are estimates for 2008, in *quads (Q)*, or quadrillions ($=10^{15}$) of british thermal units (*btu*), where $1 btu = 1055 J$. $1Q = 1.055 EJ$ or *exajoules*, each equal to 10^{18} J.

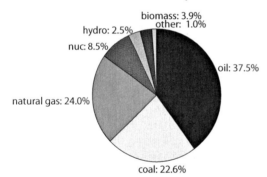

Here is a chart that shows the consumption of the various kinds of energy in the United States in 2008. The total amount is 99.2 Q or 104.7 EJ.

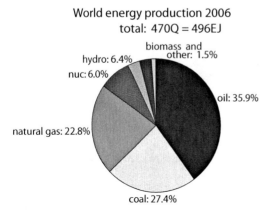

And here is a similar chart for the energy production of the world. (The numbers for both charts are from the US Energy Information Administration.)

The most drastic message of these diagrams is that about 86% of the world's energy comes from fossil fuels. The other side of the message is how little of our energy comes from other sources. We will look at the alternatives, and why the patterns of energy use have been so resistant to change.

The second message is that in the United States we use about 21% of the world's energy, while our population is only 4.6% of the total.

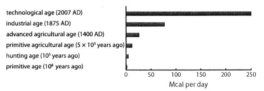

Source: Adapted from a similar figure in *Energy and Power* (W. H. Freeman and Co., 1971).

This figure shows how our appetite for energy has changed with time. At the beginning all that was consumed was the daily food, of about 2000 food calories (or 2×10^6 cal = 2 Mcal or about 8.6 MJ) per day. With hunting the amount of food increased, and wood was burned for heat and cooking. Primitive farming brought the planting of crops and the use of animal energy. In a later age animals were also used for transportation, coal was used for heating, and there were other sources of energy, including wind and water. A large increase came (in what we now call the developed world) at the time of the industrial revolution with the introduction of the steam engine and the widespread use of machinery. Finally we arrive at our own time, with modern modes of transportation and all the gadgetry of modern civilization, when we use (in the United States) about 250,000 food calories per person.

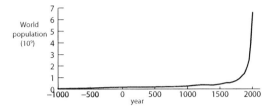

In the same period of time the number of people has increased dramatically. Already in 1798 Malthus saw that this increase could not continue indefinitely. He predicted that the population would be limited by famine, disease, war, and vice if drastic steps were not taken to limit it by other means. The world's population is now larger by a factor of about 8 than in the time of Malthus. Was he wrong? Or did he just not have the right timescale in mind when he said that disaster was then imminent? Today there are those (the neomalthusians) who believe that limiting the population is our most urgent task, while others expect that human ingenuity and technological advances will meet and resolve all challenges.

We begin with a form of energy that is not among the sources that we have listed. *Electric energy* is a *carrier* of energy. It allows us to transform energy from its primary source in one place and to use it in another.

Later we discuss another carrier of energy, one that we are not now using, but that is being considered seriously, namely hydrogen. It has to be separated from water or another substance at the cost of energy, and can then be used as fuel somewhere else.

14.2 Electric energy: what is it and what does it do for us?

Our civilization depends heavily on electric energy, on its generation, its distribution, and its eventual transformation into internal (thermal) energy, light, and mechanical energy. We also use electricity to send and receive information along wires and fibers, and through space as electromagnetic radiation.

Electric energy is often said to be *clean*. Electric motors lack the noise, the air pollution, and the waste heat of gasoline engines. Electricity is, however, not a *primary* form of energy. It has to be *generated*, i.e., transformed from some other kind of energy, today most often the internal (atomic and molecular) energy of coal, oil, or natural gas or the (also internal) nuclear energy of uranium. It is *released*, i.e., transformed to thermal energy, through the chemical reaction of burning in the first case and the nuclear reaction of fission in the second. The waste materials and unused thermal energy are still there, but they can be far from our backyards.

The question of the nature of electric energy turns out to be somewhat subtle. There is not likely to be an entry for *electric energy* in a physics textbook. We talk of *using* it, but just what are we using?

The battery stands ready to transform its internal, chemical energy to some other kind of energy, either right where it is or somewhere else. In the basic circuit the transformation is to thermal energy in the resistor. But there is an ephemeral intermediate state, before the energy is used, when it is electric *potential* energy. It is this intermediate state that allows us to transport energy from the source to the place where it is used.

We can think of a mechanical circuit with a similar sequence of energy transformations: lift some marbles in one place, transforming energy from your muscles or some device to gravitational potential energy. The marbles can now roll around at the new height, using up only the fraction of their energy necessary to overcome friction. They can drop down somewhere else and there transform their gravitational potential energy to kinetic energy. On impact there is a further transformation, ending most often in *dissipation* as thermal energy.

14.3 DC and AC: transformer and generator

Only a minute fraction of the electric energy that we use is transported via the direct current (DC) that we have been talking about so far. An even smaller fraction comes from batteries. The advantages of alternating current (AC) are overwhelming. Just what is *AC* and what are these advantages?

The first is the generator. Take a loop of wire with a magnetic field through it and rotate it about its diameter. For half a turn the magnetic flux through the loop is in one direction with respect to the loop. For the next half turn it is in the other direction. Imagine yourself standing on the loop and rotating with it. The direction of the field and of the flux will alternate, up toward your head, down toward your feet, up, down, and so on, and so will the emf that is induced, in accord with Faraday's law.

The rotating loop is an elementary AC generator. To make a DC generator, with an emf and a current that are always in the same direction, there has to be a switch (the *commutator*) that reverses the current direction after each half turn. Such a switch can be built in by having a split ring on the axle turning with the loop (see page 205). The current in the external circuit, on the other side of the commutator, will still vary in magnitude, but it will remain in one direction. The commutator is a weak part of any DC generator. All of the generated current has to go through the sliding contacts (the *brushes*), and the sparks there are a continuous source of deterioration.

The other great advantage of AC is that we can use *transformers*. Take two loops close to each other. A current in one will produce a magnetic field through both. But only a changing current in one will produce an emf and a current in the other loop.

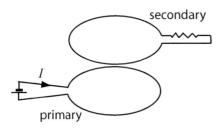

Two loops in the passive *secondary* circuit will each have the same emf induced, and they can be connected so that the emfs add up. No energy is produced, but not much is lost either. The product of the emf and the current, $\mathcal{E}I$, remains almost the same, but $\mathcal{E}$ and I can be changed at will by changing the number of turns in one or the other of the two circuits.

Why is this so important? A lower current means less loss of energy in the transmission. Wherever energy is transported through transmission lines there are transformers to increase the potential difference, the *voltage*, between the wires. The current decreases, and therefore also the energy (I^2R) that is lost to the surrounding atmosphere. On the other hand, more extensive and more costly insulation is then required. The balancing of the variable cost of the energy lost in transmission against the fixed cost of the installation at a certain voltage is an interesting problem in economics. The increasing cost of electric energy and the greater distances over which it is transported have pushed the voltages at which it is transmitted higher and higher, up to between 100,000 and a million volts.

14.4 Energy storage

We are really bad at storing energy in any form. It can be done with gravitational potential energy by pumping water to a greater height. It is also done in a pendulum clock. Flywheels, with stored kinetic energy, and superconducting coils, with energy stored in their magnetic field, are being considered, but storage as chemical energy in batteries is really the only method widely used.

Batteries

The cost of batteries and their poor ratio of stored energy to the mass that is required have made them a minor source of energy, used as backup for times when the transmission system fails or when portability is an overriding concern, as in cars and flashlights. There are also uses that require very little energy, as in watches, calculators, cameras, and hearing aids, and here batteries are used to great advantage.

With better batteries or other storage mechanisms the use of intermittent and unpredictable forms of energy, such as wind and solar energy, would, no doubt, be much more widespread.

Artifacts have been found that may have acted as batteries in antiquity, but modern electrochemistry began with the work of Luigi Galvani (1737–1798) and Alessandro Volta (1745–1827) near the end of the eighteenth century. Galvani was a physician and professor at the University of Bologna in Italy. The most cited story of his discovery is that he was using an iron scalpel to dissect a frog leg that was held by a brass hook when he saw the leg twitch as the muscle contracted.

He ascribed this phenomenon to "animal electricity," but Volta soon showed that the essential components of the experiment were the two different metals and an electrolyte (a liquid containing mobile ions). He constructed the first "voltaic cells," one of which had electrodes of copper and zinc separated by paper soaked in salt water. Later batteries, to this day, differ primarily in the materials that are used.

The first rechargeable cell was the lead-acid cell, invented in 1859 and now used in just about every car. Its electrodes are lead and lead oxide with sulfuric acid as the electrolyte. In a "dry cell" the liquid electrolyte is replaced by a paste. In "alkaline" batteries the electrodes are zinc and manganese dioxide and the electrolyte is potassium hydroxide, KOH.

There are two other kinds of rechargeable batteries that are widely used today. One is the nickel-metal hydride (NiMH) battery, invented in the 1980s. Its electrodes are hydrogen (as a metal hydride) and nickel hydroxide, $Ni(OH)_2$, and the electrolyte is KOH. Its stored energy per kg can be about twice that of the lead-acid battery. NiMH batteries are used in digital cameras and other small-scale applications. Large assemblies are used for short-term storage in hybrid cars.

The lithium-ion battery in one of its forms has electrodes of lithium and lithium cobalt oxide. It can store about four times the energy per kg of the lead-acid battery. It has the additional advantage that it keeps its charge better than other storage batteries. It loses only about 15% of its charge in a year, compared to 5% per month for the lead-acid battery and 20–25% per month for the NiMH battery. Its use beyond small-scale applications has been slow to develop because of its cost and because it is subject to thermal instabilities. With intensive research it is on its way to fulfill its promise.

EXAMPLE 1

(a) What is the amount of energy and the amount of energy per kg stored in a fully charged 12 V lead-acid battery with a mass of 18 kg and rated at 45 Ah?

(b) How does this compare to the kinetic energy of a 1.5-tonne car at 60 mph?

(c) What would be the cost of this amount of energy from a wall plug at 10 cents per kwh?

(d) What would be the cost of an equivalent amount of gasoline at 3 dollars per gallon, used in an engine with an efficiency of 25%?

Ans.:

(a) 45 ampere-hours at 12 V is 12×45 Wh or 540 Wh. The amount of energy per kg is $\frac{540}{18} = 30$ Wh/kg.

(b) 60 mph = 26.8 m/s, 1 tonne = 1000 kg, $\frac{1}{2}mv^2 = \frac{1}{2}(1500)(26.8)^2 = 0.539 \times 10^6$ J = 0.15 kwh. The battery stores 3.6 times as much energy.

(c) The cost of the 0.54 kwh stored in the battery, if it were from a wall plug at 10 cents per kwh, would be 5.4 cents. The cost of the kinetic energy of the car would be 1.5 cents.

(d) The energy density of gasoline is about 45 MJ/kg, or, since 1 kwh = 3.6 MJ, 12.5 kwh/kg. The density of gasoline is 0.7372 kg/liter, or, since 1 gallon = 3.785 liters, it is $45 \text{MJ/kg} \frac{0.7372 \text{ kg}}{1 \text{ liter}} \frac{3.785 \text{ liters}}{1 \text{ gallon}} = 125.6$ MJ/gallon = 34.9 kwh/gallon. With an efficiency of 25% we get 8.7 kwh from 1 gallon.

The kinetic energy of the car (0.15 kwh) uses 1.7×10^{-2} gallons, at a cost of 5.2 cents.

Supercapacitors

A recent development in energy storage is that of capacitors with a large effective area. The capacitance of a parallel-plate capacitor is $\epsilon_0 \frac{A}{d}$, so that a large area means a large capacitance. The use of porous carbon and carbon nanotubes has made it possible to increase the effective area and to achieve capacitances of thousands of farads. (A carbon nanotube is a cylinder whose diameter is about 1 nm, i.e., it is of the order of an atomic size.)

These "supercapacitors" have a number of advantages over batteries. They can be charged in seconds rather than hours. Since no chemical changes are involved they can go through millions of charging cycles without degradation, compared to perhaps hundreds for batteries. Very little of their energy is lost in a charging cycle. One of their first experimental large-scale applications has been in buses powered by supercapacitors that are recharged at every bus stop.

EXAMPLE 2

What is the amount of energy stored in a capacitor of 3000 f charged to 100 V and 200 V?

Ans.:
At 100 V it is $\frac{1}{2}CV^2 = (0.5)(3000)(100^2) = 1.5 \times 10^7$ J or 4.17 kwh. At 200 V it is four times as large, or 16.7 kwh.

Hydrogen

Hydrogen is not a primary fuel. It is the most abundant element, but it does not occur on earth as an independent substance. It combines too easily with other elements, principally with oxygen to form water. This is why it is a good fuel. For each molecule of water that is formed, 2.97 eV of energy is liberated, i.e., transformed from the internal energy of the separate atoms of hydrogen and oxygen to the energy of motion of the molecule that is formed. For a mole (18 g) of water the liberated thermal energy is 2.86×10^5 J.

The energy available from a kg of hydrogen is about 142 MJ/kg (142×10^6 J/kg.) This is greater than the value for gasoline, which is about 45 MJ/kg. The energy per unit volume is, however, smaller than that for gasoline. It is feasible to store hydrogen at a pressure of 70 MPa or about 700 atmospheres. At this pressure hydrogen takes up about eight times the volume of gasoline storing the same amount of energy.

To produce elemental hydrogen it has to be separated from the other elements with which it combines. Water can be decomposed by electrolysis, with an energy expenditure equal to that which is freed when the elements combine. This is not, however, how most hydrogen is produced today. Rather it is separated from natural gas, primarily from methane (CH_4). This is a less expensive process, but it does not take advantage of the freedom from carbon emission that hydrogen promises.

There has been much speculation about the possibility of using hydrogen to replace gasoline as a fuel for cars. There are, however, formidable obstacles to the development of what has been called the *hydrogen economy*. The first is, as we have seen, that the available energy per unit volume of hydrogen is about one-eighth that of gasoline. Among the technological changes that would be required is the development of ways to store and to transport large amounts of hydrogen, to do so safely, and to provide readily available fuel through a network equivalent to that of today's filling stations.

There are also some features that favor hydrogen. One is that intermittent sources of energy, such as wind and solar energy, could be used for the electrolytic separation of hydrogen from water. Another is that it can be used in *hydrogen fuel cells* that can have efficiencies about twice as large (about 50%) as those of gasoline engines.

Fuel cells

In a hydrogen fuel cell internal energy is transformed directly into electric energy. Such a cell consists of two plates separated by a barrier through which ions can pass. Hydrogen gas flows past one of the plates and, with the help of a catalyst such as platinum, dissociates into ions and electrons. The ions move to the other plate through the barrier. The electrons move through the external circuit and their current represents the electric energy produced by the cell.

14.5 Entropy and the second law of thermodynamics: the limits of energy transformation

Thermal energy

All materials have *internal energy*. Most of the time when we say "internal energy" we mean the energy of the random motion of the atoms or molecules of which the material is composed. Part of this energy is the kinetic energy of their motion and part is their mutual electric potential energy. The sum of these two energies is the material's *thermal energy*. The thermal energy is greater when the temperature is higher. It also changes when the atoms or molecules rearrange themselves to form a different structure. This happens when there is a phase change (as from solid to liquid) or a chemical reaction.

In addition to the internal thermal energy of the motion of the atoms and molecules there is also the energy inside the atoms and molecules. Each molecule, atom, and nucleus has internal energy. The atoms in each molecule, the nuclei and electrons in each atom, the protons and neutrons in each nucleus all have kinetic and mutual potential energy.

When two objects are in contact, the greater thermal motion in one is transmitted to the other by collisions between the atoms and molecules and in metals by the free electrons. This is the process of thermal conduction. The direction is from the one at higher temperature to the one at lower temperature. When the two are at the same temperture there is no heat transfer, and the two are said to be in *thermal equilibrium*.

What happens when wood, coal, oil, or any other fuel is burned? Some of the internal energy stored inside the molecules is transformed into thermal energy, i.e., into the energy of their random motion.

Similarly, in a nuclear reactor some of the internal energy stored inside the nuclei of the nuclear fuel (most often ^{235}U) is transformed to kinetic energy of the fission products. When this kinetic energy is shared among the atoms and molecules by collisions it becomes thermal energy. The fusion reactions in stars also change some of the internal energy of nuclei to kinetic energy of the reaction products, which when shared becomes thermal energy.

In some cases (as on a stove or in a furnace) the increase in thermal energy and the accompanying rise in temperature are all we want to achieve. But we may also want to transform some of the energy to mechanical energy. A device that continuously changes thermal energy to mechanical energy is called a *heat engine*. Examples are the steam engine and the gasoline engine. They achieve the required transformation, but with a characteristically low efficiency, because of the limitations imposed by the second law of thermodynamics.

The second law of thermodynamics

If we were limited only by the law of conservation of energy (which is also called the *first law of thermodynamics*), a ship could cross the ocean without fuel, just by cooling down the ocean and using its internal energy. That doesn't happen. Observation and experiment show that it can't be done. The generalized statement of this observation is called *the second law of thermodynamics*.

Thermal energy tranfers spontaneously from a hotter object to a cooler one, but not in the other direction. It is a process allowed by the first law of thermodynamics but not by the second.

There are various versions of the second law, illustrating its immense and pervasive importance. One version is about the limitation of converting internal energy to work, usually to mechanical energy. Yes, it can happen, but only if there is a hot part of the system (the steam in a steam engine, the burnt fuel in a gasoline engine) and a cooler part, to which the condensed steam or the spent fuel can go, taking with it most of the thermal energy. A second version, which can be shown to be equivalent to the first, is about cooling an object below the temperature of its surroundings. This does not use any net energy, but requires work anyway. Only if work is done on a system can thermal energy be transferred from an object at a lower temperature to one at a higher temperature. This is what happens in a refrigerator.

Entropy

To describe the effects of the second law of thermodynamics quantitatively, there is a special

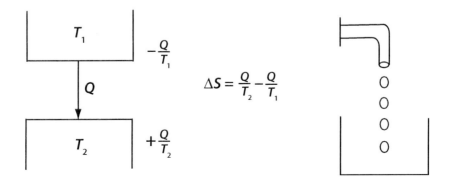

concept, the *entropy*. Unlike force, work, and energy, it has no counterpart in ordinary non-technical language.

Thermal energy can transfer from an object at a higher temperature to an object at a lower temperature, but the process is *irreversible*. We can't make it go in the other direction, at least not without expending another kind of energy. Something is lost. It's a little like pure, clean water from the faucet. Once it hits the sink, or goes down the drain, we won't want to drink it. There is just as much water, but it is no longer the same.

Let's say an amount of thermal energy, Q, is taken from an object at the high (absolute) temperature, T_1, and transferred to an object at the low temperature, T_2. The hot object loses an amount of energy, Q, which is the same as the amount gained by the cold object. We need some other quantity to describe what has changed. That's where we define the *entropy*, S, to say that an amount of entropy $\frac{Q}{T_1}$ is taken from the object at high temperature and an amount of entropy $\frac{Q}{T_2}$ is given to the object at low temperature. The two are not the same. If T_1 is the higher temperature, $\frac{Q}{T_1}$ is smaller than $\frac{Q}{T_2}$, and the net entropy has increased by the amount $\Delta S = \frac{Q}{T_2} - \frac{Q}{T_1}$.

The entropy turns out to be an excellent measure of the irreversibility of the process and of the amount of energy that, although it is still there, has become unavailable. In a system that does not gain or lose energy (an *isolated* system), the entropy can only increase. At best, it can come close to the limiting "ideal" case where it remains the same.

We can take an amount of heat Q_1 and a corresponding amount of entropy $\frac{Q_1}{T_1}$ from the hot part of the system and transfer a smaller amount of energy Q_2 and the corresponding entropy $\frac{Q_2}{T_2}$ to the cooler part of the system. The difference in energy $Q_1 - Q_2$ can be *used*, i.e., transformed to some other kind of energy, such as mechanical energy. This is what happens in a heat engine.

The best that we can imagine, the "ideal" limiting case, is that the entropy does not change, so that $\frac{Q_1}{T_1}$ is the same as $\frac{Q_2}{T_2}$. In any real situation it increases. We have seen that the entropy increases whenever there are two parts of a system at different temperatures, T_1 and T_2. Some thermal energy (Q) is then transferred from the

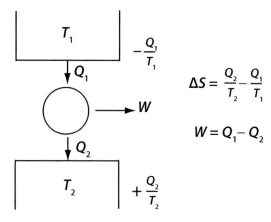

hotter part to the cooler one, and the entropy increases by $\frac{Q}{T_2} - \frac{Q}{T_1}$.

The process is *irreversible*, and the increase of entropy is a measure of the irreversibility. The amount of thermal energy that is transferred to the object at the lower temperature, T_2, remains thermal energy. It is "waste heat" that is unavailable, unless an object at a still lower temperature can be found, so that another heat engine can be used for a further energy transformation.

EXAMPLE 3

A pot of boiling soup at 100°C is put on the table, where it eventually cools to the room temperature of 22°C. 5×10^4 J leave it while it is still at 100°C.

(a) Describe the changes in energy.

(b) Describe the changes in entropy.

Ans.:

(a) The internal energy of the pot and the soup decreases. This is the energy of the motion of the atoms and molecules of which the pot and the soup are composed. It consists of both kinetic energy and the mutual potential energy of the atoms and molecules. The energy is transferred to the surroundings, i.e., to the table and to the air, primarily by collisions between the atoms and molecules. (There is also some electromagnetic radiation.)

(b) The entropy leaving the soup is $\frac{Q}{T_1}$, where T_1 is the absolute temperature of the soup, $100 + 273 = 373$ K, so that it is $\frac{5 \times 10^4}{373} = 134$ J/K. The entropy gained by the surroundings at $22 + 273 = 295$ K is $\frac{5 \times 10^4}{295} = 169.5$ J/K. The net increase of entropy is 35.3 J/K.

EXAMPLE 4

A steam engine uses steam at 100°C and cools it to 25°C as it does work at the rate of 10 kW.

(a) What is the work done in one minute?

(b) What is the maximum (ideal) efficiency?

(c) How much energy (Q_1) is taken from the steam in one minute?

(d) How much energy is given to the environment as waste heat in one minute?

(e) What can be done to increase the ideal efficiency?

Ans.:

(a) $10\,\text{kW} = 10^4\,\text{J/s} = 60 \times 10^4\,\text{J/min}$.

(b) In the ideal heat engine with the maximum efficiency there is no change in entropy, and $\frac{Q_1}{T_1} = \frac{Q_2}{T_2}$. Hence $\frac{Q_2}{Q_1} = \frac{T_2}{T_1}$, and the efficiency is $\frac{Q_1 - Q_2}{Q_1} = \frac{T_1 - T_2}{T_1} = 1 - \frac{T_2}{T_1}$. Here this is equal to $1 - \frac{298}{373} = 1 - 0.80 = 0.20$.

(c) Since the efficiency is $\eta = \frac{W}{Q_1}$, $Q_1 = \frac{W}{\eta} = \frac{60 \times 10^4}{0.20} = 3 \times 10^6$ J/min.

(d) $Q_2 = Q_1 \frac{T_2}{T_1}$, so that $Q_2 = 3 \times 10^6 \frac{298}{373} = 2.4 \times 10^6$ J/min.

(e) Even in this unattainable best case only a fifth of the energy taken from the hot part of the system is available to be used. Four times as much is rejected to the environment as waste heat or *heat pollution*. We can do better only by making T_1 larger or by making T_2 smaller. T_2 is the lowest temperature in the environment, i.e., that of the atmosphere, or of the available cooling water. The ideal efficiency can then be increased only by raising the high temperature, T_1. The temperature of the steam can be raised above 100°C. (It is then called "superheated steam.")

14.6 Our addiction to fossil fuels

Availability

Coal, oil, and natural gas have accumulated over millions of years. Over smaller time intervals they are a nonrenewable resource. Eventually they will be used up, but it is not clear when that will happen.

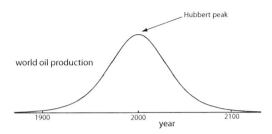

In 1956 the geologist M. King Hubbert described the exploitation of a limited natural resource through a cycle of expansion followed by decline and eventual exhaustion. He developed a formula for such a cycle and applied it to oil and coal. With the data available at that time he suggested that the U.S. production of oil would reach its maximum near 1970. This is in fact what happened. (See *Energy and Power*, W. H. Freeman, and Co., 1971, first published as the September 1971 issue of *Scientific American*.)

It is more difficult to make a similar analysis for world oil production. New sources of oil continue to be discovered, and some that were previously too expensive to exploit become economically competitive.

Hubbert's analysis leads to the conclusion that after the maximum (the "Hubbert peak") is reached, on the downward part of the curve, the supply can no longer meet the demand. There are then shortages, rising prices, and disruption of activities and processes that depend on oil.

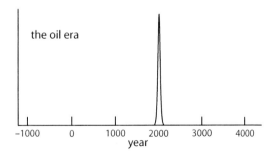

Even before the middle of the twentieth century it was widely predicted that there would be sufficient oil for only about 30 years. This prediction has derisively been called "time invariant," in that now, almost three quarters of a century later, some people think that the Hubbert peak is still 30 years away. Others suggest that the time of the peak is much closer. The figure shows a

Hubbert curve with its peak at the year 2000. Looked at from the perspective of history, over a longer time, the "petroleum era" is just a blip. The known reserves of coal will last longer, probably at least 100 to 200 years. Unfortunately, this most abundant of the fossil fuels is also the dirtiest, leading (if unchecked) to greater amounts of air pollution than oil and natural gas.

Some of the pollution comes from the part of the fuel that is not carbon. When fossil fuels burn, sulfur and nitrogen compounds are emitted, as well as small amounts of other substances, including mercury and radioactive materials. There are also solid particles, primarily of carbon, which are so small that they remain suspended in the air. Today all cars in the United States have pollution-control devices that limit the amounts of some of the combustion products that are released into the atmosphere.

It is possible to derive liquid fuel from coal. There has been exploratory research on ways to use coal cleanly, but costs have prevented the further development and use of the methods that have been found.

The greenhouse effect

Concerns about the continued availability of fossil fuels have been expressed almost since the beginning of their use. A more recent additional problem is the "greenhouse effect."

The atmosphere is transparent to electromagnetic radiation in the visible region, but absorbs radiation in the infrared portion of the spectrum. The absorption is not by the main air components, nitrogen and oxygen, but by larger molecules that have more closely spaced energy levels as a result of their rotation. The most important are CO_2, H_2O, and CH_4.

The radiation that reaches the earth from the sun in and near the visible region of the spectrum passes through the atmosphere more or less unhindered and heats the earth. The earth, in turn, radiates electromagnetic energy. The energy and frequency of the earth's radiation are determined by the earth's temperature, and are much lower than those of the radiation that comes to us. (The emitted spectrum has photon energies with a peak near kT, where k is Boltzmann's constant.) Consequently some of this radiation is absorbed by the "greenhouse gases" and transformed into internal energy of the atmosphere, which then heats the earth.

(In a real greenhouse a major fraction of the warming comes from the fact that the enclosed air is stationary, and therefore is not cooled by convection, i.e., by the movement of air.)

The most important greenhouse gas is CO_2, because as a result of the burning of fossil fuels its concentration in the atmosphere is increasing to the point where it can lead to major climate changes, principally the increase in the average temperature of the earth, commonly called "global warming." There may also be secondary effects, such as increasing severe storms, and more extreme temperature swings, but their connection with the increase of CO_2 is harder to establish.

Global warming through the greenhouse effect is expected to have such drastic consequences that it is now thought to impose the most severe limit on the use of fossil fuels. In other words, even though the resources, particularly of coal, are sufficient for our energy use in the near future, the threat of global warming is leading to reductions and modifications in the use of fossil fuels.

The rate of energy increase

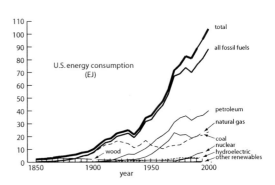

The consumption of energy in the United States has gone from 2.5 EJ in 1850, when it came mainly from burning wood, to 104 EJ in 2000. The figure (with data from the United States Energy Information Administration) shows the rise in the use of coal to about 1920, followed by the more rapid rise of oil and gas during the rest of the twentieth century, and the resulting overwhelming dependence on fossil fuels.

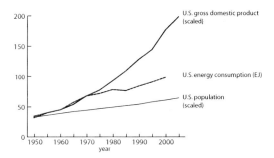

The figure correlates the rise in the use of energy with that of the gross domestic product (GDP) and that of the population. The line for the GDP as a function of time is multiplied by a scaling factor so that it coincides with the energy line in 1970. That allows us to see that in the decades before then energy use and GDP rose at about the same rate. It seemed evident that to tame the rise in the consumption of energy it would be necessary to reduce the rate of increase of the GDP. As the graph shows, this prediction proved to be incorrect. What happened?

Perhaps surprisingly, it seems that we became more efficient. Although the country's output continued to rise, the amount of energy that it took to do that rose at a considerably slower rate. The graph of population against time, scaled to coincide with the others in 1950, shows that the rate of increase in energy use became closer to the rate of population change than to the rate at which the GDP changed.

14.7 Other sources of energy

Nuclear energy

A nuclear reactor provides heat for the high-temperature part of a heat engine, just as does a boiler fueled by burning coal or oil. Let's look at the process in more detail. The nuclei of uranium barely hold together. Heavier elements don't occur in nature because they are too unstable and fall apart. The net force holding the protons and neutrons together in the nucleus is the result of a delicate balance between the attractive nuclear force that acts between the nucleons (protons and neutrons) in the nucleus and the repulsive electric force that tends to drive the protons apart.

A neutron that moves toward a uranium nucleus and collides with it causes the forces to become unbalanced. Normally the nucleus is roughly spherical, but when a neutron sticks to it the uranium nucleus becomes even more unstable. It begins to vibrate so that its surface area increases. At the surface the particles are held less strongly because there are neighboring particles on only one side. The result is that as the surface area increases during the vibration, the attractive force decreases. On the other hand the repulsive force does not change much because it doesn't act just between neighboring particles. Its range is much greater (as given by Coulomb's law) and is influenced by all the protons in the nucleus.

As a consequence, the nucleus doesn't just fall apart. It flies apart, exploding into two roughly equal pieces, the *fission fragments* or *fission products*.

After the fission fragments fly apart they collide with the other atoms. They slow down as they share their kinetic energy, leading to increased thermal energy and higher temperature of the material. Some of this energy is carried to the hot part of a heat engine by the circulation of water or some other fluid.

In the reactors used for power the neutrons are slowed down by a *moderator*. The moderator decreases the speed of the neutrons, but it increases the rate at which fission occurs, because slow neutrons are more efficient in causing fission. Of course there still has to be enough energy for the reaction to take place even if the neutron does not bring any kinetic energy. There are two substances for which this is the case. Both ^{235}U and ^{239}Pu can undergo fission when they are bombarded by slow neutrons. Plutonium does not occur in nature (although trace amounts have been identified) and ^{235}U is only 0.72% of natural uranium.

The nuclear reactors that now produce energy for the generation of electricity do so with slow neutrons on ^{235}U. The number of reactors worldwide was 439 in 2007, of which 104 were in the United States. Each produces of the order of 1 GW of energy. The country that relies on them most heavily is France, where in 2004 they produced 78% of the country's electric energy. In Germany it was 32.1%, in Japan 29.3%, and in the United States 19.9%.

It is also possible to build a reactor without a moderator that slows the neutrons. This increases the number of neutrons that are absorbed by the abundant isotope ^{238}U with the production of ^{239}Pu. They are therefore called "breeder reactors." To use a resource that is 140 times as abundant as ^{235}U is tempting, especially since with the use of only slow-neutron reactors the known uranium reserves are expected to be exhausted in less than a century. Breeder reactors have been built in France, Japan, and the United States, but have been largely abandoned. They are inherently less stable and safe than slow-neutron reactors.

All reactors produce plutonium to some extent. It can be separated from uranium chemically, which is a lot easier than to separate the uranium isotopes or to "enrich" the uranium to enhance the concentration of ^{235}U. It is, however, generally not done because of the extreme hazards associated with plutonium. The lethal dose of ingested plutonium is in the microgram range. Furthermore, an amount beyond the "critical mass" of a few kilograms undergoes a spontaneous chain reaction. In other words, it explodes as an "atomic bomb."

The possibility of abundant energy for thousands of years has given rise to discussions of a "plutonium economy" based on the widespread use of breeder reactors. It is evident that the dangers that would accompany large-scale use of plutonium would be great. In addition to the hazards associated with the chemical and radioactive properties, it would be necessary to store or dispose of large quantities of radioactive waste material. There would also be the problem of security, primarily to prevent the stealing of amounts sufficient for bombs. To some extent we live with these hazards now with the present use of nuclear reactors. It is doubtful that we could tolerate their magnification by several orders of magnitude.

Fusion

Nuclear fusion reactions are responsible for the radiation of energy from the sun and the stars, but it has been unexpectedly difficult to use the same kinds of processes in a controlled way on earth. The reactions have been studied by using accelerators that give the reacting positively charged nuclei the kinetic energy to come close in spite of the repulsive elecrostatic force between them. The experiments show how much energy is transformed from the internal energy of the nuclei to kinetic energy of the reaction products.

For any sustained energy release, however, it is necessary to bring a much larger number of nuclei together and to keep them close long enough for the reactions to occur. They still have to have sufficient kinetic energy to overcome the Coulomb barrier, and this is done by heating the gas containing the reacting nuclei to a high temperature.

Two methods have been explored extensively. One is to bombard pellets of material with high-power lasers This is called *inertial fusion*. The more widely used method is to produce a *plasma*, i.e., a gas of ions, and to contain it by using magnetic fields while it is being heated by high currents or other means. This is called *magnetic confinement fusion*.

The great attraction of using fusion reactions is that the raw material is essentially unlimited. The difficulty of confining the reacting material for sufficient times at a high enough temperature has, however, turned out to be great. Although the amount of material that needs to react is minute, the installation that is required is very large.

Solar energy

What about solar energy? No fuel, no pollution. Why are we not using more than the present small amount?

The energy that reaches the earth is huge, but it doesn't always get down to where we are, and the 1.3 kW/m^2 turns out to be rather dilute. In other words, it takes quite a bit of area to use solar energy on a large scale.

Direct heating is the simplest way to use the sun's radiation. In some parts of the world with a lot of sunshine, tanks on the roof are widely used to heat water. Solar radiation can also contribute to home heating, but its intermittent nature is an obvious drawback.

The simplest solar collection system consists of tubes with circulating water. The water is heated and carried to where it is needed. The efficiency can be increased by using a surface that maximizes the absorption of the solar radiation and at the same time minimizes its reemission. The absorber can also be surrounded by a reflecting surface to concentrate the radiation. At the cost of even greater complexity it can *track* the sun, i.e., it can be turned so as to remain at the best angle to the sun's rays. Quite elaborate large-scale systems have been built. One example is a "power tower" on top of which a container receives the radiation focused on it by an array of mirrors.

Conversion to electric energy is possible through a kind of photoelectric effect in semiconductors. We describe the process used in *solar cells* in the next chapter. Why they are not more widely used is a question of economics not physics. The reasons have to do partly with the expensive manufacturing processes and the lack of good energy-storage mechanisms as well as with the other difficulties that come from having an intermittent source. It would be necessary to devote large areas to the collection and conversion of the energy, and to devise systems for the maintenance of the vast arrays.

In 2009 the cost of a unit of energy from solar cells was at least three times that from fossil fuel sources. So far the relatively low cost, continued availability, and convenience of the fossil fuels have discouraged the investment required for the development of large-scale solar installations. In 2005 0.01% of the energy used in the United States was supplied by solar collectors and solar cells. The exhaustion of our oil resources, and, more slowly, of coal, has been long predicted, and is eventually inevitable. From time to time government subsidies have encouraged research and development, but most of the time the solar industry has been an orphan looking for its place.

Wind

Modern windmills are much more efficient than those used in the Netherlands and elsewhere for hundreds of years. They are also less picturesque. Opposition to their use shows that even the most benign sources of energy have drawbacks. "Wind farms" are installations of many windmills, often along mountain ridges or offshore near beaches. They tend to be opposed by the local population because they spoil the view. Their huge rotating blades are noisy and hazardous to birds.

Nevertheless wind energy is gaining ground. The cost of electric energy generated by wind is now close to that from fossil fuels.

Biomass

The word "biomass" refers to material of biological origin that can be burned as fuel. Wood has been used as a fuel since the time of the first man-made fire. It supplied 3.7% of the world's energy in 2004. Other fuels derived from biological material include household and agricultural waste such as straw and manure. Today there is increased attention to crops specifically planted as energy sources. In general they are not used directly, but are processed to produce ethanol and other petroleum substitutes.

Ethanol (C_2H_5OH) is the same alcohol as in alcoholic drinks. It is made by the fermentation of sugar or starch-containing plant material and subsequent distillation. The amount of energy released when 1 liter of ethanol is burned is 23.6 MJ, which is 65% of the amount (36.1 MJ) from 1 liter of gasoline. Up to 10% can be added to gasoline and used in conventional engines. "E85" is a mixture of 85% ethanol and 15% regular gasoline that is available in some parts of the United States for use in cars with specially modified engines.

In the United States ethanol for fuel use is derived primarily from corn kernels. The amount of energy that is gained this way is controversial. The sun's energy is used to grow the corn, but a good deal of additional energy is required for tractors, fertilizers, and the distillation process. Estimates differ depending on the kinds of energy that are considered and how they are counted. The consensus is that the energy available when the ethanol is burned is between 1.2 and 1.7 times as great as the energy that is used.

The burning of ethanol made from corn leads to only a small reduction (13% by one estimate) in the emission of greenhouse gases. Its most significant advantage is that it replaces gasoline made from domestic or imported oil. In 2004 the ethanol used in the United States represented the energy equivalent to 1.3% of the gasoline and 11% of the nation's corn harvest.

The balance of energy is quite different when ethanol is made from other plant materials. In Brazil ethanol from sugar cane provided 18% of the automobile fuel in 2004. The gain in energy is much greater than with corn, with the ethanol yielding about eight times the amount of fossil fuel energy input. One-half of all cars and 80% of new cars in Brazil are made to accept mixtures of ethanol and gasoline in any proportion.

The basic building block of plant material is *cellulose*, and when ethanol is made from whole plants, including stalks and leaves, it is called *cellulosic ethanol*. Estimates of the ratio of energy gain range from 4.4 to 6.6, representing a significant improvement over the values for ethanol derived from corn. The reason that cellulosic ethanol has not been made on a large scale is that different microorganisms are required for the fermentation, and processes that can be used for industrial-scale manufacture are still under development.

Energy from rivers, oceans, and the earth

The kinetic energy of the flowing water in rivers has long been exploited with water wheels, and more recently with more efficient turbines that drive electric generators. Even this seemingly very benign source of energy has detrimental side effects, primarily from the dams that are built to control the flow. Sometimes large areas are flooded, destroying animal and human habitat. Ancient monuments were submerged by the *Aswan Dam* project in Egypt, and whole villages disappeared as a result of the *Three Gorges* project in China.

The vast amounts of energy in the oceans have led to speculations on how they might be used. Demonstration projects are planned and some are underway to use the kinetic energy of tides and waves. The thermal energy is huge, but it is unavailable as long as the water is all at the same temperature. If there are temperature differences, however, energy can be extracted with heat engines between the hotter and the colder parts. For the relatively small temperature differences in the ocean the efficiency is very small, and attempts to use this energy have remained on a small scale.

There is also the thermal energy of the earth, its *geothermal* energy. The temperature gets higher as we go further down into the interior of the earth. Eventually it gets so large that the core of the earth is molten. (This is the *outer* core. The inner core is under such pressure that it is solid.) The origin of the energy is believed to be the radioactivity of the materials that are part of the composition of the earth.

We become aware of the thermal energy through the rising temperature as we drill down, and more spectacularly through volcanoes and geysers. In some regions the energy is being used in geothermal power plants.

The sustainability transition

The earth, as our habitat, faces a number of obstacles, and is often said to be *in crisis*, as at the turning point of a disease, from which the path is either toward recovery or toward death. Are the natural resources on which we have come to depend being depleted to the point where they are no longer adequate for our needs? Are we polluting the air, the water, and the ground with waste material to the extent that they can no longer sustain us? Is our use of fossil fuels producing greenhouse gases in amounts that will produce disastrous climate changes? Each one of us contributes to these effects, and their damage grows as the population of the earth increases. Was Malthus right when he said that the number of people was reaching the limit beyond which the earth cannot support it?

We are engaged in a grand experiment of which we are the subject. In time we will find out whether we are demanding more of the earth than it is able to give. But we are also in a position to influence the outcome. We can use our knowledge and ingenuity to bring about new ways to use the materials that are available to us. Our record of scientific discoveries and accomplishments is great and we may be able to push back the apparent limits to our resources and to the earth's population.

As time goes on we are, however, getting closer and closer to the limits set by nature. It is becoming more and more urgent that we limit what we use and what we waste. To avoid the catastrophe predicted by Malthus we must eventually approach a state of equilibrium, where we use only what can be renewed or replaced, and keep the population to a size that can be fed and housed and that can live with what we are given.

Will we be intelligent and resourceful enough to find new ways to use the materials at our disposal and so extend the limits to our growth? Will we be sufficiently intelligent to understand and to act on the fact that the earth's size and its resources are fixed, and that we must restore what we use?

In sum, will we humans be able to make the transition to an existence that can be sustained over time? At present we are using resources that are not replaced. We are disposing of waste in the atmosphere, in the rivers and oceans, and in the soil in ways that cannot continue indefinitely. Our sources of energy will need to change as fossil fuels are depleted. Pollution with toxic substances and greenhouse gases is increasing. The population of the earth is twice what it was 25 years ago.

The grand experiment in which we are engaged will show whether we can make the transition to sustainability, and so continue to exist. Malthus thought the time of crisis was at hand 200 years ago. Are we there now?

The search for new sources of energy and new methods of pollution control and waste disposal shows that the success of any new process depends on four different kinds or stages of support. The fundamental idea and scientific research come first, with observations and with experiments and tests in the laboratory. But it is not enough to have a good idea. It has to be followed by technological development, i.e., by finding ways to use it on a large scale. Solar energy, for example, continues to be the subject of search both for new scientific ideas and for improvements in technology. The reason why it and other alternatives are not more widely in use can be found as we look at the two other necessary pillars of support. One of these is economic competitiveness. As long as energy from solar cells is more than three times as expensive as that from other available sources, there is little chance that it will become more widespread. Finally, there must be the political will and popular support to bring about a change. This is true whether we talk about obstacles to the use of nuclear reactors or about the environmental damage that has led to opposition to the further large-scale development of other new and old sources of energy.

14.8 Summary

In spite of their drawbacks the *fossil fuels*, coal, natural gas, and oil, are the predominant sources of the energy we use.

Electric energy is a *carrier* of energy. It allows us to use energy more conveniently and in a place different from where it is generated. Another possible carrier is hydrogen.

Most electric energy is transported as *alternating current (AC)*. One advantage of AC is that it is more easily generated than DC (direct current). A second is that transformers can be used to change the voltage and current so as to reduce the loss of energy during transmission.

Batteries are used to store energy. Their use is largely restricted to applications where portability is a main consideration.

The internal energy of the fossil fuels is liberated by burning, and transformed to mechanical energy in a *heat engine*. In addition to being limited by the law of conservation of energy, the transformation is limited by the *second law of thermodynamics*, which requires that the *entropy* not decrease. It normally increases, and remains unchanged only in the limiting ideal case.

When energy ΔQ is transferred to an object at an absolute temperature T, its entropy, S, increases by $\Delta S = \frac{\Delta Q}{T}$.

The efficiency of a heat engine is $\frac{W}{Q_1} = \frac{Q_1 - Q_2}{Q_1}$. Its maximum (ideal) value, without an entropy change, is $\frac{T_1 - T_2}{T_1}$.

Another statement of the second law of thermodynamics is that energy transfers spontaneously from a hotter object to a cooler one, but a transfer in the opposite direction can happen only if work is done on the system. Similarly an object can be cooled below the temperature of its surroundings only by doing work.

All bodies radiate. The spectrum of the radiation depends on the body's temperature. The sun's high temperature causes its radiation to be strong in the visible range, and we see it as white. Except for the effect of clouds, it passes easily through the earth's atmosphere. The earth radiates also, but because of its lower temperature the radiation is not visible. It is at a lower energy and frequency and longer wavelength, in the infrared region. The atmosphere absorbs some of this radiation as a result of the presence of molecules other than nitrogen and oxygen. These more complicated molecules can absorb infrared radiation, ususally because they can rotate. The energy that gets absorbed and does not escape is trapped and leads to warming of the earth. This is called the *greenhouse effect*.

Energy is increasingly used from sources other than fossil fuels. These sources include nuclear reactors and energy from the sun. Solar energy can be used directly to produce thermal energy, and through solar cells to produce electric energy. Other sources are wind, rivers and oceans, and geothermal energy.

The fossil fuels are resources that cannot be replaced, so that their use cannot continue indefinitely. Ultimately we must make the transition to sustainable use, that is, to sources that can continue to be available, such as wind and solar energy.

14.9 Review activities and problems

Guided review

1. The battery pack of a hybrid car consists of 38 modules, each of six nickel-metal hydride (NiMH) cells, all connected in series. Each module is rated at 6.5 Ah and has a mass of 1.04 kg.

(a) What is the energy in kwh that can be stored in this battery pack? (1 cell: 1.2 V.)

(b) What is the energy density in watt-hours per kilogram?

(c) How many gallons of gasoline would deliver the same energy to the car?

2. How large a capacitor, charged to 150 V, is necessary to store an amount of energy equivalent to that delivered to a car by a gallon of gasoline?

3. Energy is transferred from a hot object to one that is cooler. Show that the net entropy increases.

4. Why can the ideal efficiency of a heat engine never be attained?

Problems and reasoning skill building

1. An alkaline AA 1.5-V battery is rated at 3000 mAh. What would be the cost of this amount of energy from an electric outlet at ten cents per kwh?

Synthesis problems and projects

1. A hybrid car cannot run on electric energy alone for more than a small distance. Design an NiMH battery pack for a "plug-in" hybrid car whose batteries are to be recharged overnight from a household electric outlet. Use the given and calculated information of the first Guided review question.

Consider how many miles you expect the car to travel on one charge, the mass of the battery pack, the cost of the battery pack compared to that of the one in Example 1, and the cost of the overnight charge.

Based on your results, discuss some of the obstacles facing the design of fully electric cars.

CHAPTER 15

Atomic Physics Pays Off: Solar Cells, Transistors, and the Silicon Age

The real solid: metals, insulators, and semiconductors
> *Electrons in a metal : not so free*
> *Bands and gaps*

Tiny changes and vast consequences: impurities in semiconductors
> *p-type, n-type*

The transistor and the information revolution
> *The transistor*
> *The p–n junction*
> *The solar cell*
> *The future*

Solid-state devices, of which the transistor is the most important, have had an enormous impact on our civilization. Their development was made possible by our knowledge of atomic structure and of the motion of electrons in solid materials. The understanding of the process of electric conduction is one of the triumphs of twentieth-century physics. It depends crucially on the quantum mechanics first developed by Schrödinger and Heisenberg in 1925. The invention and proliferation of the various types of transistors are unthinkable without this theoretical development. But it could also not have happened without the experimental advances in materials science that allowed the production of crystals of silicon and other elements with extreme purity or with impurities that are precisely controlled in quantity and spatial distribution. As so often in history, theory and experiment complemented each other as they progressed hand in hand.

15.1 The real solid: metals, insulators, and semiconductors

Electrons in a metal: not so free

The understanding of electric conduction and of metallic behavior became possible only after the development of quantum mechanics. A part of that triumph with the most profound influence was the understanding of why some materials are metals and some are not, and the control over conduction properties that then became possible. The term *semiconductor* implies a conductivity between that of a metal and that of an insulator, but that doesn't begin to deal with the essential differences between these three classes of materials. To get further we have to go beyond the free-electron model of metals.

Of course the electrons in a metal are not free. First of all they are held to the metal in accord with Coulomb's law so as to preserve the equality of positive and negative charges. You might think that they have to twist and turn past the atoms, which are now ions, and that this is the greatest impediment to their motion. This is not the case. In fact, a rigid symmetrical arrangement of ions, a *perfect lattice*, does not impede the electron motion at all, a fact that can be understood and makes sense only as a consequence of the wave nature of the electrons.

The reason is the same as that which allows light to go through a window. The question needs to be turned around: not "why does the light wave go through?" but "what prevents it from going through?" If it does not go through, if it is *absorbed*, there must be some mechanism by which it loses energy. The energy of the photons, E_{ph}, must be just equal to an energy that can be accepted by the material. It must be equal to the energy difference, ΔE, between an energy level that is "occupied" and a higher one that is not. The atom, or the whole solid, can then go from the lower level to the higher one. The photon disappears. It is absorbed. If there are no energy levels with the required difference in energy, the photons will pass through without being absorbed, and the material will be *transparent* to the light. Glass, for example, has no energy levels that allow the absorption of visible light, but it absorbs light in the ultraviolet part of the spectrum. The story is the same whether we talk about light going through a window or electrons going through a metal.

The figure shows two levels separated by the energy difference ΔE. If the lower level is occupied, the photon with just the right energy, $E_{ph} = \Delta E$, will be absorbed. But the system with these energy levels is transparent to other photons.

Bands and gaps

How is it that electrons in a piece of copper can move about freely, and so make this metal a *conductor*? To answer this question we have to see what happens when copper atoms form solid copper. Each of the electrons in an atom comes close to an electron in a similar state in the neighboring atoms. But the Pauli exclusion principle says that they can then no longer be in exactly the same state. The energies change, and what was a single level in the isolated atom becomes a *band* of energies in the metal.

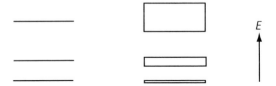

For electrons in one atom to interact with electrons in another, they have to get so close that their wave functions overlap. There is just one electron in the outermost shell of a copper atom. As atoms come together, the outermost electrons are the ones whose wave functions overlap the most. The band that they form is broader than those formed by the inner electrons. There is another difference: all of the levels in the bands formed by the inner electrons are *filled*, i.e., occupied by electrons. That is not so for the outermost band. In it is one electron from each atom. But there is room for two electrons with the same energy in each atom, one in each spin state, i.e., one with each value of m_s, namely $+\frac{1}{2}$ and $-\frac{1}{2}$. As a result the outermost band in the copper metal is only half-filled.

That means that the electrons in that band can easily be given more energy. There are other levels in the same band that they can occupy. That happens when they are accelerated by an electric field. That's why this band is called the *conduction band*. The next lower band, on the other hand, is *full*. It has no empty states. Its electrons are unable to accept small amounts of energy.

If, on the other hand, there were two electrons from each atom in the uppermost band, then all the available energy levels would be occupied by electrons, and the Pauli exclusion principle would make it impossible for any one electron to go to another energy level in that band. Each electron would be trapped, unable to accept more energy and so be able to move away, and the material would not be able to conduct. It would no longer be a metal, but would, instead, be an *insulator*.

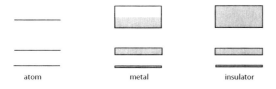

atom   metal   insulator

In the figure the shaded parts represent closely spaced energy levels that are occupied by electrons. In the conduction band of the metal there are empty levels, so that the electrons can accept small amounts of energy and occupy them. In the insulator, on the other hand, the levels in the corresponding band are completely occupied, so that none of them can accept small amounts of energy. Only an energy sufficient to move the electron to the next higher band can be absorbed.

It's a good story, but we have a problem: if two electrons from each atom cause the band to be filled so that the material is an insulator, why are the elements of the second column of the periodic table, with two electrons in their outermost shell, magnesium, calcium, barium, etc., not insulators? They are metals, although not really *good* metals. Their metallic properties of electric and thermal conduction and mechanical strength are relatively weak.

It turns out that what we said is still correct, but in these elements the uppermost energy band, the one that we would expect to be filled, overlaps with the next empty one, and electrons spill

over into it. There are now adjacent empty levels and the electrons are no longer locked into one band. They are free to accept energy, and so to move. We have to wait for the fourth column of the periodic table to get to nonmetals. Here new adventures are waiting.

Let's look at the insulator again. The band with the highest-energy electrons is filled. The next one is empty. We are here talking about the *ground state*, the state where electrons occupy the lowest energy levels that they are allowed to be in, one at a time, as limited by the Pauli exclusion principle.

An electron from a filled band can be promoted to a band with empty states if we can give it the required energy. It can be by shining light on it (with just the right energy) or by any other means that accomplishes the same purpose. In any case the minimum energy is the amount that brings the electron from the top of the filled band to the bottom of the next empty band, the energy of the *gap*, E_g, between the two bands.

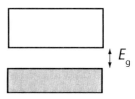

For some materials the gap is so small that even the internal (*thermal*) energy at room temperature, the energy of the shaking of the atoms of the crystal lattice, is sufficient to get some of the electrons across the gap. In this case, even when the material is an insulator as the temperature approaches absolute zero, there will be some electrons in the *conduction band*, the otherwise empty band across the gap to which the thermal energy can promote them. This is what happens in germanium and silicon, and this is why these elements are *semiconductors*.

Think of a two-floor garage, with the whole floor divided into spaces, each just big enough for one car. If all spaces on the first floor are filled,

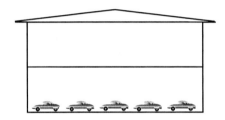

15.2 Tiny changes and vast consequences: impurities in semiconductors

If that were the whole story of semiconductors it would already be great. But there is an even easier way to affect the conduction properties, and this is the one that has made the creation of today's electronic devices possible.

Here is a representation that shows the individual atoms and their outer electrons.

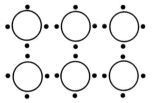

none of the cars there can move. Now let one car from the first floor be taken to the empty second floor. There it can move around at will. But in addition something else has happened: the first floor is no longer full. If forces are applied to all cars, to the left, a car to the right of the now empty spot can move to its left. A new empty space is created, and a car to its right can move one step over. The two-floor garage model has been adapted from *The Junction Transistor* by Morgan Sparks in the July 1952 issue of *Scientific American*, where it is attributed to William Shockley.

We can also take a new and marvelously fruitful point of view: focus on the empty space, the *hole* in the sea of cars. As cars, one by one, move to the left, the empty space moves to the right. Go back to the electrons. Put on an electric field to the right. Each electron experiences a force to the left, but in a filled band none can move. Take just one electron to the next, the conduction band, and it can contribute to the current, pushed by the field. But the hole left behind in the lower band also responds to the field. It moves in the opposite direction to that of the electron, in the direction of the force on a positive charge.

It is only electrons that move. There are no moving positive charges. But the hole in the electron sea moves in the direction of the electric field *as if* it had a positive charge, and is sometimes said to *have* a positive charge. When an electron is "promoted" to the conduction band there is a double reward: the electron is now able to move, and so is the hole that it leaves behind.

Put in some impurities. Silicon is in the fourth column of the periodic table, element number 14. At absolute zero all its electrons are in filled bands. Each atom contributes four electrons to the highest band. These are the electrons that are responsible for the force between atoms. Each of them forms a bond with one of the electrons in a neighboring atom. The two electrons of the bond are shared by the two atoms in what is called a *covalent* bond.

Now exchange one of the silicon atoms for an atom from the fifth column, such as phosphorus ($Z = 15$), with one more electron in its atom. Four of the electrons take the place of the silicon atom's four outer electrons in holding on to their neighbors with covalent bonds. But what happens to the fifth one? It isn't needed. It stays near "its" atom, but because it is superfluous, because it does not form part of a bond, it is very loosely attached, and can be removed with just a small amount of energy. Once liberated, it is a free electron, free to move around. It is now in the conduction band. The amount of energy needed to get it there is much smaller

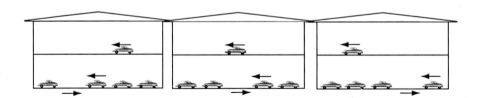

15.2 Tiny changes and vast consequences: impurities in semiconductors

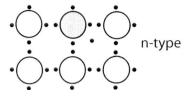
n-type

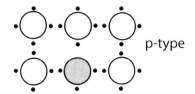
p-type

than that which would be necessary to pull up an electron across the gap. It is so small that the internal energy at room temperature is sufficient to get a significant number of the electrons there and so to contribute to the conduction process.

p-type, n-type

There is still another wrinkle to add to the versatility of semiconductors. Let the impurity atom be in the third column, like boron, with three outer electrons. The three electrons can take the place of only three of the silicon atom's four bonding electrons. The atom needs a fourth, and is ready to snare one wherever it can find one. It doesn't take much energy to steal one from a neighbor. But the neighbor is now unfulfilled, and it, in turn, pulls one away from some other neighbor. The missing place, *the hole*, travels backward, as before. The material is now called a *p-type semiconductor*, because of the holes that experience forces in the electric field as if they were positively charged. In contrast, the material with extra negatively charged electrons is called an *n-type semiconductor*.

We can use the two-floor garage model to show what happens in an n-type semiconductor: a car is moved to the second floor while the first floor remains filled, that is, an electron is promoted to the conduction band while the next lower band remains full. Similarly, in a p-type semiconductor an empty space appears on the first floor, without any other changes, that is, a hole is formed in the highest band, making conduction possible there, without any electrons in the conduction band.

Note that we have used three different representations: the energy level diagram that shows the energy bands, the two-level garage, and the lattice of individual atoms and electrons.

EXAMPLE 1

Rank the following in order of their resistivity at room temperature: (a) copper, (b) pure silicon, (c) silicon with 0.01% phosphorus.

Ans.:

The resisitivity depends primarily on the number of free electrons (the electrons in the conduction band) in a given volume of material.

(a) In copper one electron from each atom is a free electron.

(b) At absolute zero pure silicon has no free electrons and is an insulator. At higher temperatures some of the electrons have enough thermal energy to transcend the energy gap of 1.1 eV so that they are in the conduction band. Since the gap energy is much larger than the thermal energy at room temperature ($kT = 0.025$ eV), only a very small number of electrons is in the conduction band.

(c) The atomic number of phosphorus is greater than that of silicon by one. Each phosphorus atom therefore has one more electron than an atom of silicon. The phosphorus atom takes the place of a silicon atom in the crystal lattice. It takes very little energy to remove the extra electron so that it becomes a free electron. Since only 0.01% of the atoms provides free electrons, the resistivity is much greater than that of copper, but much less than that of pure silicon, where

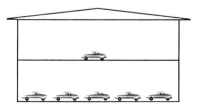

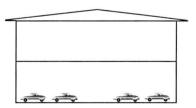

there are no impurities to provide free electrons, and the only free electrons are those that are in the conduction band because of their thermal energy.

The order of increasing resistivity is therefore (a), (c), (b).

EXAMPLE 2

How are the resistivity of a material and the resistance of a piece of wire related to the number of charges that are free to move?

Ans.:
Under the heading *Motion of a charged object in a magnetic field* on page 205 we looked at a wire in which particles with a total charge Q are moving, each with a velocity v parallel to the wire. The current in the wire is I and its length is L. We saw that $IL = Qv$.

If the charge Q consists of lots of charges, each with magnitude e, and their number is n per unit volume, then the charge per unit volume is ne. This is called the *charge density*, for which the SI unit is C/m^3. The total charge in the volume LA is $neLA$.

We see that $IL = neLAv$, which we can rewrite as $\frac{I}{A} = nev$. If we define the current density J to be $J = \frac{I}{A}$, then $J = nev$. (This is an important relation that relates the macroscopic quantitiy J, which describes the current, to the microscopic quantities n, e, and v, which describe the number per unit volume, the charge, and the average speed of the free electrons.)

Since $R = \frac{\Delta V}{I}$, and the resistivity, ρ, is given by $R = \rho \frac{L}{A}$, we see that both are inversely proportional to the charge density, ne.

15.3 The transistor and the information revolution

The history of the 1940s is that of war and suffering. The development of the nuclear reactions of fission and fusion for explosives, and later for the utilization of energy, is part of that history. So is the invention of radar. The vast human and technological consequences of these scientific developments tend to overshadow the fact that this was also the time of the *semiconductor revolution*, the beginning of what is sometimes called the *silicon age*.

It is the confluence of two tidal waves of history that led to this revolution. One is based on pure thought, the other on a very down-to-earth practical development, both at the central core of modern physics. One is quantum mechanics, the science that describes atoms and electrons. The other is materials science, the descendant of the metallurgy of the ancients, which characterized the bronze age and the iron age. It made possible the preparation of semiconducting elements in crystalline form with until then unimaginable regularity and chemical purity. The knowledge of one part in 10^8 that was achieved is analogous to knowing the number and position of everyone in the United States to ± 2 individuals.

This is history outside the mainstream of popular culture, too often ignored or misunderstood, but with the most profound consequences for that culture. It is the history that paved the way for how we communicate today. Computers and the internet, cell phones, and portable music players could not exist without it.

The transistor

In semiconductors the transfer of a small fraction of the electrons from one band to another can cause large changes in the conduction properties, and hence in an electric current. Such changes can be produced by the addition of minute amounts of impurities. They can also be brought about by electric fields, which move electrons in and out of narrow current-carrying channels.

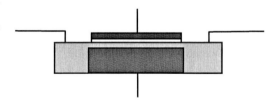

This is what happens in a *field effect transistor*. This was not the first or even the second type of transistor to be made, but it is the one that is simplest to describe, and it is the one in most widespread use today. The concept was patented in 1926 by Julius Lilienfeld, but at that time semiconductor technology was not even in its infancy, and no usable devices were constructed. The idea was reinvented much later, and the first field effect transistors were made at Bell Laboratories in about 1960. Even then it took another 15 years for them to be in common use.

In a field effect transistor the electric field is at right angles to the current in a conducting channel of either an n-type or a p-type semiconductor. The field causes electrons to be pulled in or out of that channel and can even change it from n-type to p-type or the other way around. As a result the current in the channel can change by a large amount.

Since a small change in the field can cause a large change in the current, the transistor can act as an amplifier. It can control the current and switch it on or off. A combination of two transistors in parallel can be used as a *bistable* combination, with the current in either one or the other. This is the fundamental unit of a calculator or computer, where the two possible states are usually referred to as "1" and "0." A great number of such units combine to give us the logical and computational power of today's devices.

The p-n junction

Take a piece of n-type semiconductor (with its extra electrons) and put it together with a piece of p-type material (with its extra holes). Not with glue, which would mess it up, but with the two types of impurity *grown-in*, each on one side of the border region, which is now called a p–n junction.

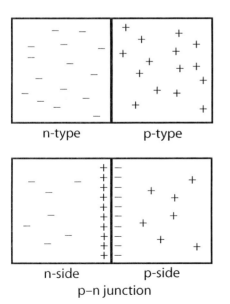

p–n junction

On the n-type side there are electrons moving about freely, aimlessly. The same is true for the holes on the p-type side. At the junction the electrons on their side are close to the holes on the other side. A hole is a place where an electron is missing, and a passing electron can drift across the border and "fall in."

When electrons do that they accumulate along the junction on the p-side where they have been captured into holes. There are then two layers of charge, negative on the p-side and positive on the n-side, where the electrons have left to cross the boundary. The two charge layers produce a built-in electric field between them. The field builds up until there is so much negative charge on the p-side that it repels further negative charges so strongly that the migration stops.

In this equilibrium configuration there is now a barrier that repels electrons as they try to drift to the p-side and holes as they try to enter the n-side. In other words the barrier is such that both kinds of charge carriers are kept "home," the electrons on the n-side and the holes on the p-side.

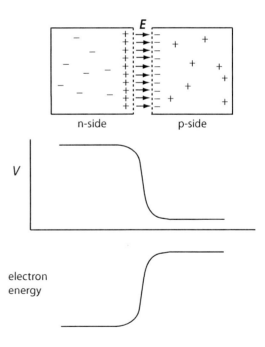

We can also look at the electric potential. The n-side, with its extra positive charge, is at a higher potential than the p-side. The field, as always, is in the direction from the higher potential to the lower potential.

A third representation is that of the energy of the electrons. Since the electrons are negatively charged, their potential energy increases as the electric potential decreases. While they "fall up" on the graph of the electric potential, they fall down on the energy diagram. The figure shows all three: the built-in field in the junction region, the variation of the electric potential, and the variation of the electron energy.

Suppose that we connect the two sides of the p–n junction to a battery, with the n-side connected to the positive terminal and the p-side to the negative termiinal. It is now even harder for the holes to get to the n-side and the electrons to get to the p-side. The height of the barrier has increased and any flow of charge is inhibited.

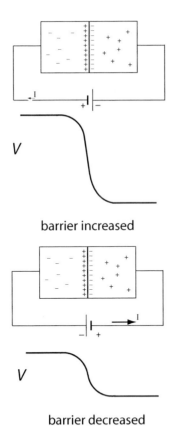

barrier increased

barrier decreased

The result is quite different if we connect the positive terminal of the battery to the p-side of the junction and the negative terminal to the n-side. The barrier is now reduced. In the equilibrium situation it just barely held the charges in their home territory. When the height of the barrier is reduced, charges will flow across it as over a waterfall.

We see that charges flow easily in one direction, but not in the other. The p–n junction acts as a rectifier.

The solar cell

Fabricate the junction so that it is near the surface. Shine a beam of light on it. By the photoelectric effect a photon can give its energy to an electron near the junction region, perhaps enough to liberate it and make it a free electron.

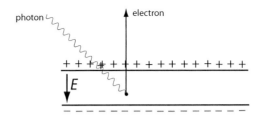

If it gains its freedom in the region of the built-in electric field, it will feel a force, and it will be pushed across the border, back to the n-side. If this happens to lots of electrons they will constitute a current, to which the energy of the photons has been transferred. If the photons come from the sun, we have transformed solar energy into electric energy.

The raw material, the silicon, costs next to nothing. The solar energy is free. The manufacturing process, however, and the auxiliary network of contacts and wires, is so expensive that up to this point solar cells are used only to a very small extent, mostly in special applications where cost is not the main factor. Predictions of a turn in the economic balance have not yet been realized, but it is widely thought that this will eventually happen.

The future

We are surrounded, at least in the developed world, by the fruits of electron science. Communications depend on it, and so does a large fraction of modern entertainment. Most recently, and still growing explosively, there is the information industry. Telephone, television, the recording of sound and image, and the computer have transformed our lives and moved the boundaries of the possible.

Some adventurous people call themselves futurists, and try to predict new technologies and the challenges that they will pose. As we look back we see that they have rarely been equal to the task. One of the pioneers of the computer age was asked what he thought computers might be good for. His answer was that they would allow accurate weather prediction. Certainly no one thought that they would displace typewriters, or give rise to message networks.

The development of science and technology has been fanciful and unpredictable. As we look to the future we must expect it to continue to be so. Our powers of prediction, on the other hand, depend on our present knowledge. There are technologies waiting to be exploited, such as solar cells and refrigerators based on solid-state science. Our understanding of the structure of matter and the behavior of electrons in solids has led to the development of new materials that promise new applications of magnetism and electricity in ways that we can only dimly foresee. Our knowledge is being applied to more and more complex systems, particularly in biology. We don't know where it will lead us, but the new developments will, no doubt, be informed by the past, even as they lead into the unknown.

Modern technology touches our lives in many ways. It is not always looked on as favoring the common good, and there are strong reactions against its pervasive presence. In some quarters science is looked on with suspicion and with hostility.

Does our educational system prepare us to live in an age in which science plays such an important role? The presence of science and technology in our lives has generally not brought about a corresponding increase in widespread knowledge about these areas. Yet citizens and their representatives are asked to make choices and judgments that channel private and public activities and resources and profoundly affect all of us.

It is hard to know whether, individually and collectively, we will bring wisdom to our choices. Surely, however, a prerequisite is that we know what we are talking about. It is toward the increase in that knowledge that this book is dedicated.

15.4 Summary

When atoms cluster together to form a solid, each electron energy level of the atom becomes a *band* of energy levels in the solid. Each energy level in the band may be occupied by an electron or it may be empty.

If all of the bands of a solid are either completely full or completely empty, the solid is an insulator. If one or more of the energy bands is only partially filled, the material is a metal. A special situation occurs when the bands are fully occupied up to a certain energy at the absolute zero of temperature, but the next band is so close in energy that thermal energies are sufficient to cause electrons to go to it. The material is then a semiconductor.

A semiconductor may have impurities with extra electrons that are easily removed so as to become *conduction electrons*, i.e., electrons that are free to move in a band that was previously empty. Such a material is called an *n-type semiconductor*.

Impurities may also have less electrons so that electrons from the semiconductor can jump to it, leaving a missing electron or *hole* in a band that was previously filled. The hole behaves like a positive electron, and the material is called a *p-type semiconductor*.

In a *field effect transistor* an electric field can be applied to change the number of electrons or holes in a current-carrying channel. A very small amount of energy can then change or control a current with a much larger energy.

A boundary where an n-type semiconductor changes to a p-type semiconductor is called a *p–n junction*. Electrons migrate to the p-side and an electric field is set up at the boundary from the n-side to the p-side. The p–n junction conducts easily in only one direction so that it acts as a rectifier. An electron liberated in the junction region by the photoelectric effect is propelled by the electric field. This is the principle of the solar cell.

Our ability to predict the future course of science and technology is limited. Based on past experience, we can expect new and surprising developments that have profound effects on the way we live.

15.5 Review activities and problems

Guided review

1. (a) Describe the difference between an insulator and a metal in terms of their energy bands.
   (b) Describe the difference between an insulator and a semiconductor in terms of their energy bands.
   (c) Describe n-type and p-type semiconductors.

2. (a) What is the number of free electrons per m^3 in copper? (You may need its atomic mass, 63.5, and its density, 8.9×10^3 kg/m^3.)
   (b) What is the average or drift velocity of the electrons in a wire with cross section 1 mm^2 carrying a current of 10 A?

Problems and reasoning skill building

1. Use the answer to *Guided review* question 2 to answer the following questions:
   (a) How many states are there in the conduction band of 1 g of copper?
   (b) What fraction of these states are occupied by electrons?

2. (a) Sulfur is an insulator. What does that tell you about its energy bands?
   (b) Calcium is a metal. What does that tell you about its energy bands?
   (c) Can you tell that calcium is a metal from its position in the periodic table of elements? What additional information do you need?

3. (a) In the ground state of silicon the electrons with the highest energy are 3p electrons, i.e., electrons with $n = 3$ and $\ell = 1$. Rank all of the electrons in the ground state of silicon in the order of their energy, from the lowest to the highest.
   (b) For which of the electrons does the orbit (the dominant part of the wave function) have the largest diameter? The smallest?

4. (a) What is the number, n, of free electrons per m^3 in silicon with 0.01% of phosphorus? (Assume that all of the extra electrons provided by the phosphorus are free. The atomic mass of silicon is 28.1 and its density is 2.3×10^3 kg/m^3.)
   (b) Assume that the drift velocity is the same as in the copper wire of *Guided review* question 2. What is then the resistivity of the silicon with the phosphorus impurity?

5. A free electron appears in the junction region of a p–n junction. Toward which side will it experience a force? Explain this with the help of a diagram.

Synthesis problems and projects

1. For silicon with phosphorus impurities we have shown two representations for the electron behavior: the lattice of atoms and electrons and the two-level garage. Now show the upper energy bands for silicon. In them show the level for the liberated conduction electron provided by the phosphorus impurity atoms.
   Then decide where the level is before the electron is free. (Do this from your knowledge that the electron is much more weakly bound than the silicon electrons.)

2. (a) A hole is created in pure silicon when an electron "jumps across the gap." Show this transition on an energy level diagram. Show the level of the electron and of the hole that it leaves behind.
   (b) It takes much less energy to create a hole in silicon with a small amount of boron impurity. Show the level of the hole that is created in this case and an arrow showing the transition as the hole is created.

CHAPTER 16

There Is No End

Bibliography

Most of the thoughts and ideas in this book are not original. We have learned from many who have come before us. Here is a list of sources that we have found particularly helpful.

Sanborn C. Brown and Benjamin Thompson. *Count Rumford* (MIT Press 1979).

Energy and Power (W. H. Freeman 1971, first published as the September 1971 issue of *Scientific American*).

Philipp Frank. *Foundations of Physics, International Encyclopedia of Unified Science*, vol.1, no. 7 (University of Chicago Press, 1946).

Gerald Holton and Stephen G. Brush. *Physics: The Human Adventure* (Rutgers University Press 2001).

Robert W. Kates. Labnotes from the Jeremiah Experiment: Hope for a Sustainable Transition. *Annals of the Association of American Geographers*, vol. 85 (December 1995), pp. 623–640.

Abraham Pais, *"Subtle is the Lord ..." The Science and the Life of Albert Einstein* (Oxford 1982).

———. *Inward Bound: Of Matter and Forces in the Physical World* (Oxford 1986).

———. *Niels Bohr's Times in Physics, Philosophy, and Polity* (Oxford 1991).

Susan Quinn. *Marie Curie: A Life* (Simon and Schuster 1995).

Robert W. Reid. *Marie Curie* (Saturday Review Press 1974).

Morgan Sparks. The Junction Transistor. *Scientific American* (July 1952), pp. 28–32.

The Teaching of Electricity and Magnetism at the College Level, Report of the Coulomb's Law Committee of the AAPT. *American Journal of Physics*, vol. 18 (1950), pp. 1–25 and 69–88.

Alan Van Heuvelen and Xueli Zou. Multiple Representations of Work-energy Processes. *American Journal of Physics*, vol. 69 (2001), pp. 184–194.

Victor Weisskopf. *Knowledge and Wonder* (Science Study Series, Doubleday and Co. 1963).

Some Constants, Astronomical Quantities, and Masses

Constant in Newton's law of gravitation	G	6.67×10^{-11} Nm²/kg²
Constant in Coulomb's law	$k \, (= \frac{1}{4\pi\epsilon_0})$	9.00×10^{9} Nm²/c²
Magnitude of charge on electron and proton	e	1.60×10^{-19} C
Speed of electromagnetic waves in a vacuum	c	3.00×10^{8} m/s
Planck's constant	h	6.63×10^{-34} Js
	$\hbar \, (= \frac{h}{2\pi})$	1.055×10^{-34} Js
Universal gas constant	R	8.31 J/mole K
Avogadro's number	N_A	6.02×10^{23} particles/mole
Boltzmann's constant	k	1.38×10^{-23} J/K
Earth: Mass		5.98×10^{24} kg
Mean radius		6.38×10^{6} m
Mean orbital radius		1.50×10^{11} m
Moon: Mass		7.35×10^{22} kg
Mean radius		1.74×10^{6} m
Mean orbital radius		3.85×10^{8} m
Sun: Mass		1.99×10^{30} kg
Mean radius		6.96×10^{8} m

Masses
    Electron: 9.11×10^{-31} kg = 0.000549 u = 0.5110 MeV
    Proton: 1.67×10^{-27} kg
       (more precisely: $1.6726233 \times 10^{-27}$ kg = 1.007276 u = 938.272 MeV)
    Neutron: 1.67×10^{-27} kg
       (more precisely: 1.674929×10^{-27} kg = 1.008665 u = 939.565 MeV)
    Deuteron: 2.013553 u
    Alpha particle: 4.001504 u
       (1 u = 1.660540×10^{-27} kg = 931.494 MeV)

Index

absolute potential energy, 184
absolute temperature, 138, 142, *142*, 144–146, 148, 154, 330, 337. *See also* temperature
absolute zero, 142
absorbed dose, 304, 317
AC. *See* alternating current
acceleration
    angular, 80, 82, 203–204
    centripetal, 77, 82, 84, 87, 119, 265
    constant, 40–46
    mass and, 60
    in mechanics, 36, 40–46, 49
action and reaction, 74
activity, 302
air track, 58
alchemy, 6
alpha decay, 298–299, *298–299*
alpha rays, 295, 304
Alpher, R. A., 314
alternating current (AC), 217, 325–326, *326*, 337
amorphous material, 4
ampere, 188
Ampere's law, 209, *209*, 234
amperian currents, 199
amplitude, 228, 252
angle of incidence, 245
angstrom, 266
angular acceleration, 80, 82, 203–204
angular displacement, 79
angular kinetic energy, 82
angular magnification, 246
angular momentum
    intrinsic, 218
    law of conservation of, 79, 82
    orbital, 81, 211, 274, 281–282, 288
    rotation and, 79–81, *80*
    spin, 281
angular motion, 82
angular velocity, 79–80
annihilation, 269–270, *269*
anode, 187
antiderivative, 47
antineutrinos, 300, 316
antinode, 230
antiparticles, 300
aperture, 245
approximation, 100
Archimedes' principle, 135, 140, *140*
area, 39, 47–49
arthroscopic surgery, 245
asymmetry term, 306
atmospheric pressure, 138–140, *138–139*
atomic mass, 142–143

atomic mass unit, 143, 297, 299, 316
atomic number, 10, 12, 295
atomic radius, 266
atomic weights, 9–10, 142
atoms. *See also* Models of the Hydrogen Atom simulation
    Bohr's model of, 2, 264–268, *264–267*, 271, 273, 277, 284, *284*, 286–287. *See also* Bohr, Niels; Bohr model
    changes in, 6
    charges in, 7–8
    home, 4
    ions, 6
    knowledge of, 91
    neutral, 6
    periodic table and, 8–11, *9–10*
    scanning tunneling microscope seeing, 11–12, *12*
    stable, 297
    structure of, 1–4, 9–10, 147, 150–152, 171, 264, 270, 277, 283–284, 294–296, 328, 339
aurora, 209
average speed, 36
Avogadro's number, 143, 153
axes, 24

background radiation, 305, 313–314, 317
Balloons and Static Electricity simulation, 172
baryons, 315
basal metabolic rate (BMR), 118
battery
    polarization of, 212–213, *212–213*
    resistor and, 187–188, *187*
    transformation of energy, 325–326
Battery-Resistor Circuit simulation, 189
Becquerel, Henri, 261, 294–295, 303
becquerel (Bq), 302
Bernoulli, Daniel, 141
Bernoulli's equation, 140–142, *140–141*, 154
beta decay, 299–301, *299–301*
beta rays (beta particles), 295, 304
big bang, 313
binding energy, 6, 12, 250, 263, 267, 297–298, *297*, 306, 310
binding-energy curve, 310
Binnig, Gerd, 11, *11*
binomial theorem, 250
biomass, 335–336
bistable combination, 345
BMR. *See* basal metabolic rate
Bohr, Niels, 2, 264–268, *264–267*, 271, 273, 277, 284, *284*, 286–287. *See also* Bohr model
Bohr model, 2, 264–268, *264–267*

Boltzmann's constant, 151
bombardment, 186
bouncing molecules, 136–138, *136–137*
boundary conditions, 231
Boyle, Robert, 142
Boyle's law, 142
Bq. *See* becquerel
Brahe, Tycho, 95
breeder reactors, 334
british thermal unit (btu), 324
brushes, 325
btu. *See* british thermal unit
buoyant force, 140, *140*

calculus, 46–48, *46*
Calculus Grapher simulation, 48
caloric, 116
calorie, 117
camera, 246, *246*
cancer, 304–305
capacitance, 236
capacitor, 171, 236, *236*
cathode-ray tube, 201, *201*
cathodes, 187
cellulosic ethanol, 336
Celsius scale, 142
centripetal acceleration, 77, 82, 84, 87, 119, 265
centripetal force, 77–79, 82, 87–88, 95, 207, 266
Chadwick, James, 309
chain reaction, 317
changing velocity, 40–46
characteristic x-rays, 270, 287
Charges and Fields simulation, 161, 164, 172, 179, 193, 196
chemical elements, 8–10. *See also* elements
chemical energy, 8, 103, 151, 159
chemical properties, 12
chromatic aberration, 246
Ci. *See* curie
Circuit Construction Kit simulation, 197
civilization
    electrical nature of, 158–159
    energy in, 322–338, *323*
classical mechanics, 33, 35, 99–101, 135, 251–252
classical physics, 98, 101, 286
coefficient of sliding friction, 67, 94
coefficient of static friction, 67
coherence, 239–240
collision, 71–72
colonoscope, 245
commutator, 217, 325
compass, 198–199
complementarity, 271
component, 118–119, *118*
composite materials, 100
Compton effect, 268–269, *268*, 287, 304, 308, 316, 320
Compton effect equation, 268
concave lenses, 243–244
concave mirrors, 243
condensation, 148
condensed matter, 149–150, *150*
conduction band, 341
conduction electrons, 347
conductor, 150, 340

conservation
    of angular momentum, 79, 82
    defined, 72, 115
    of energy, 104, 108–110, 116–118, 125
    of momentum, 71–77
constant acceleration, 40–46
constant velocity, 36–40
constructive interference, 229, 240, 273
constructs, 95–96
contact forces, 159
continuous x-rays, 287
contour lines, 178
convex lenses, 243
convex mirrors, 243
coordinate system, 24, 66
Copernicus, Nicolaus, 32, 94
cosine, 26
cosmic background radiation, 313–314, 317
cosmic rays, 209
Coulomb, Charles Augustin, 27
Coulomb barrier, 309, 311–312, 317, 334
Coulomb energy, 306
Coulomb field, 283
Coulomb's law, 27, 147, 159, 162–168
counter, 271, 307, *307*. *See also* detectors
covalent bond, 342
cross-product, 202, 217
Curie, Marie, 294–295, 303
Curie, Pierre, 294–295, 303
curie (Ci), 302
currents
    AC, 217, 325–326, *326*, 337
    amperian, 199
    DC, 217, 325–326, *326*, 337
    displacement, 234, 252
    electricity and, 187–188, 192
    induced, 214
    magnetism and, 192–205, *199*, *201–205*
    parallel, force between, 199–200, *199*
    tunneling, 11–12
cyclotron, 6, 207, 309
cyclotron frequency, 207

Dalton, John, 8–9
daughter products, 286
Davisson, Clinton, 273
DC. *See* direct current
de Broglie, Louis, 272–274
de Broglie wavelength, 272
decay
    alpha, 298–299, *298–299*
    beta, 299–301, *299–301*
    defined, 296
    equation, 302
    scheme, 299
    series, 302–303, *302–303*
dees, 208, *208*
degrees of freedom, 148, 150
Democritus, 2
derivatives, 46–47, *46*
destructive interference, 229, 237–240, 252–253, 273
detectors, 60, 233, 269, 307
determinism, 98–99
deuterium, 295–296
deuteron, 6–7, 14, 250, 296, 298, 311–313

diatomic gases, 152–153, *153*
diatomic molecules, 147
Dicke, R. H., 314
differential equation, 275
diffraction
    electron, 273
    grating, 238–239, *238*
    single slit, 239, *239*
diode, 187
diopters, 246
Dirac, P.A.M., 281
Dirac equation, 281
direct current (DC), 217, 325–326, *326*, 337
displacement, 35–36, 49, 79, 234, 252
displacement current, 234, 252
dissipation, 113–116, 124, 325
dissipative force, 114–115, 119–122, 124
distance, 35–36
domains, 212, *212*
Doppler, Christian, 232
Doppler effect, 232–233, 252
Doppler Effect simulation, 233
dose, 304, 317
dose–response relationship, 305
double slit experiment, Young's, 237–238, *237*
drift velocity, 222, 251
Dulong, Pierre, 150

earth's magnetic field, 208–209, *208*
effective dose, 304
Einstein, Albert, 62, 91, 97, 152, 248, 250–251, 262–263, 297
elastic force, 111, 113, 125
elastic potential energy, 111–113, *112*, 125
electric circuits
    battery and resistor, 187–188, *187*
    electricity and, 187–192
    resistance of wire and, 188–191, *188–189*
    resistivity and, 191–192, *191*
electric energy, 103, 325
electric field
    electricity and, 160–162, *160–162*
    energy of, 236, *236*
    flux of, 165–171, *167*, 213–217
    lines, 165–166, *165–166*, 171, 183
    uniform, 179–182, *180–181*
    as vector quantity, 171
electric flux, 165–171, *167*, 213–217
electric force, 7–8, 11–12, 14, 27–28, 59, 100, 110–111, 113, 125, 147, 162, 171, 177–179, 183–184, 192, 198–200, 217, 222, 235, 251, 265–266, 274, 286, 295–296, 298, 305, 309, 333
    electricity and, 158–160
    polarization and, 160, *160*
electric motor, 204–205
electric potential, 177–188, 192, 212, 218, 283, 345–346
electric potential energy, 110–111, 113, 125
    approximate method for, 183
    electricity and, 177–187
    moving charges and, 185–187
    point charge and, 182–185, *182–183*
    summarized, 192
    uniform fields and, 179–182, *180–181*
electrically charged particles, 7–8

electricity
    Coulomb's law and, 162–168
    currents and, 187–188, 192
    electric circuits and, 187–192
    electric field and, 160–162, *160–162*
    electric force and, 158–160
    electric potential energy, electric potential, and, 177–187
    electromagnetic waves and, 250–252, *251*
    energy transformations and, 187–192
    equipotentials and, 178–179
    field lines and, 165–166, *165–166*, 171, 183
    Gauss's law and, 167–171, *169–170*
    generating, 212–217, *212–214*
    gravitation and, 159
    gravitational field and, 168
    magnetism and, 198–199, 212–217, *212–214*, 250–252, *251*
    nature of matter and, 159
electrodes, 187
electromagnetic force, 1, 7–8
electromagnetic spectrum, 235–236, *235*, 238–239
    electric field energy and, 236, *236*
    electricity and, 250–252, *251*
    magnetism and, 250–252, *251*
    Maxwell's equations and, 233–235, *234*
    propagating, 162, 236–237, *236*
    in quantum physics, 270–272
    sound waves $v.$, 247
electromagnetic waves, 98, 109, 198, 209, 247–248, 253, 261, 262, 287
electromagnetism and relativity, 247–252
electron(s)
    Bohr model, 2, 264–268, *264–267*. See also Bohr, Niels
    capture, 301
    conduction, 347
    diffraction, 273
    free, 4, 150–151, *150*, 188, 329, 340, 342–344, 346
    high-speed, 187
    magnetism, 210–212, *210–212*
    in metals, 160, 262, 268, 329, 340, 341, 357
    pairs, 269–270, *269*, 287
    in photon-electron interactions, 268–273
    planetary model of, 264–268
    shells, 6, 284, 306
    spin, 210–211, *210*, 218
    valence, 285
    waves, 272–281, *273*, 287
    in well, 277–280, *277*
electron gun, 187
electron volt (eV), 6, 12, 138
elementary particles, 34, 100, 315
elements. *See also* periodic table of elements
    chemical, 8–10
    defined, 4, 12
    families of, 9
    order of, 284
    in quantum physics, 281–286
    rare-earth, 286
    transition, 285–286, 288
    transuranic, 11, 310
emf, 189, 212–217, *212–214*, 218

empirical laws, 93, 101
endoscope, 245
energy, 5–7, 20, 103–126, 176–192, 322–337. *See also* heat; mechanical energy; potential energy
    bands, 340–342, *340–342*
    binding, 6, 12, 250, 263, 267, 297–298, *297*, 306, 310
    from biomass, 335–336
    chemical, 8, 103, 151, 159
    in civilization , 322–338, *323*
    conservation of, 104, 108–110, 116–118, 125
    Coulomb, 306
    density, 236, 252
    dissipated, 113
    distribution, 149
    electric, 103, 325
    of electric field, 236, *236*
    flow of, 324–325, *324–325*
    fusion, 334
    gaps, 340–342, *340–342*
    geothermal, 336
    internal, 115–118, 144–145, 328
    ionization, 6, 268
    kinetic, 21, 82, 103, 105–106, 122, 125
    levels, 152, *152*, 265, 274, 286
    in mass-energy equivalence, 250
    nuclear, 297, 324–325, 333–334
    quantized, 148, 152, 262–263, 273–274
    rate of increase, 332–333, *332–333*
    rest, 249
    solar, 334–335
    stability and, 5–7
    star, 313
    stored, 113, 326–328
    surface, 306
    thermal, 125, 148, 153, 328–329
    total, 115, 297
    transfer of, 145
    transformations, 187–192
    uses of, 103, 322, 324, 334
    volume, 306
    wind, 335
    work and, 104–106, 115–125, 328–330
    zero-point, 280
Energy Skate Park simulation, 109, 117, 126–127
enriched uranium fuel, 312, 334
entropy, 329–331, 337
epicycles, 94
equal tempered scale, 232
equations, physics and, 21–25, 28
equipotentials, 178–179
equivalence, principle of, 62
ethanol, 335
ether, 247–248
eV. *See* electron volt
evaporation, process of, 149
excitation, radioactivity and, 303–304, *304*
excited states, 151, 153, 267, 299, 303, 312
experiment
    defined, 100
    double slit, 237–238, *237*
    Michelson–Morley, 240, 261
    observation and, 91, 94, 100, 329, 337
exponents, 19, 28
external force, 74, 108

extrapolation, 23
eyepiece, of microscope, 247
eyes, camera and, 246, *246*

Faraday's Electromagnetic Lab simulation, 214
Faraday's law, 212–218, *212–214*, 234
farsighted eye, 246
femtometer, 12
Fermi, Enrico, 309–310, 312
ferromagnetism, 199, 211–212
fiber optics, 245
field. *See* electric field; magnetic field
field effect transistor, 344–345, 347
first law of motion, 58–59
first law of thermodynamics, 146
field lines, 165–166, *165–166*, 171, 183
fission fragments, 310, 333
fission reactions, 286, 309–312, *309–312*, 317, 325, 329, 333, 344
flux
    electric, 165–171, *167*, 213–217
    magnetic, 213–215, 218, 220, 234, 326
focal length, 242, 244, 246–247
focus, 241–242
force(s)
    adding, 62–71
    buoyant, 140, *140*
    centripetal, 77, 82, 84, 87, 119, 265
    component of, 118–119, *118*
    contact, 159
    defined, 20, 59–60
    diagrams of, 66, 74, 82
    disadvantages of using, 176
    dissipative, 114–115, 119–122, 124
    elastic, 111, 113, 125
    electric, 7–8, 11–12, 14, 27–28, 59, 100, 110–111, 113, 125, 147, 162, 171, 177–179, 183–184, 192, 198–200, 217, 222, 235, 251, 265–266, 274, 286, 295–296, 298, 305, 309, 333
    electromagnetic, 1, 7–8
    equation, 287
    external, 74, 108
    four fundamental, 1, 7–8, 12, 26–28, 59, 100
    gravitational, 1, 7–8, 27–28, 108, 147
    internal, 74–75, 108
    magnetic, 199–200, *199*, 217
    magnitude of, 163
    net, 58–59, 66, 81, 106
    normal, 67
    nuclear, 1, 7–8, 11–12, 27–28, 147, 296, 305–306
    operational definitions and, 97
    between parallel currents, 199–200, *199*
    strong, 12
    symbol for, 58
    unbalanced, 59
    as vector quantity, 107
    weak, 8, 11–12, 27–28, 313, 315
Forces and Motion simulation, 67–68, 83
formulas, 17
fossil fuels, *323*, 324, 331–333, *331–333*, 337–338
four fundamental forces, 1, 7–8, 12, 26–28, 59, 100. *See also* electric force; gravitational force; nuclear force; weak force
Fourier: Making Waves simulation, 258

Franklin, Benjamin, 30
free electrons, 4, 150–151, *150*, 188, 329, 340, 342–344, 346
freely falling body, 40
freezing, 148–149
frequency, of waves, 227, 229–233, 235–237, 248, 252–253
friction, 58, 67–70, 113–116, *114–116*
fringing field, 171
fuel cells, 328
function, 21–25
fusion
    energy, 334
    heat of, 149
    reactions, 312–313, 317
future developments, 346–347

Galilei, 40, 91–93
Galvani, Luigi, 326
galvanometer, 205
gamma rays, 235, 295, 299, 304, 307–308
garage model of electrons in solids, 342–343
Gas Properties simulation, 146
Gauss's law
    electricity and, 167–171, *169–170*
    Maxwell's equations and, 233–234
Gay-Lussac, Joseph Louis, 117
Geiger, Hans, 307
Geiger counter, 307, *307*
general theory of relativity, 62, 97
generators, 214, 325–326
geometric optics, 241
Geometric Optics simulation, 244
geothermal energy, 336
Germer, Lester, 273
Global Positioning Systems (GPS), 62, 250
global warming, 332
GPS. *See* Global Positioning Systems
gram, 61
graphs, 21–25, 28, 38–39
gravitation
    electricity and, 159
    law of, 26, 28, 92, 97–98, 168
gravitational field, 168, 180–181, *180*
gravitational force, 1, 7–8, 27–28, 108, 147
gravitational mass, 62
gravitational potential energy, 106–108, 113, 125
graviton, 315
gray (Gy), 304, 317
greenhouse effect, 332
ground state, 6, 151, 267, 283, 341
Gy. *See* gray
gyromagnetic ratio, 211

hadrons, 315
Hahn, Otto, 310
half notes, 232
half-life, 301–302, *301–302*
Hall effect, 221–222
halogens, 9
Harris–Benedict equation, 118
heat
    capacity, 144–145, 151–152, *152*, 153
    energy, 116
    engine, 329, 337
    of fusion, 149
    pollution, 331
    second law of thermodynamics and, 116, 328–331, 337
    of vaporization, 149
    waste, 330
    work and, 116
heating, 145
heavy hydrogen, 295–296
heavy water, 296
Heisenberg, Werner, 280–281
Heisenberg uncertainty principle, 280–281, 288
Hermann, R., 314
hierarchy of size, 2–5
high-speed electrons, 187
holes, electrons and, 222, 342, 345–347
homogeneity, 168
Hooke's law, 93, 112–113, *112*, 147, 150
horse and wagon, 75–76, *75*
Hubbert, M. King, 331
Hubble's constant, 233
Hubble's law, 233
hydrogen, 3–4, 6, 9, 12, 14, 27, 143–144, 153, 250, 311, 313, 325, 327, 337
    economy, 328
    fuel cells, 328
    heavy, 295–296
    quantum physics and, 151, 264–266, 268, 270, 273–274, 276–277, 281–288
hypothesis, 93, 101

ideal gas
    first law of thermodynamics and, 146
    heat capacity and, 144–145
    heating and, 145
    internal energy and, 144–145
    law, 138, 143
    at macroscopic level, 136–138
    at microscopic level, 136–138, *136*, 153
    model, 136–146, 153–154
    pressure and, 138–140, *138–139*
    real gas *v.*, 136, 152
    temperature and, 142–146, *142*
    work and, 145–146, *145–146*
ideal system, 136
image, 241–247, 253
incompleteness, 100
index of refraction, 241
induced current, 214
induced magnetic field, 214
inertia, moment of, 80, 82
inertial fusion, 334
inertial mass, 62
information revolution, 344–347
infrared radiation, 235, *235*
initial conditions, 44
insulator, 340–341, *341*
integrals, 47–49
intensity, 228, 252, 287
interaction, 58
interference
    coherence and, 239–240
    constructive, 229, 240, 273
    destructive, 229, 237–240, 252–253, 273
    effects, 287

interference (*cont.*)
    of light, 237–241, *237–240*
        Michelson interferometer and, 240, *240*
        single slit diffraction and, 239, *239*
        thin films and, 239–240, *239*
    waves and, 226, 229, 237–241, *237–240*
interferometer, 240, *240*
internal consistency, 17
internal energy, 115–118, 144–145, 328
internal forces, 74–75, 108
interpolation, 23
interpretation, reality *v.*, 91–97
intrinsic angular momentum, 218
intrinsic magnetic moment, 211
inversely proportional, 60
iodine, 308
ion, 6. *See also* ionization
ionization
    chamber, 306–307, *307*
    defined, 6
    energy, 6, 268
    radioactivity, 303–304, *304*
iron, ferromagnetism of, 199, 211–212
irreversibility, 330
isolated system, 330
isotope, 10, 12

J. *See* joule
Joliot, Frederic, 300
Joliot-Curie, Irène, 300
Joule, James Prescott, 117
joule (J), 6
joule/second, 125

K. *See* kelvin
kelvin (K), 138, 142
Kelvin scale, 142
Kepler, Johannes, 95
kilogram, 61
kinematics, 35
kinetic energy, 21, 82, 103, 105–106, 122, 125
Kleiber's law, 5

Ladybug Revolution simulation, 78
laminar flow, 141
language, of physics, 17–26
laser, 240
lattice, 4, 222, 340–341, 343
law of conservation of angular momentum, 79, 82
law of conservation of energy, 104, 108–110, 116–118, 125
law of conservation of momentum, 74, 82, 269, 299
law of friction, 94
law of reflection, 241, *241*
law of refraction, 241, *241*
Lawrence, Ernest, 309
laws, scientific, 93–95. *See also specific laws*
Leibniz, Gottfried Wilhelm, 46
length contraction, 249
lenses
    camera, 246
    concave, 243–244
    convex, 243
    as magnifying glass, 246–247, *247*
    refraction and, 243–245, *243–244*
    thin-lens relation and, 244–245, *244*
Lenz's law, 214
leptons, 315
lifetime, of nucleus, 314
light waves
    diffraction grating and, 238–239, *238*
    interference and, 237–241, *237–240*
    particles and, 272
    in quantum physics, 270–271
    reflection, refraction, and, 241–247, *241–247*
    speed of, 247–248
    thin films and, 239–240, *239*
Lilienfeld, Julius, 344
linear momentum, 80
liquid drop model, 298, 305–306
LNT hypothesis, 305
longitudinal wave, 225
Lorentz contraction, 248, 251

macroscopic level
    defined, 115
    ideal gas at, 136–138
magic numbers, 306
magnet, 218
magnetic confinement fusion, 334
magnetic field
    of currents, 192–205, *201–205*
    earth's, 208–209, *208*
    induced, 214
    motion of charged object in, 205–208, *206*, *208*
    solenoids, 208, *208*
magnetic flux, 213–215, 218, 220, 234, 326
magnetic force, 199–200, *199*, 217
magnetic materials, 211–212, *212*
magnetic moment, 210–211, *210*
magnetic torque, 205, 211, 217
magnetism
    Ampere's law and, 209, *209*
    currents and, 192–205, *199*, *201–205*
    electricity and, 198–199, 212–217, *212–214*, 250–252, *251*
    electromagnetic waves and, 250–252, *251*
    of electrons, 210–212, *210–212*
    ferromagnetism, 199, 211–212
    history of, 198–199
    magnetic materials and, 211–212, *212*
    motional emf and, 212–213, *212–213*
    paramagnetism, 211
    poles and, 192
    summarized, 217–218
magnification, 11–12, *12*, 243, 246
magnifying glass, 246–247, *247*
magnitude, 17, 20
Malthus, Thomas, 325, 336–337
mass, 16, 20–21, 26–28, 59–63, 71–76, 79–82, 92, 96–97, 101, 105–107, 110–111, 113, 117, 122–126, 138, 141–145, 159, 168, 171, 180–183, 211, 229–230, 250, 252–254, 261, 265, 273, 295, 297–302, 304, 313–317, 326, 334
    acceleration and, 60
    atomic, 142–143
    biomass, 335–336
    gravitational, 62
    inertial, 62

in mass-energy equivalence, 250
neutrino, 301
number, 295
mass-energy equivalence, 250
Masses and Springs simulation, 113, 115, 133
mathematical description, of waves, 226–229, *227–228*
mathematical representation, 38
mathematics
    of change, 46–49
    in physics, 16–17, 28
matter
    condensed, 149–150, *150*
    electrical nature of, 159
    waves, 272
Maxwell, James Clerk, 209, 233–235, 251, 271
Maxwell's equations, 167, 233–235, *234*, 248, 251
mechanical energy
    friction and loss of, 113–116, *114–116*
    potential energy and, 108–110
mechanical waves, 225
mechanics. *See also* quantum mechanics
    acceleration in, 36–49
    approximation, 34–35
    calculus and, 46–48, *46*
    classical, 33, 35, 99–100, 135, 251–252
    derivatives and, 46–47, *46*
    displacement in, 35–36, 49
    distance in, 35–36
    heat and, 77
    integrals and, 47–49
    models in, 34–35, 49
    problem of, 35–46
    simplification, 34–35
    slopes and, 46–47
    speed in, 36, 49
    velocity in, 36–46, 49
medium, 225, 229–230, 233, 241, 245, 247–248, 252–253
Meitner, Lise, 310
melting, 150
Mendeleev, Dmitri Ivanovich, 9–10
mercury, 6, 138–139, *138*
mesons, 315
metals, 4, 150–151, *150*, 340
Michelson, A. A., 240, *240*, 248, 260–261
Michelson interferometer, 240, *240*
Michelson–Morley experiment, 240, 261
microscope
    refraction and, 247, *247*
    scanning tunneling, 11–12, *12*
microscopic level
    electric force at, 159
    ideal gas at, 115, 136–138, *136*, 153
    Newton's laws of motion at, 135
    quantum mechanics at, 135
microwave background radiation, 314, 317
microwaves, 233, 236, 314, 317
mirrors
    concave, 243
    convex, 243
    parabolic, 241, 253
    spherical, 241–242, *241*, 253

models
    Bohr's model of atoms, 2, 264–268, *264–267*, 271, 273, 277, 284, *284*, 286–287
    of chemical energy, 151
    of condensed matter, 149–150, *150*
    creation of, 101
    defined, 92, 135
    of diatomic gases, 152–153, *153*
    free electron, 4, 150–151, *150*
    of heat capacity, 152, *152*, 153
    ideal gas, 136–146, 153–154
    limits of, 152
    liquid drop, 298, 305–306
    in mechanics, 34–35, 49
    of metals, 150–151, *150*
    molecule, 147–148, *147–148*
    Newtonian, 94, 97–99
    in organization of scientific knowledge, 92–93, *92*, 97–99
    of phase changes, 148–150, *149*, 154
    planetary model of electrons, 264–268
    predictions by, 136
    quantum theory and, 151–152
Models of the Hydrogen Atom simulation, 290–291
moderator, 311, 317, 333
modern physics, 271
molar heat capacity, 144
mole, 138, 142–143
molecules
    bouncing, 136–138, *136–137*
    changing, 6
    defined, 4
    diatomic, 147
    model of, 147–148, *147–148*
moment of inertia, 80, 82
momentum. *See also* angular momentum
    conservation of, 71–77
    defined, 71, 82
    linear, 80
    orbital angular, 211
    relativistic, 249
monatomic gas, 148
monochromatic, 238, 273
moon, path of, 119–120
Morley, Edward, 248, 261
motion diagram, 37, 39, 41–42
motional emf, 212–213, *212–213*, 218
moving charges, 185–187
Moving Man simulation, 40, 43, 50–51
muon, 314
musical scales, 229–233, 252

N. *See* newton
natural scale, 232
near point, 246
nearsighted eye, 246
negative charges, 7–8, 30, 110
negative numbers, 17–18
neptunium, 11, 286, 312
net electric charge, 18
net force, 58–59, 66, 81, 106
neutral atom, 6
neutrino, 299–301
neutron excess, 311
neutron number, 295

neutron reactions, 309–312, *309–312*
neutrons, 2–3, *3*, 6–8, 10, 12, 81, 250, 295–301, 304–306, 309–312, *309–312*, 314–317, 329, 333–334
Newton, Isaac, 7, 26, 28, 46, 251–252
newton (N), 60–61
Newtonian model, 94, 97–99
Newton's law of gravitation, 26, 28, 92, 97–98, 168
Newton's laws of motion, 93, 98–99, 101, 135, 252
    first, 58–59
    in microscopic realm, 135
    overview of, 57–58
    principle of equivalence and, 62
    second, 59–61, 66, 71, 81–82, 92, 276
    third, 71–77, 82
    units and, 61–62
noble gases, 144, *144*
node, 230
normal force, 67
normal orientation, 204
notes, 232
n-type semiconductor, 343, *343*, 347
nuclear energy, 297, 324–325, 333–334
nuclear force, 1, 7–8, 11–12, 27–28, 147, 296, 305–306
nuclear physics, 294–295
nuclear reactions
    alchemy and, 6
    energy of stars and, 313
    fission, 286, 309–312, *309–312*, 317, 325, 329, 333, 344
    fusion reactions, 312–313, 317
    neutron reactions, 309–312, *309–312*
    observed, 308
    summarized, 316–317
nuclear reactor, 250, 294, 311–312, 317, 329, 333–334, 337–338
nucleons, 2–3, *3*, 295, 297, *297*, 315
nucleus
    binding energy of, 297–298, *297*
    discovery of, 2, 264–266, 294–295
    half-life, 301–302, *301–302*
    lifetime of, 314
    nuclear force and, 296, 305–306
    particles and, 314–315
    radioactivity and, 298–308, *304*
    shell structure of, 306
    structure of, 2–4, *3*, 295–296, *296*, 306
    summarized, 315–317
nuclide, 295–296, 316

objective, of microscope, 247
observation, 91, 94, 100, 329, 337
octave, 232
odd–even effect, 306
ohm, 188
Ohm's law, 188
Ohm's Law simulation, 189
"On the Electrodynamics of Moving Bodies" (Einstein), 250
one dimension, 25, 62–63
operational definitions, 96–97, 101
operations, 17
optic nerve, 246
orbital angular momentum, 81, 211, 274, 281–282, 288
orbital magnetic moment, 211
order, 1–2

order of magnitude, 20
origin, 36
oscillation, 78

pair annihilation, 269–270, *269*, 300
pair production, 269–270
parabolic mirror, 241, 253
parallel connection, 190
parallel currents, force between, 199–200, *199*
paramagnetic material, 211
paramagnetism, 211
particle accelerators, 309
particles. *See also* elementary particles
    angular momentum of, 81
    beta, 295, 304
    electrically charged, 7–8
    elementary, 34, 100, 315
    light and, 272
    nucleus and, 314–315
    synthesis with waves, 271–273
    universe made of, 98
Pascal, Blaise, 138
pascal, 138
Pauli, Wolfgang, 283, 299
Pauli exclusion principle, 283, 288
pendulum, 119–120, *120*
Pendulum Lab simulation, 120
Penzias, Arno, 314
period, 227
periodic table of elements
    atoms and, 8–11, *9–10*
    changing order in, 285–286
    defined, 4, 12
permittivity of free space, 166
perpendicular orientation, 204
Petit, Alexis, 150
phase
    changes, 148–150, *149*, 154
    difference, 228, *228*
    transition, 149
photocell, 262
photoelectric effect, 262–264, *262–263*, 268–269, 271, 286, 304, 307–308, 316, 335, 346–347
Photoelectric Effect simulation, 263
photomultiplier, 307–308
photon-electron interactions, 268–273
photons, 262–263, 268–273
    Compton effect and, 268–269, *268*
    exchange of, 314–315
    guide of, 271–272
    in new synthesis of quantum physics, 270–272
photopeak, 308, 320
physical laws, 93. *See also specific laws*
physics. *See also* quantum physics
    classical, 98, 101, 286
    equations and, 21–25, 28
    formulas and, 17
    functions and, 21–25
    graphs and, 21–25, 28
    language of, 17–26
    mathematics in, 16–17, 28
    modern, 271
    nuclear, 294–295
    powers of 10 and, 19
    ratios and, 21

right-angled triangles and, 25–26
significant figures and, 19–20, 28
symbols and, 17
tables and, 21–25, 28
units and, 20–21, 24, 28
zero and, 18–19, 24
pictorial description, of waves, 226–229, *226*
piezoelectric effect, 11–12
pion, 315
pitch, 229
Planck, Max, 152, 261, 263
Planck's constant, 210
planetary model of electrons, 264–268. *See also* Bohr model
plasma, 313, 334
plutonium, 11, 286, 312, 333–334
p–n junction, 345–347, *345–346*
point charge, 182–185, *182–183*
point mass, 183
polarization
  of batteries, 212–213, *212–213*
  defined, 171
  electric force and, 160, *160*
poles, 192
polonium, 294, 309
positive charges, 7–8, 30, 110
positive numbers, 17–18
positron, 300
postulate, 93, 101
potassium, 9–10, 285, 301, 320
potential energy. *See also* electric potential energy
  absolute, 184
  elastic, 111–113, *112*, 125
  gravitational, 106–108, 113, 125
  mechanical energy and, 108–110
  of objects, 177
  power and, 125
  reference level of, 108
  of systems, 177
pound, 61
power, 125
powers of 10, 19
precision, 19–20
pressure
  atmospheric, 138–140, *138–139*
  Boyle's law and, 142
principle of equivalence, 62
principle of relativity, 214
prism binocular, 247
probability, 99
projectile motion, 120–121, *121*, 181
propagating fields, 236–237, *236*
proper distance, 249
proper time interval, 249
proportional reasoning, 21
proportionality, 60–61
proton–proton cycle, 313
protons, 2–4, *3*, 6–12, 18–19, 79, 81, 159–160, 171, 183, 199, 250, 265–267, 273–274, 276, 282–288, 295–301, 305–306, 309–311, 313–317, 329, 333
Prout, William, 9
Ptolemy, 94
p-type semiconductor, 343, *343*, 347
pulse, 224
Pythagorean theorem, 26

Q. *See* quads
QF. *See* quality factor
quads (Q), 324
qualitative graphs, 39
quality factor (QF), 304, 317
quanta, 262–263, 272, 287
quantitative graphs, 38–39
quantities, 20–21. *See also* constructs
quantization condition, 265–267, 287
quantized energy, 148, 152, 262–263, 273–274
quantized field, 271
quantum, 286. *See also* quantized energy
quantum mechanics
  classical mechanics *v.*, 99–101
  development of, 2, 280
  electron model based on, 264
  in microscopic realm, 135
  Schrödinger and, 273–274
  Schrödinger equation and, 273–280, *275*, 287–288
quantum numbers, 266, 278, 281–285, *284*, 287–288, 306
quantum physics
  annihilation and, 269–270, *269*
  atomic structure in, 283–284
  complementarity in, 271
  Compton effect in, 268–269, *268*
  de Broglie and, 272–274
  electromagnetic waves in, 270–272
  electron waves and, 272–273, *273*
  elements in, 281–286
  Heisenberg uncertainty principle in, 280–281, 288
  hydrogen in, 151, 264–266, 268, 270, 273–274, 276–277, 281–288
  light waves in, 270–271
  new synthesis of, 270–281, *273*, *275*, *277–279*
  old, 261–270, *261–270*
  pair production and, 269–270, *269*
  Pauli exclusion principle in, 283, 288
  photoelectric effect and, 262–264, *262–263*, 286
  quantum numbers in, 266, 281–282, 284–285, *284*, 287
  radioactivity and, 286
  x-rays in, 270, *270*
quantum theory, models and, 151–152
quarks, 315

R. *See* roentgen
rad, 304, 317
radar detector, 233
radial field lines, 183
radians, 77
radiation
  background, 305, 313–314, 317
  electromagnetic, 235, 248, 261–262, 265, 270, 325, 332
  radioactive, 306–308, *307–308*
  sickness, 304
radio waves, 236
radioactive radiation, 306–308, *307–308*
radioactivity, 261, 286, 289, 298–303, 306, 308–309, 311–312
  alpha decay, 298–299, *298–299*
  beta decay, 299–301, *299–301*
  biological effects of, 303–305, *304*
  defined, 11

radioactivity (*cont.*)
  discovery of, 294
  excitation and, 303–304, *304*
  gamma rays, 235, 295, 299, 304, 307–308
  half-life and, 301–302, *301–302*
  ionization and excitation, 303–304, *304*
  nucleus and, 298–308, *304*
  of nuclides, 296
  quantum physics and, 286
  stability, decay series, and, 302–303, *302–303*
  summarized, 316
  units, 304–305
radium, 294, 303
Ramp: Forces and Motion simulation, 68, 83
rare-earth elements, 286
rate of energy increase, 332–333, *332–333*
ratios, 21
RBE. *See* relative biological effect
real gas, 136, 143–144, 148, 152–153
real image, 242, 253
reality, interpretation *v.*, 91–97, 101, 146–153, 182–183, 237
reference level, of potential energy, 108
reflection
  law of, 241, *241*
  lenses and, 243–245, *243–244*
  light waves and, 241–247, *241–247*
  magnifying glass and, 246–247, *247*
  mirrors and, 241–243, *241–243*
  resolution and, 245
  total internal, 245, *245*
refraction
  cameras and, 246, *246*
  index of, 241
  law of, 241, *241*
  lenses and, 243–245, *243–244*
  light waves and, 241–247, *241–247*
  magnifying glass and, 246–247, *247*
  microscope and, 247, *247*
  resolution and, 245
  telescope and, 247, *247*
region of stability, 296
relative biological effect (RBE), 304
relativistic momentum, 249
relativity
  electromagnetism and, 247–252
  general theory of, 62, 97
  principle of, 214
  special theory of, 248–250, *249–250*
rem, 304
representations, 22–23, 28, 37–38
Resistance in Wire simulation, 191
resistivity, 191–192, *191*, 196–197, 343–344, 348
resistor, 187–188, *187*
resolution, 245
resonance, 232
rest energy, 249
retina, 246
retrograde motion, 94
right-angled triangles, 25–26
roentgen (R), 304, 317
Röntgen (Roentgen), Wilhelm, 261
Rohrer, Heinrich, 11, *11*
Roll, Peter, 314
root-mean-square (rms) speed, 156

rotation
  angular momentum and, 79–81, *80*
  uniform circular motion, 77–79, 82
rotational kinetic energy, 122
rounding, 19
Rumford, Count, 116–117
Rutherford, Ernest, 10, 295, 298, 305, 308
Rutherford scattering formula, 305
Rutherford Scattering simulation, 291

salts, 9
scalar quantity, 35–36, 107, 137, 177, 192
scale
  calibrated, 60
  equal tempered, 232
  factor, 5
  musical, 229–233, 252
  natural, 232
  size, *3*
  temperature, 142
scanning tunneling microscope, 11–12, *12*
Schrödinger, Erwin, 273–274, 280
Schrödinger equation, 273–280, *275*, 287–288
scientific knowledge, organization of
  constructs in, 95–96
  laws in, 93–95
  mechanics and beyond, 99–100
  models in, 92–93, *92*, 97–99
  operational definitions in, 96–97, 101
  reality *v.* interpretation, 91–97
  theory in, 92–93, *92*
scientific method, 91
scintillation spectrometer, 308
scintillations, 307–308, *307–308*
second derivative, 49
second law of motion, 59–61, 66, 71, 81–82, 92, 276
second law of thermodynamics, 116, 328–331, *329–330*, 337
seismic waves, 208
semiconductor, 340, 342–344, *342–343*, 347
semiempirical binding energy relation, 306
sensor, 246
series connection, 190
shells
  electron, 6, 144, 284–285, 297, 340–341
  of nuclei, 306
SI system, 20–21
sievert (Sv), 304, 317
significant figures, 19–20, 28
silicon age, 344
simple harmonic motion, 78, 147, 150, 280
simplification, 34–35, 100
simulations
  Balloons and Static Electricity, 172
  Battery-Resistor Circuit, 189
  Calculus Grapher, 48
  Charges and Fields, 161, 164, 172, 179, 193, 196
  Circuit Construction Kit, 197
  Doppler Effect, 233
  Energy Skate Park, 109, 117, 126–127
  Faraday's Electromagnetic Lab, 214
  Forces and Motion, 67–68, 83
  Fourier: Making Waves, 258
  Gas Properties, 146
  Geometric Optics, 244

Ladybug Revolution, 78
Masses and Springs, 113, 115, 133
Models of the Hydrogen Atom, 290
Moving Man, 40, 43, 50–51
Ohm's Law, 189
Pendulum Lab, 120
Photoelectric Effect, 263
Ramp: Forces and Motion, 68, 83
Resistance in Wire, 191
Rutherford Scattering, 291
Torque, 80
Vector Addition, 64
Wave Interference, 226–227, 238, 254–255
Wave on a String, 225, 227, 231, 254
sine, 26
single slit diffraction, 239, *239*
sinusoidal variation, 78, 217, 228
size
    hierarchy of, 2–5, *3*
    scales, *3*
slope, 24–25, 28, 37–40, 42–49, 60, 113, 216, 234, 275–276, 278–279, 302
Snell's law, 241
solar cells, 335, 346, *346*
solar energy, 334–335
solenoids, 208–209, *208*, 217, 220, 236
solids, 4, 7, 95, 147, 149–153, *152*, *153*, 158–160, 235, 274, 277, 285, 347
solid-state devices, 339, 347
sound waves
    Doppler effect and, 232–233, 252
    electromagnetic waves *v.*, 247
    musical scales and, 229–233, 252
    properties of, 225
    resonance of, 232
    standing, 230, *230*
special theory of relativity, 248–250, *249–250*
specific heat capacity, 144
spectrum
    electromagnetic, 235–236, *235*, 238–239
    of energy levels, 274
speed, 36, 49, 77, 247–248
spherical aberration, 246
spherical mirrors, 241–242, *241*, 253
spherical symmetry, 93, 168–169, 185
spin, 81, 199, 210–211, *210*, 218, 281–282, 286, 306, 315, 340
spring constant, 112
springs, 111–113, 147, *147–148*
spyglass, of telescope, 247
square well, 277–280, *277*
stability
    decay series and, 302–303, *302–303*
    energy and, 5–7
    region of, 296
stable atoms, 297
standing waves, 230, *230*, 273
stars, energy of, 313
statcoulomb, 304
states
    excited, 151, 153, 267, 299, 303, 312
    ground, 151, 267, 283, 341
    of materials, 142, 149
stopping potential, 263
Strassmann, Fritz, 310

strong force, 12. *See also* nuclear force
subshells, electron, 6, 284
supercapacitors, 327–328
surface energy, 306
surface-to-volume ratio, 5
sustainability transition, 336–337
Sv. *See* sievert
symbols, physics and, 17
symmetry, 169–171, *169–170*
synchrotron, 235
synthesis of waves and particles, 271–273
systems, 4, 49, 61, 99, 135–153, 248–250, 335, 347
    choice of, 109
    coordinate, 24, 66
    ideal, 136
    isolated, 330
    of objects, 71–72, 81
    potential energy of, 177
    of units, 61, 143–144

tangent, 26
tau, 315
technique, 90
telescope, 247, *247*
temperature
    absolute, 142, *142*, 144
    ideal gas and, 142–146, *142*
    mechanics and, 77
    scales, 142
test charge, 161, 163, 184
theoretical laws, 101
theory
    creation of, 101
    general theory of relativity, 62, 97
    observation and, 94
    in organization of scientific knowledge, 92–93, *92*
    quantum, 151–152
    special theory of relativity, 248–250, *249–250*
thermal energy, 125, 148, 153, 328–329
thermal equilibrium, 145, 329
thermal motion, 150, 235
thermodynamics
    first law of, 146
    second law of, 116, 328–331, *329–330*, 337
thin films, 239–240, *239*
thin-lens relation, 244–245, *244*
third law of motion, 71–77, 82
Thompson, Benjamin, 116
Thomson, George P., 273
Thomson, J. J., 261, 273
Thomson, William (Lord Kelvin), 142
thought experiments, 248–249
threshold, 305
time dilation, 249
torque, 79–82, 204
    magnetic, 205, 211, 217
    rotation and, 79–81, *80*
Torque simulation, 80
Torricelli, Evangelista, 139
total binding energy, 297
total energy, 115, 297
total internal reflection, 245, *245*
transformers, 325–326
transistor, 339, 344–347, *344–345*
transition elements, 285–286, 288

transuranic elements, 11, 310
transverse wave, 225
traveling waves, 230
triangles, right-angled, 25–26
tritium, 296
tunneling current, 11–12
turbulent flow, 141
two dimensions, 25, 63–66

ultraviolet radiation, 235, *235*
unbalanced force, 59
uncertainty principle, 275, 280–281, 288
uniform circular motion, 77–79, 82
uniform fields, 179–182, *180–181*
uniform speed, 77
units
    Newton's laws of motion and, 61–62
    physics and, 20–21, 24, 28
    radioactivity, 304–305
    system of, 61, 143–144
universal gas constant, 138, 154
universal gas law, 143
universe
    beginning of, 313
    organization of, 1–3
    particles in, 98
    programmed, 98–99
uranium, 6, 11, 309–310, 312, 333–334

V. *See* volt
vacuum tube rectifier, 187
valence electrons, 285
Van Allen belts, 209
vaporization, heat of, 149
vector(s)
    adding forces and, 62–71
    components, 66–67
    defined, 82
    one dimension, 62–63
    product, 202
    quantity, 35, 49, 107, 171
    sum, 81
    symbols, 64
    two dimensions, 63–66
Vector Addition simulation, 64
velocity
    angular, 79–80
    changing, 40–46
    constant, 36–40
    defined, 18
    drift, 222, 251
    in mechanics, 36–46, 49
    of waves, 229–230
verbal representation, 37
vibration, 147
virtual image, 241, 253
viscosity, 141
volt (V), 177
Volta, Alessandro, 326
voltage, 325
voltage difference, 177, 188
voltage doubler, 309
volume, energy, 306

waste heat, 330
water waves, 224–225, *225*
wave(s). *See also* electromagnetic waves; light waves; sound waves
    electron, 272–281, *273*, 287
    equation, 274–276, 287
    function, 274, 276–277
    intensity of, 228
    interference and, 226, 229, 237–241, *237–240*
    longitudinal, 225
    mathematical description of, 226–229, *227–228*
    matter, 272
    mechanical, 225
    medium of, 229
    microwaves, 233, 236, 314, 317
    number, 275
    out of phase, 228
    packet, 280
    in phase, 229
    pictorial description of, 226–229, *226*
    properties of, 226–229
    Schrödinger equation, 275–276, *275*
    special theory of relativity and, 248–250, *249–250*
    standing, 230, *230*, 273
    summarized, 252–254
    synthesis with particles, 271–273
    transverse, 225
    traveling, 230
    types of, 224–226, *225*
    velocity of, 229–230
    water, 224–225, *225*
Wave Interference simulation, 226–227, 238, 254–255
Wave on a String simulation, 225, 227, 231, 254
wavelength, 227, 272
weak force, 8, 11–12, 27–28, 313, 315
weight, 20, 59, 61
Weizsäcker, C. F. von, 306
whole notes, 232
Wilkinson, David, 314
Wilson, Robert, 314
wind energy, 335
wire, resistance of, 188–191, *188–189*
work
    energy and, 104–106, 118–124
    function, 263
    heat and, 104–106, 115–125, 328–330
    ideal gas and, 145–146, *145–146*
    pendulum and, 119–120, *120*
    projectile motion and, 120–121, *121*
work-energy theorem, 106

x-ray tube, 187, 235
x-rays, 187, 235, 261, 270, *270*, 287

y-intercept, 25
Young, Thomas, 237–238, *237*
Yukawa, Hideki, 314–315

zero, physics and, 18–19, 24
zero-point energy, 280

About the Authors

Peter Lindenfeld has been a professor of physics at Rutgers University for several decades. In addition to his research in superconductivity and materials physics he has taught at all levels, including courses for physics teachers. He is a recipient of the Millikan medal of the AAPT.

After graduating from the University of Washington, Suzanne White Brahmia served as a Peace Corps physics teacher in Africa. She attended graduate school at Cornell University. Suzanne currently heads the Extended Physics Program for engineering students, and is the Associate Director for Physics of the Math and Science Learning Center at Rutgers University.